Manual of Low-Slope Roof Systems

C. W. Griffin
R. L. Fricklas
Third Edition

McGraw-Hill
New York San Francisco Washington, D.C. Auckland Bogotá
Caracas Lisbon London Madrid Mexico City Milan
Montreal New Delhi San Juan Singapore
Sydney Tokyo Toronto

Library of Congress Cataloging-in-Publication Data

Griffin, C. W. (Charles William),
    Manual of low-slope roof systems: / C. W. Griffin and R. L. Fricklas.
    3d ed.
       p.   cm.
    Rev. ed. of: Manual of built-up roof systems. 2d ed. 1982.
    Includes Index.
    ISBN 0-07-024784-6
    1. Flat roofs   1. Roofing   1. Fricklas, R. L.   II. Griffin, C.
W. (Charles William), Manual of built-up roof systems
III. Title.
TH2409.G75 1995
695   dc20   95-41566
                                                                                                                                           CIP

McGraw-Hill

A Division of The **McGraw·Hill** Companies

Copyright © 1996, 1982 by The McGraw-Hill Companies, Inc. All rights reserved. Printed in the United States of America. Except as permitted under the United States Copyright Act of 1976, no part of this publication may be reproduced or distributed in any form or by any means, or stored in a data base or retrieval system, without the prior written permission of the publisher.

2 3 4 5 6 7 8 9 0   DOC/DOC   9 0 1 0 9 8 7 6

ISBN 0-07-024784-6

The sponsoring editor for this book was Wendy Lochner, the editing supervisor was Caroline R. Levine, and the production supervisor was Pamela A. Pelton. This book was set in Century Schoolbook by Dina John of McGraw-Hill's Professional Book Group composition unit.

Printed and bound by R. R. Donnelley & Sons Company.

McGraw-Hill books are available at special quantity discounts to use as premiums and sales promotions, or for use in corporate training programs. For more information, please write to the Director of Special Sales, McGraw-Hill, 11 West 19th Street, New York, NY 10011. Or contact your local bookstore.

This book is printed on acid-free paper.

Information contained in this work has been obtained by The McGraw-Hill, Companies, Inc., ("McGraw-Hill") from sources believed to be reliable. However, neither McGraw-Hill nor its authors guarantees the accuracy or completeness of any information published herein, and neither McGraw-Hill nor its authors shall be responsible for any errors, omissions, or damages arising out of use of this information. This work is published with the understanding that McGraw-Hill and its authors are supplying information, but are not attempting to render engineering or other professional services. If such services are required, the assistance of an appropriate professional should be sought.

Contents

Preface xiii
Acknowledgments xv

Chapter 1. Introduction   1

  Why Roofs Fail   2
  Lessons from History   5
  Need for Research and Standards   7

Chapter 2. The Roof as a System   9

  Design Factors   9
  Basic Roof Components   10
  Disastrous Combinations   15
  Constructing the Roof   17
  Divided Responsibility   17
  Unified Responsibility   18
  Performance Criteria   19

Chapter 3. Draining the Roof   27

  Why Drain the Roof?   30
  Life-cycle Costing Example   33
  How to Drain the Roof   35
  Drainage Design   37
  Draining an Existing Roof   40
  Alerts   43
    Design   43
    Maintenance   44

Chapter 4. Structural Deck   45

  Deck Materials   45
  Design Factors   48
    Slope   50

Deflection	50
Dimensional Stability	52
Moisture Absorption	52
Anchorage of Components	54
Deck Surface and Joints	55
Steel Decks	57
Recommendations for Steel Decks	58
Alerts	59
Design	59
Field	60

Chapter 5. Thermal Insulation — 63

Thermal Insulating Materials	64
Rigid Board Insulation	65
Poured-in-Place Insulating Concrete Fill	66
Design Factors	70
Strength Requirements	71
Moisture Absorption	72
Venting	74
Dimensional Stability	76
Tapered Insulation	78
Anchoring Insulation to the Deck	78
Solid Bearing for Insulation Boards	79
Refrigerated Interiors	79
Effects of Increased Insulation Thickness	80
Joint Taping	81
Double-Layered Insulation	82
Principles of Thermal Radiation	84
Moisture Reduction of Insulating Value	85
Heat-Flow Calculation	89
Alerts	91
Design	91
Field	92

Chapter 6. Vapor Control — 95

Fundamentals of Vapor Flow	95
Techniques for Controlling Moisture	99
General Ventilation	99
Vapor Retarder	100
Self-drying Roof System	108
The Futility of Topside Venting	112
Theory of Vapor Migration	113
Diffusion	113
Air Leakage	115
Sample Vapor-Retarder Calculation	118
Alerts	121
Design	121
Field	122

Chapter 7. Wind Uplift — 123

Historical Background	125

Mechanics of Wind-Uplift Pressure	130
Basic Wind-Speed Determinants	131
Building Shape, Size, and Enclosure	137
Determining Final Wind-Uplift Pressure	139
Design Example for Anchored System	139
Perimeter Flashing Failures	145
Wind-Uplift Mechanics on Ballasted Roofs	147
Gravel Blowoff and Scour	148
Wind-Uplift Design for Ballasted, Single-ply Roof Assemblies	153
Basic Ballasted Wind-Design Systems	155
PMR Wind Design	162
Retroactive Redesign	165
Design Example for Ballasted, Loose-Laid System	165
Wind-Uplift Testing	165
Laboratory Tests	166
Field Wind-Uplift Tests	168
Repair Procedures after Wind-Uplift Failure	172
Mechanical Fasteners	173
Alerts	179
General	179
Design	179
Field	179

Chapter 8. Fire Resistance 181

Nature of Fire Hazards	181
External Fire Resistance	182
Internal Fire Hazards	184
Below-Deck Fire Tests and Standards	185
Flame-Spread Test	185
FM Fire Classification	186
Time-Temperature Ratings	188
Fire Rating vs. Insulating Efficiency	189
Building Code Provisions	189
Alerts	190

Chapter 9. Historical Background of Contemporary Roof Systems 193

Built-up Bituminous Membranes	194
The First-Generation Single-ply Pioneers	195
The Cold-Process Dead End	197
The Second-Generation Success	197
New Roof Systems	200
Modified Bitumens	202

Chapter 10. Elements of Built-up Membranes 205

Membrane Materials	207
Roofing Bitumens	207
Incompatibility of Asphalt and Coal Tar Pitch	214
Coal Tar Pitch vs. Asphalt	215
Felts	215

Surfacing	219
Smooth-Surfaced Membrane	225
Mineral-Surfaced Roll Roofing	226
Heat-Reflective Surfacings	226
Built-up Membrane Specifications	227
Shingling of Felts	228
Coated Base Sheet	230
Phased Application	231
Temporary Roofs	234
Joints in Built-up Membranes	235
Membrane Performance Standards	237
BUR Failure Modes	240
Blisters	241
Splitting	252
Ridging	257
Membrane Slippage	259
Alerts	265
General	265
Design	265
Field	266

Chapter 11. Modified-Bitumen Membranes 269

Historical Background	270
Modified-Bitumen Materials	272
Rubberizing Polymers	272
Reinforcement	275
Surfacing	278
Modified-Bitumen Specifications	280
Modified-Bitumen Flashings	283
Performance Criteria for Modified Bitumens	284
Modified-Bitumen Failure Modes	288
Defective Lap Seams	288
Checking	290
Shrinkage	291
Blistering	291
Delamination	294
Slippage	294
Splitting	296
Alerts	297
General	297
Design	297
Field	298

Chapter 12. Elastomeric Membranes 299

Basic Material	299
Field Seaming	302
Flashings	304
Mechanically Fastened Systems	307
Elastomeric Failure Modes	309

Chapter 13. Weldable Thermoplastics — 315

- Materials — 315
- Historical Background — 318
- Flashings — 321
- Failure Modes — 322
- Alerts — 323

Chapter 14. Flashings — 325

- Flashing Functions and Requirements — 326
- Materials — 327
  - Base Flashing Materials — 329
  - Counterflashing Materials — 332
- Principles of Flashing Design — 333
  - Eliminating Penetrations — 333
  - Reducing Openings — 334
  - Flashing Elevation — 334
  - Differential Movement — 336
  - Flashing Contours and Cants — 336
  - Flashing Connections — 336
- Specific Flashing Conditions — 337
  - Edge Details — 337
  - Vertical Flashings — 340
  - Roof Penetrations — 341
  - Expansion-Contraction Joints — 346
  - Water Conductors — 348
- Flashing Failures — 349
  - Sagging — 350
  - Ponding Leakage — 350
  - Diagonal Wrinkling — 351
  - Post-construction Damage — 352
- Alerts — 352
  - General — 352
  - Design — 353
  - Field — 354
- Standard Details — 355

Chapter 15. Protected Membrane Roofs and Waterproofed Decks — 373

- Why the Protected Membrane Roof? — 375
- How PMR Changes Component Requirements — 379
- PMR Design and Construction — 380
  - Drain Location — 381
- PMR Flashings — 381
- PMR Insulation Performance Requirements — 381
  - Thermal Quality of PMR Insulation — 384
  - Refinements in Locating PMR Insulation — 385
- Surfacing Protected Membrane Roofs — 386
- Life-cycle Costing — 387
- Waterproofed Deck Systems — 389

Standard Decks	390
Membranes	391
Protection Boards	393
Drainage Design	393
Aggregate Percolation vs. Open Subsurface Drainage	394
Insulation	396
Flashing	397
Wearing Surface Design	398
Plaza "Furniture"	399
Detailing	400
Flood Testing	400
Alerts (for PMRs)	400
Alerts (for Waterproofed Decks)	401
Design	401
Field	402

Chapter 16. Sprayed Polyurethane — 405

Historical Background	406
Materials	409
Urethane Foam	409
Membrane Coating	411
Surfacing Aggregate	418
Detailing SPF Systems	420
Application Requirements	420
Foam Application	427
Coating Application	428
Quality Assurance	430
Maintenance	430
Alerts	431
General	431
Design	432
Application	432
Maintenance	432

Chapter 17. Metal Roof Systems — 435

Historical Background	436
Metal Roof Concepts and Terminology	437
Design Factors	440
Thermal Movement	440
Drainage	440
Condensation	441
Penetrations	442
Wind-Uplift Resistance	443
Corrosion Protection	444
Metal Roofing Accessories	446
Thermal Insulation	451
Underlayments	451
Vapor Retarders	452
Detailing Metal Roofing	455
Metal-Roof-System Defects	458

Aesthetic Problems	458
Functional Problems	458

Chapter 18. Field Inspections — 461

Quality Assurance	461
QA Inspections	462
Material Storage Checklist	465
Deck Preparation	468
Insulation Application	469
Membrane Application	471
Visual Roof Surveys	476
Nondestructive Moisture Detection	479
Benefits of Nondestructive Moisture Detection	480
Limitations	481
Principles of Nondestructive Moisture Detection	481
Other Moisture-Detection Techniques	484

Chapter 19. Reroofing and Repair — 487

Re-covering or Tearoff-Replacement?	489
Investigation and Analysis	490
Preliminary Investigation	491
Leak-detection Technique	491
Visual Inspection and Analysis	492
Field-Test Cuts	493
Large-Scale Moisture Surveys	494
Assessment for Remedial Action	494
New Performance Criteria	495
Reroofing Specifications	496
Isolating the New Membrane	497
Flashings	497
Surface Preparation of Existing Membrane	498
Alerts	499
For Replacement Bituminous Roof Membranes	499
For Loose-Laid Synthetic Single-ply Membranes	501
For Sprayed Polyurethane Foam (Plus Fluid-Applied Coating)	502

Chapter 20. Roof-System Specifications — 505

Specifications and Drawings	506
Drawing Requirements	506
Specification Requirements	507
Writing Style	507
Specifying Methods	508
Master Specification	509
Division of Responsibility	509
Temporary Roofing	510
Special Requirements	511
Specifying New Roofing Products	512
Quality Assurance	513

Coordination	515
Test Cuts	516
Alerts	518
General	518
New Products	518
General Field	518
Technical	519
Chapter 21. Roofing Guarantees and Warranties	**521**
Manufacturer's Bond	521
Manufacturer's Warranty	524
Analyzing the Warranty	525
Warranty Checklist	528
Roofer's Guarantee	530
Roof-Maintenance Program	531
Glossary of Roofing-Related Terms	**533**
Appendix A	**567**
Appendix B	**571**
Index 591	

Preface

Haunted by the pessimistic prophecies propounded by the Reverend Thomas Malthus and David Ricardo in the early nineteenth century, the great historian Thomas Carlyle labeled the emerging study of economics "the dismal science." Carlyle, of course, knew nothing about low-slope roof technology and the monumental problems of wind blowoffs; blistered, split, and ridged membranes; water-soaked insulation; and other maladies resulting in leaking roofs and energy waste. In this third edition of the roof manual, we consider this dismal history as well as the changes in roof technology that necessitate this revision. "Those who can't remember the past are doomed to repeat it." Santayana's well-worn maxim applies to roof technology as well as politics.

The roofing industry's history is littered with the wreckage of materials and systems introduced into the marketplace without adequate research and analysis. Some were marketed for years after early warnings clearly pointed to big problems ahead. On their introduction into the roofing market in the late 1960s, two-ply, coated-felt membranes appeared to be a labor-saving solution to ridging and several other defects afflicting built-up membrane systems. By the mid-1970s, an unnecessarily large host of split and blistered roofs attested to their general failure, and two-ply, coated-feld membranes were belatedly rejected as an unmitigated disaster.

Disasters of this magnitude should not recur in today's more sophisticated roofing industry. The introduction of totally new materials and systems has slowed over the past decade after frantic innovation during the 1980s. This third edition focuses on improvements in products, systems design, and performance requirements and refinements in application technique in this slowly evolving, but nonetheless increasingly science-based field.

Among the notable products given their own chapter in this edition are weldable thermoplastic membranes—plasticized PVC, Hypalon, chlorinated polyethylene, and various polyethylene blends and copolymers. The application techniques and properties of these weldable systems require that they be considered separately from vulcanized elastomeric systems, such as EPDM.

EPDM now accounts for fully a third of currently installed membrane roofing systems. Changes in seaming and anchorage techniques have simultaneously solved the problems of earlier generations of materials and complied with ever-stricter environmental regulations.

Metal roofing systems are no longer restricted to preengineered buildings. They can provide ventilation, slope to drain, and give a new look to existing buildings. A new chapter (17) discusses these systems.

Another new chapter (16) covers sprayed-in-place polyurethane foam. New guidelines and specifications are available through the Polyurethane Foam Contractors' Division of the Society of the Plastics Industry. Better guidelines on surface preparation, density, compressive strength, and coating selection should make this material perform even better.

Chapter 11 on Modified Bitumens, looks at the fastest-growing segment of the commercial roofing business. Two distinct polymer types of modifiers are discussed, the polypropylene plastic type and the "block" copolymers of the styrene-butadiene family. Application methods vary from conventional hot-mopping and cold process to torch fusion.

Lest we be accused of ignoring built-up roofing, the king is not dead. It has evolved, with glass-fiber felts solving many of the problems highlighted in the last two editions. Flashings are now hybrids, frequently incorporating the versatility and toughness of modified bitumens. Membranes for BUR continue to use the multiple-ply concept, providing redundancy for those that feel single-ply and MB systems are too vulnerable to typical roof abuse.

This book has attempted to assimilate the excellent work presented as roofing conferences and symposia since the last edition. These include NIST/NRCA programs, ASTM publications on roofing research, and ORNL conferences on wind and reroofing. Quality publications from NIST and other research organizations have been included and referenced where appropriate.

Acknowledgments

There is no practicable way to acknowledge all the roofing industry experts who have contributed, one way or another, to a book that has undergone three revisions in a quarter century. For this third edition, however, we owe especial debts to consulting architect Justin Henshell, AIA, who provided major portions of text incorporated in Chaps. 15 and 21 like a reporter working under a deadline, and Wayne Tobiasson, U.S. Army Cold Regions Research and Engineering Laboratory, who contributed a meticulous critique of revised Chap. 6.

Others who have helped with illustrations and/or technical input include Dan Benedict, Polyurethane Foam Contractors Division of SPI; Peter Butler, W. R. Grace; P. Cogneau, Performance Roof Systems; Bob Carothers, Partek; T. W. Freeman, Dow Chemical; Smith Funk, Celotex; Rob Haddock, Metal Roof Advisory Group; Helene Hardy-Pierce and Kent Blanchard, Tamko; Riaz Hasan, Buildex; Stephen Phillips Esq., Hendrick, Phillips, Schemm & Saltzman, P.C.; Jon Peterka, Colorado State University; Dave Roodvoets, T-Clear Corp.; Walter Rossiter, National Institute of Standards and Technology; Tom Smith, Jim Carson, Jack Robinson and Bill Cullen, National Roofing Contractors Association, George and Phil Smith, Factory Mutual Research Corp.; Marc Somers, Eastman Chemical; Ed Stewart, Gale Assoc.; and Patricia Wood-Shields, Roofing Industry Committee on Wind Issues.

Particular thanks is extended to The Roofing Industry Educational Institute for allowing Dick the time to work on this text and for providing many illustrations as well as to Susan Kaminski who prepared many of the special charts and tables.

C. W. Griffin
Dick Fricklas

Chapter

1

Introduction

Viewed from the perspective of 1982, when the second edition of this manual was published, the low-slope roofing industry has experienced such drastic change that the book's title required a change. For roughly 140 years, from its invention in the mid-1840s until the mid-1980s, built-up bituminous roofing (BUR) systems dominated the low-slope roofing market. That dominance is long gone. Other systems—notably single-ply elastomers and thermoplastic weldables, modified bitumens, and even metal roofing—have eroded BUR's formerly dominant market share, which has dwindled to roughly one-third of the low-slope market in the mid-1990s.

While roof materials and systems have changed, the tide of low-slope roof construction rolls serenely on. Roughly $10 billion is spent annually constructing (and reconstructing) low-slope roof systems in the United States, according to the National Roofing Contractors Association (NRCA).[1] In area, this vast volume would cover Washington, D.C., twice. Low-slope roof systems retain their popularity despite several decades of rampant litigation over roof failures. According to some estimates, the number of lawsuits involving low-slope roof systems equals (or exceeds) the total number of lawsuits filed over all other building systems combined. Regardless of whether this estimate is strictly accurate, it conveys an incontrovertible fact: No other building system can approach the roof system as a source of litigation.

Yet despite their many recurring problems, low-slope roofs maintain their predominance. Steeply sloped roofs on the large, sprawling buildings that dominate today's construction would drastically cut the costs of reroofing, repair, and litigation for problem-prone low-slope roofs. But they would raise construction costs by a far greater amount through the costs for additional building volume. Low-sloped roofs will maintain their predominance in commercial roofing for a simple economic reason: the costs of steeply sloped roofs over the vast acreages covered by shopping centers and other modern buildings are simply too high a price to pay to avoid the problems posed by low-slope roof systems. Roofing a single-story, 200-ft-wide building with 100-ft-span trusses sloped 25 percent (i.e., 3 in 12) would increase building volume by roughly 40 percent.*

Why Roofs Fail

Premature roofing failures are caused by both economic and technical factors. Economically, a building's roof system normally lags far behind the more architecturally glamorous building subsystems competing for the building owner's money. Pennywise, dollar-foolish decisions underlie many premature roofing failures: Whether through ignorance, laxity, or sheer perversity, many roof designers and building owners refuse to pay the slight additional cost for sloping the roof to avoid the ponding of rainwater (see Fig. 1.1).

Technically, the factors contributing to premature roof failures can be listed as follows:

- The extraordinary rigors of roof-performance requirements

---

*If the building were 15 ft high, the additional space required for the 25 percent slope would increase building volume by $2 \times \frac{1}{2} (12.5 \times 100)/(15 \times 200) = 42$ percent.

Figure 1.1 This scene is not an irrigated wheat field; it is a level built-up roof system with ponded water. The roots of the flourishing vegetation can force their way through the membrane into the insulation, ultimately producing widespread leaks. (*Sellers & Marquis Roofing Co.*)

- Proliferation of new materials
- Complexity of roof-system design
- Expanding roof dimensions
- Field-application problems
- The modern trend toward more flexible buildings

Roofs must withstand a much broader attack from natural forces than other building components. In some parts of the continental United States, roof surfaces experience annual temperature changes exceeding 200°F and daily changes exceeding 100°F. These temperature changes can occur rapidly, as when a summer shower suddenly cools the sun-baked membrane surface. Solar radiation heats the roof to extraordinary temperatures, up to 180°F for black surfaces; this heat greatly accelerates photochemical deterioration (see Fig. 1.2). Rain, snow, sleet, and hailstones pound the roof; acid mists and other airborne pollutants—even fungus—attack the roof.

The proliferation of new roof materials—new deck, insulation, vapor retarder, membrane, and flashing materials, used in countless combinations—has complicated the field-manufacturing process and the designers' job of evaluating durability. Until a material has been field-tested in service, its durability remains highly unpredictable. Accelerated laboratory tests are better than nothing, but in-service performance is essential for proving a material's durability. Some materials enter the roofing market lacking even laboratory testing.

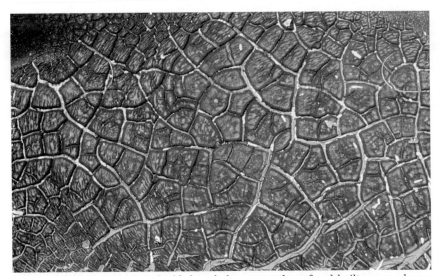

Figure 1.2 Alligatoring of unshielded asphalt on smooth-surfaced built-up membrane betrays the hazards of overweight mopping. Alligatoring is shrinkage cracking resulting from ultraviolet radiation, erosion, and consequent embrittlement of the asphalt. (*RIEI*)

Figure 1.3 New Orleans Superdome is roofed with sprayed polyurethane foam, over an area measured in acres. (*RIEI*)

Many materials that are satisfactory for some uses prove disastrous when incorporated in the wrong system. Extruded polystyrene foam is a remarkably good material in its exposed position in protected membrane roofs, where its resistance to water preserves a high proportion of its thermal resistance (see Fig. 1.3 and Chapter 15). But it was a disaster in conventional roof assemblies, sandwiched between deck and membrane. (It was impossible to adhere it dependably to the deck with hot-mopped asphalt.)

Thermal insulation is a vitally important component in roof systems, serving several ancillary functions in addition to its primary purpose of cutting heating and cooling costs. But it increases the probability of entrapping water within the roof assembly. This threat of condensation of water vapor creates the possible need for another roof-system component, a vapor retarder designed to intercept the flow of water vapor into the insulation, where it can reduce the insulation's thermal resistance, form a source of leakwater, and even destroy the insulation itself. But the vapor retarder creates a kind of moisture trap, preventing the venting of vapor downward into the building interior during hot weather. (See Chapter 6 for an extended discussion.)

Thus roof designers must never consider a component in isolation; they must always investigate its compatibility with other materials and its effects on the whole system. Far more important than the quality of the individual materials are their design and installation as compatible components of an integrated system.

Expanding roof-plan dimensions are a source of roofing troubles. Its greater size alone makes a large roof a more complex technical problem than a small roof. A roof more than 400 ft long generally should have an expansion joint to accommodate thermal contraction and expansion. On the other hand, a roof only 200 ft long should require no expansion joint. Membrane splitting from movement of unanchored insulation boards, the major cause of splitting, occurs more frequently on large roof surfaces because the boards have greater area to move and produce membrane stress concentrations.

Large, level built-up roofs also run a greater risk of inadequate drainage—a major cause of roofing failure. Long structural framing spans deflect more than short spans, and these deflections increase the probability of ponding. Large plan dimensions increase the inevitable humps and depressions stemming from construction inaccuracies in column-base-plate elevations and fabricated column heights; dimensional variations in deck, insulation, and membrane thickness; and so on. Ponded water creates a host of hazards, ranging from the threat of structural collapse from roofs overloaded with deep ponds caused by clogged drains to leakage threats through defective membrane lap joints and flashings. (See Chapter 3 for a detailed discussion of reasons for draining the roof.) Yet the lure of first-cost economy, achieved through the simpler fabrication of dead-level framing, often seduces owners into accepting a "dead-level" roof, which inevitably contains depressions that pond water.

Despite problems of bad roof design, poor fieldwork nonetheless accounts for most roof failures, according to the majority of roof experts. Some roofing contractors ignore the roof specifications, if they take the trouble to read them. Design errors rank next, with structural deficiencies and material failure further down the list.

Financial pressures cause much faulty fieldwork. Under the threat of liquidated damages extracted by an owner if the project is not completed on schedule, the general contractor often forces the roofing subcontractor to install the roof before the deck is ready, or in damp, rainy, or severely cold weather.

Lessons from History

The roofing industry's conventional wisdom, represented by its more experienced and skeptical members, has been validated over the past decade. Research from NRCA's continuing Project Pinpoint, reported by the dean of U.S. roofing experts, W. C. Cullen, has confirmed several traditional roofing-industry maxims:

- There are no panaceas.
- Re-covering existing roof systems is risky.

The *no-panacea* lesson springs from Project Pinpoint data on roof problems involving various membrane materials. The single-ply materials that appeared to be at least a partial panacea several decades ago have shown that they, too, are only erringly "human." Judged by the experience over the decade stretching from 1983 to 1992, the three most popular membrane materials—EPDM, modified bitumens, and BUR—experienced a similar level of problem projects. The data are too imprecise to draw accurately quantified conclusions, but they point to something quite surprising—that of the three materials, BUR has possibly outperformed the other two. What might be called a *normal* problem ratio of 1.0 (computed as the proportion of problem projects experienced by a given membrane type divided by the proportion of total projects represented by the given membrane type) is considerably higher than the 0.72 ratio reported for BUR.[2] That ratio is better than that reported for EPDM (0.97) and for modified bitumens (0.93). When readers recall that EPDM and modified bitumens appeared as solutions to the apparently intractable problems of BUR systems—the blistering, splitting, ridging, and slippage that harassed building owners a decade ago—they may well ask, "What's going on here?"

The apparent paradox is not difficult to explain. The contemporary BUR system has undergone a beneficent transformation. The troublesome organic and asbestos felts that caused so many problems in the past have been replaced by improved, inorganic glass-fiber felts. Moreover, the problem ratios cited above do not account for other determinants of problem roofs—notably, whether the roofing project was a re-covering, new construction, or reroofing (i.e., tearoff-replacement). Re-covering of existing problem systems is a much more hazardous gamble than new construction. Both EPDM and modified bitumens are more likely than BUR to be used for re-covering, and this factor may well account for their higher problem ratios.

This subject leads naturally into the second roofing-industry maxim, concerning the greater risks involved in re-covering existing roof systems. That maxim was spectacularly validated by the Project Pinpoint data.

Re-covering carries a roughly 50 percent greater risk of roof problems than constructing a new (or replacement) system.[3]

There are many reasons why re-covering an existing system poses additional risks. Anchoring through the old system requires much longer fasteners, which reduces their lateral stability and prospects for dependable anchorage into the deck. Moisture retained in the old system can corrode the fasteners. Flashings are much more complicated on re-covering projects than on new or tearoff-replacement projects. (See Chapter 19, "Reroofing and Repair," for a fuller discussion of re-covering hazards.)

Project Pinpoint's data also show that we can learn from experience. The problem ratio for polyvinyl chloride (PVC) membranes was 3.24, indicating that these membranes failed at a rate more than three times the average failure rate. This high failure rate for PVC is misleading in today's context, as it includes many *unreinforced* PVC projects. Today's PVC membranes are all reinforced, with far better prospects for survival than their unreinforced predecessors. Unreinforced PVC sheets have followed the trail of such roofing dinosaurs as two-ply coated-felt membranes, uncured neoprene flashing, butyl rubber sheets, asbestos felts, and a host of other materials consigned to oblivion by poor performance.

Need for Research and Standards

The roofing industry has lagged in the promulgation of installation standards and test methods. Instead of focusing on the whole field-manufactured roof system, the true guide to roofing-system performance, the industry has concentrated on component-material quality. There are appropriate American Society for Testing and Materials (ASTM) and federal standards for testing important properties of surfacing aggregates, felts, bitumens, insulation, vapor retarders, and structural decks. But except for fire and wind-uplift resistance, there are no generally accepted tests for performance of the entire built-up roof system assembled from these components.

The Europeans have progressed further in this respect than the United States. (See Chapter 2, "The Roof as a System," for further discussion of this topic.) It is an astonishing fact that the pioneering National Bureau of Standards research report, "Preliminary Performance Criteria for Bituminous Membrane Roofing," published in 1974, remains *preliminary* more than two decades later! In that same time span, computer technology has propelled us deep into the Information Age, with marvels that make 1974's computers seem like relics from the Dark Ages.

Yet despite its comparatively elephantine pace, the roofing industry has made discernible progress over the past decade or two. The catastrophic damage from hurricane Hugo in 1989, followed by the even greater wind damage from hurricanes Andrew and Iniki in 1992, spurred cooperative efforts among industry groups in the formation of the Roofing Industry Committee on Wind Issues (RICOWI). RICOWI's narrower goals could serve as wider goals for the entire industry:

- To promote and coordinate research
- To accelerate the promulgation of new consensus standards for design and testing
- To educate the building industry about roof problems

To whatever contribution they can make to promoting these worthy goals, the following pages of this manual are dedicated.

References

1. W. C. Cullen, *Project Pinpoint Analysis: Ten-Year Performance Experience of Commercial Roofing 1983–1992,* NRCA, 1993, p. 2.
2. Ibid., p. 12.
3. Ibid., p. 5.

Chapter

2

The Roof as a System

The low-slope roof system is an assembly of interacting components designed, as part of the building envelope, to protect the building interior, its contents, and its human occupants from the weather. It is one of many other building subsystems, e.g., curtain walls; structural framing; heating, ventilating, and airconditioning (HVAC); ceiling lighting; and internal electronic communications, each similarly designed for a specific function.

Design Factors

For each specific project, the best roof design is a synthesis of many factors. The most obvious and basic requirement is that the roof must satisfy the local building code. The roof probably will also have to satisfy an insurance company's requirements for wind and fire resistance. Beyond these mandatory requirements, the designer should pursue at least an informal life-cycle cost analysis. Ultimately, the roof specification should consider the following design factors:

- First cost and life-cycle (long-term) cost
- Energy conservation opportunities
- Value and vulnerability of building contents
- Required service life
- Type of roof deck
- Climate
- Maintenance

- Availability of materials and component applicators
- Local practices
- Environmental factors

Designing the roof system requires consultation with other members of the design team. The architect must confer with the mechanical engineer about heating and cooling loads to design the insulation and to keep roof penetrations to a minimum, thereby reducing the chances for flashing leaks. The architect must work with the structural engineer to ensure slope and framing stiffness that will avoid ponding and must advise the owner to institute a periodic maintenance and inspection program.

A major purpose of this maintenance and inspection program is to avoid clogged drains, which may pond rainwater to excessive depths and, in some well-publicized incidents, collapse the roof. Another purpose is to check the integrity of flashings and membrane, to ensure that splits, bare spots, blisters, lap-seam openings, and other repairable defects are corrected before they worsen and allow water to enter the building or degenerate to a point where expensive tearoff-replacement is necessary.

Perhaps the most important basic problem confronting the designer concerns roof slope. Through long, sad experience, roof experts have learned that ponded water is a major threat to the integrity of the roof system. Deliberately ponded roofs, those kept perpetually flooded (except in winter), once enjoyed a minor vogue as a means of cooling the roof surface. Roof experts today reject this strategy; there are far less hazardous ways of saving cooling energy than deliberately ponded roofs. Moreover, even deliberately ponded roofs should be sloped for drainage in cold weather.

Basic Roof Components

A modern low-slope roof system generally has three basic components: structural deck, thermal insulation, and membrane. A fourth component, vapor retarder, is sometimes required for roofs over humid interiors in northern climates. Flashing, although not a basic component of the roof system, is an indispensable accessory. It seals joints wherever the membrane is either pierced or terminated—at gravel stops, walls, curbs, expansion joints, vents, and drains. Metal roof panels serve as both deck and membrane.

The roof assembly, including flashing, functions as a system in which each component depends on the satisfactory performance of the other components. The integrity of the waterproof membrane depends on secure anchorage of all components, plus adequate shear strength between the deck and the vapor retarder, between the vapor retarder and the insulation, and between the insulation and the membrane. The insulation's thermal resistance, which can be drastically reduced by liquid moisture, depends on the effectiveness of the vapor retarder and the membrane. The integrity of the vapor retarder, insulation, and membrane depends on the stability of the structural deck (see Fig. 2.1).

Each component has its own unique primary function and also secondary functions that it must serve in conjunction with other component materials. The *structural deck* transmits gravity, earthquake, and wind forces to the roof framing. Its four major design factors are

- Deflection
- Component anchorage

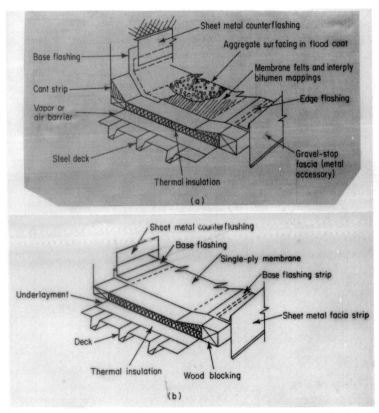

Figure 2.1 BUR-system components include cant strip at base flashings (top), whereas single-ply systems (bottom) require no cant strip.

- Dimensional stability
- Fire resistance

Decks can be classified as nailable or nonnailable (sometimes both) for purposes of anchoring the vapor retarder, insulation, or membrane to the deck (see Fig. 2.2). Some decks, e.g., timber or plywood, should only be nailed because of the threat of heated bitumen dripping through the joints. Poured concrete is generally limited to nonnailed anchorages, except for some mechanically fastened single-ply systems incorporating predrilled anchors.

Vapor retarders can be made of various materials. A common vapor retarder, often known as a *vapor seal,* comprises three bituminous moppings with two plies of saturated felt or two bituminous moppings enclosing an asphalt-coated base sheet. Vapor-retarder materials also include various types of plastic sheets, aluminum foil, and laminated kraft paper sheets with bitumen sandwich filler, or bitumen-coated kraft paper.

Like insulation, a vapor retarder can cause problems, but unlike insulation, the vapor retarder may cause problems that outweigh its benefits. If the insulation contains moisture when installed, the vapor retarder helps prevent its escape because the retarder forms the bottom of a sandwich whose top is the membrane. Moreover, under the

Figure 2.2 Electrically powered equipment drives insulation board fasteners into steel deck flanges. (*ITW Buildex*)

many field threats to its integrity—punctures, cutting of new roof openings, and so on—a vapor retarder almost always admits some water vapor. Most roofing experts today design for vapor release during the hot, summer months, following guidelines from Wayne Tobiasson of the Cold Regions Research and Engineering Laboratory of the Corps of Engineers (CRREL). These experts specify vapor retarders only when CRREL critical conditions are exceeded. (See Chapter 6, "Vapor Control," for expanded discussion of CRREL guidelines.) The best advice appears to be this: "If you need a vapor retarder, make sure it's a good one."

Thermal insulation has many benefits: It cuts heating and cooling costs, enhances interior comfort, and prevents condensation on interior building surfaces. Its secondary functions are also vital. Through its horizontal shearing strength, insulation can relieve concentrated stresses in the membrane, distributing them over wider areas and limiting membrane strain to below the breaking point. Insulation also provides a level substrate for application of membranes on steel decks.

Insulation comes in many types: rigid insulation prefabricated into boards, poured insulating concrete fills (sometimes topped with another, more efficient rigid board insulation), dual-purpose structural deck and insulating plank, flexible batts generally installed under metal panel or plywood decks, and sprayed-in-place foam. Fiberboard insulations are generally the most vulnerable to moisture, which eventually rots and weakens any fibrous organic vegetable board or organic plastic binder. But all insulation materials are vulnerable to some degree to moisture or freeze-thaw damage.

Roof insulation should have four structural properties:

- Good shearing strength, to distribute tensile stresses in the membrane and prevent splitting
- Compressive strength, to withstand traffic loads and (especially in the midwestern states) hailstone impact
- Cohesive strength, to resist delamination under wind uplift
- Dimensional stability under thermal and moisture changes

Because of these demanding requirements, the design of thermal insulation and choice of materials is one of the roof designer's most complex tasks (see Fig. 2.3).

The roof membrane, the weatherproofing component of the system, may comprise several elements. In a built-up roof membrane, there are three: *felts* and *bitumen* alternated like a multideck sandwich, and a *surfacing,* normally of mineral aggregate. The membrane forms

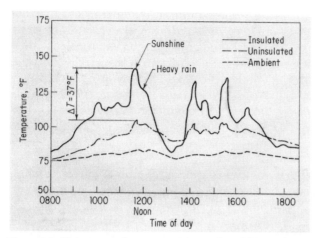

Figure 2.3 Extreme roof temperatures produced by insulation sandwiched between deck and built-up roof membrane can make an insulated roof surface 40°F hotter in sunlight and 10°F colder at night than an uninsulated roof membrane. This extreme temperature cycling accelerates the membrane's deterioration, but in an era of constantly escalating energy prices, well-insulated roofs are essential. Moreover, the thinnest insulation produces nearly as extreme an effect on membrane temperature as the thickest, most efficient insulation (see Chapter 5, "Thermal Insulation"). (*National Bureau of Standards Tech. Note 231*)

Figure 2.4 Contemporary three-ply shingled BUR membrane requires no base sheet, since ridging problems have been eliminated by substitution of inorganic glass-fiber felts, which do not absorb moisture from interior, for organic and asbestos felts. (*RIEI*)

a semiflexible covering, with as few as one or as many as five plies of felt, custom-built to fit the deck (see Fig. 2.4). Polymer-modified bituminous membranes combine bitumen, reinforcement, and surfacing into single- or two-ply components. Polymeric membranes may provide surfacing and membrane in single-layer, prefabricated sheets.

In a built-up membrane, the bitumen—coal tar pitch or asphalt—is the waterproofing agent. The felts stabilize and strengthen the bitumen, prevent excessive flow when warm bitumen is semifluid, and distribute contractive tensile stress in cold, glasslike bitumen.

Mineral aggregate surfacing, normally gravel, crushed rock, or blast-furnace slag, protects the bitumen flood coat from life-shortening solar radiation. Through its damming action, aggregate permits use of heavy, uniform pourings of bitumen (up to 75 lb/square), with consequently better waterproofing and longer membrane service life. It also serves as a fire-resistive skin, preventing flame spread and protecting the bitumen from erosion and abrasion from foot traffic.

Surfacing aggregate may sometimes be omitted on smooth-surfaced asphalt roofs with glass-fiber felts and on some unsurfaced modified-bitumen products. As its chief advantage, a smooth-surfaced membrane facilitates the detection and repair of leaks through membrane fissures obscured by aggregate. As an offsetting disadvantage, however, smooth-surfaced membranes are generally less durable, especially when they are poorly drained.

For the newer membrane materials, weatherproofing protection is more varied. Elastomeric membranes get their ultraviolet resistance from carbon black, which screens the radiation, or from ultraviolet-absorbing chemicals included in the polymer mixture. Modified bitumens have a variety of shielding surfacings: factory-embedded mineral granules, factory-bonded metal foils, or field-applied reflective coatings.

Membrane design requires a knowledge of what the membrane can't do as well as what it can do. No membrane is strong enough to resist large movements in the substrate material. All membranes are threatened by puncture from sharp objects. Wherever more than occasional light foot traffic is anticipated on roofs, and especially around roof-mounted equipment where workers may drop heavy tools, the designer should provide walkways or other means of puncture protection.

Flashings are classified as *base flashings,* which protect and seal the upturned edges of the membrane where it is pierced or terminated, and *counterflashings* or *cap flashings,* which shield the exposed edges of base flashing. Base flashings are normally made of bitumen-impregnated felts or fabrics, elastomerics, plastics, or other nonmetallic material. Counterflashings are often made of sheet metal: aluminum, copper, lead, or stainless or galvanized steel. Generally recognized as the major source of roof leaks, flashing demands as much of the designer's attention as the basic roof components.

Disastrous Combinations

The characteristic problems of roof-system designs are the result of a combination of incompatible materials rather than isolated failures of single components. Two or more components may satisfy their individual material requirements to perfection and yet, in combination, fail disastrously. The art of roof-system design has lagged far behind the introduction of new materials. Because of this lag, some new materials have left a wake of litigation pressed by building owners plagued with failed roofs.

One of the worst system failures stemmed from the use of extruded polystyrene board insulation in conventional built-up roof systems, i.e., with the membrane on top of the insulation. These preformed board insulations offer excellent thermal resistance. But as noted in the previous section, insulation has vital secondary functions as well as primary functions. Through its failure to perform these secondary functions, extruded polystyrene caused roof splits and even pulled flashings from their wall anchorages.

Polystyrene board is unsuitable as a substrate material for conventional built-up roof systems with felt-bitumen membranes mainly because it is virtually impossible to bond it to a deck with hot asphalt. It also has an extremely high thermal coefficient, roughly twice the thermal coefficient of a built-up membrane. Yet expanded polystyrene has proved to be an excellent substrate for ballasted elastomeric systems, where shear resistance and dimensional stability are less critical factors.

Fire resistance also requires compatibility of materials. Through no intrinsic fault of the individual materials, certain combinations can be disastrous. When testing the fire resistance of a roof-ceiling assembly to qualify it for a given fire rating (in hours), one criterion is the assembly's resistance to heat flow. As a safeguard against the roof covering igniting, the average surface temperature rise during the furnace test must not exceed 250°F above the initial temperature. Because insulation retards heat flow, the designer might assume that adding more insulation to a rated roof-ceiling assembly must improve its fire performance.

The designer could be disastrously mistaken. Added insulation indeed depresses the roof-surface temperature, but if heat loss through the roof is excessively retarded, it can cause a structural collapse. A lower roof-surface temperature means a *higher* ceiling plenum temperature (resulting from undissipated heat). This higher plenum temperature could buckle steel joists or ignite combustible structural members that otherwise would continue to carry their loads. The system designer, like a juggler, must watch more than one ball and must never assume that you cannot have too much of a good thing.

A third illustration of the system approach to roof design concerns the method of anchorage. Wind-uplift failures, in which all or part of a roof system is ripped off the deck, make anchorage of the roof components an especially important problem. But when balancing the advantages against the handicaps of the various methods of anchoring roof components, the designer can never focus solely on one troublesome factor, but must always consider several factors. When choosing among the various anchoring methods—nails or other mechanical anchors, hot-mopped bitumen, or cold-applied adhesives—the designer must consider fire resistance and membrane splitting as well as wind-uplift resistance.

Constructing the Roof

Under the present organization on a normal roofing project, the work ideally proceeds as follows:

- The architect specifies the roofing components and installation procedures and submits progress and final inspection reports before acceptance.
- The roofing manufacturer furnishes products complying with the specifications and cooperates with the roofer to ensure that materials are dry when delivered to the site. Under the normal manufacturer's guaranteed roof program, the manufacturer inspects the work of the roofing subcontractor, who has previously demonstrated ability to install the manufacturer's materials.
- The general contractor schedules and coordinates the work of the roofing subcontractor and other subcontractors working on the deck (plumbers, electricians, HVAC workers, and so on), and makes sure that the stored materials are kept dry. The general contractor also provides the roofing subcontractor with a satisfactory deck surface.
- The roofing subcontractor performs the actual field work, coordinated by the general contractor, and installs the vapor retarder, insulation, membrane, and, usually, flashing.

Along with this apportionment of responsibility, a roofing job may be covered by a roofing materials manufacturer's *guarantee* or *warranty*. Warranties, however, sometimes prove worthless when the manufacturer or purveyor of the roof system declares bankruptcy. (For more detailed discussion of warranties and guarantees, see Chapter 21.)

Divided Responsibility

Preceding a roofing failure on a typical job, where the manufacturer's warranty may lull the architect and owner into a false sense of security because they have not studied the warranty provisions, the following sequence of events may occur:

- The designer relies on the integrity of the prequalified roofer and the manufacturer and skimps on roofing-system specifications and details.
- The roof subcontractor exercises the option to select a cheaper roof specification by a manufacturer whose product still qualifies for the specified warranty period.
- The general contractor, disregarding the roofer's qualifications and ignoring the application technique, selects the low roofing bid and relies on the manufacturer's inspection required under the warranty.
- The manufacturer's inspector, often the sales representative who sold the materials to the roofer, is charged with inspecting the work of a customer on whose continued good will the sales representative depends for future material sales.

Assigning responsibility for roof leaks is a formidable challenge. Did the trouble stem from faulty design, careless fieldwork by the roofing subcontractor, defective materials supplied by the manufacturer, abuse by the owner, poor design by the architect, or an inadequate deck installed by the general contractor or another subcontractor? The technical complexity of modern roofing systems, with proliferating material combinations creating new and sometimes unforeseen interactions between components, puts new stress on the creaking structure of responsibilities. Their diffused apportionment encourages a round or two of buck-passing.

Unified Responsibility

For intricate mechanical or electrical systems, a simpler, more rational apportionment of responsibilities has evolved. For an elevator subcontract, for example, the architect sets the general performance standards. The manufacturer advises the architect on hatchway and door clearances, access, power outlets, and other requirements. As the building rises, the elevator company's construction superintendent arranges for the installation of the rail brackets or inserts.

The elevator contract is essentially complete when the architect or owner accepts the installation, following the manufacturer's testing program. Under a contract provision, the elevator manufacturer maintains the equipment, generally for a 3-month period. This more unified responsibility works better than the splintered responsibility of a typical roofing subcontract.

Several short steps toward unified responsibility in the roofing industry include the advent of "total" roof-system guarantees, which include insulation as well as the traditionally guaranteed membrane and flashing components. But the roof system's intimate physical association with the rest of the building precludes the clear, isolated responsibilities assignable under an elevator subcontract.

The current fragmented state of the roofing industry manifests an anomalous disintegration from a more integrated state a century ago. In the late nineteenth century, the manufacturer was also the applicator. When the volume of built-up roof systems increased with the industrial expansion of those times, the manufacturers found their dual role as fabricator and applicator economically impractical. Independent roofing contractors entered the industry, and what was formerly an integrated responsibility split in two. This split has proved a benefit to trial lawyers, a detriment to building owners.

Performance Criteria

We all know the busy doctor's all-purpose prescription for a patient's complaint received late in the working day: "Take two aspirin and call me in the morning." As a quick analogy, the two aspirin represent the prescriptive approach; the patient's call on the following morning initiates the performance approach.

Viewed in a more technical light, performance criteria present another facet of systems design, in which roof-system requirements receive more rigorous analysis and are tested to assure satisfactory service. Since publication of this manual's second edition in 1982, progress in the development of performance criteria has continued, but at a tortuously slow pace. More than two decades after the 1974 publication of the pioneering work BSS55, "Preliminary Performance Criteria for Bituminous Membrane Roofing," by NBS (now NIST) researchers W. C. Cullen and R. G. Mathey, the criteria remain preliminary. And there have been no generally accepted performance criteria promulgated for the new single-ply and modified-bitumen membranes.

Performance criteria constitute a literally radical (back-to-roots) advance over the traditional method of specifying building systems via *prescriptive criteria*. Prescriptive criteria establish requirements that, to some extent, may be arbitrary. They usually represent an *implicit* idea of what is required for satisfactory performance of a building system, but this idea often lacks clear, concentrated focus. With performance criteria, what is merely implicit in prescriptive criteria becomes *explicit*. You start with a general statement of the system's (or system component's) required performance—i.e., to resist hurricane wind-uplift forces. The performance criterion then quantifies the general requirement—e.g., to resist wind-uplift pressures produced by a wind velocity of 120 mph. Next comes a test method to assure the components' ability to pass the test.

The performance approach has two basic goals:

- To provide a precise description of required performance, so that the building system (or system component) will perform more dependably than a system (or component) that merely satisfies some traditional prescriptive requirement
- To allow the system (or component) manufacturer more freedom in providing a product that may be more economical or more suitable than a product restricted by narrow prescriptive criteria

An extreme illustration of the stultifying effects of prescriptive criteria comes from the building industry's historic treatment of masonry bearing-wall design. Prescriptive criteria in the nation's building codes have often degenerated into regulatory, progress-obstructing taboos, delaying for years the introduction of such historic innovations as prestressed concrete, rationally designed timber structures, and the like. With masonry bearing-wall design, the situation became ludicrous. Until the late 1960s, masonry bearing-wall design was governed by prescriptive criteria originally promulgated in 1881. These nineteenth-century criteria ignored the modern technique of reinforcing masonry bearing walls with steel bars. As a consequence, U.S. building codes required masonry bearing walls several feet thick for multistory structures that were perfectly safe with reinforced brick only 8 in thick. These obsolete prescriptive criteria economically eliminated masonry bearing walls from consideration for multistory structures until updated criteria setting rational structural requirements replaced the obsolete prescriptive criteria.

For roof systems, however, code requirements have traditionally been too lax rather than too restrictive. The catastrophic damage inflicted on Kauai by hurricane Iniki in September 1992 was partially attributable to a drastically inadequate design wind velocity of 80 mph, far below hurricane velocity of 120 mph. Until recently, the

South Florida Building Code required only 40 lb/square flood-coat weight for aggregate-surfaced asphalt built-up membranes, compared with the 60 lb/square unanimously recommended by major manufacturers for at least half a century. Lawyers defending contractors and developers often argue that local conditions justify departures from nationally recognized standards. That argument collapses with regard to waterproofing requirements in Florida, where rainfall, both in intensity and in quantity, equals or exceeds that recorded in the other 49 states.

The performance approach avoids both problems created by the prescriptive approach—i.e., obstruction of technological progress, on the one hand, and overly lax requirements, on the other. To avoid the technological stagnation promoted by taboo-style prescriptive criteria, the performance approach is designed to incorporate a format for promulgating the criteria *plus* further steps for reconsideration, via feedback and revision.

As presented in the pioneering publication BSS55, "Preliminary Performance Criteria for Bituminous Membrane Roofing," by NIST (formerly NBS) researchers W. C. Cullen and R. G. Mathey, the format for setting performance criteria comprises four steps:

1. Qualitative statement of the requirement
2. Criterion (quantifying the qualitative statement in Step 1)
3. Test method designed to assure that the system (component) satisfies the criterion
4. Commentary explaining the reason for, or intent of, the criterion

The entire process can be diagramed as an algorithm, as shown in Fig. 2.5.

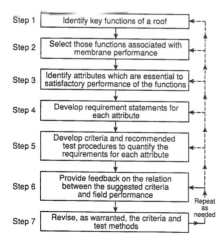

Figure 2.5 Diagram shows a seven-step procedure for low-slope roof-system criteria, including feedback and revision as field performance is monitored. (*NIST*)

As a simple illustration how the performance criteria format works, here is a requirement for a modified bitumen membrane's dimensional stability (from NIST BSS167):

Requirement: The roof membrane shall not exhibit large, irreversible dimensional change when exposed to temperatures encountered under normally expected service conditions.

Criterion: The linear dimensional change shall not exceed 1 percent.

Test: ASTM proposed test method, Section 10, which subjects the test specimens to 24-h oven heating at 176°F (80°C), after which the cooled specimens are measured.[1]

Commentary: Dimensionally unstable membrane materials pose several hazards that can drastically reduce service life. Shrinkage can stress membrane material and cause cracking, as well as loss of adhesion between components. Expansion can create ridges or buckles that can crack or split. Heat exposure is a major environmental factor that may cause irreversible dimensional change.

If, in the future, field observation and laboratory investigation indicate that the 1 percent limit is either too high or too low, this performance criterion can be changed to reflect new data.

With the introduction of elastomeric and modified-bitumen membranes, lap-seam failures have supplanted the prime failure modes of built-up membranes (blistering and splitting). This change obviously required a new emphasis on lap-seam failure, with expanded criteria for lap-seam performance. (Fully half of elastomeric single-ply membrane failures and more than one-third of modified-bitumen failures are lap-seam failures, according to W. C. Cullen.[2])

These expanded performance criteria for lap seams, promulgated in NISTIR 4638, "A Performance Approach to the Development of Criteria for Low-Sloped Roof Membranes" (July 1991), comprise four basic attributes:

- Watertightness
- Puncture resistance
- Water transmission
- Environmental attack

Criteria for each of these attributes (six for watertightness, three for puncture resistance, one for water transmission, and five for environmental attack) are included in Chapter 12, "Elastomeric Membranes."

At their current incomplete stage of development, performance criteria require complementary prescriptive requirements, which pose no obstacle to progress. Again, to take a simple illustration, prescriptive criteria for modified bitumens include a minimum 95 percent of the nominal thickness listed by the manufacturer. Purists can argue that this requirement violates the principles of performance criteria. If a membrane only half as thick as the listed value can perform satisfactorily—resisting hailstone impact, resisting fire exposure, satisfying strain-energy limits, and meeting all the other criteria—then a purist could argue that this membrane, regardless of its thickness, should be acceptable. As a practical matter, however, a few prescriptive criteria holding a manufacturer to its published data help to assure that the purchaser gets the product paid for. Tests for these prescriptive criteria can help to quickly identify or characterize the material.

As indicated in the foregoing discussion, performance criteria can govern the entire roof system or merely a component included in the system. The prototypical criteria governing the entire system involve wind-uplift and fire resistance. For several decades, Underwriters' Laboratories and Factory Mutual System have been testing the performance of entire roof assemblies in resisting fire and wind-uplift exposures designed to simulate actual field exposure. For reasons discussed earlier in this chapter, approvals for various roof assemblies are restricted to the precise combination of components tested. Performance criteria for individual components of the roof system (notably the membrane) can be isolated for readily definable attributes—watertightness, dimensional stability, weathering resistance, impact resistance, strain-energy absorption—all of which are normally essential for the membrane's, and consequently the roof system's, overall performance.

With the proliferation of different roof systems and materials over the past decade, specialized performance criteria have similarly proliferated (see Tables 2.1 and 2.2). Progress in achieving complete sets of performance criteria has nonetheless been slow. Sometimes the obstacle is in setting the performance criterion, sometimes there is a problem devising a satisfactory test, and sometimes both of these obstacles exist. A major deficiency in modified-bitumen criteria, for example, is lack of a generally accepted criterion or test method for seam strength. Since, as previously indicated, defective lap seams are the most common source of modified-bitumen failures, this is an urgent matter.

Indications that the industrialized nations' adoption of performance criteria has beneficial effects come from French researchers Chaize and Fabvier.[3] France's FIT system, introduced in 1989, classifies both system requirements and performance by numerical indexes (with numbers increasing with severity of requirements and levels of performance). For each index (F = fatigue, I = indentation, T = temperature), there are levels of required performance and performance levels

TABLE 2.1 Documents Applying the Performance Approach to Membrane Roofing

Designation[a]	Title	Crit[b]
BRAB (1964)	Study of Roof Systems and Constituent Materials and Components	No
BSS 55 (1974)	Preliminary Performance Criteria for Bituminous Membrane Roofing	Yes
MRCA/PVC (1981)	Recommended Performance Criteria for PVC Single Ply Roof Membrane Systems	Yes
MRCA/ELAS (1982)	Recommended Performance Criteria for Elastomeric Single Ply Roof Membrane Systems	Yes
MRCA/MB (1983)	Recommended Performance Criteria for Modified Bitumen Roof Membrane Systems	Yes
BSS 167 (1989)	Interim Criteria for Polymer-Modified Bituminous Roofing Membrane Materials	Yes
SIA 280 (1981)	Plastic Waterproofing Sheets: Requirements and Test Methods	Yes
UEAtc (1983)	General Directive for the Assessment of Roof Waterproofing Systems	Yes

[a]The year of publication is given in parentheses.
[b]This column indicates whether the document contains recommended criteria.
Key to source acronyms:
BRAB = Building Research Advisory Board
BSS = Building Science Series
MRCA = Midwest Roofing Contractors Association
SIA = Swiss Society of Engineers and Architects
UEAtc = European Union of Agreement
SOURCE: "A Performance Approach to the Development of Criteria for Low-Sloped Roof Membranes," NISTIR 4638, July 1991.

measured by test results to qualify roof membranes and membrane-insulation assemblies.

For fatigue, as one example, there are five risk classes (with 5 denoting the most rigorous and 1 denoting the least rigorous). Fatigue tests cyclically stress the membrane for different amplitudes of joint movement, initial joint width, and temperature. Indentation tests measure puncture resistance (static and dynamic) for varying loads, and temperature tests measure slippage resistance under varying conditions. The designer selects a membrane (or membrane-insulation assembly) with a rating at least equal to the required performance level.

The French researchers report remarkable progress following the institution of national research programs based on the performance approach over the past two decades. Since the 1970s, the proportion of claims for low-slope roof damages have declined from 18 to 5 percent of roof projects, according to Chaize and Fabvier.[4]

TABLE 2.2 Summary of Requirements in Documents Setting Performance Criteria for Membrane Roofing

Requirement	BRAB	UEAtc Gen	SIA 280	BSS 55	BSS 167	MRCA PVC	MRCA ELAS	MRCA MB
Mechanical Loads								
Abrasion resistance	x	·	·	·	·	x	x	x
Adhesion	x	·	·	·	·	·	·	·
Cyclic movement resistance	x	x	·	·	·	x	x	x
Dimensional stability	·	x	x	·	x	x	x	x
Elongation	x	·	x	·	·	x	x	x
Fatigue strength	·	·	·	x	·	·	·	·
Flexural strength	x	·	·	x	·	·	·	·
Flow resistance	x	x	·	·	x	·	·	x
Impact resistance	·	·	x	x	x	x	x	x
Pliability (low-temperature flexibility)	·	x	x	·	x	x	x	x
Puncture resistance—dynamic	x	x	x	x	·	x	x	x
Puncture resistance—static	·	x	x	·	·	x	x	x
Tear resistance	x	·	·	·	·	·	x	·
Temperature-induced load	·	·	·	·	·	x	x	x
Tensile strength	·	·	x	·	·	x	x	x
Thermal expansion	·	·	·	x	·	·	·	·
Thermal shock resistance	x	x	·	x	·	·	·	x
Seam strength	·	x	x	·	·	x	x	x
Strain energy	·	·	·	·	x	·	·	x
Environmental Loads								
Chemical resistance	·	·	x	·	·	x	x	x
Moisture absorption	·	·	·	·	x	·	·	·
Moisture resistance	x	x	x	·	·	x	x	x
Permeability	x	x	x	·	·	x	x	x
Seam leakage	·	x	·	·	·	·	·	·
Surfacing durability	·	·	·	·	·	·	·	x
Weathering resistance (durability)[b]	x	x	x	·	x	x	x	x
Biological Loads								
Resistance to microbial attack	x	·	x	·	·	x	·	·
Resistance to plant growth	x	x	·	·	·	·	·	·
Wind Loads								
Peel (wind)	·	x	·	·	·	·	·	·
Uplift (wind)	x	x	·	x	x	x	x	x
Fire								
Fire resistance	·	x	x	x	x	x	x	x
Number of Requirements	15	16	13	9	9	18	18	21

[a]Document titles appear in Table 2.1 Except for the BRAB report, all documents include suggested criteria for requirements indicated by "x."
[b]This requirement generally includes reference to long term resistance to heat, ultraviolet radiation, ozone, and other pollutants.

SOURCE: "A Performance Approach to the Development of Criteria for Low-Sloped Roof Membranes," NISTIR 4638, July 1991.

Though this section on performance criteria has been totally rewritten for this updated revision, the concluding paragraph of the second edition remains as accurate today as it was in the early 1980s:

Improved roof performance depends less on purely technological progress, manifested in the perennial search for new miracle materials, than on a deeper understanding of the roof as a complex system of interacting components. The correct combination of materials and good field application is as important as material quality as ingredients in a durable, weathertight roof.

References

1. ASTM Draft, "Standard Test Methods for Sampling and Testing Modified Bituminous Sheet Materials Used in Roofing and Waterproofing," November 1987.
2. W. C. Cullen, "Project Pinpoint: Database Continues to Grow," *Professional Roofing,* April 1990, p. 28ff.
3. Alan Chaize and Bruno Fabvier, "FIT Classifications for Roofing Systems," *Proc. Third Intern. Symp. Roofing Technol.,* 1991, p. 356ff.
4. Ibid., p. 356.

Chapter 3

Draining the Roof

If there is any single point on which roofing industry experts are unanimous, it is the necessity of draining the roof. Several decades ago, ponded roofs enjoyed a minor vogue as a means of cooling the roof surface in hot summer weather. But that vogue has long since disappeared. There are now far safer ways of conserving cooling energy. Today's roofing industry consensus can be starkly stated: *If at all practicable, drain the roof. Take every practicable precaution to ensure that no accidental ponds are left after rainfall.*

Well-drained roofs' superior performances are confirmed by the industry's general experience. In a study of 86 randomly selected roofs, Montreal building consultant Donald J. Smith found that 67 buildings with roof slopes less than 2 percent (that is, $\frac{1}{4}$ in./ft) had a 58 percent leak rate, compared with an 11 percent leak rate for 19 roofs sloped at 2 percent or more. A nationwide survey of military authorities responsible for built-up roofs generally paralleled this finding. Although these general findings may lack scientific rigor, they represent an overwhelming empirical consensus. Owners and designers who ignore this experience proceed at their own peril.

Despite the long-standing industry advice to drain the roof, the nation's roof designers still display a durable obstinacy to assimilating this basic lesson of good roof design. To whatever it can contribute to the cause of eradicating this residual resistance to the fundamental rule of good design, this chapter is dedicated.

Ironically, the deliberately ponded, water-cooled roof argument boomerangs against those who cite it as a case against draining the roof. Since 1938, one major manufacturer has published a double-pour, aggregate-surfaced, built-up roof specification for roofs that must resist ponded water (for water cooling the roof surface, for roofs with retarded drainage systems designed to relieve storm sewers, or for roofs subjected to periodic discharges from cooling towers or industrial processes). This specification contains the following requirements:

- A 75 lb/square coal-tar-pitch flood coat, applied with 400 lb/square gravel or 300 lb/square slag, swept free of loose aggregate
- A second 85 lb/square coal-tar-pitch flood coat with 300 lb/square gravel or 200 lb/square slag, thoroughly broomed to remove loose aggregate
- Drainage of the roof prior to freezing weather (to avert the hazards of ice damage)
- Design of the roof structure for minimum design live load of 50 psf
- Recommended use of structural concrete deck, not wood or steel
- Deflection limit of 1/360 span
- Use of a rigid, dimensionally stable insulation with high compressive strength

Only a tiny number of existing roofs can satisfy these rigorous requirements. Note, for example, that slope is required to fulfill the third requirement. (Otherwise how can the water drain?) This collapses the defense of unsloped roofs via the deliberately ponded, water-cooled-roof argument.

In 1963 (and reiterated in its 1989 *BUR Systems Design Guide*), with the unanimous concurrence of all member companies, the Asphalt Roofing Manufacturers Association (ARMA) Builtup Roofing Committee adopted a resolution recommending minimum ¼ in./ft slope. The ¼-in. figure contains a tolerance for the many inherent building industry imperfections. Flat roof surfaces inevitably contain humps and depressions resulting from a host of factors: column-foundation settlement; structural framing and deck deflection (elastic and long term creep); variations from plane surfaces in the deck and top flanges of structural members from inaccurate fabrication or inaccurate field finishing of slabs and topping; dimensional variations in insulation thickness; even variations in interply mopping thickness and flood coat, plus the irregularities of aggregate surfacing, forming dams and pockets impeding rainwater drainage.

Concurring with the ARMA recommendation of ¼-in. minimum slope are the National Roofing Contractors Association (NRCA), the Midwest Roofing Contractors Association (MRCA), and the American Institute of Architects (AIA). Even more important, ¼-in. slope is now required in all major building codes—notably, the Uniform Building Code (UBC) and the Building Officials Conference of America (BOCA) code. Such unanimity among manufacturers, contractors, designers, and, at long last, building officials is rare in the roofing industry.

In addition to being a safe figure for most low-slope roofs, the minimum ¼-in./ft (2 percent) also appears to be an economically optimum slope, according to research prepared for the Air Force.[1] The slight (1½ percent) increase in membrane service life achieved by increasing roof slope to 1 in./ft is economically unjustified, despite some additional reduction in maintenance costs (see Fig. 3.1). A 1-in. (8 percent) slope can substantially increase a building's construction cost.

Despite the importance of drainage and the overwhelming industry consensus, the rule has been casually violated. The nation's high proportion of unsloped roofs is indicated by the ponded roofs observable from a jetliner landing at almost any urban airport in the United States. It is doubtful that half of the nation's flat roofs are properly drained. The prevalence of unsloped roofs has, in fact, been cited during roofing litigation as evidence that an architect who fails to slope a roof (even a smooth-surfaced roof) is not guilty of professional negligence because many of the architect's colleagues pursue the same benighted policy. With the Looking Glass logic characteristic of the

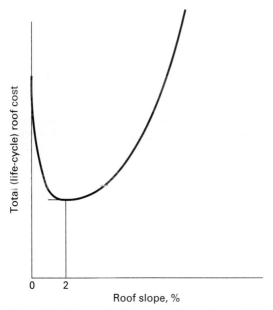

Figure 3.1 The qualitative curve above shows minimum total roof cost (estimated on a life-cycle cost basis) at an industry-recommended 2 percent (¼ in./ft) slope. Steeply sloped roofs raise construction costs, whereas inadequately sloped roofs raise maintenance, repair, reroofing, and litigation costs.

legal world when it defends the indefensible, a lawyer defending an architect client attempts to establish incompetence as the prevailing community standard of competence by which to judge the client's performance.

Why, if this rule to drain the roof is so important, has it been so frequently flouted? The most obvious explanation is first-cost economy. It is almost always cheaper (but only slightly) to build a "dead-level" roof than to slope the roof. A more respectable reason for a dead-level roof is to provide for future expansion requiring conversion of the roof into a floor. Perhaps an even more common explanation for undrained, unsloped roofs is designer ignorance and carelessness. Not only is it *cheaper* to design and construct a dead-level roof, it is *easier*. A nominally dead-level roof is of course never dead-level, but warped, humped, and depressed, with drains normally located near columns, which are usually high points (owing to the increasing deflection of roof framing members in proportion to their distance from columns). It takes care to ensure that openings, walls, expansion joints, and other flashed components are located at high points, or at least elevated above roof low points. Many designers and owners apparently rely on the slovenly logic that just as some cigarette smokers do not get lung cancer, so some ponded roofs do not fail prematurely.

As an additional point about ponding, note that ponding area and depth are important factors when assessing the hazard to the roof system in particular and to the building in general. A roof with 50 percent of its area chronically ponded is about 500 times worse than a roof with only a 1 percent chronically ponded area. Ponding furnishes a source of leakwater proportional to the volume of the contributory ponded area, and volume varies exponentially with ponding depth. If the contributory area of ponded water flowing through a membrane-leak opening is pyramidal, a doubling of ponding depth—say from $\frac{1}{2}$ to 1 in. maximum—multiplies the leakwater volume by a factor of 8.*

Why Drain the Roof?

Treatises on roof design usually attempt to convince readers of the need for positive drainage by citing long lists of reasons. Through their cumulative impact, these long lists are designed to mentally bludgeon the reader into acquiescence. This manual does the same.

---

*A pyramid's volume = $\frac{1}{3}$ × base area × altitude. Thus for the same side slope, an inverted pyramidally shaped ponded area with $\frac{1}{2}$-in. depth has only one-quarter of the base area of a pyramid with 1-in. depth.

But before we get to this recitation, consider an anecdote that cuts to the heart of the issue.

Back in the early 1940s, during World War II, when such novelties were more impressive, one of my Navy colleagues always wore his "waterproof" wristwatch in the shower and the swimming pool. Asked to explain this cavalier attitude toward the hazards of water, he casually replied, "This watch is waterproof."

Some days later, when my friend appeared sans wristwatch, I inquired, "Where's the watch?"

"At the jeweler's," he replied. "It stopped when the works got wet."

The elementary lesson here springs from the question, "Why take unnecessary risks when the prospective losses far outweigh the gains?" Like the "waterproof" watch, a low-slope roof membrane is commonly supposed to be waterproof. And so it is, under favorable conditions. If the design is perfect, the materials perfect, and the application perfect, or nearly so, then you can ignore slope in the roof without a worry in the world. You need worry only if any one of these critical factors—design, materials, or application technique—falls short of perfection. On the other hand, if you cannot count on a combination of perfect roof design with perfect materials and perfect application, you can trust in luck.

Now, to shift to a more technical viewpoint, the basic reason why you should drain the roof is the superiority of *water shedding* over *water resistance* as a means of keeping water out of the building. As anyone who has ever observed roofing application should know, practicable field techniques of applying a roof fall considerably short of laboratory precision. A thin spot in the top coat of bitumen; a fishmouth at a lap edge; a split caused by drying shrinkage of a wet felt, thermal contraction of unanchored insulation boards, or cracking of the substrate material; an interply void that later expands into a blister; or simply a puncture from a dropped tool—any of the foregoing can ultimately destroy the waterproof membrane. Because of its vulnerability to relative movement, a flashing joint submerged in ponded water poses an especially grave leak threat. However, if the roof sheds water, it can survive some imperfections without leaking. The advice to drain the roof is basically as simple as that.

For those unconvinced by this central argument, here is the previously promised list of ancillary reasons:

1. Structural roof collapses, sometimes with fatalities, are periodically caused by ponded water following heavy rains.
2. Ponded water is a source of moisture invasion into the membrane through any membrane imperfections: fishmouths, splits, cracks,

bare felts, or even via an unsealed lap. Infiltrating liquid moisture can invade the insulation or leak into the building. Heated by the summer sun, entrapped moisture can accelerate growth of an interply void into a blister. As ice, moisture can delaminate the membrane as the freezing water expands, and subsequent freeze-thaw cycles can progressively enlarge these delaminated areas.

3. Perennial cycles of ponding and evaporation, corresponding with weather cycles of rainfall and sunshine, accelerate degradation of asphalt and polymeric materials. Ultraviolet radiation degrades exposed asphalt chemically through photo-oxidation, which increases the number of high-molecular-weight hydrocarbons (asphaltenes). It also increases the water-soluble products (OH groups on the HC chains, for which the H_2O molecule has a chemical affinity). Physically, the combination of heat and ultraviolet radiation is manifested in (*a*) migration of oily constituents to the surface (hence the slickness of wet asphalt pavements) and (*b*) hardening. Because it dissolves the photo-oxidized constituents and washes them away, water accelerates asphalt degradation by exposing fresh surfaces to photo-oxidative degradation.[2]

Like asphaltic materials, PVC can suffer accelerated loss of plasticizer from these perennial cycles of ponding and evaporation. This accelerated aging process results in embrittlement and severe shrinkage.

4. Moisture in insulation can rot organic fibers, weaken binders, and drastically reduce the insulation's thermal resistance.

5. Frozen ponded water moves with changing temperature. (Ice has an extremely high coefficient of thermal expansion-contraction.) Thermal movement of ice can erode the aggregate surfacing.

6. Standing water promotes the growth of vegetation and fungi, creating breeding places for insects and producing objectionable odors. Plant roots can penetrate the membrane and spread into insulation. When the plant dies, it leaves a large opening for entry.

7. Wide variations in roof-surface temperature between ponded and dry areas of the roof can range up to 60°F or more in summer, and these temperature differences can promote a warping pattern of surface elongation and contraction, possibly wrinkling the membrane. Lateral migration of entrapped water from hot (high-pressure) to cold (low-pressure) areas can promote condensation that will enlarge the areas of wet insulation.

8. Especially in warm, humid climates, water can penetrate even a sound BUR membrane via the following process: When the sur-

face at 100% RH is warmer than the interior, exterior vapor pressure must exceed interior vapor pressure, and water vapor will migrate into the membrane (a condition that would not exist if the membrane were dry). With daily and seasonal temperature changes, the predominantly inward vapor flow will supply moisture that eventually condenses (when membrane temperature drops below the dew point). Eventually this process wets the felts through the entire membrane cross section.[3]

9. Randomly ponded water often occurs at flashed joints, which may provide easy access for standing water to enter the roof assembly.
10. Evidence of ponded water after a rainfall nullifies some manufacturers' roofing guarantees.

The collective weight of the foregoing items should convince the reader that any kind of dead-level roof multiplies the odds on failure. But the risk also depends on another factor: the type of roof surfacing. If it is imprudent to specify an undrained aggregate-surfaced roof, it is stupid to specify a dead-level or very low-slope smooth-surfaced or mineral-surfaced cap-sheet surfaced roof. No statistical data are available to confirm these estimates, but in a locality with average United States rainfall, say, 30 in./year, it is highly doubtful that an undrained smooth-surfaced roof has better than an even chance of averting premature failure. According to a National Bureau of Standards (NBS)–conducted survey of military roofs at bases in the Northeast, middle Atlantic, Midwest, Alaska, and offshore Pacific islands (Hawaii and Guam), few built-up roofs, even aggregate-surfaced ones, that ponded water gave more than 10 to 15 years' service without major repairs and very expensive corrective maintenance. That finding applied to both asphalt and coal-tar-pitch membranes.

Regardless of the built-up roof's surfacing, even for the most favorable odds offered by an aggregate-surfaced roof, the no-drain policy becomes a ridiculously bad risk if you rationally weigh the initial cost savings gained from a no-drain policy against the probable cost of tearoff-replacement.

Life-cycle Costing Example

Let us examine the economic implications of gambling on a ponded roof. Assume that the probability of a tearoff-replacement rises from 5 to 15 percent (i.e., an increased failure probability of 10 percent). Assume further that this tearoff-replacement cost is incurred after the fourth year. Other assumptions: tearoff-replacement costs 50 percent more than original roof-construction cost of $3 psf with roof-cost escalation rate 5 percent, interest rate 10 percent, and original cost of providing ¼-in. slope for drainage $0.15 psf. With failure probability rising by 10 percent, for 10 roofs there is a probable cost of 1.5×$3.00 = $4.50 (current price) for one tearoff-replacement in exchange for a saving of 10×$0.15 = $1.50 (cost of sloping 10 roofs).

Because this replacement roof has a salvage value (assumed straight-line depreciation) for 4 years of remaining service life at the end of the original roof's 20-year projected service life, this salvage value must be subtracted from the replacement roof's cost for accurate estimation of the future pretax return on investment in slope. This salvage value is the value 4 years from now of the replacement roof's remaining value when it is 16 years old.

Accordingly, for the left-hand term of our equation, we have the Future Worth (4 years from now, when the assumed tearoff-replacement becomes necessary) of the investment in sloping the 10 roofs. For the right-hand term, we have the projected cost of the tearoff-replacement, *minus* the replacement roof's Future Worth (4 years from now) when it is 16 years old. With straight-line depreciation, this value equals the projected tearoff-replacement cost×⁴⁄₂₀ (to account for straight-line depreciation) divided by 1.10 raised to the 16th power, which discounts the projected 20-year Future Worth to a 4-year Future Worth.

Pretax return on investment is thus computed as follows:

$$\$1.50\,(1+r)^4 = \$4.50 \times 1.05^4 - \frac{4}{20} \times \frac{\$4.50 \times 1.05^4}{1.10^{16}}$$

$$1.50\,(1+r)^4 = 5.47 - 0.24$$

$$= 5.23$$

$$r = \left(\frac{5.23}{1.5}\right)^{0.25} - 1$$

$$= 37\%$$

A 37 percent return on investment is probably too conservative because the assumed $0.15 psf cost of sloping the deck is almost certainly too high.

There is, moreover, another dismal fact about tearoff-reroofing expenditures that tends to raise the rate of return on an investment in good drainage in the original design: A dead-level deck may be essentially uncorrectable. Providing proper slope for drainage may require raising peripheral or wall flashings to economically impracticable elevations, thus forcing a compromise on proper drainage. It is not merely more economical to do things right originally; it may be the only practicable time to do them right.

Still another factor may raise the pretax return of an original investment in a properly drained roof. If ponding promotes water invasion into the insulation, the moisture can substantially raise heating and cooling bills because water absorption can triple heat loss through the roof system.

If there are any countervailing positive factors economically favoring randomly ponded roofs, they remain a closely guarded secret.

How to Drain the Roof

For the basic decision on how to drain the roof, the roof designer must decide between *interior* and *peripheral* drainage systems. In an interior drainage system, rainwater flows from elevated peripheral areas to interior roof drains. Leaders conduct the rainwater down through the building interior. Leaders in an interior drainage system are almost always located at columns. In a peripheral drainage system, rainwater flows in the opposite direction, from elevated interior areas to peripheral low points, to scuppers and leaders located outside the building.

Interior drainage has several notable advantages over peripheral drainage. Interior drain pipes, heated by the building interior, continue to conduct rainwater or melting snow through the cold winter weather. Peripheral drainage systems are totally subject to outside temperatures, freezing up and not functioning during cold winter weather. They also require more elaborate flashings, to protect gutter and scupper areas from erosion. Moreover, these peripheral areas are vulnerable to ice damming and metal distortion from freeze-thaw cycles. Peripheral drainage is obviously less troublesome in mild climates where winter temperatures remain above freezing.

Scuppers can, however, provide a safety feature on roofs with interior drainage systems and the threat of clogged drains (see Fig. 3.2). On roofs with parapets that could retain water to hazardous depths, peripheral scuppers can act as overflow valves. They should be located no more than 4 in. above general roof elevation. (A 4-in. average water depth results in an average uniform roof load of 21 psf.)

An alternative to scuppers is to incorporate overflow drains. These are placed in close proximity to interior drains, elevated 2 to 4 in. above the roof surface. If the primary drains clog, the overflow drains carry away excess water. Many plumbing codes require that the leaders from overflow drains be independently plumbed. (See Fig. 3.3.)

For roofs divisible into rectangular areas, the most positive drainage geometry comprises inverted pyramids, with four-way slope into each

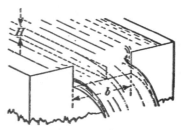

Figure 3.2 A scupper drain is essentially a "contracted" weir with drain capacity calculable by the Francis formula (for water with 0 approach velocity).

$$Q = 3.33(b-0.2H)H^{1.5}$$

where Q = scupper capacity (ft³/s)

b = scupper width (ft)

H = hydraulic head (ft) (assumed equal to scupper height)

Example: Check a 6-in.-wide by 2½-in.-high scupper with 2,500-ft² contributory area in Fort Myers, Florida (maximum rainfall intensity = 4.3 in./h):

$$\text{Required scupper capacity} = \frac{4.3}{12} \times 2,500 = 896 \text{ ft}^3/\text{h}$$

$$Q = 3.33[0.5-(0.2 \times 0.21)]0.21^{1.5} = 0.147 \text{ ft}^3/\text{s}$$

$$Q = 0.147 \times 60^2 = 528 \text{ ft}^3/\text{h}$$

$$\text{Deficiency in scupper capacity} = \frac{896-528}{896} = 41\%$$

Figure 3.3 Overflow drain (right) is at a higher elevation than regular drain (left) as a safeguard. Thin, tapered insulation boards lead to regular drain.

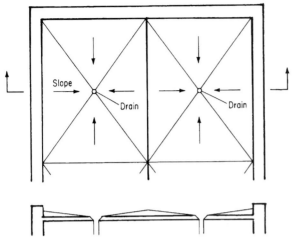

Figure 3.4 An inverted pyramidal pattern with interior drains provides the most dependable roof drainage.

drain (see Fig. 3.4). Inverted pyramids can be created with field-built roof framing systems—wood and cast-in-place concrete. Tapered board insulation, wet insulating concrete fill, or dry-mixed, thermosetting asphaltic fill can also create these sloping pyramidal contours.

With prefabricated roof framing systems, e.g., structural steel or precast concrete, one-way slopes may prove more practicable and economical. Saddles and crickets, formed with tapered board insulation or poured fill, can complement one-way slope to provide positive drainage (see Figs. 3.5 and 3.6).

An important but often neglected aspect of roof design is flashing elevations. Locate flashings at a roof's high points, if practicable. In any event, keep flashings out of a roof's low points. (See Fig. 3.7 and Chap. 14, "Flashings," for further discussion of this topic.)

Drainage Design

To ensure good drainage, specify a minimum of two drains for a total roof area less than 10,000 ft^2. Add at least one additional drain for each additional 10,000 ft^2 of roof area, and limit the maximum spacing of drains in any direction to 75 ft. (That rule limits the contributing area to 5625 ft^2 in large interior areas.)

Irregularly shaped roofs, with penthouses or other obstructions, require additional drains for good drainage. Do not expect water to turn corners as it flows toward drains. Pond-free drainage normally requires straight water flow from high points to a drain. The slight

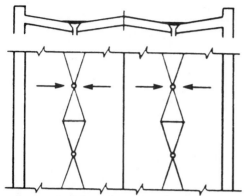

Figure 3.5 Saddles, built by tapered insulation sections, can provide slope in valleys formed by decks only in one direction.

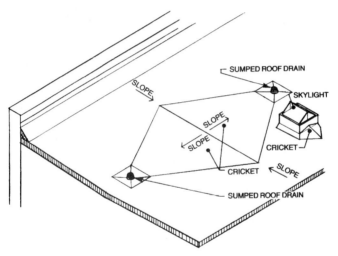

Figure 3.6 Crickets are required on the high side of curbed openings to divert water flow around such obstacles. (*National Roofing Contractors Association*)

additional cost for several extra drains is cheap insurance against ponding. The drain's original installation costs much less than a later addition to relieve ponding. This slight additional drainage-assurance cost is trivial compared with the cost of tearoff-replacement to rectify a condition caused by ponding.

With uncambered roof framing, a drain's best location is near midspan—ideally at the point of the roof framing's maximum deflection (see Fig. 3.8). Avoid drain locations near columns or bearing walls, which are often high points regardless of the designer's intent. The extra cost for additional lateral leaders back to a column is well justi-

Figure 3.7 The scupper drain shown above is located at a higher elevation than the surrounding membrane. This reversal of proper slope direction is more common in exterior drainage systems than in interior drainage. It is also more hazardous. Ponded water at the roof perimeter is more likely to result in leakage than ponded water in interior sections, because flashing is more vulnerable than membrane.

fied. Leaders should start at a 90° angle with the drain head, to prevent deck deflection from dislodging drain rings installed over a rigid vertical drain leader. (Expansion joints can also accommodate this movement.)

Provide sumps at drains, recessing drains below immediately adjacent roof surfaces to prevent local ponding around the drain.

Required drainpipe size depends on three basic factors:

- Contributory area (maximum 10,000 ft^2)
- Rainfall rate (in./h)
- Roof slope (in./ft)

After you determine the contributing roof area per drain, consult local code authorities, the nearest rainfall recording station, or Table 3.1 for maximum rainfall rate.

A 1 in./h rainfall rate builds up at a 0.0104 gpm/ft^2 rate. [Because 1 gal = 231 in.^{3}, gpm/ft^2 = $^{144}\!/_{231}$ × 60 = 0.0104 gpm/(ft^2 · in. · h) rainfall.] For Chicago, maximum rainfall rate = 3 in./h (from Table 3.1) and gpm/ft^2 = 3 × 0.0104 = 0.0312. For a 10,000-ft^2 contributing drain area, you need drain capacity = 0.0312 × 10,000 = 312 gpm. From Table 3.2, read drainpipe diameter of 5 in. (capacity = 360 gpm). Note also from this same table that you need a 6-in.-diameter horizontal drainage pipe (capacity = 315 gpm) for $\frac{1}{4}$-in. slope.

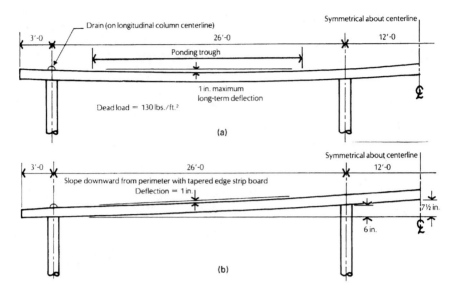

Figure 3.8 Cross section A (top) shows a typical double-loaded corridor school wing designed dead-level. Drains are placed on the centerline of exterior columns, midspan between them. This design made ponding inevitable. The long-term 1-in. dead-load deflection of the 9½-in. concrete flat-plate slab down each line of classrooms forms a ponding trough. It is deepest near the center point of each flat-plate bay (intersection of diagonals between columns at the corner of each bay). This ponding was theoretically about 1 in. lower than the high points at drain locations. (Because of the inevitable construction errors and inaccuracies, the actual elevation difference between drains and low points probably exceeds 1 in. in some locations.)

Cross-section B (bottom) shows how flat-plate structures can be sloped to assure positive drainage. The center of the classroom bay would be the best location for drains. But that would require the expense of drop ceilings to hide the horizontal piping. Note that the minimum recommended slope of ¼ in./ft is necessary to get a slope of considerably less. The 1-in. deflection at a point roughly 11 ft from the exterior column line reduces the slope from ¼ in./ft (2¾ in. in 11 ft) to 1¾ in. in 11 ft or ⅙ in./ft. Construction inaccuracies can reduce this slope still further, possibly to ⅛ in./ft.

Draining an Existing Roof

The toughest roof-drainage problem occurs with an existing roof that ponds water. This is an extremely common problem faced by many building owners with failed roofs and dead-level or inadequately sloped decks. Besides tapered insulation and sloped insulating fills (see Chapter 5, "Thermal Insulation"), another alternative for the existing deck is the *siphon drainage system*; it is, however, less satisfactory than sloping the roof.

The siphon-drainage system is a set of pumps placed in the ponded areas of a roof, with plastic hoses leading to roof drains. The system recycles every hour, pumping only when ponds reach a ⅝-in. depth, when the units prime the drain lines. A siphoning action starts water

TABLE 3.1 Maximum Rainfall Rates (In./h)

State & city	Rainfall	State & city	Rainfall	State & city	Rainfall	State & city	Rainfall
Alabama:		Macon	3.7	Escanaba	2.3	**North Carolina:**	
Anniston	3.6	Savannah	4.0	Grand Rapids	2.6	Asheville	3.2
Birmingham	3.6	Thomasville	4.0	Marquette	2.2	Charlotte	3.4
Mobile	4.2	**Hawaii:**		Port Huron	2.4	Greensboro	3.4
Montgomery	3.8	Hilo	3.0	Saginaw	2.4	Raleigh	3.7
Alaska:		Honolulu	3.2	Sault Ste Marie	2.0	Wilmington	4.0
Anchorage	1.0	**Idaho:**		**Minnesota:**		**North Dakota:**	
Fairbanks	1.2	Boise	1.1	Duluth	2.6	Bismarck	2.8
Juneau	1.0	Lewiston	1.1	Minneapolis	3.0	Fargo	2.8
Arizona:		Pocatello	1.3	St. Paul	3.0	**Ohio:**	
Phoenix	2.5	**Illinois:**		**Mississippi:**		Cincinnati	2.7
Tucson	3.0	Cairo	3.4	Jackson	3.8	Cleveland	3.0
Arkansas:		Chicago	3.0	Meridian	3.8	Columbus	2.7
Bentonville	3.8	Peoria	3.1	Vicksburg	3.7	Dayton	2.6
Fort Smith	3.9	Springfield	3.3	**Missouri:**		Sandusky	3.0
Little Rock	3.7	**Indiana:**		Columbia	3.6	Toledo	3.0
California:		Evansville	3.0	Hannibal	3.4	Youngstown	2.6
Bakersfield	1.5	Fort Wayne	2.6	Kansas City	3.7	**Oklahoma:**	
Eureka	1.6	Indianapolis	2.8	St. Joseph	3.6	Oklahoma City	3.9
Fresno	1.5	South Bend	2.8	St. Louis	3.4	Tulsa	3.9
Los Angeles	2.0	Terre Haute	2.9	Springfield	3.7	**Oregon:**	
Sacramento	1.4	**Iowa:**		**Montana:**		Medford	1.4
San Diego	1.5	Burlington	3.3	Billings	2.0	Pendleton	1.0
San Francisco	1.6	Davenport	3.2	Havre	1.8	Portland	1.4

State & city	Rainfall
Memphis	3.6
Nashville	3.1
Texas:	
Abilene	3.5
Amarillo	3.4
Austin	4.0
Brownsville	4.5
Corpus Christi	4.5
Dallas	4.0
Del Rio	4.5
El Paso	2.4
Fort Worth	4.0
Galveston	4.6
Houston	4.5
Port Arthur	4.5
San Antonio	4.0
Wichita Falls	3.8
Utah:	
Salt Lake City	1.3
Vermont:	
Burlington	2.0
Virginia:	
Lynchburg	3.4
Norfolk	3.8

TABLE 3.1 Maximum Rainfall Rates (in./h)*(Continued)

Location	Rate	Location	Rate	Location	Rate
San Jose	1.6	Des Moines	3.3	Helena	1.5
Colorado:		Dubuque	3.1	Missoula	1.3
Denver	2.5	Sioux City	3.3	**Nebraska:**	
Durango	1.7	**Kansas:**		Lincoln	3.6
Grand Junction	1.6	Concordia	3.7	North Platte	3.3
Pueblo	2.6	Dodge City	3.6	Omaha	3.5
Connecticut:		Topeka	3.7	**Nevada:**	
Hartford	3.0	Wichita	3.7	Reno	1.2
New Haven	3.0	**Kentucky:**		**New Hampshire:**	
Delaware:		Lexington	2.8	Concord	2.4
Wilmington	3.5	Louisville	2.9	**New Jersey**	
District of Columbia:		**Louisiana**		Atlantic City	3.5
Washington	3.4	Lake Charles	4.4	Newark	3.0
Florida:		New Orleans	4.5	Trenton	3.2
Apalachicola	4.3	Shreveport	4.0	**New Mexico:**	
Fort Myers	4.3	**Maine:**		Albuquerque	2.1
Jacksonville	4.0	Portland	2.3	Roswell	2.6
Key West	4.5	**Maryland:**		Santa Fe	2.2
Miami	4.6	Baltimore	3.4	**New York:**	
Pensacola	4.3	**Massachusetts:**		Albany	2.5
Tampa	4.2	Boston	2.6	Binghamton	2.5
Georgia:		Nantucket	3.0	Buffalo	2.4
Atlanta	3.5	**Michigan:**		New York	3.2
Augusta	3.6	Alpena	2.2	Rochester	2.4
Columbus	3.7	Detroit	3.0	Syracuse	2.3

Location	Rate	Location	Rate
Pennsylvania:		Richmond	3.6
Allentown	3.0	Roanoke	3.3
Erie	3.0	**Washington:**	
Harrisburg	2.9	Port Angeles	1.0
Philadelphia	3.3	Seattle	1.0
Pittsburgh	2.7	Spokane	1.0
Reading	3.0	Tacoma	1.1
Scranton	2.6	Walla Walla	1.0
Puerto Rico:		Yakima	1.0
San Juan	4.0	**West Virginia:**	
Rhode Island:		Charleston	2.8
Block Island	3.0	Huntington	2.8
Providence	3.0	Parkersburg	2.6
South Carolina:		**Wisconsin:**	
Charleston	4.1	Green Bay	2.6
Columbia	3.6	La Crosse	2.9
Greenville	3.4	Madison	3.0
South Dakota:		Milwaukee	2.7
Pierre	2.9	**Wyoming:**	
Rapid City	2.7	Casper	2.1
Sioux Falls	3.3	Cheyenne	2.5
Tennessee:		Sheridan	2.1
Chattanooga	3.3	Yellowstone Park	1.5
Knoxville	3.2		

*Tabulated data from Josam Manufacturing Co. (100-year recurrence)

TABLE 3.2 Pipe-Sizing Data

Pipe diameter, in.	Flow Capacity for storm drainage systems, gpm.			
	Roof drain and vertical leaders	Horiz. storm drainage piping slope, in./ft		
		$1/8$	$1/4$	$1/2$
2	30			
2½	54			
3	92	34	46	69
4	192	78	110	157
5	360	139	197	278
6	563	223	315	446
8	1208	479	679	958
10		863	1217	1725
12		1388	1958	2775
15		2479	3500	4958

Tabulated data from Josam Manufacturing Company.

flow down existing drains until each pond is reduced to a ⅛-in. depth. Evaporation then takes over, draining the small remnant of ponded water. (One obvious drawback to this electrically powered siphon-drainage system is that power failure during a rainstorm renders it temporarily useless.)

The *solar-powered drain* is still another technique for draining ponded roof areas and pumping the water to a roof drain or over the building's edge. Cycling radiant solar energy creates a siphon pumping action. Solar radiation absorbed on a black surface within the solar drain's transparent plastic dome increases the internal pressure within the solar drain and exhausts air through an outlet valve. Cooling reduces internal pressure, but the closed exhaust valve prevents pressure-equalizing outside air from entering. Vacuum pressure initiating each new cooling-heating cycle siphons water from the intake hose and pumps it through an exhaust hose.

Alerts

Design

1. Provide minimum ¼ in./ft (2%) slope generally for low-slope roofs, ½ in./ft (4%) for smooth-surfaced BUR roofs.
2. Locate drains at low points (an obvious rule often overlooked).

3. Provide sumps at drains, or other means of recessing drain heads below the adjacent roof level.
4. Favor interior drainage systems over peripheral drainage, especially in cold climates.
5. Provide drain inlets with strainers, to prevent debris from clogging leaders.
6. Provide scupper-type receivers with flanges wide enough for anchorage to wood nailers.
7. When gutters form part of peripheral drainage systems, detail the gutters' outside vertical section at least 1 in. below roof elevation (to prevent water from standing on roof if downspouts are clogged).
8. Use tapered insulation boards to build slopes or crickets around rooftop equipment to keep water away from flashings, moving toward drains.

Maintenance

1. Inspect and clean drain screens regularly.
2. Keep roof clean. (Rubbish impedes drainage.)

References

1. T. R. Sharpe, R. L. Wendt, and J. E. McCorkle, *Design Guide for Roof Slope Selection,* ORNL-6520, Oak Ridge National Laboratory, Oak Ridge, TN, October 1988.
2. P. G. Campbell, J. R. Wright, and P. B. Bowman, "The Effects of Temperature and Humidity on the Oxidation of Air-Blown Asphalt," *Mater. Res. Stand., ASTM,* vol. 2, no. 12, 1962, pp. 988ff.
3. E. C. Shuman, "Some Effects of Moisture Migration and Persistence in Building Materials," *Moisture Migration in Buildings,* STP 779, ASTM, 1982, pp. 66, 67.

Chapter 4

Structural Deck

The structural deck's major purpose—to resist gravity loads and lateral loading from wind and seismic forces—is beyond the scope of this manual. But in addition to its structural function as the base for the roof system, the structural deck must satisfy other design requirements:

- Deflection
- Component-anchorage technique
- Dimensional stability
- Fire resistance (see Chapter 8, "Fire Resistance")
- Surface character (continuous or jointed)

Rational deflection limits that prevent the formation of rainwater ponds can maintain positive drainage. Anchorage of the component layers above the deck is essential to prevent both delamination of the roof assembly by wind uplift or horizontal movement and membrane splitting. Deck dimensional stability depends on the coefficient of thermal expansion-contraction and, in organic fibrous materials, the degree of swelling accompanying moisture absorption. Excessive movement of the substrate can wrinkle or split a bituminous membrane. Dimensional stability is especially important when thermal insulation is either omitted or placed below the deck. And the structural deck must carry its design loads through any fire-resistance tests.

Deck Materials

The basic roof decks commonly used with commercial membrane roofing systems are

- Steel—lightgage, cold-rolled sections
- Wood sheathing—sawed lumber, plywood or OSB (oriented stranded board)
- Concrete—poured-in-place or precast
- Gypsum—precast or poured-in-place
- Preformed, mineralized wood fiber
- Composite decks of lightweight insulating concrete on corrugated steel or formboards

Steel decks, by far the most popular, usually come in a narrow range of gages—22 gage (0.028 in.), 20 gage (0.034 in.), and 18 gage (0.045 in.) thick (down to 28-gage for centering for lightweight insulating concrete fill). In cross section, steel deck is ribbed, with ribs generally spaced at 6 in. (on centers) and $1\frac{1}{2}$ or 2 in. deep. The slope-sided ribs measure 1 in. across at the top for *narrow-rib* deck, $1\frac{3}{4}$ in. for *intermediate* deck, and $2\frac{1}{2}$ in. for *wide-rib* deck. Prefabricated units are 18 in. to 3 ft wide, in 6-in. increments, up to 20 ft or more long. Spans are typically $5\frac{1}{2}$ to 7 ft (see Fig. 4.1).

Wood decks come in three basic types:

1. Board sheathing (up to 1-in. nominal thickness)
2. Plank (to 5-in. nominal thickness)
3. Plywood (minimum $\frac{15}{32}$ in. (12 mm) thick, 4×8 ft panels) (see Fig. 4.2)

Underlayments are generally nailed or stapled to wood decks.

Concrete decks come in two forms: cast-in-place and precast. A cast-in-place structural deck is continuous, except where interrupted by an expansion joint or another building component.

Precast concrete decks come in a variety of cross sections: single T, double T, inverted channels, solid or cored plank. Single or double Ts vary in width from about 2 to 8 ft, with spans up to 60 ft or so. Inverted channels range from shallow, 3-in. depths to 12 in., with spans ranging up to 20 ft or so. Planks, solid or cored, range in depth from 2 to 10 in., in 16- to 24-in. widths, and spans up to 25 ft or so. Because joints between precast deck units are usually uneven, they may require mortar troweling to smooth the surface for insulation.

Gypsum decks, like concrete, are either precast or cast-in-place. A cast-in-place gypsum deck is poured on formboards spanning flanges of closely spaced steel bulb tees. Like a poured concrete deck, a cast-in-place gypsum deck presents large, seamless expanses of roof surface

Structural Deck   47

Figure 4.1 Lightgage steel deck sections come in sections of 18- to 36-in. width, generally around 30-ft length. (*Roofing Industry Educational Institute [RIEI]*)

(except where it cracks from thermal contraction or drying shrinkage).

Precast gypsum planks come in 2-in.-thick panels, with metal-bound tongue-and-groove edges. Because gypsum decks are nailable and noncombustible, insulation can be either nailed or bonded with hot bitumen.

Preformed, mineralized wood fiber, in $1\frac{1}{2}$- to $3\frac{1}{2}$-in -thick planks, comprises long wood fibers bonded with a cement or resinous binder and formed under a combination of heat and pressure. Some preformed, mineralized wood-fiber planks have an integrally bonded foam layer for more efficient combination of this deck's dual structural-insulating function.

Decks can be classified as *nailable* or *nonnailable*. Nailable decks include wood, structural wood fiber, gypsum, and lightweight insulating concrete (see Fig. 4.3). Structural concrete is generally considered nonnailable, though there are specially developed mechanical fasteners designed for anchorage into structural concrete (see Fig. 4.4). Lightgage steel decks require a mechanically fastened overlay (i.e., a thin insulation board).

Figure 4.2 Prefabricated plywood decks are popular on the West Coast. (*RIEI*)

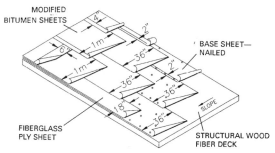

Figure 4.3 Structural wood fiber deck provides a nailable substrate for base sheet. (*Tamko*)

Decks are also classified for fire resistance—combustible or noncombustible. Wood and preformed wood fiber are combustible, whereas steel, concrete, and gypsum are noncombustible.

Design Factors

Design factors, considered in order, are

- Sloped framing for drainage

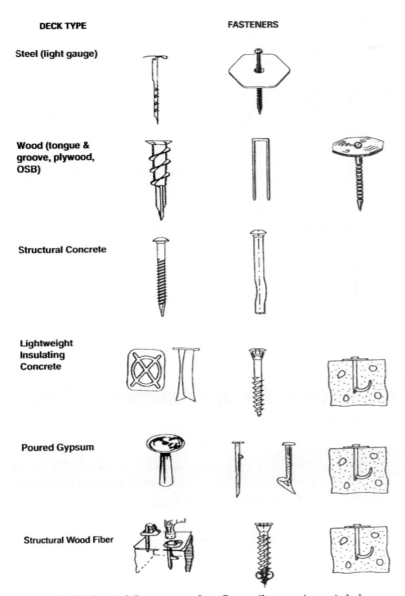

Figure 4.4 Mechanical fasteners and roofing nails come in varied shapes, sizes, and anchorage techniques. (*RIEI*)

- Deflection
- Dimensional stability
- Moisture absorption
- Anchorage of components
- Deck surface and joints

Slope

The most dependable and often most economical way to provide slope on a new building's roof is to slope the structural framing and deck. This is relatively simple with cast-in-place concrete and wood-framed structures, which lend themselves to ready field sloping. Structural steel supporting steel decks is slightly more difficult to slope. It is very difficult to slope the dead-level plane of a ponded, failed roof. As demonstrated in Chapter 3, the economic returns from draining the roof and avoiding ponded-water problems should far outweigh the trivial cost of planning and building slope into the structure. This is usually better accomplished by sloping the framing than by using tapered insulation or tapered fills.

Compared with sloped framing and constant insulation thickness, tapered insulation has the following four disadvantages:

- Difficult design coordination for flashing heights and other roof details affected by varying insulation thickness
- Fairly complex field operations, involving special field drawings and coded insulation pieces for most tapered insulation systems
- Hazards of excessive insulation thickness—6, 7, or even 8 in. is sometimes required for adequate slope, with possible reduced membrane restraint and consequently increased splitting risk
- Complication of longer fasteners needed to penetrate deck

Adding sloped insulating concrete fills sandwiched between the deck and the membrane to provide slope carries two unique liabilities:

- Increased risk of entrapped moisture within the roof assembly, especially in ordinary steel decks that lack underside vent openings or over concrete
- Requirement for dry, warm weather to pour wet fill

Deflection

Deflection limits normally set in building codes, manufacturers' bulletins, and other guides or recommendations have little or no relevance to roof problems. The typical deflection limits—$\frac{1}{180}$, $\frac{1}{240}$, or even $\frac{1}{360}$ of the span for live, or even total load—may be excessive for long-span roofs, which are especially vulnerable to ponding, and too conservative for short-span decks. Structural failures, including sudden and total collapse, have occurred in some poorly designed level roofs. Most of these roofs satisfied the normal code design provisions.

Ponding caused by faulty design can be explained as follows: The deflection curve produced by an accumulating weight of rainwater may form a shallow basin in a roof. If the outflow does not prevent the capacity of the basin from increasing faster than the influx of rainwater, the roof is unstable, and a long, continued rainfall is structurally hazardous. Even if the structural deck withstands the ponding load, the standing water threatens the membrane, the insulation, and ultimately, if the roof leaks, the building contents.

As a rough safeguard against ponding, for level roofs designed with *less* than the minimum recommended slope of $\frac{1}{4}$ in./ft, the *sum* of the deflections of the supporting deck, purlins, girder, or truss under a 1-in. depth of water (5-psf load) should not exceed $\frac{1}{2}$ in.

A proper code provision would specify a minimum stiffness, such as maximum deflection = $\frac{1}{2}$ in. for live load = 5 psf (the weight of 1 in. of water). Thus the deflection would increase at roughly half the rate of the rainwater buildup. If the roof structure does not satisfy this requirement, the designer should present computations substantiating the safety of the slope used.[1]

When computing deflections, the designer must consider plastic flow in concrete and wood. Plastic flow (increasing deflection under constant prolonged loading) is inevitable in materials like reinforced or prestressed concrete. Laboratory tests indicate plastic strains ranging from $2\frac{1}{2}$ to 7 times elastic strains in structural concrete. Because these larger extremes seldom occur in practice, the normal assumption is that plastic flow adds an increment of $2\frac{1}{2}$ to 3 times elastic deflection. Thus the total deflection attributable to elastic strain + creep = 3 or 4×instantaneous elastic deflection.[2]

Wood is similarly subject to long term plastic deformation. For glued laminated members (including plywood) and seasoned sawed members, residual plastic deformation is about 50 percent of elastic deformation. For unseasoned sawed members, plastic deformation rises to 100 percent, thus requiring a doubling of computed elastic deflection to approximate long term deflection.

In highly humidified interiors, preformed structural wood-fiber plank sometimes exhibits excessive, and erratically unpredictable, permanent inelastic deflection. These deflections are not a true creep or plastic flow, attributable solely to load-produced strain. A combination of thermal deformation and moisture absorption, which weakens certain cementitious binders used in wood-fiber materials, causes some preformed structural fiber planks to sag—up to ½ in. in a 4-ft span. Such excessive deflection threatens the entire roof assembly, increasing the chances of membrane splitting. Portland cement binder makes the preformed wood-fiber plank generally immune to excessive deflection.

Deflection and deck movement can also be aggravated by foundation settlement. In a roof designed for the minimum recommended ¼-in. slope, foundation settlement is normally insignificant. (A differential settlement of 1 in. in adjacent column footings 30 ft apart reduces a ¼-in. roof slope by only 13 percent.) Roof-deck vibration from seismic forces, traffic, or vibrating machinery may also require attention.

Changes in deck-span direction also require attention, as they produce stress concentrations (see Fig. 4.5).

Dimensional stability

The dimensional stability of a roof deck is determined largely by its coefficient of thermal expansion and dimensional change with changing moisture content. These factors vary greatly with different materials.

Wood has a relatively low thermal coefficient of expansion longitudinally, from 1.7×10^{-6} to $2.5 \times 10^{-6}/°F$ for sawed members of different species and about $3 \times 10^{-6}/°F$ for plywood. The average value is thus about one-sixth the value for aluminum, less than one-third the value for steel. Moisture is the greatest threat to a wood deck's dimensional stability. Under the extreme moisture variation normally anticipated in service, the expansion of plywood roughly equals the expansion of steel or reinforced concrete under a 150°F temperature rise.

Moisture absorption

Sooner or later, as an expanding gas, expanding freezing liquid, or contracting solid, moisture retained in materials like wood, concrete, and gypsum damages the roof.

Opinions vary about the time required for poured decks to dry. If the roof system is mechanically fastened, base sheets are normally installed when the deck is sufficiently cured to withstand construction traffic. Proper dryness to assure adhesion can be simply resolved, under normal conditions, by the U.S. Army Corps of Engineers tests for dry deck. Pour a small amount of bitumen on the deck. Watch for frothing or bubbling. If none occurs, the deck is probably dry enough for hot-bitumen application. After the bitumen cools, try another test. If you can readily remove it with your fingernails and hands, reject the deck as too wet for application. If the cooled bitumen sticks to the deck too tightly to be manually removed, then accept the deck as dry enough for insulation or roofing application.

Stress concentrations at deck-span change

There are four components of stress concentration at changes in deck-span direction:
1. Axial tensile stress concentration
2. Additional tensile stress from bending rotation at end of deck support
3. Horizontal shearing stress
4. Vertical shearing stress

At the typical support condition, with both adjacent deck sections supported, the first three stress components will occur. Where one deck section is cantilevered to the stress-concentration line, all four stress components can occur.

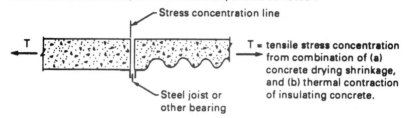

① Axial tension at change in deck-span direction

② Additional tensile stress concentration from bending

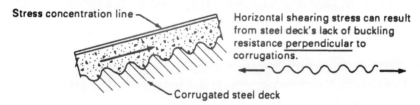

③ Horizontal shearing stress along stress-concentration line

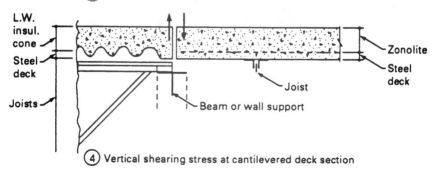

④ Vertical shearing stress at cantilevered deck section

Figure 4.5 At change of deck span an expansion joint is generally required to prevent cracking of lightweight insulating concrete (see Chapter 14, "Flashings," for expansion joint details).

Because of their high absorptivity, poured gypsum and cement-fiber board are not recommended for roof decks in buildings whose uses involve high temperature and high relative humidity (e.g., laundries, bakeries, and textile mills). For buildings with highly humidified interiors, concrete and properly pressure-treated wood are the most generally satisfactory roof-deck materials. When properly maintained to control corrosion, galvanized or painted steel decks are also satisfactory.

Wood decks can rot when exposed to high, unrelieved humidity. In the normal airconditioning range of 40 to 50% relative humidity (RH), wood's moisture content generally runs 8 or 9%, far below the 35 to 50% range required for active rot-fungus growth. But in unvented ceiling plenums, where periodic condensation can keep relative humidity close to 100%, wood decks can rot.[3]

Anchorage of components

For anchoring roofing or insulation to the structural deck, you have a choice of two basic techniques:

- Nailing or mechanical fastening
- Adhesives (hot or cold applied asphalt or foam)

Nailing or mechanical fastening is generally the most dependable method. The alternatives to mechanical fastening are hot asphalt and foam adhesives. When the application is properly carried out on a dry deck surface, hot asphalt generally provides the strongest adhesion.

Mechanical fastening, however, has a critical advantage over hot asphalt as a dependable anchorage technique because of its vastly greater tolerance for cool and even windy weather. Because of their extremely high heat conductivity, steel and structural concrete decks congeal hot asphalt within a few seconds, even under highly favorable conditions (i.e., 70°F, 0-mph wind). Steep asphalt cools from 500°F to 300°F (minimum temperature for assuring dependable adhesion) within 6 seconds for both steel and structural concrete, according to cooling curves calculated by NIST researchers (see Fig. 4.6). Lower temperature accompanied by wind would reduce this already short time interval even further. In cold weather with any significant wind velocity, hot-asphalt adhesion is, if not impossible, highly improbable, to say the very least. Cold adhesives and foam may also pose difficult problems in cold weather, as there may be condensed moisture or a thin layer of ice on the deck.

Other considerations can rule out mopping, especially solid mopping, as an anchoring method. On deck surfaces subject to shrinkage cracking, solid mopping increases the hazard of membrane splitting. By bonding two components throughout their contact area, solid mopping intensifies local stress concentration in a membrane directly

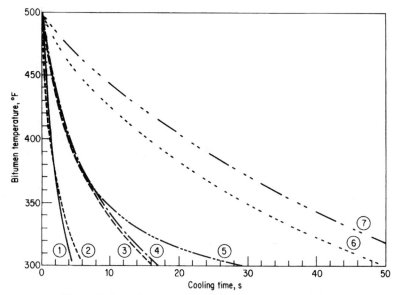

Figure 4.6 Hot-asphalt cooling curves (calculated by NIST researchers) show the tremendous variation in cooling time from 500 to 300°F for hot asphalt mopped to substrates of varying thermal resistance and heat capacity. Air and substrate temperatures are 70°F; wind speed is 0 mph. Substrates are (1) concrete, (2) steel, (3) insulating concrete, (4) plywood, (5) felt/PUF insulation, (6) fiberglass insulation, and (7) PUF insulation. Concrete, which promotes the fastest cooling, has low thermal resistance and high heat capacity. Urethane, which promotes slow cooling, has high thermal resistance and low heat capacity. (*National Institute for Standards and Technology.*)

over the deck crack. Solid mopping also seals in entrapped moisture, thereby promoting blisters and membrane wrinkling.

Strip or spot mopping alleviates both problems. By leaving areas unbonded to the membrane to distribute cracking strains, strip or spot mopping reduces membrane stress concentrations and possible splitting. Intermittent mopping provides lateral avenues of escape for water-vapor-pressure relief, which reduces the hazard of membrane blistering or wrinkling.

For lightweight insulating concrete and preformed, mineralized wood-fiber decks, which are subject to both shrinkage and high moisture absorption, nailing is mandatory. It provides better stress distribution in the membrane and better access for water-vapor-pressure relief to vents.

Deck surface and joints

As a working surface for the application of insulation membrane, poured-in-place decks have an advantage that at least partly offsets the moisture problems: They provide large deck areas without joints. Horizontal gaps between adjacent precast (or precut) units may allow bitumen to drip through, creating a potential fire hazard. Vertical misalignment between adjacent units can ruin the surface as a substrate for applying the insulation.

There are several techniques for closing the horizontal gaps between prefabricated units and leveling the ridges formed at joints. For wood sheathing, tongue-and-grooved instead of square-edged boards help in both respects: The tongue-and-grooved joint is much tighter than a square-edged butt joint, and the tongue-and-grooved boards do not warp as easily as square-edged boards. For plywood decks, when there is no supporting purlin or other structural member under the joint, H-shaped metal clips between adjacent units prevent unequal deflection and thus avoid the formation of vertical irregularities.

As precautions against bitumen drippage through the joints of wood decks, plywood can be stripped with felt, and board sheathing can be covered with 5-lb rosin-sized paper or No. 15 asphalt-saturated felt. For glass-fiber-felt membranes on nailed wood decks, one manufacturer recommends a glass-fiber "combination" sheet: an asphalt-coated base sheet with kraft paper laminated to the underside.

Joints in preformed, mineralized wood-fiber or precast concrete decks should be pointed with cement mortar, which is itself subject to shrinkage cracking. Prestressed concrete units bent in a convex deflection curve by the prestressing may require a concrete topping to level a surface marred by extreme joint irregularities. In less serious instances, precast, prestressed units may require only local grouting along joint lines.

Low spots on concrete deck surfaces (more than $\frac{1}{2}$ in. below level) should be filled with portland cement mortar. High spots should be ground down.

Before insulation, membrane, or vapor retarder is applied, poured concrete decks should receive a surface primer or cutback bitumen. The primer is dual-purpose:

- To absorb the dust inevitably remaining even after the surface is cleaned
- To penetrate into the concrete pores, thereby improving adhesion via increased level surface area and reduced surface tension

Uninsulated bulb tees used to support the gypsum and structural wood-fiber decks can create problems. Bulb tees without covering insulation act as thermal bridges, transferring heat through the highly conductive steel. (This thermal-bridge heat transfer can melt snow in parallel strips.) By creating varying membrane surface temperatures (lower in the exposed strips than in the insulated snow-covered areas), this melted-snow pattern can cause different thermal stresses in the membrane.

Thermal contraction or expansion of uninsulated steel bulb tees and their supporting joists can promote structural cracking of a poured gypsum deck, which is also subject to drying shrinkage. These thermal movements heighten the risk of membrane splitting in either a gypsum or structural preformed wood-fiber deck.

To avoid these hazards with uninsulated bulb tees, always specify a layer of insulation over the gypsum or structural wood-fiber decks. In this energy-conscious era, you will almost always need additional insulation for both its thermal resistance and its membrane-protecting qualities.

Steel Decks

As the most popular deck material by far, used for about 65 percent of new industrial buildings and many other building types, lightgage, cold-rolled steel merits special attention. The steel deck's fluted (i.e., nonplanar) surface has a marked disadvantage compared with other deck materials as a roof-system substrate, and the steel deck's flexibility, its often distorted joints and its general vulnerability to localized distortion make it a special problem for securely anchoring insulation.

Most problems with steel decks concern excessive deflection. Steel decks are subject not only to longitudinal but to transverse deflection (dishing or "rolling"). Live-load deflection in service *after* the insulation is in place can break the original bond if the deck and insulation do not deflect in perfect congruity.

Construction loads or handling stresses can cause even greater problems. Permanent deflection or distortion (beyond the steel's elastic limit) can prevent adhesive bonding of the insulation and deck at the time of application. Because construction loads normally exceed loads imposed during the building's service life, Factory Mutual researchers tested decks with four 250-lb stationary concentrated loads, each representing a 200-lb man carrying roofing felts or other material, located at midspan of one end span and the center span of a three-span continuous deck spanning up to 7 ft.

Steel-deck design requires special attention to three items:

- Side lap fastening
- End lap detail
- Span and thickness

Side lap fastening is vital to prevent differential deflection between adjacent steel-deck units. This differential deflection can result in:

TABLE 4.1 Maximum Spans for Lightgage Steel Deck, per FMRC Recommendations

	Steel deck gage		
Rib type	22	20	18
Narrow (1-in.)	4'10" (1.5 m)	5'3" (1.6 m)	6'10" (1.8 m)
Intermediate (1¾-in.)	4'11" (1.5 m)	5'5" (1.6 m)	6'3" (2.0 m)
Wide (2½-in.)	6'10" (1.8 m)	6'6" (2.0 m)	7'5" (2.3 m)

- Permanent deck distortion (because the steel is locally stressed beyond the elastic limit)
- Breaking of adhesion with the insulation
- Fracture of insulation board and membrane

Correct end lapping of the steel sections is similarly necessary to avert differential deflection. If the ends are cantilevered, they can deflect differentially under concentrated loads, with the same consequences as inadequate side lap fastening.

Deck spans should be limited by stress loading calculations and maxima set by Factory Mutual Research Corporation (FMRC). (See Table 4.1.)

Recommendations for steel decks

1. Deck thickness must at least equal the value upon which the manufacturer's moment of inertia calculation is based. Minimum acceptable thickness for published gages are:

    18 gage = 0.045 in. (1.143 mm)

    20 gage = 0.034 in. (0.864 mm)

    22 gage = 0.028 in. (0.711 mm)

2. Provide structural steel framing (angles) around openings that interrupt more than one rib.

3. Space side lap anchors (No. 8 self-drilling screws) at maximum 3-ft spacing (i.e., midspan location for spans up to 6 ft, two fasteners for spans over 6 ft, at one-third points).

4. Anchor deck to bar joists with ½-in.-diameter welds or screw fasteners at maximum 12-in. spacing. At roof perimeter, anchor at 6-in. maximum spacing (i.e., at every rib).

5. End laps = 2-in. minimum, located over supports, with one end crimped to facilitate nesting. Maximum dimension between end lap surfaces = $\frac{1}{16}$ in.

6. Live-load deflection limit = $\frac{1}{240}$ span, computed by the following formulas:

$$\text{Single deck span: } D = 0.015 \frac{KPL^3}{EI}$$

$$\text{Two or more spans: } D = \frac{KPL^3}{48\,EI}$$

where $K = 1$ (unless substantiated as less than 1)
$I =$ moment of inertia (in.^{4})
$E = 29.5 \times 10^6$
$L =$ span (in.)
$P = 300$ lb*

7. Provide factory galvanizing or factory coating with aluminum zinc alloy.

Alerts

Design

1. Check deck and supporting structure for ponding deflection. Include creep or plastic flow factor when computing deflection of concrete or wood members. Also check for deflection under equipment wheel loads (see Fig. 4.7).
2. Design the roof to drain, with $\frac{1}{4}$-in. minimum slope, preferably by sloping deck and framing.
3. Locate drains near midspan, not near columns (unless beams are cambered to slope toward columns).
4. Check the method of attachment of next component above deck with insulation manufacturer, roofing manufacturer, and insur-

---

*To represent a 300-lb *dynamic* load, simulating a 200-lb man carrying 100 lb of material, as opposed to a *static* load, P should be a minimum 600 lb because a dynamic load at least doubles the static-load deflection, in accordance with the following formula:

$$\Delta = \Delta_{st} + \sqrt{\Delta_{st}^2 + 2\,\Delta_{st} h}$$

where $\Delta =$ deflection under *dynamic* load
$\Delta_{st} =$ deflection under *static* load
$h =$ height (in.) from which dynamic load falls

Thus Factory Mutual's deflection formulas are not conservative enough to constitute a true representation of the dynamic construction loads.

Figure 4.7 Decks should be checked for equipment wheel loads, which may deflect steel decks in particular and break the bond between deck and insulation. (*RIEI*)

ance agent (see Chapters 7 and 8, "Wind Uplift" and "Fire Resistance," respectively.)

5. To reduce the threat of bitumen drippage when insulation, vapor retarder, or base sheet is hot-mopped to the deck, specify steep asphalt. Specify rosin-sized paper, felt, or tape over joints in wood sheathing, plywood, or other prefabricated units.
6. Check the fire resistance of the entire roof-deck-ceiling assembly (see Chapter 8, "Fire Resistance").
7. Limit steel decks to minimum thickness of 22 gage.
8. Specify minimum $\frac{3}{8}$-in.-diameter nail heads, driven through metal caps of minimum 1-in. diameter and 30-gage thickness.
9. Specify insulation over bulb tee steel-deck supports.
10. Specify a vapor barrier over a poured gypsum (or lightweight insulating concrete) deck, to prevent vapor from migrating into the insulation.

Field

1. Require a smooth, plane deck surface, with proper slope. For prefabricated deck units, tolerances are

Vertical joints: $\frac{1}{8}$-in. gap

Horizontal joints: $\frac{1}{4}$-in. gap

Flat surfaces of steel decks: $\frac{1}{16}$ in. (between adjacent ribs)

When the vertical joint tolerance is exceeded, require leveling of the deck surface on a maximum 1 in./ft slope with grout or other approved fill materials.

When the membrane is applied directly to the deck, require caulking or stripping of all joints. Where insulation is applied on monolithic poured-in-place decks, fill the low spots (more than $\frac{1}{2}$ in. below true grade) and grind high spots. Cover openings over $\frac{1}{4}$ in. in wood decks with nailed sheet metal.

2. Permit no deformed side laps or broken or omitted side welds in metal decks.
3. Limit moisture content in deck materials to satisfactory levels—for example, 19 percent for wood and plywood.
4. Require waterproofed tarpaulin coverings over moisture-absorptive deck materials (wood, preformed wood fiber, precast concrete, precast gypsum) stockpiled at the site. Also require that these materials be blocked above the ground.
5. Prohibit placing of vapor retarder, base sheet, insulation, or roofing on a deck containing water, snow, or ice. Decks must be clean and dry (including flute openings of metal decks) before next roofing component is applied. (Refer to the moisture test described under "Moisture Absorption.")
6. Place plywood walkways on a vulnerable deck surface, e.g., light-gage steel.
7. Forbid stacking materials on the deck in piles that exceed design live load. Also check for equipment wheel loads.
8. Set close tolerances for steel-column base plate elevations (because they determine roof framing and deck slope).

References

1. For a more elaborate analysis of required roof slope and framing stiffness, see the *Timber Construction Manual,* John Wiley & Sons, Inc., New York, 1966, pp. 4–153ff and "Roof Deflection Caused by Rainwater Pools," *Civ. Eng.,* October 1962, p. 58.
2. For more detailed analysis of the part played in the deflection of concrete structural members by shrinkage and temperature deformation as well as plastic flow, see *Proc. Am. Concr. Inst.,* vol. 57, 1960–1961, pp. 29–50; also see W. G. Plewes and G. K. Garden, "Deflections of Horizontal Structural Members," *Can. Build. Dig.,* no. 54, June 1964.
3. M. C. Baker, "Designing Wood Roofs to Prevent Decay," *Can. Build. Dig.,* no. 112, April 1975, pp. 112-1, 112-3.

Chapter 5

Thermal Insulation

Thermal insulation in an airconditioned or merely heated building offers the greatest return on initial investment of any building material. Substantial thermal insulation is indispensable for occupied buildings.

Besides reducing energy costs, insulation also enhances building occupants' comfort. In cold weather it raises interior surface temperatures, thus reducing radiative losses that chill the occupants. In summer it reduces interior surface temperatures that retard occupants' heat rejection and make them feel uncomfortably warm.

Roof insulation has three other benefits:

- It can prevent condensation on interior surfaces.
- It furnishes an acceptable substrate for the membrane over steel decks.
- It stabilizes deck components by reducing their temperature variations and consequent expansion-contraction.

For a conventional roof system (membrane on top of insulation), insulation has three drawbacks:

- Although it reduces the probability of condensation on exposed interior surfaces, insulation increases the probability of condensation *within* the roof system. (Insulation creates a wider temperature range between the roof assembly's interior and exterior surfaces. During the heating season, insulation warms the interior surface and cools the exterior surface; it thus increases the chances of water vapor infiltrating from the warm side condensing within the roof's cross section.)

- Insulation raises a roof's surface temperatures in the summer, thereby accelerating the oxidizing chemical reactions that harden bitumen and make it more brittle and more subject to alligatoring, wrinkle cracking, blister growth, and general degradation.
- Insulation produces more rapid membrane thermal contraction, thus increasing the hazard of membrane splitting.

The protected membrane roof (PMR) concept, which places the insulation above the membrane, averts these last two problems.

There are many problems, apart from thermal efficiency, associated with the selection of an insulation material, including water absorption, drying rate, dimensional stability, treatment of board joints, impact resistance, and compressive and shearing strength. Despite these problems, thermal insulation is indispensable, and, as demonstrated in this chapter, many fears expressed about increased thermal efficiency are grossly exaggerated. The first $\frac{1}{2}$-in. thickness of insulation in a roof system exerts a far greater effect on the surface temperature than does the next 4 in. Regardless of the problems created by insulation, few owners can afford to eliminate or skimp on thermal insulation.

Thermal Insulating Materials

The insulation used in low-slope roof systems falls into four categories:

- Rigid insulation, prefabricated into boards applied directly to the deck surface
- Dual-purpose structural deck and insulating planks
- Poured-in-place insulating concrete fills, or lightweight fills with embedded foam insulation
- Sprayed-in-place plastic foam (see Chapter 16)

Soft blanket and batt insulations obviously cannot serve as the substrate for a low-slope roofing membrane. They must be located below the structural deck, normally on the ceiling, below a ventilated space. This arrangement is especially good in arid climates with daily temperature extremes. The ventilated space interposes an additional insulating medium that modulates the ceiling temperature, thus increasing occupants' comfort. It can also dissipate water vapor (see Chapter 6, "Vapor Control").

Below-deck insulation offers better acoustical control than conventional above-deck insulation. And low-density batt insulation costs much less than high-density above-deck roof insulation of equal insulating value. However, below-deck insulation has several drawbacks that normally outweigh its advantages. Except in arid climates, below-deck insulation creates the risk of wetting from migrating water vapor condensing on a cold roof deck and dripping back into the ceiling-supported insulation. Below-deck insulation cannot substitute for above-deck insulation in a steel-deck roof assembly; a steel deck requires board insulation as a substrate for the membrane.

Combinations of above- and below-deck insulation are practicable, but the designer should check the dew-point location under design conditions.[1]

Winter venting of the ceiling plenum solves the moisture problem, but it sacrifices the thermal-insulating value of above-plenum roof components: deck, above-deck insulation, membrane, and outside air film. Whether these above-deck components' insulating value is totally lost remains an unanswered question, requiring extensive research. And although ventilation of a plenum space is a winter-heating energy loser, it is a summer-cooling energy winner.

In addition to the moisture problem, below-deck insulation has three other drawbacks:

- It increases building construction cost if additional building height is required for ventilating space.
- It does not protect utilities in a ventilated space from freezing.
- It does not restrict thermal expansion and contraction in the deck, whose movement may create problems.

Rigid board insulation

Far more common than ceiling insulation in modern low-slope roofing systems is rigid board insulation, normally applied to the deck's top surface. Such insulation is classified chemically as *organic* or *inorganic*. Organic insulation includes the various vegetable-fiber boards and foamed plastics. Inorganic insulation includes glass fibers, perlite board, cellular glass, mineral fiberboard, and poured insulating concretes with lightweight aggregates.

The newest roof insulation materials, foamed plastics, include polystyrenes, phenolics, and isocyanurates. Air or other gas introduced into the material expands it as much as 40 times. Cells form in various patterns: open (interconnected) or closed (unconnected). Most rigid foams are expanded with one of the hydrogenated chlorofluorocarbons (HCFCs). Because of their extremely low thermal conductivity (about one-third that of air), the foaming agents give isoboards a high insulating quality. Air gradually diffuses into the cells, replacing some of the original foaming gas and eventually lowering thermal resistance by about 20 percent. The isoboards' thermal resistance nonetheless remains extremely high (see Table 5.1).

Fibrous insulation includes various fiberboards, made of wood, cane, or vegetable fibers, and sometimes impregnated, or later coated, with asphalt to enhance moisture resistance. Fibrous glass insulation consists of nonabsorbent fibers formed into boards with phenolic binders. It is surfaced with an asphalt-laminated kraft facer. Basal-fiber insulations are frequently faced with glass-fiber mats.

Perlite board, another fibrous insulation, contains both inorganic (expanded silicaceous volcanic glass) and organic (wood fibers) materials bonded with asphaltic binders.

Composites comprising polystyrene/wood fiber, isoboard/OSB, etc., are sometimes offered for their combination of properties required for special uses.

Poured-in-place insulating concrete fill

The perlite or vermiculite aggregates normally used in lightweight insulating concrete contain varied-sized cells, which raise the concrete's thermal resistance to 14 to 20 times that of ordinary structural concrete. This is still a relatively inefficient insulation. Composites, constructed by embedding polystyrene foam into the lightweight fill, can vastly increase their thermal resistance.

Lightweight insulating concrete comes as a cellular, foamed material, formed by adding a liquid-concentrate foaming agent to the mix (see Fig. 5.1).

Perlite is silicaceous volcanic glass; vermiculite is expanded mica. Both materials expand 15 to 20 times at high temperatures (somewhere between 1400 and 2000°F). Normally, the concrete made from these aggregates has a portland cement binder, but perlite aggregate sometimes has an asphalt binder, a thermosetting fill discussed later.

Lightweight insulating concrete also comes with other aggregates (e.g., expanded polystyrene).

Still another type of insulating fill is thermosetting, produced by mixing a perlite aggregate with a hot, steep (Type III or IV) asphalt binder. Mixed and heated at the jobsite, thermosetting insulating fill is placed dry on a primed surface, screeded to the proper thickness (usually variable, to provide slope), and compacted with a hand tamper or roller to the specified density (about 20 pcf) (see Fig. 5.2).

Poured insulating concrete fills economically provide both insulation and a sloped roof surface for drainage. They also form a smooth

TABLE 5.1 Roof Insulation Material Properties*

Material	Density pcf	Density kg/m³	Thermal conductivity† k, Btu/(in.·h·ft²·°F)	Thermal resistivity† R, in.·h·ft²·°F/Btu*	Vapor permeance, perm-in.	Coefficient of thermal expansion, in./(in.·°F · 10⁻⁶)	Moisture expansion, %(50-97% RH)	Compressive strength psi‡	Fire resistance
Board insulations:									
Perlite	10	160	0.36	2.8 (0.49)	25	10-13	0.2	35 (241)	Excellent
Glass fiber	12	192	0.24	4.2 (0.74)	40	5-8	<0.05	20 (138)	Excellent
Wood fiber	17	272	0.36	2.8 (0.49)	25	7-10	0.5	80 (551)	Poor
Foamglass	9	144	0.38	2.6 (0.46)	0	4.6	0	100 (689)	Excellent
Isocyanurate foam	2.0	32	0.16	6.2 (1.1)	<1.0	42	<4.0	16 (111)	Good
Polystyrene foam (extruded)	2.3	3.7	0.20	5.0 (0.88)	0.4	35	0.4	45 (310)	Poor
Polystyrene foam (beadboard)	1.0-1.3	16-2?	0.26	3.8 (0.69)	3.0	—	—	12 (83)	Poor
Phenolic (Resol)	1.5	24	0.21	7.5 (1.32)	—	17	—	25 (172)	Excellent
Basalt wool	—		—	4.2 (0.74)	—	—	low	6.5 (45)	Excellent
Composite boards:									
Iso-wood fiber									Fair/good
Iso-OSB									Fair/good
Iso-basalt-wool									
EPS-wood fiber									Poor

TABLE 5.1 Roof Insulation Material Properties* (Continued)

Material	Density pcf kg/m³		Thermal conductivity† k, Btu/ (in.·h·ft²·°F)	Thermal resistivity†, R, in.·h·ft²·°F/Btu	Vapor permeance, perm-in.	Coefficient of thermal expansion, in./ (in.·°F·10^{-6})	Moisture expansion, % (50–97% RH)	Compressive strength psi†‡		Fire resistance
Poured-in-place: Lightweight concrete, perlite aggregate	27	(432)	0.64	1.5 (0.26)	13	4.8	low	140	(965)	Excellent
Lightweight concrete, vermiculite aggregate	25	(400)	0.80	1.3 (0.23)	22	3.0	low	125	(860)	Excellent
Cellular concrete compacted	27	(432)	0.70	1.6 (0.28)	—	5.6	low	160	(1100)	Excellent
Thermosetting fill	22	(352)	0.40	2.5 (0.44)	—	—	—	40	(276)	Good
Sprayed-in-place urethane foam	2.5–3.0	(37–48)	0.16	6.2 (1.1)	1.2	30	—	45	(310)	Good/excellent¶

*Tabulated data come from varied sources. Check with manufacturers and other sources for more complete, precise data.
†Thermal resistance and thermal resistivity are for oven-dry material at 75°F. Water-absorptive materials may suffer substantial losses in insulating value under typical service conditions. Figures in parentheses are in metric units: m²·°C/W.
‡Values for compressive strength may not be comparable because they vary in deformation at which ultimate strength is recorded. Figures in parentheses are metric (kPA).
§Because of varying thicknesses in composite boards, properties vary with varying proportions of each material.
¶Depends upon surfacing.

Figure 5.1 Lightweight insulating concrete pumped to deck is leveled with steel-framed screed. (*Roofing Industry Educational Institute.*)

Figure 5.2 Thermosetting insulating fill, comprising perlite aggregate with asphaltic binder, mixed and heated at the jobsite, can provide slope for dead-level decks. Rolling compacts fill to about 20-pcf density. (*F. J. A. Christiansen Roofing Co., Inc.*)

substrate on top of steel, precast concrete, and other decks for application of the membrane or supplementary insulation boards, if required. Unlike some other insulating materials, they offer excellent compressive strength and fire resistance.

As one notable advantage over lighter materials, massive insulating materials like lightweight concrete fill (which may weigh 50 times as much as foamed plastic of an equivalent thermal resistance) stabilize the extreme fluctuations of an insulated roof's surface temperatures. A roof membrane directly on top of a light, highly efficient insulation reacts to the sudden heat of the sun like an empty pan on a burner. The metal in contact with a good insulator—in this case air—heats up faster than the metal of a pan filled with a poor insulator like water. A massive insulating material, like the water in the heated pan, has greater heat capacity than a lighter, more efficient insulator. Because they store and release heat more slowly than thinner, lighter insulations do, these heavier materials tend to stabilize roof-surface temperatures. Thus membrane temperature reacts less quickly to the sudden heat of the sun emerging from behind a cloud or to the chill of a sudden rain shower. Massive structural decks—notably concrete or poured gypsum—stabilize roof-surface temperatures even more than insulating concrete (Fig. 5.3).

Design Factors

The growing popularity of loose, ballasted and protected membrane roof (PMR) systems complicates the task of setting performance requirements for roof insulation. Compressive strength, to resist traffic loads and hailstone impact, is a universal requirement for insulation in all above-deck locations. In its exposed location above the membrane in a PMR, however, insulation requires several properties—notably moisture resistance and freeze-thaw cycling resistance—that eliminate most insulating materials currently used in the conventional location under the membrane (see Chapter 15, "Protected Membrane Roofs and Waterproof Decks," for detailed discussion of roof-insulation requirements in a PMR). In contrast with a PMR, a loose, ballasted system sets less rigorous requirements for insulation than does a conventional system with insulation sandwiched between the deck and the membrane.

In the conventional sandwich-style roof assembly with an adhered membrane, insulation requires

- Compressive strength
- Cohesive strength to resist delamination under wind uplift

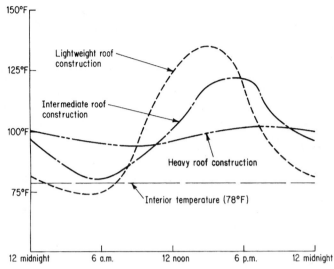

Figure 5.3 These cooling-load temperature curves (for interior temperature of 78°F) show how increased roof-system mass reduces peak cooling load and smooths the cooling-load curves from solar heat gain. Heavy roof construction retards solar heat gain by absorbing the heat and slowly releasing it. Lightweight roof construction = steel deck + 1-in. (or 2-in.) insulation + suspended ceiling. Intermediate roof construction = 1-in. wood deck + 1-in. insulation + suspended ceiling. Heavy roof construction = 6-in. concrete deck + 1-in. (or 2-in.) insulation + suspended ceiling. (*Plotting data for the curves from* ASHRAE Handbook and Product Directory, 1977 Fundamentals, *table 5, p. 25.7 for summer day with maximum 95°F air temperature.*)

- Horizontal shear strength to maintain dimensional stability of the roof membrane under tensile contraction stress
- Resistance to damage from moisture absorption
- Dimensional stability under thermal changes and moisture absorption
- A surface absorptive enough to adhere the bituminous mopping, or liquid-applied adhesive, but not so absorptive that it soaks up the adhesive

Strength requirements

A *compressive strength* of 30 psi minimum is generally recommended in insulation to resist traffic loads, hailstones, dropped tools, and miscellaneous missiles' impact on the roof.

The *cohesive strength* within the insulation must at least equal the required wind-uplift resistance to prevent the insulation from breaking loose from the deck in high winds.

Horizontal shearing strength helps distribute membrane tensile stress, thereby preventing stress (or strain) concentrations that can split the membrane. Insulations in general use usually provide this horizontal shearing strength. However, thickened layers may create a problem with some materials.

Moisture absorption

Water entrapped in insulation (during roof application, through leaks, or via condensation of infiltrating water vapor) can destroy the thermal-insulating value of some insulating materials. Water vapor can generally flow wherever air can flow—between fibers, through interconnected open cells, or where a closed-cell structure breaks down. Water can easily penetrate the larger enclosures of insulating concrete. Wherever water replaces air, the insulating value drops drastically because water's thermal conductivity exceeds that of air by roughly 20 times.

Fibrous organic insulations are especially vulnerable to moisture damage. Free water eventually damages any fibrous organic material or organic plastic binder. Fiberboard long exposed to moisture may warp or buckle and eventually decay. Expansion and contraction accompanying changing moisture content can seriously damage any fully adhered membrane.

Some preformed structural wood-fiber planks are subject to inelastic sag, attributable chiefly to moisture weakening of some (not all) cementitious binders. For that reason these planks are not recommended over highly humidified interiors.

Although less vulnerable to moisture, inorganic materials are not immune. Water penetrating into fiberglass insulation not only impairs the insulating value but may also degrade the binder-fiber interface. Foamed cellular glass neither absorbs water in its closed cells nor allows the passage of water vapor. But if moisture is trapped between the roofing and the top of the insulation, water accumulates in open-surface cells. With roof-surface temperature alternating above and below freezing temperatures, ice formation can break down the walls between open-surface cells and the interior closed cells. Repeated freeze-thaw cycles progressively destroy the foamed glass, leaving a water-saturated gray-black dust. (Laboratory tests have indicated complete breakdown under as few as 20 such cycles.)

Lightweight insulating concrete poses an especially severe moisture-absorption problem because it may retain high moisture content for years. When laboratory tested, insulating concrete cores cut from some roofs in place for 10 or more years had more than 100 percent water content, measured on the so-called dry basis.*

A 100 percent dry-basis moisture content is the ideal moisture-content range for wet vermiculite concrete mix ready for placing. (Wet density ranges from 44 to 60 pcf vs. 22 to 28 pcf for dry density.) Unless it is properly vented, lightweight concrete fill can thus maintain a water content nearly equal to its original mixing water content throughout its service life.

A properly vented lightweight insulating concrete should attain a 10 to 25 percent minimum dry-basis moisture content after 2 to 5 years, with the low (10 percent) figure in arid places like Tucson, Arizona, and the 25 percent figure in the humid climates of the eastern United States. On its way to this 10 to 25 percent equilibrium moisture range, insulating concrete should lose from 25 to 40 percent of its moisture content from its drying period prior to membrane application, according to Tony Clapperton, manager of W. R. Grace's Zonolite Roof Applications. That slightly exceeds the 20 to 30 percent drying-period loss estimated by Professor C. E. Lund.[2]

Water entrapped within insulating concrete can result from a combination of the following three sources:

- Long term retention of original mixing water
- Condensate formed within the fill from vapor migration from the building interior
- Roof leaks

Although it can relieve vapor pressure within the insulating concrete, venting cannot significantly reduce moisture content. This is the conclusion of NBS researchers, citing both experimental evidence and theoretical calculations.[3] This NBS research makes it difficult to reconcile claims that lightweight insulating concrete loses about 50 percent of its original moisture content within several years following membrane application. A prudent designer thus assumes greater moisture retention than the more optimistic estimates claimed for lightweight insulating concrete.

---

*Dry basis, % moisture $= \dfrac{\text{wt. of moisture}}{\text{oven-dry wt.}} = \dfrac{\text{total wt.} - \text{oven-dry wt.}}{\text{oven-dry wt.}}$

Wet basis, % moisture $= \dfrac{\text{wt. of moisture}}{\text{total wt.}} = \dfrac{\text{total wt.} - \text{oven-dry wt.}}{\text{total wt.}}$

100% moisture content, dry basis = 50% on wet basis.

Excessive water entrapment in lightweight concrete fill has two harmful consequences:

- Shortened membrane life from increased probability of blistering (from vapor migrating upward into membrane voids) and splitting (from water weakening the membrane felts).
- Potentially drastic reduction in the roof system's thermal resistance and consequently increased heating and cooling bills. (With a dry-basis moisture content of 50 percent, a percentage sometimes exceeded in laboratory core samples cut from existing roof systems, vermiculite aggregate concrete can lose 75 percent of its dry-state thermal-insulating value.)

Moisture entrapped within insulation undergoes cyclic seasonal change. In winter, when the vapor pressure promotes upward vapor migration, water vapor starts condensing at an obstructive plane below the dew point, usually the underside of the membrane. So long as this condensed moisture is not excessive, it can remain in the upper levels of the insulation. The upward vapor-pressure differential tends to keep the moisture from dripping back through the insulation into the interior.

In summer, conditions change drastically. With the roof-surface temperature soaring to 150°F or thereabouts, the trapped water evaporates. Vapor migration is then downward, impelled by the higher outdoor temperature and absolute humidity. Except in roof assemblies over refrigerated interiors, there is no condensation plane anywhere within the roof assembly; the temperatures are too high. Thus a roof assembly with a properly vented soffit tends to dry out in summer, from heat energy supplied by solar radiation.

Venting

Although backed by the conventional wisdom of the roofing industry, topside venting of insulation via stack vents may do more harm than good. Infrared moisture-detection surveys conducted by Wayne Tobiasson, research engineer with the U.S. Army Cold Regions Research and Engineering Laboratory (CRREL), have detected wet areas around the overwhelming majority of breather vents. In many investigated cold-region roofs, Tobiasson found that almost all the wet insulation was attributable to exterior water entry at flashings and roof penetrations rather than to the condensation of water vapor flowing up from the building interior. These topside vents are apparently worthless as a means of releasing previously accumulated moisture or preventing moisture accumulation. And, on the negative side, they add vulnerable flashed openings for exterior water penetration and objects that can be kicked over or otherwise damaged. Stack vents, mindlessly used as anchors for wind-bracing cable, have been found overturned, with complete rupture of the surrounding membrane.)

If lightweight insulating concrete fill is poured onto a monolithic structural concrete deck, venting is mandatory even without a vapor retarder. A poured-in-place structural concrete deck is a poor substrate for lightweight insulating concrete, generally unacceptable to roofing manufacturers. Lightweight insulating concrete should be restricted to decks with underside venting: slotted steel decks, permeable formboards (e.g., fiberglass), or precast concrete sections with venting joints. If it is nonetheless impracticable to avoid using lightweight insulating concrete over a nonventing deck, nail a layer of permeable isulation (e.g., fiberglass) to the insulating concrete's top surface, or specify a venting base sheet (i.e., a heavy-coated felt with an underside grid of lateral venting grooves) between the fill's top surface and the membrane.

Underside venting of lightweight insulating concrete is superior to topside venting for the following reasons:

- Solar heat normally creates a steeper downward vapor-pressure gradient during the summer than the upward vapor-pressure gradient during the winter.
- A properly vented underside normally offers a larger, better distributed surface area than the top surface, with its widely spaced vents.
- The topside is initially drier because it is open to the ambient air for several days before membrane application.

Major roofing manufacturers require both underside and topside venting of lightweight insulating concrete for warranted roofs.

Edge venting. Edge venting creates horizontal escape paths through grooved edge boards open to the edge of the insulation and placed around the roof periphery.

Underside ventilation. Underside ventilation for lightweight insulating concrete on steel decks consists of slotted holes in the flutes, providing a 2.5 percent open soffit area. Slotted holes are preferable to metal vent clips at longitudinal deck joints because they drastically reduce the distance, to one-third or less, that the vapor must travel through the insulating concrete to escape to the building interior. Dimpled side lap venting is totally ineffective.

Dimensional stability

As the roofing sandwich filler between deck and roof membrane, insulation must retain stable dimensions or it may damage the membrane. Contraction and expansion with changing moisture content can also distort some insulation materials beyond tolerable limits. Repeated expansion and contraction of insulation boards promotes membrane splitting.

Insulation made of cellulosic fibers swells with moisture absorption and contracts on drying as much as 0.2 to 0.5 percent, with 50 to 90% RH changes. Under such a humidity change, a 4-ft-long cellulosic fiberboard may contract $\frac{1}{4}$ in., thus opening a joint.

Use of smaller insulation boards—2×4 ft instead of 4×8 ft—is recommended by Robert A. LaCosse, former technical services manager of the National Roofing Contractors Association (NRCA). This practice can reduce joint width dimensions and also alleviate the hazards of insulation buckling and bowing.

Insulation materials with a high coefficient of thermal expansion can contribute to the splitting of roof membranes subjected to extreme temperature changes. Membrane splitting can occur when a roof membrane shrinks as temperature drops, or as wet felts dry out. Because the flexible membrane buckles under compressive stress, the membrane cannot expand with rising temperature, and contraction can proceed with a sort of ratchet action resulting in accumulating contraction, closing open joints between the insulation boards.

Molded expanded polystyrene (MEPS) boards display a combination of long term shrinkage and high thermal expansion-contraction coefficient. According to researchers R. M. Dupuis and J. G. Dees, $\frac{3}{8}$-in. gaps were found in 8-ft board dimensions of expanded polystyrene in the field. In laboratory tests, aging at 158°F, expanded polystyrene shrank more than 0.5 percent, a figure that held even after 200 days' storage at 70°F.*

Largely because of its dimensional instability, coupled with the difficulty of getting good adhesion with hot-mopped asphalt, extruded polystyrene insulation (XEPS) board is no longer used in conventional built-up roof systems (i.e., membrane on top of insulation) in the United States. However, this material is successfully marketed as insulation for protected membrane roofs.

This polystyrene foam insulation was involved in numerous BUR membrane splitting failures (discussed in Chapter 10). These failures

---

*Research Report, "Expanded Polystyrene Insulation for Use in Built-up and Single-ply Roofing Systems," Dupuis and Dees, Structural Research, Inc., 1984.

emphasized the vital importance of anchoring insulation to the deck, an extremely difficult task with this insulation. (Its cell structure starts to collapse at 175 to 180°F, well below the temperature required for good adhesion with hot-mopped steep asphalt. Under hot asphalt at 400°F, it melts and collapses into craters.)

Isoboard insulation can pose similar dimensional-stability problems. Polyisocyanate insulation—prefabricated boards and sprayed-in-place polyisocyanate—exhibits long term growth that results from inward diffusion of small atmospheric gaseous molecules—oxygen, nitrogen, water vapor, carbon dioxide, and so on—that is faster than the outward diffusion of the large HCFC molecules. Under the consequently increased internal pressure, the cellular walls yield plastically, with slow, permanent cell-enlarging deformations.

Warm, humid climates present the most favorable conditions for isoboard growth and consequently the worst conditions for its use. Water vapor, experiencing condensation-evaporation cycles with changing temperature, is the most active gas promoting long-term growth. At constant pressure, evaporation of liquid water entails a 1500-fold volume increase. Trapped within foam cells, this water-vapor pressure increases exponentially with rising temperature as more liquid water evaporates. The foam's cellular structure expands under rising temperature and pressure, but it does not contract elastically with falling temperature and pressure.

Swelling also varies inversely with density, providing yet another reason for setting minimum foam core density at 2.0 pcf. As a performance criterion for urethane board expansion under the "humid-aging" (ASTM D2126) test, which subjects the board to 158°F temperatures at 97% RH for 14 days, researcher Duane Davis, of GAF Corporation, had proposed the following limits: 1 percent increase in length and width, 7 percent increase in thickness. For a 4-ft-long board, a 1 percent increase in length is nearly ½ in. Even though the humid-aging test should be tougher than any roof system would normally endure in service, it nonetheless indicates the scope of the problem. It further indicates the greater importance of moisture than of temperature as a factor in dimensional stability. Even with its high coefficient of thermal expansion-contraction [30×10^{-6} in./(in. · °F)], a board expands only 0.3 percent under a 100°F temperature rise, one-third of the 1 percent allowable expansion under the humid-aging test.

The warping-expansion sometimes observed between adjacent insulation boards may result from moisture and thermal expansion coupled with top-surface contraction under high-temperature drying. The warping can also be explained simply as buckling under the compressive stress of adjacent moisture-expanded boards.

Tapered insulation

With the increasing recognition of the importance of positive drainage, tapered roof insulation has become popular for reroofing projects, where it may provide the only practicable means of obtaining slope. There are two basic types of tapered insulation: (1) lightweight insulating concrete or asphaltic fills poured in place and screeded to sloped contour and (2) preformed insulation boards (cellular glass, perlite, plastic foam) with a tapered cross section designed to provide slopes of $\frac{1}{16}$ in., $\frac{1}{8}$ in., or $\frac{1}{4}$ in./ft.

A third method of tapering insulation is field grinding perlite board insulation placed in a terraced pattern of different thicknesses. Tapered-insulation-board systems require an extra step in the construction process: preparation of shop drawings indicating roof slope, drain location, drain height, and tapered block location. Extra care is necessary to place the coded blocks in the proper pattern, as shown on the shop drawings.

Anchoring insulation to the deck

In conventional sandwich-style bituminous roof systems, insulation must be firmly anchored to the deck to resist wind-uplift stresses and to prevent membrane splitting, especially in cold climates, distributing stresses that otherwise (i.e., with unanchored insulation) would become concentrated.

Of the three basic methods for adhering insulation—cold adhesives, hot-mopped bitumen, and mechanical fastening—mechanical fastening is by far the most generally dependable, especially on steel decks. Cold-applied adhesives are so unreliable that in 1978 three major manufacturers withdrew their approval of this method for bonded roofs. Hot-mopped steep asphalt, correctly applied with the insulation boards tamped firmly into the hot, fluid bitumen, provides excellent adhesion. But getting the insulation boards down into the hot asphalt in the few seconds available before it congeals into a nonbonding solid makes this a highly risky method, especially on steel decks, where the mopping requires extra care to limit its weight and keep it centered on the steel-deck flange. Hot-mopped asphalt adhesive is especially risky in cold, windy weather.

Overlooking the need for good anchorage at the deck-insulation interface, some designers of buildings planned for future expansion have specified insulation boards loosely laid on the deck, for easy removal when the roof is later converted to a floor. This first-cost economy may prove costly. Unanchored insulation boards vastly increase the risk of membrane splitting. Through internal stresses produced by thermal and moisture changes, a bituminous membrane exerts a ratchet action on a poorly anchored insulation. (The flexible membrane expands, compresses, and buckles in heat, and contracts and pulls in the cold, thus producing a cumulative ratchet action toward the center of the roof area.) Compounded by aging and moisture changes in the felts, this ratchet action sometimes pulls the insulation 2 or 3 in. from the roof edges, destroying the edge flashing.

Solid bearing for insulation boards

Solid bearing for insulation boards on steel-deck flanges is another anchorage rule sometimes violated in the field. Cantilevering insulation boards over deck flutes, especially wide-ribbed decks, can fracture brittle insulating materials, destroying the vital substrate support for the membrane and making it vulnerable to puncture. This practice can also promote membrane splitting by opening a joint between adjacent boards. The joint is subject to an extreme degree of separation and thermal warping (see Chapter 10, "Elements of Built-up Membranes"). An unsupported cantilevered section of insulation board is subjected to both additional longitudinal contraction and upward deflection from low top-surface temperature. This upward deflection can add membrane flexural stress to tensile stress.

To avoid this hazard, most insulation boards should have a minimum $1\frac{1}{2}$-in. bearing on the steel-deck flange. Whenever the installation pattern reduces this dimension, the roofer should trim the board.

Refrigerated interiors

The architect should not attempt to economize by insulating a refrigerated interior with insulation placed on top of the structural deck, for two main reasons:

- Placing the roof insulation between a cold, dry interior and (in summer) a hot, humid exterior increases the pressure of water-vapor migration down through the membrane. If the refrigerated space is maintained at 0°F, this pressure may always be downward. Constant downward vapor migration almost inevitably saturates the insulation and, at the least, aggravates membrane blistering, icing delamination, and the other problems associated with entrapped moisture.

- The thicker insulation required for a refrigerated space creates a less stable substrate for the membrane than does the normally thinner roof insulation. The inevitably wider joints and greater movement in a thickened substrate magnify the hazard of membrane splitting.

Providing a ventilated ceiling air space between the separately insulated refrigerated space and the roof deck relieves this threat to the roof. It greatly reduces the vapor-pressure differential between the roof and the interior, and it also reduces the chances of water penetrating into the refrigerated space, where it will freeze, reduce insulating efficiency, and raise operating costs.

Effects of increased insulation thickness

Thickened roof insulation has raised fears, well publicized in the technical press, about deleterious effects on the roof membrane. These fears, especially those concerning the effect of thickened insulation on accelerated degradation of the membrane, have been greatly exaggerated. As the following section shows, roof color is a much more important determinant of roof-surface temperature than insulation thickness, and roof-surface temperature is a prime determinant of the built-up membrane's degradation. The trivial effect of insulation on roof-surface temperature is established by both theoretical and empirical research.

Here, in summary, is the case against thickened insulation, which can expose the roof membrane to several specific life-shortening effects:

- Accelerated chemical degradation
- Increased risk of splitting
- Reduced impact resistance
- Increased risk of slippage

Note that most of these hazards result from the greater temperature range—hotter in summer, colder in winter—experienced by a heavily insulated membrane.

Chemical degradation of bitumen accelerates with the higher summer surface temperature resulting from thickened insulation because oxidation rate, chief agent of this chemical degradation, rises exponentially with temperature. But the temperature rise from increased insulation thickness is slight. An NBS study calculates a maximum temperature increase of 4°F, from 153°F for a black surface with 0.25 U factor (1-in. fiberboard insulation) to 157°F with a 0.066 U factor (5-in. fiberboard insulation).[4] Another theoretical study by Carl G. Cash and W. H. Gumpertz confirms the Rossiter and Mathey study's conclusions.[5] And a study based on field tests of membranes over varying thicknesses of fiberglass insulation (from $\frac{3}{4}$ to $7\frac{1}{4}$ in. in three layers) by D. E. Richards and Ed Mirra of Owens-Corning Corporation concludes that heavily insulated membranes experience only slightly higher membrane temperatures during hot weather.[6]

Roof color has a much greater effect on membrane temperature than insulation thickness. For roofs of equal insulation thickness, maximum calculated temperature difference between a black and a gray surface is 15°F; between black and white it is 27°F, compared with the 4°F temperature difference between 1- and 5-in.-thick insulation. A black-surfaced membrane over 1-in. insulation gets 10°F hotter than a gray-surfaced membrane over 5-in. insulation.[7]

Increased splitting risk results from several factors associated with thickened insulation. Most important is thickened insulation's reduced horizontal shearing resistance to membrane contraction, which can result from either a temperature drop or drying shrinkage of the felts. Horizontal shearing resistance can be assumed to be inversely proportional to insulation thickness; i.e., doubled insulation thickness reduces horizontal shearing resistance by half.

Remedies for this indeterminate increased splitting risk include mechanically anchoring insulation to the deck; specifying an additional felt ply to strengthen the membrane; more closely spacing expansion-contraction joints; and mechanically anchoring the base sheet to insulation, which can help reduce the slippage risk.

The splitting and impact hazards especially call for research into the structural properties of insulation materials. Insulation materials with low horizontal shear strength may require limitation in depth because the insulation must transmit membrane stresses to the deck.

Joint taping

Taping of insulation joints is recommended by Owens-Corning Corporation. This practice, with 6-in.-wide glass-fiber tapes bonded with steep asphalt at continuous joints between fiberglass insulation boards, has several benefits:

- Prevention of asphalt drippage (and consequent bitumen loss) from membrane felt moppings through insulation-board joints (see Fig. 5.4)
- Reduction in ridging over insulation joints
- Reduction in splitting hazard

The protection against membrane splitting results from the elimination—or at least attenuation—of stress concentration at the vulnerable joint lines, where joint taping can (1) reduce differential movement between adjacent insulation boards and (2) ensure a more uniform membrane cross section (free of increased thermal contraction coefficient from membrane thickening at the insulation joint).

Figure 5.4 Joint taping along continuous insulation-board joints can alleviate membrane stress concentrations and prevent bitumen drippage into insulation joints. (*Roofing Industry Educational Institute.*)

One disadvantage of joint taping is its requirement for aligning the insulation in both directions instead of staggering at least one line of joints. Staggered joints break lines of potential stress concentration in the membranes above.

Double-layered insulation

Double-layered insulation is a better method for reducing membrane stress concentration than joint taping, according to many roofing experts. According to C. G. Cash, of Simpson, Gumpertz & Heger, consulting engineers, of Cambridge, Massachusetts, double-layered insulation placed with vertically offset joints forms an almost continuous plane plate as opposed to the more likely warped surface of single-layered insulation. Double-layered insulation has the following advantages:

- Smoother top surface as membrane substrate because of better accommodation to deck irregularities
- Elimination of thermal bridges and consequent heating- and cooling-energy leakage at insulation joints
- Reduced ridging hazard (from reduced tendency of insulation-

board warping because it is generally easier to get good adhesion between insulation boards than between insulation and deck)
- Reduced membrane splitting hazard because (1) there is less chance of insulation-board warpage; (2) it is easier to get tight joints with thinner boards; and (3) there is reduced stress concentration from continuity in the insulation "plate," which according to one study reduces thermal contraction stress by 10 percent.[8]
- Blockage of vertical paths followed by upward-moving water vapor condensing on the cold underside of the membrane, where the felts can swell with the absorbed moisture, creating ridges
- Better adaptability to mechanical anchorage

Mechanical anchorage works better with double- rather than single-layered insulation for two reasons: (1) It reduces the required shank length by half the total insulation thickness because the fastener penetrates only the bottom layer of double-layered insulation (see Fig. 5.5). (Hot mopping to the bottom insulation layer can then provide good adhesion for the top insulation layer because insulation provides a better heat-retaining surface as well as an easier one for application than a deck, especially a steel deck.) (2) By limiting the fastener penetration to half the insulation depth, you also insulate

Figure 5.5 Mechanical fasteners are the only FM-approved method for anchoring the bottom layer of insulation boards to steel decks. (*Roofing Industry Educational Institute.*)

the fastener, preventing its action as a possible thermal bridge that, at subfreezing outdoor temperatures, might cause condensation at its projection into the warm interior.

One caution concerning the mopped-asphalt adhesive film between the two insulation-board layers is the possibility of condensation forming at this asphalt plane, as a consequence of upwardly migrating water vapor being impeded at this plane. The adhesive film could function as a vapor retarder, and in cold weather, under many common conditions, its temperature may fall below dew-point temperature. A simple calculation, based on assumed interior and exterior relative humidity and temperature, can indicate whether such condensation poses a threat.

Principles of Thermal Insulation

Heat is transferred through (1) conduction, (2) convection, and (3) radiation. *Conduction* depends on direct contact between vibrating molecules transmitting kinetic (or internal heat) energy through a material medium. *Convection* requires an air or liquid current, or some moving medium, to transfer heat physically from one place to another. *Radiation* transmits heat through electromagnetic waves emitted by all bodies, at an intensity varying with the fourth power of the absolute temperature.

Radiation accounts for the extremes in roof-surface temperatures, above and below atmospheric temperature. The sun's rays can raise the surface temperature of an insulated roof 75°F above air temperature, and on a clear night, without cloud cover to reflect radiated heat back to earth, roof-surface temperature can drop 10°F or more below the air temperature. (See Fig. 5.6.) Thus, in a climate with design temperature varying from a summer maximum of 95°F to a winter low of 0°F, the annual variation in roof temperature may be 160°F or more. Because it conducts less heat to or from the roof surface, insulation within the roof sandwich *increases* the extremes in roof-surface temperature; for a black roof surface the annual temperature differential could exceed 200°F. But insulation greatly reduces the daily or annual temperature difference experienced by the deck.

Heat flows through building materials primarily by conduction, less so by convection. Through an air space all three modes of heat transfer are at work. Thermal insulation resists all three, but primarily conduction.

Reflective insulation—normally with an aluminum foil or aluminum-pigmented heat-reflective coating—is common in walls with an enclosed air space. But the difficulty of accommodating such a space under a roof, plus the need to pierce the foil with numerous openings for pipes, conduit, ducts, and other mechanical items, makes

Figure 5.6 White acrylic coating reduces cooling load temperature difference (CLTD) attributable to solar radiation by a substantial amount. CLTD reduction can substantially cut cooling-energy costs in warm, sunny climates. (*Roofing Industry Educational Institute.*)

reflective insulation generally impracticable for roofs. So does the difficulty of keeping reflective insulation from contacting other materials and thereby losing its effectiveness.

Good thermal insulators generally depend on the entrapment of air, a poor heat conductor, in millions of small cells or pockets; these small cells arrest the transfer for heat by convection. Carbon steel, a good heat conductor, transfers heat nearly 2000 times as fast as an equal thickness of air. Cellular insulations, e.g., foamed glass or plastic, acquire their thermal-insulating value by establishing a temperature gradient through the cross section, with each tiny cell of entrapped air making its contribution to the total resistance to heat flow. Fibrous insulations exploit thin pockets of air between the fibers.

The efficiency of thermal insulation also depends on its capacity for impeding air flow, thus resisting heat flow by convection, and on the low thermal conductivity of its basic materials.

The quantity of heat transferred through a building component varies directly with the temperature differential, the exposed area, and the time during which the transfer takes place. It varies inversely with thickness. A good thermal-insulating material is a poor thermal conductor (and vice versa); it also blocks the passage of air, thus resisting convective heat flow. If it is sufficiently opaque, it resists the penetration of heat radiation.

Rising temperature reduces insulation's thermal resistance (i.e., increases its thermal conductivity). The R value of glass-fiber insulation, for example, drops from about 6.5°F (in. · h · ft^2)/Btu at -50°F to about 4 at 150°F, representing a loss of about 40 percent over the extreme temperature range that roofs experience.[9] This phenomenon is explained by the kinetic theory of gases: Thermal conductivity increases with faster movement of gaseous molecules transmitting heat energy from one surface to another. Temperature is an index of this internal molecular movement.

Moisture reduction of insulating value

Published values for thermal conductivity or resistance are for *dry* materials. This policy permits accurate measurements of thermal conductivity because moisture within the tested sample would distort the conductivity measurements (chiefly by adding latent heat transfer resulting from condensation or evaporation to the purely conductive heat transfer that is desired). Dry thermal-conductivity values are thus an accurate basis for comparing different materials' thermal resistances under ideal conditions.

In the field, however, where most insulations inevitably contain significant quantities of liquid moisture, the insulation's dry thermal resistance (i.e., its R factor) is reduced via two independent mechanisms:

- Sensible heat transfer, chiefly by conduction
- Latent heat transfer (i.e., changes of state from vapor to liquid or solid, or vice versa, which depend on the insulation's vapor permeability)[10]

Water filling an insulating material's interstices generally replaces air [thermal conductivity = 0.17 Btu/(in. · h · ft^2 · °F) at 40°F. Water's thermal conductivity is more than 20 times as much as air's at above-freezing temperatures [when $k = 4$ Btu/(in. · h · ft^2 · °F)] and nearly 100 times as much at subfreezing temperatures [the k value of ice = 15.6 Btu/(in. · h · ft^2 · °F)].[11]

Vulnerability to moisture penetration is the chief determinant of an insulation's thermal resistance in the presence of moisture. Foamglass, the least vulnerable of roof-insulating materials in common use, has a water-resistant, closed-cell structure that protects it from significant moisture penetration. Even after being subjected to a simulated leak, foamglass retains nearly all (98 percent) its dry thermal resistance. But poured lightweight insulating concrete fills lose about half their dry thermal-resistance values at 25 percent (dry basis) moisture content, up to 70 percent or more at 50 percent moisture content.[12]

Tobiasson's studies of wet insulation reveal great disparities among chemically identical materials. *Extruded* polystyrene (XEPS) is only slightly inferior to foamglass in its resistance to moisture. *Expanded* polystyrene (MEPS), however, is highly vulnerable, especially in the lower densities (i.e., 1 or 2 pcf). Whereas extruded polystyrene retains about 90 percent of its "dry" thermal resistance after 400 days of wetting test, the lowest-density expanded polystyrene retains only 25 percent (see Fig. 5.7).

A material is classified as "wet" by Tobiasson when its *thermal resistance ratio* (TRR = wet thermal resistivity divided by dry thermal resistivity) falls below 80 percent.[13] (See Table 5.2.)

Perlite board, fiberglass, and fiberboard insulation also suffer substantial losses from moisture absorption. Designers should consider these losses when computing total R factors through roof systems. At 100 percent moisture content (by weight), most common insulations (e.g., perlite board, fiberboard, fiberglass, and felt-faced urethane) apparently lose from 40 to 60 percent of their insulating value.

Even when designers lack precise knowledge of the insulation's moisture content (the normal situation when designing a new building or even a reroofing project), they should estimate moisture-caused losses in thermal resistance, possibly based on equilibrium moisture content.

Trapped moisture in vapor-permeable insulations—fiberglass, fiberboard, perlite board, and lightweight insulating concrete—reduces the insulation's thermal resistance more during the summer-cooling season than during the winter-heating season. Latent heat transfer is the probable explanation for this empirically discovered phenomenon. Condensation of water vapor flowing upward toward a colder, drier exterior or downward toward a cooler, drier interior adds its high

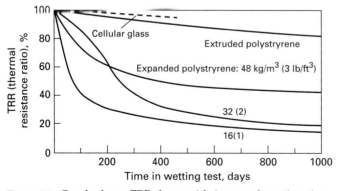

Figure 5.7 Graph shows TRR decay with increased wetting time. (*From Wayne Tobiasson, Alan Greatorex, and Doris Van Pelt, "New Wetting Curves for Common Roof Insulations,"* Intern. Symp. Roofing Technol., *1991.*)

TABLE 5.2 Tabulated Data (Top) Show Equilibrium Moisture Content and Moisture Content at 80 Percent TRR; Bottom Table Shows Moisture Content at which TRR Equals 80 Percent

Comparison of Equilibrium Moisture Contents and Those at 80 Percent TRR for Insulations without Facers

Insulation	Equilibrium moisture content (% of dry weight) at 45% RH	at 90% RH	Moisture content (% of dry weight) at 80% TRR
Cellular glass	0.1	0.2	23
Expanded polystyrene 16 kg/m^3 (1 pcf)	1.9	2.0	383
Extruded polystyrene	0.5	0.8	185
Fibrous glass	0.6	1.1	42
Isocyanurate	1.4	3.0	262
Perlite	1.7	5.0	17
Phenolic	6.4	23.4	25
Urethane	2.0	6.0	262

Moisture Content (% of dry weight) at 80% TRR

Material	Moisture content % of dry weight	% of volume
Cork	39	9.9
Fiberboard	15	4.4
Perlite	17	2.7
Fibrous glass	42	6.2
Cellular glass	23	3.1
Gypsum	8	7.0
Lightweight concrete 369 kg/m^3 (23 pcf)	10	3.7
Lightweight concrete 594 kg/m^3 (37 pcf)	9	5.3
Expanded polystyrene 16 kg/m^3 (1 pcf)	383	6.1
Expanded polystyrene 32 kg/m^3 (2 pcf)	248	7.2
Expanded polystyrene 48 kg/m^3 (3 pcf)	82	4.3
Extruded polystyrene	185	5.9
Urethane/isocyanurate	262	8.8
Foamed-in-place urethane	130	6.5
Phenolic	25	1.0

heat of evaporation (972 Btu/lb at standard atmospheric pressure) to the conductive heat loss through the roof. In winter, water-vapor flow remains upward, and condensate tends to remain in the upper, colder sections of insulation. However, in summer, vapor flow often reverses direction: downward from a solar-heated membrane during the day, upward at night, from an interior at 78°F toward an exterior at 65°F or so. Daily, or nearly daily, reversals in vapor-flow direction make the water vapor available for latent heat transfer in both directions, augmenting the already increased thermal conductivity resulting from the presence of liquid water.

Heat-flow calculation

The simplified heat-transfer calculations normally used for building design require a knowledge of four indexes of heat transmission.

1. Thermal conductivity k = heat (Btu) transferred per hour through a 1-in.-thick, 1-ft^2 area of homogeneous material per °F temperature difference from surface to surface. The unit for k is Btu/(h · ft^2 · °F · in.) To qualify as thermal insulation, a material must have a k value of 0.5 or less.

2. Conductance $C = k$/thickness is the corresponding unit for a material of given thickness. (The unit for C is Btu/(h · ft^2 · °F). For a 2-in.-thick plank of material whose k = 0.20, C = 0.10.

3. Thermal resistance $R = 1/C$ indicates a material's resistance to conductive heat flow. (For a material with C = 0.20, R = 5.0.) The unit for R is °F · h · ft^2/Btu. That is, for a 5°F temperature difference surface-to-surface, 1 Btu/h flows through a 1-ft^2 specimen.

4. Overall coefficient of transmission U is a unit like k and C, measured in Btu transmitted per hour (Btu/h) through 1 ft^2 of construction per °F from air on one side to air on the other. However, it relates to the several component materials in a wall or roof. U is calculated from the following formula:

$$U = \frac{1}{\Sigma R} \quad (5\text{-}1)$$

where ΣR = sum of the thermal resistances of the components, plus the resistances of the inside and outside air films. Designers using SI units will find conversion factors in Table A.1, Appendix A. U, in watts transmitted through 1 m^2 per °C (or K) requires multiplication of U in British units by 5.678.

To calculate the insulation's required thermal resistance R_p, the designer usually starts with a target U factor set by the mechanical engineer. For the other components, the designer merely tabulates the resistances R for all the materials, including inside and outside air films. (For the summer condition, when roof temperatures often rise to 60°F or more above outside air temperature, it is prudent to assume a roof-surface temperature of 150°F or so and omit the outside air-film resistance. See Fig. 5.8.) Data for conductances of various materials are available from the *ASHRAE Handbook and Product Directory*. If not available in general tables, data for proprietary insulating materials should be furnished by manufacturers.

Consider the roof system shown in cross section on the next page, with target U factor of 0.085.

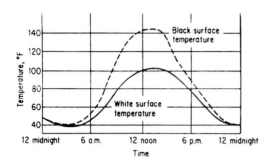

Figure 5.8 Graph shows roof-surface temperatures recorded for 25-mm (3-in.)-thick sprayed-in-place urethane roof on sunny day. Heat flow lagged by about 1 h. (*Graph from C. P. Hedlin, "Some Design Characteristics of Insulation in Flat Roofs Related to Temperature and Moisture." Proc. 5th Conf. Roofing Technol., NBS-NRCA, April 1979, p. 21.*)

By Eq. (5-1), $U = \dfrac{1}{1.67 + R_i} = 0.085$

$$R_i = \dfrac{1}{0.085} - 1.67 = 10.09$$

Two layers of 1 5/16-in.-thick fiberglass boards ($R = 10.52$) are satisfactory. (It might be prudent to reduce the insulation R factor by 10 to 15 percent to allow for moisture content.)

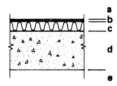

	R value
a = Outside air film (7-1/2) mph wind	0.25
b = 3/8-in. BUR membrane	0.33
c = Insulation	R_I
d = 4-in. concrete slab	0.48
e = Inside air film (heat-flow up)	0.61
ΣR = 1.67 + R_I	

Heat gain or loss

Total rate of heat gain or loss Q (in Btuh) through a roof is computed by:

$Q = U \times$ roof area $\times$ temperature difference between inside and outside air (5.2)

To calculate the temperature at any parallel plane through the roof, use one of the following formulas:

$$T_x = \begin{cases} T_i - \dfrac{\Sigma R_x}{\Sigma R}(T_i - T_o) & \text{for winter conditions} \\ T_o - \dfrac{\Sigma R_x}{\Sigma R}(T_o - T_i) & \text{for summer conditions} \end{cases} \quad (5.3)$$

where T_x = temperature at plane X
T_i = inside temperature
T_o = outside *roof-surface* temperature
ΣR_x = sum of R values between warm side and plane X

Alerts

Design

1. Select two layers in preference to one. For multiple layers, specify staggered joints.

2. Investigate the need for taping insulation-board joints (to relieve membrane stresses).

3. Require bituminous impregnation of wood fiberboard and other organic fiberboard insulations.

4. Install insulation for a refrigerated space below the roof deck and wrap with a continuous vapor retarder.

5. Investigate the need for a vapor retarder to shield the insulation from water vapor migrating from a warm, humid interior (see Chapter 6, "Vapor Control").

6. To reduce the threat of bitumen drippage when insulation is hot-mopped to the deck, specify ASTM Type II or III asphalt. Specify rosin-sized paper, felt, or tape over joints in wood sheathing, plywood, or other prefabricated deck units to be nailed.

7. Check local building code, Factory Mutual, and Underwriters' Laboratories listings for insulation's conformance with wind-uplift and fire-resistance requirements.

For lightweight insulating concrete:

1. Do not specify wet-fill materials (including perlite, vermiculite, or foamed concrete) over monolithic structural concrete, unvented metal deck, vapor retarders, or other substrates that do not permit underside venting.

2. Specify stack venting (one vent per maximum 900-ft^2 area) *and* perimeter edge venting (notched wood nailers or isolated flashing).

3. Specify slotted vent holes in steel deck used with insulating concrete.

4. Require minimum 2-in. thickness for lightweight insulating concrete (to allow embedment of mechanical fasteners).

5. Over lightweight insulating concrete fill, specify a coated base sheet for initial membrane ply, *nailed* to the fill.

6. If impracticable to avoid a nonventing substrate under a wet-fill material, specify either a layer of porous insulation or a venting base sheet as the bottom membrane ply to facilitate lateral venting.

7. Use asphalt-coated base sheets to resist alkaline attack.

8. Require a minimum 28-day compressive strength of 100 psi.

Field

1. Require that stockpiled insulation stored at the jobsite be covered with waterproofed tarpaulins and raised above ground.

2. Require a smooth, plane deck surface [For specific tolerances, see Chapter 4, "Structural Deck" (Field Alert No. 1).]

3. Prohibit placing insulation on wet decks or snow- or ice-covered decks. Decks must be cleaned and dried (including flute openings of metal decks) before insulation is installed. Normally, test for dry deck surface as follows: Pour a small amount of hot bitumen on the deck. If, after the bitumen cools, you can remove it with your fingernails, reject this deck for application of insulation; the moisture content is too high. If the cooled bitumen sticks to the deck and cannot be removed by your fingers, the deck is dry enough to receive the mopping.

    An alternative test for properly cured and dried cast-in-place decks (e.g., concrete, gypsum) is to apply hot bitumen as described and observe for frothing or bubbling. If none occurs, the deck is dry enough.

4. Limit moisture content to satisfactory levels, not to exceed equilibrium moisture content.

5. Limit daily application of insulation to an area that can be covered on the same day with roofing membrane.

6. Place plywood walkways on insulation vulnerable to traffic damage.

7. Set adjacent units of prefabricated insulation with tightest possible joints. *Trim or discard units with broken corners or similar defects.* (See Fig 5.9.)

8. Provide water cutoffs at the end of the day's work on all decks with insulation boards, cut along vertical face, and remove when work resumes (see Fig. 5.10).

9. Use mechanical anchors for attaching insulation boards to steel decks. For nonnailable decks, e.g., concrete, specify solid-mopped Type II or III asphalt.

10. Prohibit cantilevering of insulation boards over steel-deck ribs. Require solid bearing, $1\frac{1}{2}$-in. minimum.

Thermal Insulation   93

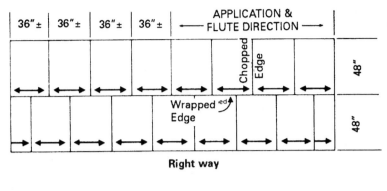
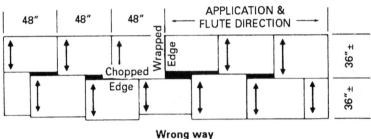

Figure 5.9 Glass-fiber insulation boards, 4×3 ft in plan dimensions, should have *wrapped* edges oriented in a continuous joint line parallel to steel-deck ribs (top), because the factory-controlled 48-in. dimension is more precisely controlled than the 36-in. dimension. If the *chopped* edges form this continuous joint, irregular gaps occur (shown as black areas of bottom diagram). These irregular gaps result from the greater variation (± ½ in.) in the 36-in. dimension of the insulation boards. (*Owens-Corning Corp.*)

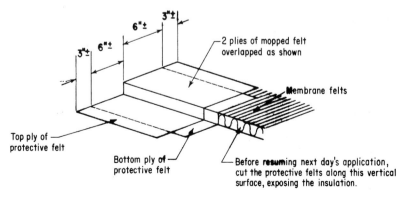

Figure 5.10 Temporary insulation-board cutoff detail (to be used at end of day's work).

For lightweight insulating concrete:

1. Prohibit placing of wet insulating concrete fill if temperature is expected to drop below freezing within 48 h.
2. Let lightweight insulating concrete dry for at least the minimum time recommended by the manufacturer—preferably 2 days, plus an additional day for each rainy day occurring during the drying period.

References

1. For detailed discussion, see R. L. Fricklas, "Insulation: Does Below-Deck Placement Increase Value?" *Specifying Engineer,* February 1977, pp. 66ff.
2. C. E. Lund, *The Performance of Perlite and Vermiculite Concrete Roof Decks,* paper delivered at 12th annual Midwest Roofing Contractors Association convention, Excelsior Springs, Mo., November 1961.
3. Frank J. Powell and Henry E. Robinson, "The Effect of Moisture on the Heat Transfer Performance of Insulated Flat-Roof Constructions," *Build. Sci. Ser.* 37, NBS, October 1971, pp. 12, 13.
4. W. J. Rossiter and R. G. Mathey, "Effects of Insulation on the Surface Temperature of Roof Membranes," NBS Rept. 76-987, 1976, p. 11.
5. Carl G. Cash and W. H. Gumpertz, "Economic and Performance Aspects of Increasing Insulation on the Temperature of Builtup Roofing Membranes," *J. Test. Eval.,* March 1977, pp. 124ff.
6. D. E. Richards and E. J. Mirra, "Does More Roof Insulation Cause Premature Roofing Membrane Failure or Are Roofing Membranes Adequate?" *Proc. Symp. Roofing Technol.,* NBS-NRCA, September 1977, pp. 193ff.
7. Rossiter and Mathey, op. cit.
8. Owens-Corning Fiberglass Corporation, *The Whys of Double Layer,* March 1980.
9. C. J. Shirtcliffe, "Thermal Resistance of Building Insulation," *Can. Build. Dig.,* no. 149, April 1975, p. 149-4.
10. F. A. Joy, "Thermal Conductivity of Insulation Containing Moisture," ASTM STP217, 1957, p. 66.
11. R. H. Perry, *Engineering Manual,* 2d ed., McGraw-Hill Book Company, New York, 1967, pp. 3-25, 3-27, 3-28.
12. Powell and Robinson, op. cit., table 2, pp. 6, 7.
13. Wayne Tobiasson, Alan Greatorex, and Doris Van Pelt, "New Wetting Curves for Common Roof Insulations," *Third Intern. Symp. Roofing Technol.,* 1991.

Chapter 6

Vapor Control

Water vapor flowing upward into the insulation from the building interior may create severe hazards for an insulated roof assembly. Condensation of this vapor impairs the insulation's thermal resistance, and it can ultimately destroy the insulation material itself. Liquid moisture may collect at the underside of the insulation until it flows into cracks or joints in the deck and leaks onto a suspended ceiling or directly into a room. Condensate trapped within voids at the membrane-insulation interface can freeze and expand, breaking the bond between insulation and base ply. Other hazards of entrapped liquid moisture include rotted wood deck and blocking material; metal corrosion; and leaching, efflorescence, and spalling of concrete and masonry.

Paradoxically, more efficient insulation heightens the problems of condensation. Insulation shifts the dew point (the surface temperature at which water vapor condenses) from under the roof system to within it (see Table 6.1 and Fig. 6.1). Thus, other factors being equal, the more efficient the insulation, the more need for a vapor retarder, subroof ventilation, or other means of preventing migrating water vapor from condensing within the roof system.

Fundamentals of Vapor Flow

In the generally temperate climate of the United States, water vapor normally flows *upward* through the roof, from a heated interior toward a colder, drier exterior (i.e., from high to low vapor pressure, along a vapor-pressure gradient). As the curves of Fig. 6.2 depict, this water-vapor pressure depends on two variables: temperature and relative humidity (RH).

The most important variable is temperature. At 0°F (−18°C), even at 100% RH, vapor pressure is insignificant. But at 100°F (38°C),

TABLE 6.1(a) Dew-Point Temperature, °F*

RH, %	Dry-bulb temperature, °F														
	32	35	40	45	50	55	60	65	70	75	80	85	90	95	100
90	30	33	37	42	47	52	57	62	67	72	77	82	87	92	97
80	27	30	34	39	44	49	54	58	64	68	73	78	83	88	93
70	24	27	31	36	40	45	50	55	60	64	69	74	79	84	88
60	20	24	28	32	36	41	46	51	55	60	65	69	74	79	83
50	16	20	24	28	33	36	41	46	50	55	60	64	69	73	78
40	12	15	18	23	27	31	35	40	45	49	53	58	62	67	71
30	8	10	14	16	21	25	29	33	37	42	46	50	54	59	62

*For intermediate, untabulated combinations of dry-bulb temperatures and RH, dew-point temperature can be interpolated on direct (i.e., linear) proportionality. For example, for 70°F, 35% RH interior temperature, dew point = (37 + 45)/2 = 41°F.

TABLE 6.1(b) Dew-Point Temperature, °C

RH, %	Dry-bulb temperature, °C								
	0	5	10	15	20	25	30	35	40
90	−1	4	8	14	18	23	28	33	38
80	−3	2	7	12	17	21	26	31	36
70	−4	0	5	10	14	19	24	29	34
60	−6	−2	3	7	12	17	21	26	31
50	−8	−4	0	5	9	14	19	23	28
40	−11	−7	−4	0	6	10	15	19	24
30	−14	−12	−6	−2	2	6	10	15	19

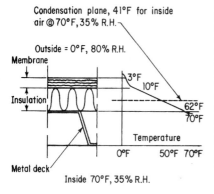

Figure 6.1 The introduction of insulation between deck and membrane usually shifts the dew point from *below* the roof system to *within* the roof system. The dew point at interior conditions (70°F, 35% RH) is 41°F. Thus migrating water vapor will condense somewhere above the 41°F temperature plane, probably at the underside of the membrane. Without insulation, water vapor would condense on the steel-deck surface at about 35°F.

even at 20% RH, vapor pressure is more than 10 times as high as vapor pressure at 0°F, 100% RH. Mathematically, vapor pressure increases *exponentially* with increasing temperature, but only *linearly* with increasing relative humidity.

For roof-system designers, interested more in vapor flow than in vapor quantities, unlike heating, ventilating, and airconditioning (HVAC) system designers, relative humidity is most meaningfully

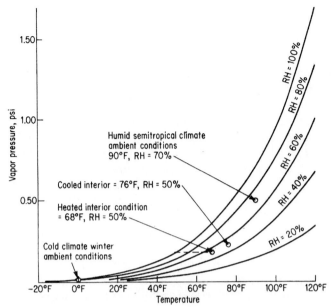

Figure 6.2 This graph, which plots vapor pressure for a given temperature and relative humidity (RH), is a handy technique for determining the direction of water-vapor flow through a roof system. When the ordinate for outside ambient conditions is *below* the ordinate representing interior conditions, the vapor-flow direction is *upward*. (The vapor-pressure differential for a heated interior at 68°F, 50% RH with 0°F, 50% RH ambient is about 0.16 psi.) Conversely, if the ordinate for outside ambient conditions is *above* the ordinate representing interior conditions, the vapor flows *downward*. (The downward vapor-pressure differential for a cooled interior at 76°F, 50% RH with 90°F, 70% RH outside is about 0.27 psi.)

defined as the ratio of actual vapor pressure to the vapor pressure of a saturated (that is, 100% RH) air-vapor mixture at constant temperature and overall atmospheric pressure. (For more detailed technical discussion, see the "Theory of Vapor Migration" section later in this chapter.)

These phenomena—exponential vapor-pressure increase with rising temperature, linear vapor-pressure increase with rising RH—explain why cold climates are characterized by low exterior vapor pressure. When the outside temperature drops to 0°F (−18°C), regardless of whether outside RH is 100% or 0%, the vapor pressure is insignificant. If interior conditions are 68°F (20°C), 50% RH, the vapor-flow problem is essentially the same in cold, dry Montana as along the cold, humid Maine seacoast. The colder the climate, the lower the outside vapor pressure and the greater the pressure differential for a given interior temperature and RH. Note in Fig. 6.2 that the vapor-pressure differential = 0 at 49°F (9°C), 100% RH, outside conditions. But at any outside temperature lower than 49°F (9°C), even at 100% RH, the outside vapor pressure drops below the interior vapor pressure, impelling vapor flow upward through the roof.

The normal direction of vapor flow reverses in the warm, humid climates of the southeastern or south central United States and in Hawaii, where vapor flow is normally downward through a roof toward an airconditioned interior. Referring again to Fig. 6.2, note that for the common Corpus Christi, Texas, ambient conditions of 90°F (32°C), 70% RH, the outside vapor pressure greatly exceeds the interior vapor pressure corresponding to a typical cooled airconditioned interior [76°F (24°C), 50% RH]. Note further that for normal occupancies—residences, office buildings, or commercial buildings—the vapor pressure impelling downward vapor flow in humid, tropical, or semitropical climates may be unidirectional, or nearly so, year-round. In northern locations with climatic extremes, like our north central states, vapor flow changes direction, to predominantly downward in summer.

Although a vapor retarder may be required in a cold climate, there is no such requirement in warm, humid climates. In these locations, the roof membrane itself must perform double duty, resisting the downward flow of vapor as well as liquid moisture. (Low membrane permeability is thus beneficial in a warm, humid climate, but it may be detrimental in a cold climate.) A vapor retarder in its conventional location between deck and insulation is detrimental in a warm, humid climate. It traps water vapor within the insulation, where it may condense and impair insulation efficiency, instead of letting the vapor escape into the interior, where it can be vented or condensed out by airconditioning equipment.

Water vapor penetrates a compact roof system via air leakage and diffusion. Depending on local conditions, air leakage and diffusion vary in relative importance, but they normally reinforce one another; i.e., both tend to force the vapor in the same direction. In the normal case—heated, humid interior and cold, dry exterior—the atmospheric pressure under the roof exceeds the atmospheric pressure on the roof, and the air escaping through deck joints and other small openings conveys water vapor into the roof system from below. Because warm air can hold more water vapor than cold outside air, the water-vapor pressure of a heated, humidified interior exceeds the outside vapor pressure and thus promotes outward, upward vapor migration.

Yet the two mechanisms are physically independent. For example, it is possible to have an inward air leakage associated with outward vapor diffusion. On a hot, windless day in an arid climate, a cooled, highly humidified plant might produce this phenomenon. Because of the reverse "chimney effect" created by the cooled interior, the outside atmospheric pressure could exceed the interior pressure, and the water-vapor-pressure imbalance from the humidified interior could produce upward water-vapor diffusion. (For a more detailed discussion, plus vapor-migration calculation, see the section on "Theory of Vapor Migration" at the end of chapter.)

Techniques for Controlling Moisture

As the basic strategy for limiting moisture within the roof assembly, roof designers can choose one of four methods:

- General ventilation, designed to carry interior moisture out on air drafts circulating through a space left under the roof
- Vapor retarders, designed as low-permeability membranes blocking the passage of water vapor into the roof system
- The "self-drying" concept, which relies on heat energy supplied by the spring and summer sun to evaporate winter-accumulated moisture and drive it downward through a vapor-permeable deck into the building interior
- Seal against air leakage

General ventilation

Properly designed subroof ventilation is the most certain technique for preventing water-vapor infiltration into a roof system. But for low-sloped roofs, it poses the most difficult set of design conditions. Because natural convection decreases with diminishing roof height, ventilation is far less effective for low-sloped roofs than for steeply sloped roofs. Thus under a level roof, the mechanisms left for dissipating moisture are diffusion and wind-induced ventilation. Exterior wind-produced pressure differentials tend to either force moist air out of the loft space or displace it with colder, drier air, but with minimal benefits, particularly for a large level roof area.

There are, moreover, practical obstacles to general ventilation in commercial, industrial, and residential apartment construction. In these buildings, ducts and pipes located in loft space pierce the ceiling, thus making an effective air seal difficult. Because warm, humid air will probably infiltrate the cold loft space and condense on cold surfaces, ducts and pipes must be insulated. Especially in multistory buildings, where the chimney effect increases pressures in upper parts of the building and promotes upward air leakage, the ceiling air seal becomes essential. A further practical objection to general ventilation is the increased building height required to provide the ventilation space. General ventilation is thus the highest first-cost method of controlling vapor migration. As a compensating advantage, it is the most dependable method.

Vapor retarder

According to conventional vapor-retarder theory, the vapor retarder forms an essentially impermeable surface on the warm, humid side of the roof sandwich, blocking the entry of water vapor. Vapor retarders are used only in roofing systems where the insulation is sandwiched between the structural deck and the roof membrane. A properly designed and installed vapor retarder can prevent condensation from forming within the roofing system (see Fig. 6.3).

Water-vapor transfer. To qualify as a vapor retarder, a material's vapor permeance rating should not exceed 0.1 perm. A material rated at 1 perm admits 1 grain of water (= 1/7,000 lb = 0.065 g) per hour through 1 ft^2 of material under a pressure differential of 1 in.Hg [0.491 psi, 57.2 ng/(s · m^2 · Pa)]. Vapor resistance is the reciprocal of permeance.

Typical winter pressure differentials are about 0.5 psi (3.44 kPa).

For perm ratings of selected vapor-retarder materials, see Table 6.2. Because of variations in some composite materials and other difficulties in measuring vapor-transmission rates, perm ratings are generally imprecise. Thus sophisticated designers do not rely on overly refined calculations, but design vapor retarders conservatively.

Another index of a vapor retarder's permeability, the *perm-inch*, or permeance of unit thickness, sometimes enables the designer to make

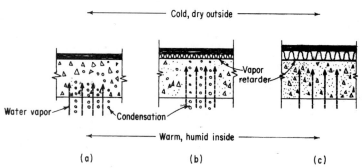

Figure 6.3 How a vapor retarder prevents condensation. In (*a*), an uninsulated concrete roof system, migrating water vapor condenses when it reaches the dew-point temperature somewhere in the roof cross section. In (*b*), an inadequately insulated system in which the vapor retarder is at or below the dew-point temperature, migrating water vapor again condenses and drips. In (*c*), an adequately insulated system, the vapor-retarder temperature is above the dew point, preventing condensation.

TABLE 6.2 Permeance of Some Roofing Components

Material	SI Units kg/(Pa · s · m^2)	IP Units Perm*
Bituminous built-up membrane	0.0	0.0
1 mm (45 mil) EPDM	0.17×10^{-11}	0.04
1.5 mm (60 mil) EPDM	0.11×10^{-11}	0.03
Aluminum foil (no holes, no laps)	0.0	0.0
0.1 mm (4 mil) polyethylene	0.29×10^{-11}	0.08
0.15 mm (6 mil) polyethylene	0.23×10^{-11}	0.06
0.1 mm (4 mil) polyvinyl chloride (PVC)	about 5×10^{-11}	about 1.2
Kraft paper laminates	less than 1.1×10^{-11}	less than 0.3
No. 15 asphalt-saturated felt	4.0×10^{-11}	1.0
No. 43 asphalt-saturated and coated felt	1.1×10^{-11}	0.3
Steel deck (forgetting seams)	0.0	0.0
Steel deck (considering seams)	more than 4×10^{-11}	more than 1.0
Uncracked concrete, 150 mm (6 in.) thick	about 2×10^{-11}	about 0.5
Plywood, 6 mm (¼ in.) thick, exterior glue	2.9×10^{-11}	0.7
Gypsum wall board, 10 mm (⅜ in.) thick	189×10^{-11}	50.0
Insulation Boards		
Cellular glass, 25 mm (1 in.) thick	about 0.0	about 0.0
Polyurethane, 25 mm (1 in.) thick	about 4×10^{-11}	about 1.0
Polystyrene, extruded, 25 mm (1 in.) thick	about 5×10^{-11}	about 1.2
Polystyrene, expanded, 25 mm (1 in.) thick	about 15×10^{-11}	about 4.0

*Grains/h · sq ft · in. mercury.

a direct comparison between materials of equal thickness. (For films less than 0.020 in. thick, the perm-inch index is impracticable.)

For many uses the 1-perm rating formerly set as the maximum permeance permissible for a vapor retarder is not good enough. Under a high differential vapor pressure, or even under low differential pressure over long time intervals (as in a refrigerated warehouse), a vapor retarder rated at 0.1 perm admits too much water vapor. For certain kinds of industrial buildings subjected to continual high temperatures and high relative humidity, vapor retarders of less than 0.1-perm rating are recommended. In severely cold northern locations, where the vapor-pressure differential may persist in the same upward direction for months, a virtually impermeable vapor barrier is needed to prevent a destructive moisture buildup within a compact roof system.

Vapor-retarder materials. An essentially 0-perm rating is attainable by the most common vapor retarder: two or three moppings of bitumen and two plies of saturated felt. A modern version of this vapor retarder uses one coated base sheet in combination with one or two bituminous moppings. Other materials that may qualify as vapor retarders include certain plastic sheets (vinyl, polyethylene film, polyvinyl chloride sheets), black vulcanized rubber, glass, aluminum foil, and laminated kraft paper sheets with bitumen sandwich filler or bitumen-coated kraft paper. Steel decks with caulked joints may also qualify as vapor retarders (see Table 6.2).

As another layer in the multideck roofing sandwich, the vapor retarder must be solidly anchored to the deck below and to the insulation above. It must resist wind-uplift stresses and horizontal shearing stresses produced by thermal stresses in the membrane and transferred through the insulation or produced by thermal or moisture-induced swelling or shrinkage in the insulation itself.

The vapor-retarder material must satisfy secondary requirements, such as practicable installation. For example, although they have low permeance ratings, plastic sheets create field problems that impair their overall performance as components in the roof system. On windy days, these light, flexible sheets are difficult to install. Billowing and fluttering heighten the risk of tearing and make it difficult to flatten the sheet on the deck. Once in place, these sheets are vulnerable to traffic damage.

For the most popular deck, lightgage steel, a vapor retarder should not be placed directly atop the deck, where it must span deck ribs. It should instead be placed on top of a thin bottom layer of insulation (usually $\frac{3}{4}$-in. perlite or $\frac{1}{4}$- or $\frac{1}{2}$-in. gypsum board) mechanically anchored to the deck (see Figs. 6.4 and 6.5). (Since 1984, Factory Mutual has required mechanical anchorage of insulation to qualify the system for a Class 1 rating; see Chapter 7, "Wind Uplift.") Mechanical fasteners driven through a vapor retarder inevitably jeopardize its effectiveness.

To maintain the vapor retarder above the dew-point temperature, a thicker layer of insulation goes atop the vapor retarder. A top insulation layer of considerably greater R factor than the bottom layer will almost always be economically justified. When you really need a

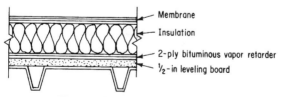

Figure 6.4 This steel-deck roof assembly is a far more dependable roof system than one with a plastic sheet vapor retarder applied directly to the steel deck with cold adhesive. When a vapor retarder is really needed in a steel-deck roof assembly, this detail should be used.

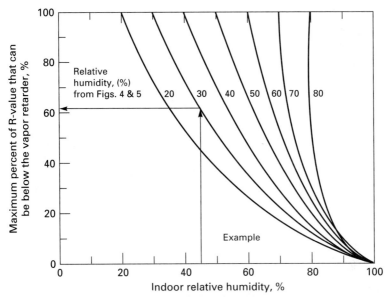

Figure 6.5 This graph shows how to maintain a vapor retarder above the dew-point temperature. In the example, with interior RH of 45%, you must maintain roughly 60 percent of the maximum total roof-system R value below the vapor retarder. (If a greater percentage is below the vapor retarder, there is a threat of condensation on the vapor retarder's underside.) A good rule of thumb is to place no more than one-third of the total insulation R value below the vapor retarder. (*Roofing Industry Educational Institute.*)

vapor retarder, the additional expense for this system, with superior permeability rating and wind-uplift resistance and reduced vulnerability to field damage, can pay off its investment many times over.

Flashing the vapor retarder. Field punctures and unsealed joints are not the only weak spots in a vapor retarder. To effectively block vapor flow into the roof system, a vapor retarder must be flashed and enveloped at roof openings and penetrations. Flashing the vapor retarder exemplifies an effort to prevent air leakage, as opposed to vapor diffusion, into the insulation. For a hot-mopped vapor retarder, set the vapor retarder's felt edges in plastic cement and flash the edges to roof-penetrating components with a felt collar sandwiched between two plastic-cement sealer trowelings. Extend the vapor retarder at all vertical surfaces.

Vapor-retarder controversy. In the past, vapor retarders were recommended in (1) areas where the January temperature averaged 40°F (4°C) or less or (2) wherever building occupancy or use created high relative humidity (see Tables 6.3 and 6.4).

TABLE 6.3 Winter Relative Humidities

Dry occupancies: Relative humidity expected to range under 20%, closely related to prevailing outdoor relative humidity and indoor temperature	Aircraft hangars and assembly plants (except paint shops) Automobile display rooms, assembly shops (except paint shops) Factories, millwork, furniture (except plywoods and finishing units) Foundries Garages, service and storage Shops, machine, metalworking (except pickling and finishing) Stores, dry goods, electrical supplies and hardware Warehouses, dry goods, furniture, hardware, machinery and metals
Medium moisture occupancies: Relative humidity expected to range from 20 to 45%, varying with outdoor relative humidity, with moisture content increased by indoor activities, equipment, and operations	Auditoriums, gymnasiums, theaters Bakeries, confectioners, lunchrooms Churches, schools, hospitals Dwellings, including houses, apartments, and hotels (highest relative humidity in kitchens, baths, laundries) Factories, general manufacturing, except wet processes Offices and banks Stores, general and department
High moisture occupancies: Relative humidity over 45%, determined primarily by processes, not climate	Chemical, pharmaceutical plants Breweries, bakeries, food processing, and food storage Kitchens, laundries Paint and finishing shops Plating, pickling, finishing of metals Public bath and shower rooms, club locker rooms, swimming pools Textile mills, paper mills, synthetic fiber processing plants Photographic printing, cigar manufacturing

This approach is oversimplified; each project requires its own analysis. Based on much sad experience, the efficacy of vapor retarders has been attacked by researchers and engineers administering vast building construction programs. A new consensus developing among many roofing experts reverses the conventional policy for vapor retarders. According to conventional vapor retarder theory, the advice is, "If in doubt, use a vapor retarder." But according to the more sophisticated modern view, the advice is, "If in doubt, think it out."

To replace the cruder traditional approach, researcher Wayne Tobiasson produced vapor-drive maps for the United States.[1] Figure 6.6 depicts this more analytical approach. The top map shows areas

TABLE 6.4 Recommended Relative Humidities and Temperatures for Various Industries*

Industry	Temperature, °F	RH, %	Industry	Temperature, °F	RH, %
Baking:			Laboratories	As required	
Cake mixing	70–75	65	Leather:		
Crackers and biscuits	60–65	50	Chrome tanned		
Fermenting	75–80	70–75	(drying)	120	45
Flour storage	65–80	50–65	Vegetable tanned		
Loaf cooling	70	60–70	(drying)	70	75
Makeup room	75–80	65–70	Storage	50–60	40–60
Mixer-bread dough	75–80	40–50	Paint application:		
Proof box	90–95	80–90	Air-drying lacquers	70–90	60
Yeast storage	32–45	60–75	Air-drying oil paint	60–90	60
Cereal packaging	75–80	45–50	Paper products:		
Confectionary:			Binding	70	50–65
Chocolate covering	62–65	50–65	Folding	75	50–65
Centers for coating	80–85	40–50	Printing	75–80	45–55
Hard candy	70–80	30–50	Storage	75–80	40–60
Storage	60–68	50–65	Textiles:		
Cork processing	80	45	Cotton carding	75–80	50–55
Data processing	75	45–55	Cotton spinning	60–80	50–70
Electrical:			Cotton weaving	68–75	85
Manufacturing			Rayon spinning	80	50–60
cotton-covered wire	60–80	60–70	Rayon throwing	70	85
Storage, general	60–80	35–50	Rayon	70	60
Food:			Silk processing	75–80	65–70
Apple storage	30–32	75–85	Woolens—carding	80–85	65–70
Banana ripening	68	90–95	Woolens—spinning	80–85	50–60
Banana storage	60	85–89	Woolens—weaving	80–85	60
Citrus fruit storage	60	85	Tobacco:		
Egg storage	35–55	75–80	Cigar and Cigarette	70–75	55–65
Grain storage	60	35–45	Other processing and		
Meat, beef aging	40	80	storage	75	70–75
Mushroom storage	32–35	80–85	Casing room	90	88–95
Mushroom, growing			Woodworking:		
stages	Various	60–85	Finished products	65–70	35–40
Potato storage	40–60	85–90	Gluing	70–75	40–50
Produce	Various	70–95	Manufacture	65–75	35–40
Sugar	80	30	Painting lacquer		
Tomato storage	34	85	(static control)	70–90	60 min.
Tomato ripening	70	85			

*From A. L. Kaschub, "Industrial Humidification—Psychometric Considerations and Effects of Humidification," *Plant Eng.*, Mar. 22, 1979.

where average January temperature falls below 40°F (5°C). In the traditional approach, a vapor retarder is recommended for compact roof systems when interior RH is 45% or higher. The bottom map shows interior RH at 68°F (20°C) above which a vapor retarder is needed. This bottom map obviously gives a more sensitive indication of the need for a vapor retarder than the top map, which ignores the variable interior RH resulting from different building uses. [See Fig.

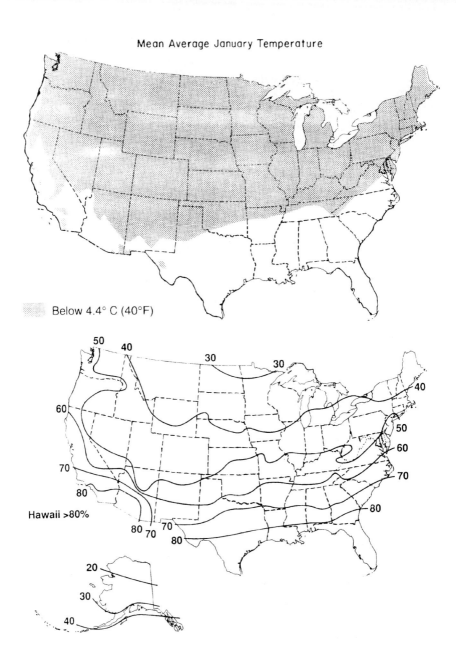

Figure 6.6 The shaded area in the top map shows where the average January temperature is below 40°F (5°C). The traditional rule recommended vapor retarders for compact roof systems with interior RH of 45% (or higher) in this sub-40°F zone. As a more sophisticated approach, the bottom shows the interior RH above which a vapor retarder is needed in membrane roofing systems. Vapor retarders are thus required for buildings with an indoor temperature of 68°F (20°C) to limit winter wetting potential to a prescribed value established as an interim industry consensus. In Chicago, vapor retarders would be required for buildings heated to 68°F if interior RH exceeds 42%. In Boston, the limiting RH would be 45%; in Fairbanks, 22%.

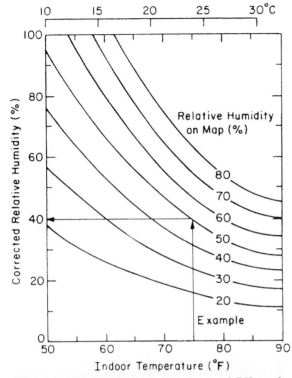

Figure 6.7 This graph shows the corrected RH on the vapor-drive map (Fig. 6.6) when the interior temperature is *not* 68°F (20°C). For example, interior conditions of 75°F, 40% RH, produce the same vapor drive as 68°F, 50% RH. (*Graph from Tobiasson and Harrington,* Vapor Drive Maps for the USA.)

6.7 for a graph correcting mapped RH for indoor temperatures varying from 68°F (20°C).]

Viewed in the context of the decades-old vapor-retarder controversy, Tobiasson's approach is a compromise. On one hand, there is the crude traditional approach: use a vapor retarder wherever average January temperature is 40°F or lower and interior RH is 45% or higher. On the other hand, there is the extreme engineering approach, exemplified by the American Society of Heating, Refrigerating and Airconditioning Engineers (ASHRAE) method, which almost always indicates the need for a vapor retarder.

In summary, here is the case *for* using a vapor retarder:

- It can protect the insulation from disintegration, delamination from freeze-thaw cycling, and loss of thermal resistance resulting from condensation of migrating vapor invading the roof system.

- It can protect deck materials and wood blocking from rot or corrosion promoted by condensation.
- It provides a good safeguard against the above-stated hazards if a building's use changes from "dry" to "wet."

Here is the basic case *against* using a vapor retarder:

- A vapor retarder is a disadvantage in summer, when vapor migration is generally downward through the roof. (Hot, humid air can infiltrate the roofing sandwich through vents or through diffusion through the membrane. It may condense on the vapor retarder itself.)
- In the event of roof leaks, the vapor retarder traps water below the insulation and may release it through openings that may be some lateral distance from the roof surface leak point, thus making it more difficult to locate the leak source at roof level. Since the vapor retarder delays discovery of the leak, a large area of insulation may be saturated before the punctured membrane is repaired.

Like the use of pitch pans or inadequately drained roof surfaces, the use of a vapor retarder as a temporary roof violates the basic principles of good roofing practice. Under the best conditions, installation of an effective vapor retarder is difficult. Subjecting it to the punishment of days (or weeks) of roof traffic and field operations exponentially increases the difficulty. If, however, cold, rainy weather or material delivery delay necessitates its use as a temporary roof, the designer should take precautions to assure the vapor retarder's integrity in the completed roof system. These measures should include (*a*) requiring plywood walkways to protect the vapor retarder from roof traffic, and (*b*) requiring a moisture survey or other means of testing insulation below the vapor retarder for moisture. Wet insulation should, of course, be removed before completing roof-system application.

Self-drying roof system

Self-drying roof systems rely on summer heat to evaporate winter-accumulated moisture and drive it down through a vapor-permeable deck into the building interior. Their feasibility has been demonstrated not only by experience with actual buildings, but by research designed to simulate the field process and to provide at least crude criteria for design of such systems. Tobiasson's research corroborates earlier research by former NBS researcher Frank Powell. This research has established the practicability of designing compact roof systems that can release small amounts of winter-accumulated moisture fast enough to prevent a perennial buildup and ultimately a premature failure of the roof system itself. In a satisfactorily functioning self-drying roof system, *all* winter-accumulated moisture is expelled during the following summer. And the system's average thermal resistance will satisfy some reasonable standard of moisture-caused loss. In contrast, an unsatisfactory self-drying system will suffer perennial moisture buildup—i.e., successive winter accumulations will exceed summer dispersals. Steadily declining thermal resistance, with consequently rising energy bills, will add a secondary insult to primary injury.

Tobiasson's study compared a building in Washington, D.C., with a building located in the more severe climate of Minneapolis. In Washington, D.C., with interior conditions of 68°F (20°C), 45% RH, the wetting/drying ratio is 0.20, with the drying period roughly 0.70 of the year (see Fig. 6.8). This system will dispel winter-accumulated moisture during the following summer. Under more severe interior conditions (i.e., 75% instead of 45% RH), the calculated wetting/drying ratio exceeds 1.0, thus indicating a potential for perennial moisture accumulation (see Fig. 6.9). Note in the graphs for Minneapolis that the wetting/drying ratio there is nearly as high for 45% interior RH as it is for 75% interior RH in Washington, D.C. With the 75% RH interior condition in Minneapolis, the wetting/drying ratio is 3.7. Unless provided with an effective vapor retarder, this system will accumulate moisture throughout a service life destined for shortening.

Earlier research, conducted in the 1960s by former researcher Frank Powell, demonstrated how scientifically designed self-drying roof systems can dissipate moisture gained during construction and winter vapor migration. For this research, a self-drying roof had to expel accumulated moisture during a laboratory-simulated summer season fast enough to (1) reestablish its equilibrium moisture (i.e., show no trend toward long term moisture gain) and (2) suffer only minor loss in thermal resistance.

Among the quantitative criteria set for a self-drying roof were the following.

- Minimum 80 percent average thermal resistance during the simulated year for insulation materials with normal hygroscopic moisture
- Moisture content at the end of the *second* summer simulation no greater than that at the end of the *first* summer (a safeguard against perennial moisture accumulation)
- Average rate of regaining moisture under simulated winter test conditions no greater than 0.05 lb/ft^2 (0.24 kg/m^2) per week

To derive the full benefits of a self-drying roof requires attention to field conditions that can cause excessive moisture in even a well-designed self-drying roof system. Powell cites a case history illustrat-

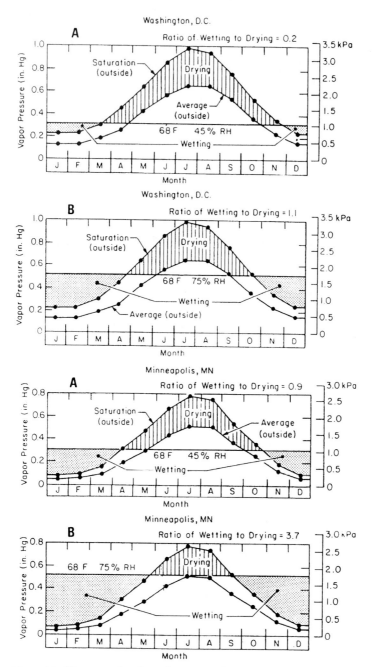

Figure 6.8 These graphs show the seasonal wetting-drying cycles for roof systems in Washington, D.C., (top) and Minneapolis (bottom). Liquid moisture accumulates in the system during cold weather and evaporates into the building interior in warm weather. The wetting-drying ratio indicates the annual quantity of moisture entering and leaving the roof system. Moisture quantity varies with vapor-flow *intensity,* measured by the exterior-interior vapor-pressure difference. It also varies with the *duration* of vapor flow. (*Graph from Tobiasson, "General Considerations for Roofs,"* Moisture Control in Buildings, *ASTM Manual 18, 1994.*)

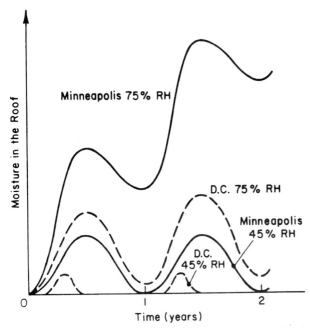

Figure 6.9 This graph shows how moisture can accumulate in roof systems. Note that at 68°F (20°C), 45% RH, there is no perennial accumulation either in mild Washington, D.C., or in colder Minneapolis. At high (75%) interior RH, there is a threat of such accumulation, especially in Minneapolis. (*Graph from Tobiasson and Harrington,* Vapor Drive Maps for the USA.)

ing field moisture problems. The roof system comprised an aggregate-surfaced built-up membrane on poured gypsum deck over fiberglass formboard supported by open web steel joists. In the laboratory, this system ranked as one of the best tested assemblies, requiring only 5 weeks drying time to dissipate initial moisture, compared with 16 or more weeks required by half the tested assemblies. During this roof's first spring, however, a layer of water appeared near the lower surface of the fiberglass formboard. The explanation was as follows: After pouring the gypsum concrete during the previous winter, the contractor used portable fuel-burning heaters within a weather enclosure during completion of the building's finishing operations. Initial moisture, plus moisture generated within the weather-enclosed building, accumulated within the built-up roof system. With the advent of spring, solar radiation, reversing the upward vapor-pressure gradient maintained during the winter by the heating of the humid enclosure, caused a downward migration of moisture to the lower surface of the fiberglass formboard. Areas with unventilated suspended ceilings suffered water damage.

Ventilation of the interior in combination with solar heating soon dissipated the accumulated moisture. But all moisture damage could have been completely prevented by simply ventilating the enclosed space during winter construction operations.

The tremendous quantities of moisture generated by hydrating construction materials cured by fuel-burning heaters are often overlooked. Combustion in an oil or propane heater produces more water as a by-product than the weight of the consumed fuel—1 gal water for each gallon of heating oil and 30 gal water for a 200-lb tank of propane. Cementitious materials release vast quantities of water vapor—roughly 1 ton water for each 1000 ft^2 wall ceiling plaster or 4-in.-thick structural concrete slab. Ventilation of enclosures where such materials are curing is absolutely essential to prevent excessive wetting of the roof system.

Unfortunately, as noted, the NBS researchers found the problem of self-drying roof design too complex to be solved by formula. Laboratory tests of unlisted systems are required to assess self-drying performance during the roof's service life. Slight perennially accumulating moisture gains could build up to hazardous levels, destroying the roof's thermal performance and ending its useful service life far short of the generally demanded 20 years.

The futility of topside venting

NBS research in the 1960s, confirmed by later research conducted by Tobiasson, demonstrated the futility of topside and edge venting, traditionally promoted as aids to expelling accumulated moisture from compact roof systems. "Breather" vents installed on NBS test specimens had no significant effect on the insulation's moisture content and R value. Canadian researchers also confirmed these results.

Tobiasson's research carries the NBS and Canadian researchers' negative findings a step further. Not only are breather vents essentially worthless as a means of expelling entrapped moisture, but they can even be a positive source of entrapped moisture, gained through leakage at the vulnerable flashed joints of the vents. Here, with this still widely recommended practice (one breather vent per 1000 ft^2 of roof surface), we see another triumph of unverified, rationalized conventional wisdom persisting long after its refutation by more scientifically based evidence.

The fear that loose-laid EPDM membranes admit moisture in hot weather has been rebutted by research by Rene Dupuis. EPDM permeability does increase in hot weather. But according to Dupuis's research, moisture accumulation through a loose-laid EPDM membrane is insignificant.

Theory of Vapor Migration

Water vapor infiltrates a roof through two mechanisms: *diffusion* and *air leakage*. Compared with heat transmission, diffusion corresponds roughly to conduction; air leakage corresponds precisely to convection.

Diffusion

Diffusion can be explained by the kinetic theory of gases—a theory whose basic principles account for changes of state from liquids (or solids) to gases and the accompanying energy changes. In a sealed container of air with water in the bottom, at any temperature above absolute zero, some liquid (or solid) molecules tear loose from the surface and escape into the air above. If constant pressure is maintained within the sealed container, then for each temperature a different quantity of water vapor (gaseous water molecules) escapes from the liquid and diffuses through the air.

So long as the water supply holds out, the liquid surface reaches equilibrium soon after the mixture reaches a stable temperature, with as many water molecules plunging back into the liquid as those escaping from its surface. When equilibrium is reached, the atmosphere is "saturated"; that is, the air-vapor mixture contains as much water vapor as it can retain at that temperature. The graph in Fig. 6.2 displays this capacity of air at constant pressure to "hold" varying quantities of water vapor at different temperatures. At normal atmospheric pressure of 14.7 psi (101.3 kPa) and temperature of $-10°F$ ($-23°C$), a cubic foot of air can hold 0.3 grain of water vapor; at 70°F (21°C) it can hold 27 times that amount, or 8.1 grain.

So long as the liquid water holds out, the air-vapor mixture remains saturated. However, if all the water evaporates and the temperature rises, the water-vapor content of the air mixture represents something less than the saturation limit. The ratio of the volume of water vapor actually diffused through the mixture to the volume of water vapor in saturated condition is the *relative humidity* (RH). But because vapor pressure is of greatest interest to the roof-system designer, the most meaningful definition for our purposes is: Relative humidity is the ratio of actual vapor pressure to the vapor pressure of a vapor-saturated-air–vapor mixture at constant temperature and pressure.

When the air-vapor mixture is saturated with vapor, the air temperature and the dew point are the same. But whenever the relative humidity is less than 100%, the dew point is, of course, lower than the air temperature. For example, at 70°F, 10% RH, the dew point is 13°F ($-11°C$).

Besides dew-point temperature, another condition is necessary to produce condensation. Water vapor flowing through permeable materials or air spaces passes through the theoretical dew-point plane into colder regions without condensing. (Condensation may actually occur, but it is followed by instantaneous reevaporation.) Migrating water vapor normally condenses only when it reaches an obstructive surface that stops or drastically retards its movement. This phenomenon is attributable to the kinetic energy of flow, which maintains the water vapor in a gaseous state at a supercooled temperature until it collides with a sub-dew-point obstruction.

Another phenomenon associated with the diffusion of water vapor throughout air explains the penetration of water vapor into roofing. The diffused water vapor is actually low-pressure steam, possessing its high latent heat energy. Like any other diffused gas, water vapor mixed with air exerts a pressure independent of the pressure exerted by the gases constituting the air (oxygen, nitrogen, carbon dioxide, and so on). In accordance with Dalton's law of partial pressures, the partial vapor pressure is directly proportional to the volume occupied by the water vapor.

If water vapor is supplied to a space, the vapor pressure rises rapidly with increasing temperature. Absolutely dry air (i.e., air with no suspended water vapor whatever) confined in a space of constant volume gains 2.2 psi (from standard atmospheric pressure of 14.7 to 16.9 psi) under a temperature rise from 70 to 150°F. (In accordance with Charles' law, the pressure increases in direct proportion with the rise in *absolute* temperature—from 530 to 610°R.) If to each cubic foot of dry air we add 76 grains of water vapor, the total pressure increases 5.6 psi under the same 80°F temperature rise. The water vapor exerts a partial pressure of 3.4 psi, or 490 psf, more than half again as much as the partial-pressure increase in the heated dry air. Thus warm, humid inside air exerts unbalanced vapor pressure, impelling vapor migration *from* the warm *toward* the cold side of the roof.

A curious consequence of the law of partial pressures is the water vapor's tendency to flow from a region of high vapor pressure toward a region of low vapor pressure, *regardless of the total atmospheric pressure.* Thus even if the outside air pressure exceeds the inside air pressure, the water vapor tends to diffuse through the roofing system from a warm, humid interior toward a cold, dry exterior. The roof merely impedes the natural tendency of both the air and the water vapor to diffuse throughout the atmosphere in equal proportions. In our vast atmospheric system, buildings are a mere local accident, and in seeking to achieve a perfectly proportioned mixture, the gaseous molecules in the atmosphere may move through walls and roofs opposite directions—with inward-bound air molecules passing outward-bound water-vapor molecules—as they seek their own partial-pressure level.

As an example of the pressure differentials encountered, consider a heated building interior within the normal range of heating and humidification—say, at 70°F (21°C) and 40% RH. Outside temperature is −10°F (−23°C), 100% RH (winter design conditions for Chicago). Under these conditions 1 lb indoor air contains 0.0063 lb (44 grain) water vapor and 1 lb outdoor air contains 0.0005 lb (about 3 grain) water vapor. (These data are obtained from a psychrometric table or chart.) According to Dalton's law, the partial pressure of the water vapor is directly proportional to the percent by *volume* (not weight) of vapor. Because the air–water-vapor mixture closely approximates an "ideal" gas, the vapor volume varies inversely with the ratio of the molecular weights of the water vapor (18.02) and the air (28.96). Assuming the pressure of the air-vapor mixture p_m = standard atmospheric pressure of 14.70 psi, we compute the vapor pressure p_v as a percentage of the air pressure p_a as follows:

$$p_v = \frac{28.96 \text{ (molecular weight of air)}}{18.02 \text{ (molecular weight of vapor)}} \times 0.0063 = 0.01\, p_a$$

$$p_m = p_v + p_a = 14.70 \text{ psi}$$

$$= 1.01\, p_a$$

$$p_a = \frac{14.70}{1.01} = 14.5 \text{ psi}$$

$$p_v = 0.01 \times 14.5 = 0.145 \text{ psi} \times 144 = 21.1 \text{ psf}$$

We can thus assume for this example a vapor migration attributable to diffusion from the warm interior toward the cold exterior—a migration impelled under a pressure imbalance of 19.4 psf (obtained by subtracting the exterior vapor pressure of 1.7 psf from the 21.1-psf interior vapor pressure). As explained earlier, this vapor-pressure imbalance induces an inside-out diffusion regardless of any possible atmospheric pressure imbalance forcing dry air toward the interior. In addition to the partial-pressure difference, the rate of diffusion depends on the length of the flow path and of course the permeance of the media through which it passes (deck, vapor retarder, insulation, and so forth).

Air leakage

In compact roof systems with the conventional arrangement of insulation sandwiched between deck and membrane, air leakage is a minor problem, accounting for less vapor transfer than diffusion.[2] This is especially true where the system components are adhered with solid moppings of hot bitumen, which seal the system against air leakage. In compact systems with mechanically anchored single-ply membranes, however, air leakage can become a big problem. The light, flexible membrane can billow upward and oscillate in response to wind-uplift pressure, and the resulting atmospheric "pumping" can draw air up from the building interior. Air leakage is a still bigger problem in roofs with plenum spaces and in roofs with permeable below-deck insulation (e.g., fiberglass batts). (See Fig. 6.10.)

Under the normally small pressure differences existing above and below the roof, the volume of air moved through these small openings is insignificant so far as it affects building heating and ventilation. But a relatively small volume of air can transport a troublesome volume of water vapor. At any point in the path where this migrating water vapor is stopped by a surface at or below the dew point, it condenses.

Several factors make the existence of atmospheric pressure differentials between inside and outside almost inevitable. The major factors are temperature differences and wind. Pressure differentials created by mechanical ventilation or exhaust systems are normally less important.

Chimney effect. In a heated building, the chimney effect produces higher pressures and consequent exfiltration in the upper part of the building and lower pressures and consequent infiltration in the lower part of the building. (Like a chimney, the heated building sets up a convection current, with cold air moving into the low-pressure space constantly vacated by the expanding heated air, which rises and raises the pressure in the upper parts.) In a cooled building, the opposite

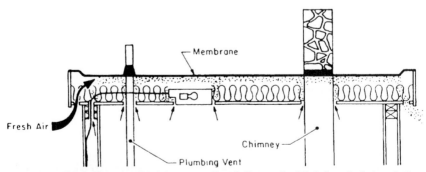

Figure 6.10 Air-leakage paths into a ventilated flat roof with below-deck insulation. (*From Tobiasson, "General Considerations for Roofs,"* Moisture Control in Buildings, *ASTM Manual 18, 1994.*)

occurs, with infiltration occurring at the upper levels and exfiltration at lower levels as the cooled, dense lower air presses out and sets up a downward convection current.

Wind suction. Wind creates a positive static pressure on a windward wall in proportion to the wind's kinetic energy (velocity squared). In conformance with the Bernoulli principle, the wind across a level or slightly pitched roof produces an uplift varying from a factor of 2.0 p_s or more at the leading edge to 0.2 p_s in some other region of the roof (p_s is the static pressure against a vertical surface).

As a hypothetical example of how pressure difference is created above and below the roof, consider a 400-ft-high building with a steady 20-mph wind blowing across the roof and a temperature differential of 60°F (70°F inside, 10°F outside).

At any level in the building, the pressure differential accompanying temperature difference can be computed from the following formula:

$$p_c = 7.6h \left(\frac{1}{T_i + 460} - \frac{1}{T_o + 460} \right)$$

$$= 7.6 \times 200 \left(\frac{1}{70 + 460} - \frac{1}{10 + 460} \right) = -0.365 \text{ in. water}$$

where p_c = theoretical pressure (in. water) attributable to chimney effect
h = distance (ft) from "neutral zone" (level of equal inside-outside pressure, estimated at midheight for multistory building without air-sealed floors)
T_o = outside temperature (°F)
T_i = inside temperature (°F)

Because static pressure for a 20-mph wind equals 0.193 in. water, an estimate of 0.5p_s (for average roof suction) yields a static pressure of about −0.1 in. water. Adding the two components of the pressure differential and multiplying by a conversion factor of 5.2 yields about 2.4 psf exfiltration pressure *upward* through the roof.

The practice of mechanically pressurizing airconditioned buildings through a 10 to 20 percent excess of fresh supply air over exhaust air can add another small increment of pressure imbalance to that produced by chimney effect and wind, further aggravating the problem of vapor migration. In arriving at a decision, the designer should weigh the benefits of mechanical pressurization against its liabilities. In humidified buildings the designer may even want to provide a small interior suction to reduce air leakage instead of deliberately increasing it.

Sample vapor-retarder calculation

Winter condition. To illustrate a vapor-migration problem for the normal (winter) condition, assume a roof construction as shown in the cross section and the following data: outside temperature $T_o = 0°F$, 90% RH, and inside temperature $T_i = 70°F$, 35% RH (See Fig. 6.11).

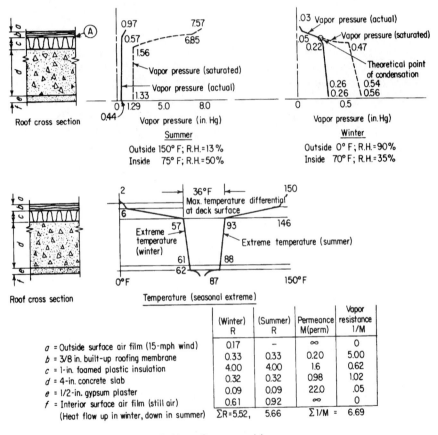

Figure 6.11 Vapor-retarder calculations, although imprecise, indicate the difficulty of solving the moisture problem with a vapor retarder. They are, however, far too conservative, because they ignore the roof's self-drying capacity.

Because air-leakage computations can be made only for known, measured cracks and other openings—a clearly impossible task—the following computations consider only diffusion. It is naïve to assume great accuracy in such computations, and so pressures (in in.Hg) are carried only to the second decimal place. Because of the many imponderables, vapor-migration computations are far less accurate than ordinary structural or mechanical design computations for, say, axial column stresses or heat losses. Thus vapor-migration computations are more an index to a problem than a precise design technique. They also ignore the self-drying capacity of roofs, which allows them to absorb limited amounts of moisture and expel them during hot summer weather.

The designer first tabulates thermal resistance (R values), perm values (M), and vapor resistances (perm-value reciprocals) for the various roof components and air layers (see Fig. 6.11). For a thermal steady state, the temperature loss upward through the roof cross section is proportional to the sum of the R values of the components between the interior and a given plane, divided by the total sum of the R values. Thus

$$T_x = T_i - \frac{\Sigma R_x}{\Sigma R}(T_i - T_o)$$

where T_x = temperature at plane X
ΣR_x = sum of thermal resistances from interior to plane X
ΣR = total thermal resistances

The vapor-pressure gradient through the roof cross section can be computed in similar fashion. (It is analogous to drop in hydrostatic head in a series of canal locks.) Thus

$$p_x = p_i - \frac{\Sigma(1/M_x)}{\Sigma(1/M)}(p_i - p_o)$$

where p_x = vapor pressure at plane X
p_i = interior vapor pressure
p_o = exterior vapor pressure
$\Sigma(1/M_x)$ = sum of vapor resistances from interior to plane X
$\Sigma(1/M)$ = total vapor resistances

The critical plane is obviously A, the interface between the insulation and the membrane (see Fig. 6.11). Because it lies on the far side of the relatively permeable insulation, it represents the worst possible combination of low temperature and high humidity and thus poses the greatest risk for condensation.

$$T_a = 70 - \left(\frac{5.02}{5.52} \times 70\right) = 6°F$$

$$p_a = 0.26 - \left(\frac{1.69}{6.69} \times 0.027\right) = 0.19 \text{ in.Hg}$$

We can now calculate the vapor flow to and from plane A. However, when we check the saturated vapor pressure at A, we find that it is only 0.05 in.Hg, which is far below the calculated 0.19 in.Hg. Thus we know that condensation must take place at (or possibly below) plane A.

To calculate the vapor flow to plane A, we divide the pressure differential by the vapor resistance:

$$\text{Vapor flow } to \text{ plane } A = \frac{0.26 - 0.05}{1.69} = 0.12 \text{ grain/(ft}^2 \cdot \text{h)}$$

$$\text{Vapor flow } from \text{ plane } A = \frac{0.05 - 0.03}{5} = 0.004 \text{ grain/(ft}^2 \cdot \text{h)}$$

These grains of water vapor are in transit through the roof, representing a more or less "steady state" of vapor migration, which varies with changing temperature and humidity. Because the air-vapor mixture at plane A is saturated, all the migrating water vapor must condense. Thus we can compute the condensation rate merely as the difference between the vapor flow to plane A and the vapor flow from plane $A \approx 0.12$ grain/(ft$^2 \cdot$ h).

To prevent condensation at plane A (the insulation-membrane interface), a vapor retarder must add sufficient vapor resistance on the high-pressure (warm) side of plane A to reduce the rate of vapor migration at plane A by at least 0.12 grain/(ft$^2 \cdot$ h). Because the condensation rate (the excess of vapor flow *to* plane A over the vapor flow *from* plane A) must equal 0, we set the vapor flow to plane A equal to the vapor flow from plane A.

$$\text{Vapor flow to plane } A = \frac{0.26 - 0.05}{1.69 + X} = 0.004 \text{ grain/(ft}^2 \cdot \text{h)}$$

where X = required vapor resistance added by vapor barrier

$$0.21 = 0.004 \,(1.69 + X)$$
$$X = 53.5$$

$$\text{Required per rating} = \frac{1}{X} = 0.02$$

To ensure this virtually 0 perm rating is, to say the least, difficult.

Summer condition. As the preceding discussion has demonstrated, vapor migration changes direction during the hottest summer months of a temperate climate. The membrane then doubles as vapor retarder, and the vapor retarder itself becomes, at best, useless. There should be no problem of condensation, because the saturation vapor pressure should remain well above the computed (or actual) vapor pressure throughout the entire roof cross section. (See Fig. 6.11, in which severe summer conditions of 100°F, 50% RH, are equivalent to a roof surface temperature of 150°F, 13% RH.)

Alerts

Design

1. As a basis for deciding whether to use a vapor retarder, calculate the location of the dew point and the rate of vapor migration under the worst winter condition (see example at end of chapter). If vapor-retarder location is below dew point, do not specify it.

2. A vapor retarder on a roof destined for future penetrations is highly vulnerable to damage and likely to fail.

3. If in doubt, do not specify a vapor retarder. Use a vapor retarder only if a study of your conditions indicates a positive need for it. (See Fig. 6.6 for critical interior RH as a guideline.)

4. If unhumidified exterior air is drawn into the building, a vapor retarder is normally not required, unless the interior relative humidity is 60% or more.

5. Never specify a vapor retarder between a poured structural deck and poured insulating concrete fill. (It will seal in moisture.)

6. Do not specify a vapor retarder in the roof over an unheated interior.

7. Require flashing and enveloping of a vapor retarder at all roof penetrations.

8. Check the vapor retarder's ability to take nailing or other anchorage punctures and still maintain a satisfactory perm rating.

9. Check a vapor retarder for its effect on the roof assembly's fire rating (see Chapter 8, "Fire Resistance").

10. For a vapor retarder hot-mopped to a deck, specify ASTM Type II (flat) or III (steep) asphalt. Specify rosin-sized paper, felt, or tape over joints in wood sheathing, plywood, or other prefabricated units to be nailed.

Field

1. Do not use the vapor retarder as a temporary roof.

2. Prohibit installation of light, flexible plastic vapor retarders on windy days.

References

1. Wayne Tobiasson, "General Considerations for Roofs," *Moisture Control in Buildings,* ASTM Manual 18, 1994, pp. 291ff.
2. Tobiasson, op. cit., p. 292.

Chapter 7

Wind Uplift

Unlike leaking roofs, afflicted with a chronic ailment that continually harasses a building owner with constant dismal reminders, a defectively anchored roof system normally inspires complacency—until disaster strikes. Disaster struck in 1989 with hurricane Hugo in South Carolina, and later, in the summer of 1992, in the gigantically destructive hurricane Andrew in south Florida and hurricane Iniki on Kauai. Total estimated cost of these three hurricanes exceeds $20 billion. By a conservative estimate, this cost would have been a minor fraction of $20 billion if we had devoted the required effort to assuring construction quality through competently drafted and, even more important, rigorously enforced wind-pressure design provisions in our building codes.

The complacency noted above is manifest in the cavalier attitude of many developers and trial lawyers. Some believe—or at least profess to believe—that a blowoff is to be expected if wind pressures reach the design pressure. In other words, they confuse the concept of *design* load, which a structure's reserve strength should easily resist, with *ultimate* load, i.e., the load at which the structure (or building component) should fail. This confusion is often displayed in courtrooms by defense attorneys, who argue that a failure was an act of God or otherwise unforeseeable occurrence for which the defendant cannot be held responsible.

The key fact about wind pressure is its exponential increase with wind velocity: doubling the wind velocity *quadruples* its pressure. This is an elusive fact, seldom recognized even by those who habitually experience it. (Veteran golfers used to making allowances for wind usually underclub when hitting against a 20- to 25-mph breeze, not realizing that wind resistance to a golf ball's flight varies with the wind velocity squared.) This general failure to maintain a decent respect for the exponential increase in wind pressure stems at least partly from the conventional use of wind speed to describe gales and hurricane winds. Few persons are conditioned to make the mental computations necessary to translate speed into uplift *pressure*—to recognize, for example, that a 140-mph wind exerts twice the uplift pressure of a 100-mph wind.

Of particular concern in this regard is the absurdly low 65-mph maximum wind speed set by most manufacturers in their roof-system warranties. Under this limitation, damage claims attributable to winds exceeding 65 mph are rejected. Windspeeds within the 50-year recurrence category exceed 65-mph throughout the United States. In hurricane belts, home to a sizable proportion of the U.S. population, maximum design wind speeds now range up to 150 mph. Hurricanes produce wind-uplift pressures more than *five times* the uplift pressure associated with a 65-mph wind speed. Warranty provisions limiting liability to a maximum 65 mph wind speed are thus essentially worthless in hurricane areas, and close to worthless in many others.

Recent developments in roofing technology have increased the complexity of an already complex problem by introducing new stresses on roof systems. Loose-laid, ballasted systems resist wind uplift by a totally different mechanism from traditionally adhered or mechanically anchored systems. The advent of single-ply battened or individually fastened systems, which allow the flexible membranes to flutter at moderate or even relatively low wind speeds, has made fatigue failure a major wind-uplift concern. The roof designer must now consider dynamic wind-induced oscillations as well as the simpler static equivalent designs to resist ultimate wind loading.

As a final prefatory note before launching into the detailed technical aspects of this chapter, let us emphasize that roof designers and building owners should heed this warning. Conservative wind-uplift design is a good supplemental insurance policy. In view of the industry's past experience, measured in billions of dollars lost through wind-caused damage, plus the inevitable uncertainties associated with wind-uplift design, the relatively slight cost of providing safe wind-uplift resistance is a sound investment. Among the horror statistics presented by Factory Mutual engineers is the multiplication of damage claims in wind blowoffs. The damage to building contents averages 5 to 10 times the cost of damages to the building itself. Prudent designers will thus convince their clients that a few extra dollars spent on wind-uplift resistance is a cheap premium for additional insurance against this worst of roof-failure modes.

The revised American Society of Civil Engineers (ASCE) Standard 7-95, the model for most U.S. building codes in setting design loading for buildings, will probably require more conservative wind design than the 1988 version. Exemplifying these changes is an increase in design wind speeds, which will raise the 50-year recurrence design speed in southern Florida's tip from 130 mph to 150 mph and raise the minimum U.S. design speed by more than 20 percent (from 70 to 85 mph) (see Fig. 7.1). This wind-speed increase results from a new criterion for recording wind speed. Peak gusts (2- to 3-s duration) replace the so-called fastest-mile wind speed. This new approach will simplify the currently complex design procedure by reducing the gust-correction factor formerly applied to fastest-mile wind speeds. (The fastest-mile wind speed is the average wind speed over the time required for the moving air to travel 1 mi; thus the time required for a fastest-mile wind speed of 90 mph is 40s.)

Historical Background

The advent of lightgage steel deck, which became the most popular deck material for low-sloped roofs in the post-World War II era, inaugurated the era of roof blowoffs. For the decade of the 1970s, Factory Mutual engineers reported some 1250 separate blowoffs involving insulated steel decks. This experience was virtually duplicated in the 1980s, at an average cost per blowoff of roughly $400,000 (1995 dollars).

Lightgage metal construction is associated with roof blowoffs involving not only membrane material but perimeter flashing, fascia strips, gravel stops, and parapet copings. A large majority of roof blowoffs (roughly 80 percent, according to FM engineers) start with failure at the perimeter flashing.

Metal standing-seam roofs also pose a special wind problem, according to George A. Smith of FMRC. The poor performance of these metal roofs stems from use of excessively thin lightgage metal, excessive spans, and weakness in sliding clips, according to Smith.

Though wind damage to low-sloped roofs is not limited to hurricane zones, they are the sites of the most numerous, most catastrophic, and most costly blowoffs. (The one possible exception concerns tornado damage, but it is generally restricted to much smaller areas.) The billions of dollars of wind-inflicted roof damage caused by hurricanes Hugo, Andrew, and Iniki exposed defective practices long recognized by knowledgeable observers of Sunbelt construction practices. According to one investigator, 90 percent of the buildings damaged in hurricane Andrew violated the government building code. Developers and their defense attorneys naturally blame the extraordinary winds of hurricane Andrew rather than defective construction. True, these winds were extraordinary, with gusts clocked at 163 mph and sustained winds at 140 mph. With a safety factor of 2, however, hurri-

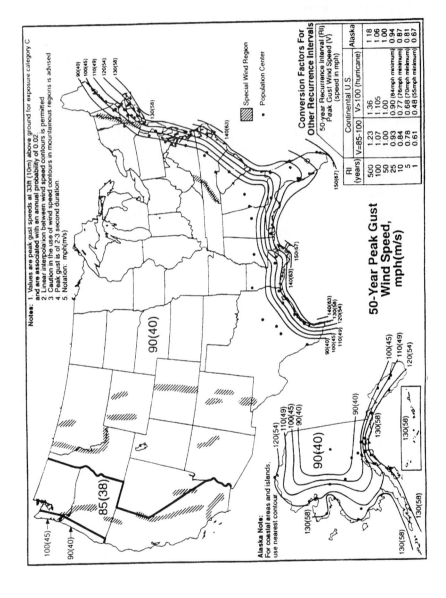

Figure 7.1 New ASCE wind map replaces old fastest-mile wind speed with peak gust (2- to 3-s) wind speeds, for 50-year recurrence. (*American Society of Civil Engineers.*)

cane-zone structures should resist winds of roughly 180 mph $[= (2\times 130^2)^{1/2}]$. In hurricane Iniki, which struck the Hawaiian island of Kauai in September 1992, maximum fastest-mile wind speeds were only 90 to 100 mph. They thus produced uplift pressures only half those created by the faster winds of hurricane Andrew. But the building code governing Kauai was also much less rigorous than south Florida's, with design wind speed = 80 mph.

Among the examples of defective construction discovered in the investigation of hurricane Andrew were:

- Grossly deficient nailing of roof sheathing to supporting wood trusses (with spacing sometimes three times the code-mandated 6 in. and nails often missing their substrate)
- Substitution of weaker staples for nails to anchor sheathing
- Omission of code-mandated "hurricane anchors" (i.e., steel straps anchoring roof trusses to their bearing walls)

The 1992 hurricanes thus exposed the sordid facts that engineers have known for years about the quality of hurricane-zone construction. Cost-chiseling construction tactics cut perhaps $100 from a house's construction cost, but magnify blowoff risk exponentially.

One coauthor, merely one witness to the slovenly construction practices prevailing in south Florida, has observed some astonishing examples of substandard anchorage, notably:

- A large project of 26 condominium buildings with the majority of its roofs *totally unanchored*—i.e., with insulation boards simply laid on concrete decks, with no attempt to provide code-mandated mopping (or fastener) anchorage
- Numerous projects with half the code-mandated mechanical fasteners, applied in random patterns, anchoring the insulation (or base sheet) to plywood or steel decks
- Substitution of defective, substandard fasteners for code-mandated fasteners

In addition to simple violations of good practice, the problems of wind-uplift resistance offer another of the bountiful illustrations of the interdependence of roof-system components, the subject of Chapter 2, "The Roof as a System." The solution of one problem often leads to the creation of another. Back in 1953, the industry learned an expensive lesson about the fire-feeding hazards of hot-mopped asphalt on steel decks from the disastrous fire at a gigantic GM plant in Livonia, Michigan. As a consequence of the Livonia fire, cold-applied adhesives and mechanical fasteners began to replace hot-mopped steep asphalt as a method of anchoring insulation to steel decks. Cold-applied adhesives offered one advantage over mechanical fasteners in steel deck assemblies including vapor retarders. For vapor retarders placed directly on the deck, mechanical fasteners jeopardized their integrity by puncturing them.

As an alternative to mechanical fasteners and limited hot-asphalt mopping, cold-applied adhesives soon betrayed their weakness as a dependable wind-anchorage technique. Irregular deck surfaces can prevent proper contact between the adhesive and the adhered component. Moreover, their slow attainment of full adhesive strength makes cold adhesives vulnerable during construction, when loadings are at their maximum. These loads can deflect the deck and break the adhesive bond between deck and insulation. (Use of a cold adhesive aged beyond its shelf life can also result in deficient adhesive strength. Aged beyond its shelf life, a cold adhesive loses volatiles required to maintain its fluidity or forms a gel. This leaves the material too viscous to attain proper contact between roof components and to cure properly.)

Hot-mopped steep asphalt is undependable for different reasons. Dependable adhesion with hot-mopped asphalt depends upon application timing that is, to say the least, difficult, and under some conditions (e.g., low temperature, high wind) may be impossible (see Fig. 7.2). Applied to steel or structural concrete decks, ASTM Type III (steep) asphalt cools from 500°F (240°C) to 300°F (150°C), the minimum temperature for dependable adhesion, in 6 s or less under highly favorable weather conditions [70°F (21°C) ambient temperature, 0-mph wind]. Structural concrete, which promotes even faster cooling than steel decks, combines high thermal conductance with high heat capacity. Tamping insulation boards in place within the few seconds available to assure dependable adhesion on a concrete deck could tax any applicator's skill. Note, however, that hot mopping is a satisfactory method of anchoring a top layer of insulation to a lower insulation layer, or anchoring a membrane base sheet to insulation. On a fiberglass insulation substrate, for example, the applicator has nearly 10 times as much time (nearly a minute under favorable conditions) before the asphalt cools below dependable adhesion temperature (see Fig. 7.3). (This vastly delayed cooling of the asphalt occurs because of the insulation substrate's low thermal conductance which retards heat-energy flow from its heated surface. It is the same phenomenon that, on sunny days, raises the membrane temperature of a compact, insulated roof system far above that of an uninsulated roof system.)

Spurred by the catastrophic damage of hurricane Hugo and the proliferating technical problems attending the introduction of new roof systems, the Roofing Industry Committee on Wind Issues (RICOWI) was created in November 1989 to pursue the following goals:

Figure 7.2 Blowoffs from deficient anchorage can result from chilled hot-asphalt adhesive (top) or from an inadequate number of mechanical fasteners (bottom). There is less excuse for the latter, as the rapid cooling rate of hot asphalt on a steel deck allows the roofer only a few seconds to embed the insulation, even under favorable conditions.

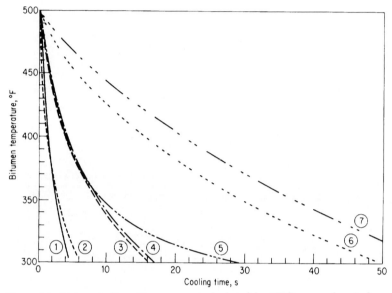

Figure 7.3 Hot-asphalt cooling curves (calculated by NBS researchers) show the tremendous variation in cooling time from 500 to 300°F for hot asphalt mopped to substrates of varying thermal resistance and heat capacity. Air and substrate temperatures are 70°F; wind speed is 0 mph. Substrates are (1) concrete, (2) steel, (3) insulating concrete, (4) plywood, (5) felt/PUF insulation, (6) fiberglass insulation, and (7) PUF insulation. Concrete, which promotes the fastest cooling, has low thermal resistance and high heat capacity. Urethane, which promotes slow cooling, has high thermal resistance and low heat capacity. (*National Bureau of Standards.*)

- To promote and coordinate research
- To accelerate the promulgation of new consensus standards for wind design and testing
- To educate the building industry about wind problems

Of these three goals, the last is probably the most important.

Mechanics of Wind-Uplift Pressure

The exponential increase in wind pressure with velocity is readily demonstrated. Wind blowing against a vertical surface exerts a positive pressure (known as *velocity, static,* or *stagnation* pressure). This positive pressure accompanies the wind's total loss of momentum—i.e., velocity loss from wind speed v to 0. It is proportional to (*a*) wind velocity squared and (*b*) air density.

Though air density varies with temperature, elevation, and moisture content, it is conventionally assumed to be 0.0765 pcf, its density at 59°F (15°C), sea level, standard atmospheric pressure, in calculating velocity pressure. This basic velocity pressure formula is:

$$q = 0.00256V^2 \qquad (7.1)$$

where q = pressure (psf)
V = wind velocity (mph)

(See Fig. 7.4) for derivation.)

Uplift pressure is correlated with this basic static pressure by negative coefficients. The minus sign denotes uplift (i.e., vacuum) pressure. Thus an uplift pressure coefficient of -1.3 for a 90-mph wind denotes an uplift pressure of -27 psf = $-1.3 \times 0.00256 \times 90^2$.

Wind blowing across a roof plane exerts uplift pressure because of a phenomenon known as the Bernoulli principle: The pressure exerted *perpendicular* to flow direction by a moving fluid drops with increased velocity and rises with decreased velocity to maximum pressure when the fluid is stationary. (As a simple demonstration of the Bernoulli principle, hold two vertically suspended sheets of paper in parallel planes about 2 in. apart and blow between them. The two sheets will come together, demonstrating the reduced pressure accompanying faster air flow.)

The final design wind-uplift pressures depend upon two basic variables: wind velocity and building size, shape, and wall openings. In keeping with the complexity of the topic, each of these primary factors depends upon secondary variables.

Basic wind-speed determinants

Wind speed is determined by the following factors:

- Geographic location
- Local topography
- Elevation

Geographic location establishes the basic wind speed used for wind-uplift design pressure. This is the primary determinant of wind velocity, dependent on the site location in the complex atmospheric system encompassing the earth. Wind is created by differences in pressure accompanying differences in solar heating (intense at equatorial regions, tenuous at polar regions) and by the earth's rotation, which induces prevailing westerlies in middle latitudes. In the continental United States, this basic, geographically determined wind speed varies by a factor of nearly 2, which, translated into wind pressure, is nearly 4. Wind speed is highest at the southern tip of Florida: 150

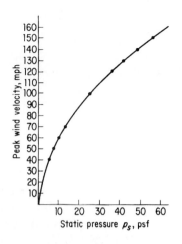

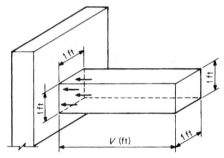

Figure 7.4 Static pressure q exerted against a wall by a prism of moving air can be computed by a formula derived as follows:

$$E_k = qs = \int \frac{M\,dv}{dt} \cdot v\,dt = \int Mv\,dv$$

For a velocity change from v to =:

$$E_k = \int_v^o Mv\,dv = \frac{Mv^2}{2} = \frac{Wv^2}{2g}$$

where E_k = kinetic energy of moving air mass (ft · lb)
q = static pressure (psf)
W = 0.0765 (lb/ft^3) @ 59°F (15°C)×v (ft)
g = gravitational acceleration = 32.17 ft/s^2
v = distance traveled by air prism in 1 s

For a 1-s interval:

$$E_k = qv$$

$$q = \frac{E_k}{v} = \frac{0.0765\,v \cdot v^2}{2g \cdot v} = \frac{0.0765 v^2}{2 \times 32.17} = 0.00119 v^2$$

Convert wind velocity to mph:

$$1 \text{ mph} = 5280/60^2 = 1.4667 \text{ ft/s}$$

$$q = 1.4667^2 \times 0.00119 v^2$$

$$= 0.00256 V^2$$

mph at 50-year recurrence in the ASCE 7-95 wind map. It is lowest in the three West Coast states. As previously noted, this new ASCE 7-95 wind map shows peak gusts for a 2- to 3-s duration, replacing the previous fastest-mile wind speed (i.e., average wind speed over the time required for the wind to travel 1 mile). Fastest-mile wind speeds are obviously lower than peak gusts, since the required time intervals are at least 10 times the duration of peak-gust measurements.

The 50-year (or 100-year) recurrence interval serving as the basis for wind maps introduces the concept of probability into wind-uplift design. A wind speed of 50-year recurrence is, by definition, a wind speed with an annual probability of $1/50 = 2$ percent. (By parallel reasoning, a 100-year-recurrence wind obviously carries a 1 percent annual probability.) Note from Fig. 7.5 that a 50-year wind is not certain to occur within a 50-year interval; its probability of occurring at least once in 50 years is 64 percent, and its probability of occurring at least once in 20 years, a normal anticipated roof-system service life, is only 33 percent. There is no basic difference, however, between design for wind and for any other assumed structural live loading. Structures are designed not for normal daily loading, but for extraordinary loads, the maximum loads they are likely to carry. A 100-psf corridor live load, for example, may be as improbable as a 50-year wind-uplift pressure during the structure's service life.

Local topography affects wind speed by varying the surface friction and turbulence caused by obstacles to air flow across the earth's surface. Like a liquid flowing through a pipe or water in a stream, wind is slowed at and near the ground by surface frictional resistance and by trees, hills, buildings, and other obstacles to its unimpeded movement. As a consequence, there are four so-called ground roughness factors in ASCE 7-88 (to be revised in ASCE 7-95). In descending order of frictional resistance, they are classified as follows:

- Exposure A (large urban centers or highly irregular terrain)
- Exposure B (suburbs, towns, wooded areas, rolling terrain)
- Exposure C (flat open country and grasslands)
- Exposure D (flat, unobstructed coastal areas, exposed to wind over water)

Though these categories suffice in the overwhelming majority of instances, there are situations where mountainous terrain, gorges, ocean promontories, and other unusual topographic features create higher wind velocities than can be accommodated by a standardized procedure. As one example of the attempt to incorporate more of these unusual features in its procedure, ASCE 7-95 provides for accelerated wind at ridges and escarpments. It also contains demarcated "special

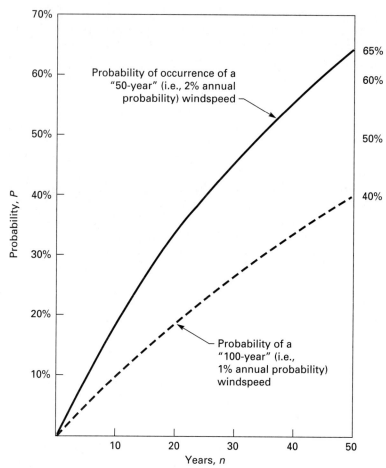

Figure 7.5 Curves show the probability that a wind speed with annual probability of 1 or 2 percent will occur in a given number of years, from 0 to 50. Note that an annual occurrence probability of 2 percent does *not* equal a probability of 1 (i.e., certainty) in 50 years. This probability is 64 percent. The probability of an occurrence at least once within 20 years (normal anticipated roof service life) is 33 percent. The formula for calculating these values, $P = 1-(1-p)^n$, is derived as follows: Let p = probability of annual occurrence (2 percent), n = years, P = probability of at least one occurrence within n years. The probability that the wind speed will *not* occur in n years = $(1-p)^n$. Then the probability that it *will* occur, P, equals $1-(1-p)^n$.

wind regions" on the lee side of mountain ranges, where winds can accelerate. Unique site conditions may even require wind-tunnel testing or other specifically targeted analytical techniques instead of a standardized procedure.

Gusting, which is largely dependent on topography, refers to the intensifying of wind speed over short time intervals (see Fig. 7.6).

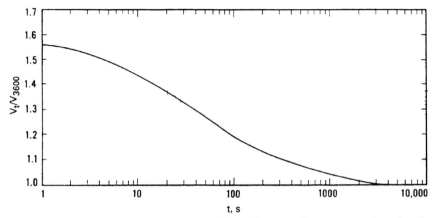

Figure 7.6 Ratio of probable maximum wind speed averaged over t seconds to hourly mean speed. (*From ASCE Standard 7-88.*)

Gusts, or lulls, are more prevalent in cities than in open country. Average wind speed in the city is generally lower than in the country, but higher gust factors occur in the city because of the complex, irregular airflow patterns around buildings and other man-made obstacles. Turbulence, created by temperature differentials at the surface, also affects gusting. Warming of the surface increases turbulence, from updrafts of heated air, whereas surface cooling, producing so-called atmospheric inversions, decreases turbulence. High surface wind speeds, however, tend to decrease the effect of turbulence on gusting.

Elevation above the earth's surface is a third determinant of wind velocity. This determinant is obviously dependent on the previously discussed determinant, topography. Frictional drag at the surface extends upward through a *boundary layer*, whose height depends on ground roughness. Boundary-layer height varies from a maximum of roughly 1500 ft (460 m) above exposure A to 700 ft (215 m) for exposure D. Above the boundary layer, the *gradient wind* blows at constant, unimpeded velocity (see Fig. 7.7).

At elevations within the boundary layer, wind velocity can be estimated by an exponential formula:

$$v_z = v_1 \left(\frac{z}{z_1} \right)^\alpha$$

where v_z = wind velocity at elevation z
v_1 = wind velocity at reference height (z_1 = 33 ft)
α = exponent, from 0.33 for exposure A to 0.1 for exposure D

Calculating the wind velocity for exposures other than C involves a convoluted procedure. The data used for the wind-contour maps come

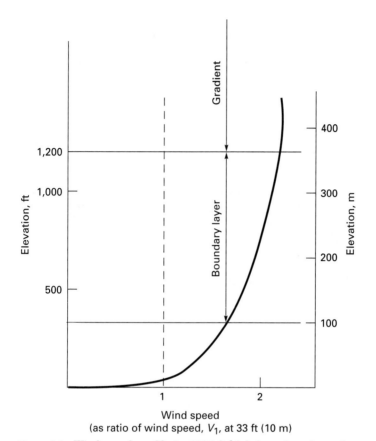

Figure 7.7 Wind-speed profile in 1200-ft-high boundary layer for ground roughness B. Wind speed increases by a factor of 2.2 at gradient-wind elevation from datum wind speed at 33-ft elevation. Wind-speed ratios are calculated from the formula $(z/z_1)^\alpha$, where z = elevation of computed wind speed, z_1 = wind speed at 33-ft elevation, and α = exponent dependent on ground roughness (= 0.22 for ground roughness B).

almost exclusively from airports, exposure C (flat, open country). As a consequence, calculating wind velocity for other exposures (i.e., A, B, and D) requires conversion of the wind-map velocity into a gradient wind velocity (i.e., at an elevation of 900 ft for exposure C), which is then converted into an appropriate wind velocity for the site elevation (i.e., building height) and exposure rating (i.e., A, B, or D). This procedure yields the following equation:

$$v_z = \left[v_{33}\left(\frac{900}{33}\right)^{0.143}\right]\left(\frac{z}{z_g}\right)^\alpha$$

where z = building height (elevation)

z_g = gradient height for the particular exposure (= 1500 ft for exposure D, 1200 ft for B, and 700 for D)

α = 0.20 for exposure A, 0.143 for B, 0.105 for C, 0.087 for D)

By computing the bracketed exponential function, we can simplify the above equation:

$$v_z = 1.6\, v_1 \left(\frac{z}{z_g}\right)^\alpha$$

Building shape, size, and enclosure

After determining the basic wind velocity prevailing at the site, the designer confronts a second basic problem: determining the influence of building size, shape, and wall openings on wind-pressure coefficients. A building in a moving air mass behaves like an ungainly airfoil. A level roof is analogous to the upper surface of an airplane wing, which, largely through the negative pressure produced by the fast-flowing air, creates an uplift force that keeps the plane airborne. Unlike an airplane wing, however, a sharply angled building is not designed for smooth, laminar flow; uplift pressures can vary widely because of irregular airflow patterns. At the windward roof edge, uplift forces can reach two times the static wind pressure. Over the remaining roof areas, uplift may equal the static pressure. At the windward corners, the uplift pressure can rise to three times the static pressure (see Fig. 7.8).

Uplift forces depend basically on wind angle and slope. For a conventionally shaped building with rectangular plan, the worst wind direction is 45°, diagonally across the building. The 45° angle produces strong vortices along the roof's windward edges (see Fig. 7.9). A level roof also maximizes uplift pressure. As roof slope increases, suction over the windward roof plane decreases to zero at about 30°. On steeper slopes, the wind exerts a positive pressure on windward roof planes and suction on leeward planes. Wind blowing perpendicular to a gable end wall produces suction over both sloping roof planes.

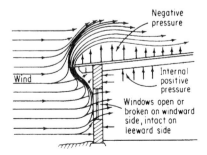

Figure 7.8 Wind uplift is severe at the roof perimeter, especially at corners, where it exceeds the normal static pressure against the wall. (*Factory Mutual Engineering Corporation.*)

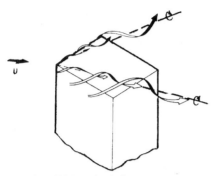

Figure 7.9 Diagonal wind sweeping across roof creates worst uplift and scouring condition at windward corner. (*R. L. Wardlaw, National Research Council of Canada.*)

Less important factors affecting wind pressure are height, size, and shape. Because wind gusts may engulf a small building, but only partially engulf a large building, small buildings should be designed for larger wind pressures than large buildings.

Parapets introduce still another complication. Tall parapets can drastically reduce wind-uplift pressure—e.g., a 5-ft parapet can cut the uplift coefficient by a factor of nearly 3, according to wind-tunnel tests conducted at the University of Toronto.[1] A low parapet, however, in the 18-in. range, may increase uplift pressures. As a further complication, the parapet's effect depends on other building dimensions: height, length, and width.[2]

A final major factor affecting wind uplift on the roof system is building enclosure. *Positive* internal pressure within the building can combine with negative external pressure to produce extraordinarily high uplift pressure. A prime example is an airplane hangar open to wind penetration. Under the more conservative internal pressure provisions in ASCE 7-95, internal pressure coefficients have been increased to + 0.80 and −0.30 for buildings in hurricane zones with glazed openings (*a*) located in the lower 60 ft and (*b*) *not* designed to resist windborne debris. For all other buildings, these coefficients are + 0.18 and −0.18.

Wind-induced internal pressure within a building will affect the roof system if the deck has open joints. Joints in steel, plywood, and other jointed-panel decks open the underside of the roof system—normally at the deck-insulation interface in conventional compact roof systems—to positive air pressure from the building interior. Roof systems with monolithic, airtight decks—e.g., structural, cast-in-place concrete—are subject to considerably less total uplift pressure. In airtight decks, any lifting of the insulation above the deck surface in response to external vacuum pressure will produce a partial vacuum in the expanded space at the deck-insulation interface, and this partial vacuum pressure will resist the external uplift pressure. With jointed decks, however, any expansion of this space at the deck-insulation interface will *add* to total uplift pressure, especially if there is positive interior pressure from an invasion of moving air through wall openings.

A 15 percent reduction in uplift pressure is allowed by FMRC for monolithic decks in its publication DS 1-28S.[3]

Determining final wind-uplift pressure

Using the variables discussed in the preceding sections, the designer can determine final wind-uplift design pressure from two basic formulas in the ASCE 7-95 approach. These formulas quantify the previously discussed independent variables affecting final design uplift pressures:

- The basic wind velocity prevailing at the site
- The particular building's response to this prevailing wind velocity

The site wind-pressure formula modifies the basic, static wind-velocity formula (7.1) with two additional factors as follows:

$$q = 0.00256 K I V^2 \quad (7.2)$$

where q = basic velocity pressure at site
V = wind velocity (mph, 3-s gust)
K = pressure coefficient that corrects the basic velocity for specific site conditions (i.e., ground roughness and elevation)
I = a so-called "importance factor," designed to increase effective velocity pressure and thus increase the safety factor for design of buildings with high occupancy or significant hazard to life and property

Having determined q (design wind velocity), the designer next calculates basic design uplift pressure p (psf) from the following formula:

$$p = qGC \quad (7.3)$$

where G = gust response factor
C = pressure coefficient values for various portions of the roof (i.e., corners, perimeters, interior)

Design example for anchored system

Continual updating of wind-design procedures as new research refines basic wind data makes it impossible to present perennially up-to-date, universally applicable design procedures in a manual revised only every 12 years or so. Accordingly, this section presents two parallel methods yielding similar results: the 1991 FM method and the ASCE Standard 7-95 method.

Consider a three-story (40-ft-high) office building in Atlanta, Georgia. Since FM uses the more conservative 100-year recurrence interval (rather than the 50-year recurrence used by ASCE, SPRI, and most codes), we located Atlanta in the 90-mph zone on the wind map. Ground roughness for the building is assumed as exposure C (unprotected grassland).

For the FM procedure, from Table 7.1 (from FM *Loss Prevention Data Sheet, 1-28S,* September 1991), basic interpolated velocity pressure (corrected for gust factor and interpolated for elevation) is 28 psf. With a safety factor of 2, this velocity pressure places the Atlanta building within Zone 1 (Class 1-60, designed for ultimate wind-uplift pressure of 60 psf).

For a steel deck in Zone 1 (Class 1-60), covered by FM *Loss Prevention Data Sheet 1-28,* there is a requirement for deck welds spaced at 6 in. for building corners (10-ft-square area). [For Zone 2, Class 1-90, there is an additional deck-welding requirement, with 6-in. spacing also required around a 6-ft (1.8-m) perimeter (or two deck panel widths, whichever is greater).]

If a vapor retarder is specified, it should be located above a mechanically anchored bottom layer of insulation (minimum 1-in.-thick perlite board or $\frac{5}{8}$-in. moisture-resistant, Type X core gypsum board) followed by a one- or two-ply vapor retarder, a hot-mopped top layer of insulation, and a BUR, MB, or approved single-ply membrane.

Fastener patterns should follow the diagrams in FM *Loss Prevention Data Sheet 1-28* (unless the FM *Approval Guide* differs). (See Fig. 7.10 for sample diagrams for 4×4 ft insulation boards.) Fasteners should be driven in accordance with the following rules:

- Maximum 12 in. (305 mm), minimum 5 in. (152 mm) distance from board edge.
- Even distribution over the board area.
- Fastener engagement with deck flange (since fasteners driven into deck flutes are ineffective). Inspect deck soffit during installation to assure proper fastener driving.

In accordance with the higher uplift coefficients in corners and perimeters, FM requires additional anchorage for these locations even for Zone 1 (Class 1-60). Roofs with parapets less than 3 ft (0.9 m) high require extra anchorage in corners (50 percent for Zone 1, 100 percent for Zone 2). In Zone 1, no additional anchorage is required in

TABLE 7.1 Velocity Pressure (P_h), Ground Roughness C, Flat, Open Country, Open Coastal Belts >1500 ft (457 m) from Coastline

Height above ground, ft	Velocity pressure in pounds per square foot					
	Wind isotach (from map, Figs. 1, 2, and 3), mph					
	70	80	90	100	120	130
0–15	14	18	23	29	35	41
30	16	21	27	33	40	48
50	18[1]	24	30	37	44	53
75	20	26	33	40	49	58
100	21	28	35	43	52	62
200	25	32[2]	41	50	61	72
300	27	35	44	55[3]	66	79
400	29	37	47	58	71	84
500	30	39	50	62	74	89

Height above ground, m	Velocity pressure in kilopascals						
	Wind isotach (from map, Figs. 1, 2, and 3), mph						
	70	80	90	100	110	120	130
0–5	0.67	0.86	1.10	1.39	1.68	1.96	
9	0.77	1.00	1.29	1.58	1.92	2.30	
15	0.86[1]	1.15	1.44	1.77	2.11	2.54	
23	0.96	1.24	1.58	1.92	2.35	2.78	
30	1.00	1.34	1.68	2.06	2.49	2.97	
61	1.20	1.53[2]	1.96	2.39	2.92	3.45	
91	1.29	1.68	2.11	2.63[3]	3.16	3.78	
122	1.39	1.77	2.25	2.78	3.40	4.02	
152	1.44	1.87	2.39	2.97	3.54	4.26	

Note:
[1] Zone 1, $P_h \leq 30$ psf
[2] Zone 2, 31 psf $\leq P_h \leq 45$ psf
[3] Zone 3, $P_h \leq 46$ psf
Wind uplift on a flat roof (excluding perimeter and corners)
SOURCE: FM *Loss Prevention Data Sheet 1-28S,* September 1991.

perimeters; roofs in Zone 2 require an additional 50 percent (see Table 7.2).

The Atlanta building is assumed to have no parapet and thus requires 50 percent additional fasteners at building corners, or six instead of the four fasteners generally required per 4×4 ft-insulation board.

Corner and perimeter dimensions are as follows:

- Corner dimension = 8 ft square (Zone 1), 10 ft square (Zone 2)

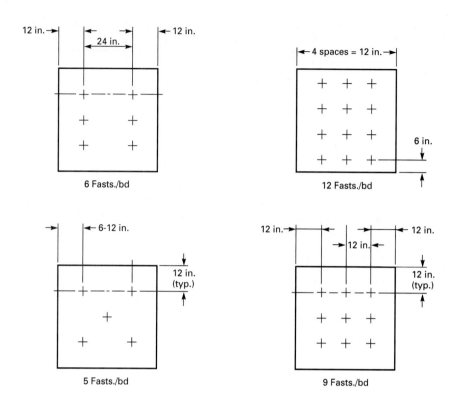

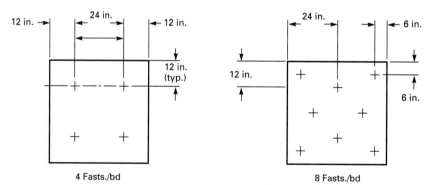

Figure 7.10 Diagrams above show fastener pattern for 4×4 ft insulation boards, per FM guidelines. There are similar diagrams for 2×4, 3×4, and 4×8 ft boards. (*Factory Mutual Research Corporation.*)

TABLE 7.2 Insulation Fastener Requirements for FM Zones 1 and 2 (FMRC)

Wind uplift zone	Parapet height	Insulation fastening needed		
		Field	Perimeter	Corner
Zone 1	<3 ft (0.3 m)	1-60	1-60	1-60 plus 50%
	≥3 ft (0.3 m)	1-60	1-60	1-60
Zone 2	<3 ft (0.3 m)	1-90	1-90 plus 50%	1-90 plus 100%
	≥3 ft (0.3 m)	1-90	1-90 plus 50%	1-90 plus 50%

Note: 1-60/1-90 refers to the FMRC-approved fastener density for the particular fastener/insulation/roof cover combination for Class 1-60/1-90 wind uplift resistance.

- Perimeter dimension = 4 ft for both zones

FM literature shows diagrams for insulation boards of various sizes (2×4 ft, 3×4 ft, 4 ft square, and 4×8 ft) for corner and perimeter requirements. (See Fig. 7.10, for example, using 4-ft-square boards.)

Now compare the FM method with the ASCE 7-95 method. From the ASCE 7-95 wind map, Atlanta's 50-year wind speed is 95 mph, from which uplift pressure is calculated as follows:

$$P = q_z (GC_p - GC_{pi})$$

and

$$q_z = 0.00256 \, K_z \, IV^2$$

where q_z = velocity pressure at height (psf)
K_z = velocity pressure exposure coefficient that accounts for terrain and height above ground (See Table 7.3)
I = importance factor (ranges from 0.87 to 1.15 depending upon probability, occupancy, and proximity to hurricane oceanline)
V = basic design wind speed at height Z
GC_{pi} = product of pressure coefficient and gust response factor

If our building had unusual openings facing the wind, we would use 0.25 as the GC_{pi} (internal pressure).

We can now solve for the uplift.

$$P = (21.6)(-2.8) - (21.6)(0.18) \quad = -64 \text{ psf for corners}$$

$$= (21.6)(-1.8) - 3.9 \quad = -43 \text{ psf for perimeter}$$

TABLE 7.3 Velocity Pressure Exposure Coefficient K_h and K_z

Height above ground level, z (ft)	A	B	C	D
0–15	0.32	0.57	0.85	1.03
20	0.36	0.62	0.90	1.08
25	0.39	0.66	0.94	1.12
30	0.42	0.70	0.98	1.16
40	0.47	0.76	1.04	1.22
50	0.52	0.81	1.09	1.27
60	0.55	0.85	1.13	1.31
70	0.59	0.89	1.17	1.34
80	0.62	0.93	1.21	1.38
90	0.65	0.96	1.24	1.40
100	0.68	0.99	1.26	1.43
120	0.73	1.04	1.31	1.48
140	0.78	1.09	1.36	1.52
160	0.82	1.13	1.39	1.55
180	0.86	1.17	1.43	1.58
200	0.90	1.20	1.46	1.61
250	0.98	1.28	1.53	1.68
300	1.05	1.35	1.59	1.73
350	1.12	1.41	1.64	1.78
400	1.18	1.47	1.69	1.82
450	1.24	1.52	1.73	1.86
500	1.29	1.56	1.77	1.89

Notes:
1. Linear interpolation for intermediate values of height z is acceptable.
2. For values of height z greater than 500 ft, K_z shall be calculated as defined on p. 135.
3. Exposure categories are defined on p. 135.

$$= (21.6)(-1.0) - 3.9 = -26 \text{ psf for roof interior area}$$

How does this compare with the FM method? FM gives pressure coefficients in Table 2 of *Loss Prevention Data Sheet 1-48* as:

$$\text{Corners} = -2.60$$

$$\text{Perimeter} = -1.75$$

$$\text{Body of roof} = -1.0$$

Computed via the FM method, uplift pressures are as follows:

$$\text{Corners} = -2.6 \times 28 = -73$$

$$\text{Perimeter} = -1.75 \times 28 = -49$$

$$\text{Field} = -1.0 \times 28 = -28$$

Though the FM method yields more conservative results than the ASCE method—notably in 13 percent higher uplift pressure at corners—the two methods are nonetheless quite close.

Perimeter Flashing Failures

Most roof blowoffs (about 85 percent, according to FM data) start with failure at the perimeter fascia–gravel stop (see Fig. 7.11). The metal fascia strip is bent outward and upward, exposing more of its area to the wind. Wind forces then pry off the fascia strip and/or the wood nailer-cant assembly. Failure of the cant-nailer assembly opens the roof assembly to peeling and suction forces, which can roll back the membrane sandwich. Positive pressure from wind entering steel-deck ribs supplements uplift forces at the roof surface. (See Fig. 7.11 for faulty fascia–gravel-stop details that led to wind-uplift failures investigated by FM engineers.)

Hurricane Hugo, which struck Charleston, South Carolina, in September 1989, confirmed the FM experience with edge-initiated roof blowoffs. Typical failures, reported by NRCA technical director Thomas Smith, included the following:

- Loss of inadequately anchored lightgage metal parapet copings
- Blowoff of inadequately strapped lightgage metal gutters
- Uplift of nailers inadequately anchored to masonry walls[4]

The wind bending of a *cleated* stainless steel fascia strip reported by Smith demonstrates the care required to secure edge flashing against intensified perimeter uplift pressure. This failure occurred despite the use of cleats designed to stabilize the fascia strip along its bottom edge (see Fig. 7.12). The 24-gage stainless steel fascia strip was stressed beyond its yield point because the continuous cleat fasteners were located near the *top* of the cleat. This faulty location increased the unbraced, cantilevered depth of the fascia strip, exponentially multiplying the bending stress exerted by the wall-deflected wind rising along the eave line.

Adequate wind-design perimeter flashing requires close attention to details by a designer cognizant of the need to eliminate weak links. Cleats must be detailed to brace fascia strips at their lower edges, to avoid prying cantilever bending moment applied by upward-deflected wind. Gravel-stop flanges must be adequately nailed into securely anchored wood nailers. Wood nailers, in turn, must be anchored to masonry with bolts adequately embedded in grout to resist wind-uplift pressure (see Fig. 7.13).

FM *Loss Prevention Data Sheet 1-49* shows recommended perime-

146   Chapter Seven

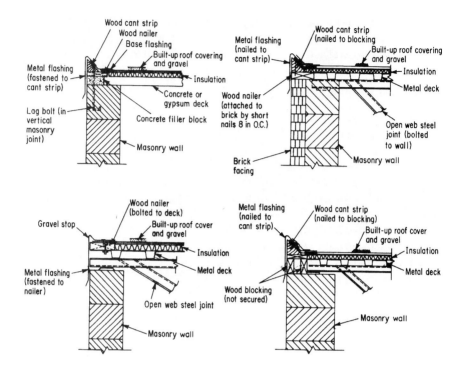

Figure 7.11 These edge-flashing details were found most vulnerable to wind-uplift damage by Factory Mutual engineers. Note that they all omit a continuous cleat, which provides bottom anchorage for the fascia strip. (*Factory Mutual Research Corporation.*)

ter flashing details for a whole spectrum of roof systems: BUR and adhered and ballasted single plies. Figure 7.14 shows one such detail, and Table 7.4 shows the minimum thickness for metal fascia flashings (steel, aluminum, and copper) for different flashing leg dimensions.

Also in FM *Loss Prevention Data Sheet 1-49* is a list of 13 recommendations concerning edge-flashing details designed for wind-uplift resistance. These recommendations prescribe minimum anchor-bolt sizes and spacing, minimum gage thickness for continuous hook strips (cleats), minimum nail (or screw) sizes and spacing for anchorage into nailers, provisions for concrete filler in hollow masonry walls, and other data essential to safe edge detailing.

Poor flashing details probably indicate ignorance more than economizing via cheap construction, though the latter has some effect. Developers normally attempt to chisel on construction costs because mortgage lenders seldom consider quality in making their loans. As a consequence, incremental costs, even small ones, can reduce developers' profits through a perverse kind of reverse leveraging. Petty econo-

Wind Uplift 147

Figure 7.12 The stainless steel fascia strip shown above bent upward under wind-lift pressure despite the use of a cleat designed to stabilize it. (*National Roofing Contractors Association.*)

mizing on perimeter flashing can be enormously costly for the ultimate owners, the ultimate losers. The cost of this folly is especially evident in a high-rise building, where the perimeter footage may average out to 1 or 2 ft per unit owner. To risk a blowoff for the sake of such trivial savings, a few dollars for each unit owner, is economic insanity. Yet it is not an uncommon occurrence.

Wind-Uplift Mechanics on Ballasted Roofs

Ballasted, loose-laid roof systems are a radically different wind-uplift problem from adhered or anchored roof systems. The normal 10- to 15-psf aggregate ballast weight on loose-laid systems falls far short of the 30- to 60-psf wind-uplift design pressures often required for adhered roof systems. Judged by field experience, however, ballasted, loose-laid systems resist wind-uplift pressure by an entirely different mechanism from that of adhered systems. In an adhered system, an area of lost adhesion no longer contributes to the roof's uplift resistance. A small area of lost adhesion can expand rapidly because of the low peeling resistance of the remaining adhered sections. An adhered system usually suffers a blowoff from local failure, like the tensile failure of the weakest link in a chain. However, in a ballasted system,

Figure 7.13 Exposed top course of hollow masonry wall units reveals the loss of a wood nailer peeled off the wall because of inadequate anchorage. (*National Roofing Contractors Association.*)

any wind-lifted area shifts its ballast to an adjacent area, whose uplift resistance is consequently increased.

Gravel blowoff and scour

The problem of loose gravel aggregate scour and blowoff has accompanied the rising popularity of loose-laid membranes for both conventional and protected membrane (i.e., "upside-down") roofs. Blowing aggregate threatens surrounding buildings' windows, which have been broken by these flying refugees from neighboring roofs. Added to the more obvious hazards to property and persons from wind-launched gravel missiles is the direct exposure of protected membrane roof (PMR) insulation or a conventional system's membrane to sunlight, traffic punctures, and other forces against which aggregate surfacing provides protection.

Some 20 years ago, two Canadian researchers, Prof. R. J. Kind of Ottawa's Carleton University and R. L. Wardlaw of the National Research Council, published a pioneering design procedure for preventing loose aggregate blowoff and scour.[5]

Basic design strategy against aggregate blowoff and scour comprises one or more of the following techniques:

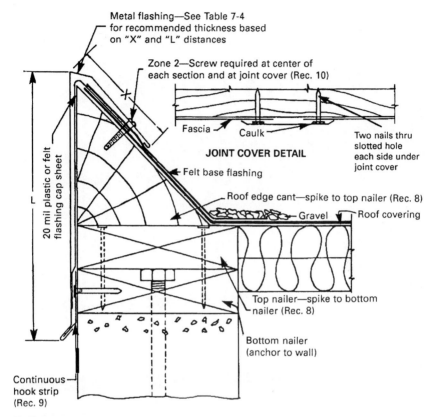

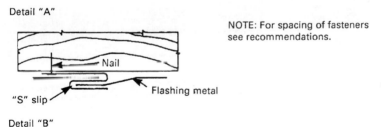

Figure 7.14 FM-recommended detail for metal edge flashing contains references to FM *Loss Prevention Data Sheet 1-49,* with requirements for nailer size and quality, nail size and spacing, etc. (*Factory Mutual Research Corporation.*)

TABLE 7.4 Maximum Recommended L and X Dimensions and Metal Thicknesses for Roof Edge Flashing and Hook Strip

Note: Not applicable for metal gravel guard and fascia—(see text in FM 1-49)

Type of metal		Maximum L dimension						Hook strip		Maximum X dimension					
		Velocity pressure (pounds per square ft)								Velocity pressure (pounds per square ft)					
		Zone 1				Zone 2				Zone 1				*Zone 2	
		10–20		21–30		31–45				10–20		21–30		31–45	
Galv. Iron or Soft Stainless Steel		(0.48–0.95 kPa)		(1.0–1.44 kPa)		(1.48–2.15 kPa)				(1.48–0.95 kPa)		(1.0–1.44 kPa)		(1.48–2.15 kPa)	
Ga.	(mm)	in.	(mm)	in.	(mm)	in.	(mm)	Ga.	(mm)	in.	(mm)	in.	(mm)	in.	(mm)
26	(0.45)	6	(150)	6	(150)	4	(100)	24	(0.61)	3	(75)	2½	(65)	use thicker metal	
24	(0.61)	8	(200)	8	(200)	6	(150)	22	(0.76)	4	(100)	3	(75)	3	(75)
22	(0.76)	10	(250)	10	(250)	8	(200)	20	(0.91)	4	(100)	4	(100)	3½	(88)
20	(0.91)	12	(300)	12	(300)	10	(250)	18	(1.06)	5	(130)	4½	(115)	4	(100)
Aluminum†															
in.	(mm)	in.	(mm)	in.	(mm)	in.	(mm)	in.	(mm)	in.	(mm)	in.	(mm)	in.	(mm)
0.040	(1.02)	6	(150)	6	(150)	4	(100)	0.050	(1.27)	4	(100)	3	(75)	2½	(65)
0.050	(1.27)	8	(200)	8	(200)	6	(150)	0.060	(1.62)	4½	(115)	4	(100)	3	(75)
0.060	(1.62)	10	(250)	10	(250)	8	(200)	0.070	(1.78)	5	(130)	4½	(115)	3½	(88)
0.070	(1.78)					10	(250)	0.080	(1.93)	5½	(140)	5	(130)	4	(100)
Copper															
oz	(mm)	in.	(mm)	in.	(mm)	in.	(mm)	oz	(mm)	in.	(mm)	in.	(mm)	in.	(mm)
16	(0.55)	8	(200)	6	(150)	4	(100)	20	(0.69)	3½	(88)	3	(75)	2½	(65)
20	(0.69)	10	(250)	8	(200)	6	(150)	24	(0.82)	4	(100)	3½	(88)	3	(75)
24	(0.82)			10	(250)	8	(200)	32	(1.10)	4½	(115)	4	(100)	3½	(88)
32	(1.10)					10	(250)	48	(1.64)	5½	(130)	5	(130)	4	(100)

*Additional fasteners are required near top of cant. One screw with washer at center of each section of fascia and one through joint cover. See Detail A in FM 1-49.

†Temper "O" aluminum, although easily formed, has a low bending strength. High tempers are advised when using aluminum.

- Increasing stone size
- Increasing parapet height
- Substituting concrete pavers for gravel

Increased stone size increases the aerodynamic force required to move or lift the loose stones. Wind exerts an aerodynamic drag force on an exposed stone. At a certain critical value, it topples the stone away from its nested position. Through a complex interaction of forces, upward deflection of a rolling stone combined with increased aerodynamic wind force, faster wind speed can blow some stones off the roof.

Critical wind speed increases with particle size because the weight of the particle varies with its diameter cubed, whereas the frontal area exposed to wind force increases with the particle's diameter squared. Aerodynamic force rises proportionately with wind velocity squared (i.e., a doubling of wind velocity quadruples aerodynamic force). It follows that the ratio of aerodynamic force to gravitational force is proportional to $V^2 d^2/d^3 = V^2/d$ (see Fig. 7.15). Critical wind velocity is thus expressed by Eq. (7.4):

$$V_c = k\sqrt{d} \qquad (7.4)$$

where V_c = critical wind velocity
d = stone aggregate size
k = proportionality constant

Parapet height is a major design variable. Parapets shield the stones from the wind; increased parapet height reduces aerodynamic forces on the stones. When the wind angle is roughly 45° to a rectangular building's axes (the worst angle for gravel blowoff), a parapet elevates the cores of the vortices, thus reducing rooftop suction.

Concrete pavers substituted for aggregate in the critical roof corners can prevent scour and blowoff in these most vulnerable areas. In plan, these corner areas are either square- or L-shaped, designed to match scour patterns.

Kind and Wardlaw derived their design data from wind-tunnel testing of scale-model buildings, with model dimensions, including gravel size, one-tenth full scale. Models and prototype buildings are related by the so-called Froude number ($V\sqrt{hg}$; V = wind velocity, h = building height, g = gravitational acceleration). Because g is constant, prototype and model wind speeds V_p and V_m are related as in Eq. (7.5):

$$V_p = \sqrt{\frac{h_p}{h_m}}\, V_m = \sqrt{10}\, V_m = 3.16\, V_m \qquad (7.5)$$

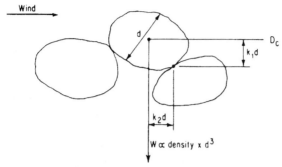

Figure 7.15 Critical wind velocities, at which loose aggregate unit rolls over or flies off roof, vary with the square root of the unit diameters. Sand grains would obviously blow off more easily than boulders, and the tendency to do so is a continuum between these two extremes (and even beyond them). The aerodynamic lifting force is a function of wind velocity squared, V^2, and d^2. Resisting force is a function of d^3. Thus

$$\frac{\text{Aerodynamic force}}{\text{gravitational force}} = \frac{V^2 d^2}{d^3} = \frac{V^2}{d}$$

$$V^2 \, \alpha \, d$$

$$V_c = k\sqrt{d}$$

where V_c = critical wind velocity. *Note:* k_1 and k_2 are proportionality constants depending on stone shape. (*R. L. Wardlaw, National Research Council of Canada.*)

Thus, a 25-mph wind speed in the wind tunnel corresponds to a full-scale wind speed of $3.16 \times 25 = 79$ mph.

In PMR roof assemblies (see Chapter 15), permeable fabric sandwiched between the light plastic foam insulation and the gravel ballast reduces wind-scouring damage and limits the gravel and insulation blowoff at any given wind velocity. With increasing wind velocity, the permeable fabric induces three staged failure modes:

- In F1, gravel scour at upwind corners leaves most of the displaced gravel in deep piles near the perimeter parapet (although some leaves the roof).

- In F2, at a higher velocity, the plastic foam insulation boards in the bare gravel-scoured portions of the roof oscillate vertically, but remain trapped under the fabric. Increased wind velocity pulls the fabric out from under the accumulated gravel piles, and the now unrestrained insulation boards blow away.

- In F3, at still higher wind velocity, the fabric peels back farther until a new stabilized condition is reached, with gravel accumulated in the fold between lifted and unlifted fabric.[6]

Wind-Uplift Design for Ballasted, Single-ply Roof Assemblies

Design of ballasted, single-ply roof systems should follow the criteria promulgated by the *Wind Design Guide for Ballasted Single-ply Roofing Systems,* ANSI/SPRI RP-4 (latest edition) and published by the Single-ply Roofing Institute. Used in conjunction with the product manufacturers' recommendations, this guide considers a whole spectrum of special conditions that can affect wind-uplift design of ballasted systems. The following summary of this guide outlines the general procedure. Designers should, however, consult the document itself, plus manufacturers' recommendations, in designing their particular system for wind-uplift resistance.

Starting with an obvious reminder, Part II of the wind-design guide proceeds as follows:

- Design a new building structurally for the additional dead load of a ballasted roof system. Check the structural design of an existing building to assure its capacity for the additional load of a ballasted system (usually from 10 to 22 psf).
- Limit roof slope to a maximum of 2 in./ft (17 percent).
- Select design wind speed (*a*) from ASCE 7-95 (which superseded American National Standard A58.1), latest edition, or (*b*) from a comparable standard by an insurance underwriter (e.g., Factory Mutual System), or (*c*) from the governing code authority. Consider intensification of wind speed by valleys.
- Take building height as the distance from the ground to the roof surface, considering special site topographic features—e.g., hills—in this calculation. If building height exceeds 150 ft (46 m) or design wind speed exceeds 120 mph (193 km/h), review special design factors with the owner, the membrane manufacturer, and the building official.
- Divide roof areas into (*a*) corners (the most highly stressed areas), (*b*) perimeter, and (*c*) field (the most lightly stressed area). *Corner* dimensions are 40 percent of building height, with a minimum of 8.5 ft (2.6 m). The *perimeter* is a rectangular strip of a minimum width of 8.5 ft, and the *field* is the interior area bounded by corners and perimeter.
- Favor parapets as perimeter terminations, to reduce roof wind loading. If the edge termination is a gravel stop, set its height at an absolute minimum of 2 in. (51 mm), higher if required to retain ballast. If the lowest parapet height outside the corner areas is *less* than 30 percent of the parapet height within the corner, use this

lower parapet height for the design. If the lowest parapet height outside the corner areas *exceeds* 30 percent of the corner parapet height, use the minimum corner parapet height (see Fig. 7.16).

- At roof perimeters, penthouses, and other terminations of the membrane for curbs, parapets, etc., anchor the membrane to resist a diagonal uniform tensile stress of 75 lb/ft (1.1 kN/m) in accordance with the test shown and described in Fig. 7.17. The substrate into which the membrane termination is anchored must be designed to resist a minimum tensile stress of 90 lb/ft (1.3 kN/m).

- Since topography at the building site affects wind exposure, identify this exposure as one of four conditions. Unprotected exposures include exposure A, large urban areas with concentrations of buildings over 60 ft (18 m) high. (Buildings in such locations are either too tall for shielding by adjacent buildings or low enough to be affected by wind chanelling between them.) Exposure C is open, level terrain—e.g., plains or shoreline—with little or no wind shelter. Exposure B is protected—suburban or small urban areas with buildings (and trees) under 60 ft (18 m) high. (ASCE/ANSI exposure D is a special case requiring review by the owner, the membrane manufacturer, and building officials.)

- Buildings with large wall openings that could be left open in a storm—e.g., a warehouse with large overhead doors—may require increased wind resistance. The roof section requiring additional wind-uplift resistance is a rectangle 1½ times the wall opening width by 2 times the wall opening length (in the perpendicular direction; see Fig. 7.18). Design this vulnerable area for the same wind loading as corner areas. Upgrade a System 1 design to System 2 for a building with large wall openings.

- When there is positive interior air pressure (greater than 0.5 in. water), compensate for the consequently reduced roof uplift resistance by increasing the wind loading. If this calculated positive pressure is between 0.5 and 1.0 in. water, upgrade the roof uplift resistance to the next level in the design table (see Table 7.5). Under this criterion, a System 1 design goes to System 2, System 2 to System 3, and System 3 to an upgraded special category with 90 psf (4.3 kPa) uplift resistance (per FM4470, UL580, or other recognized test procedure). For positive interior pressure exceeding 1.0 in. water (249 Pa), the design requires special considerations resolved by the owner, membrane manufacturer, and local building officials.

- Anchor the 4-ft roof strips around rooftop projections greater than 4 ft in width and extending more than 2 ft above parapet height, such as corner sections.

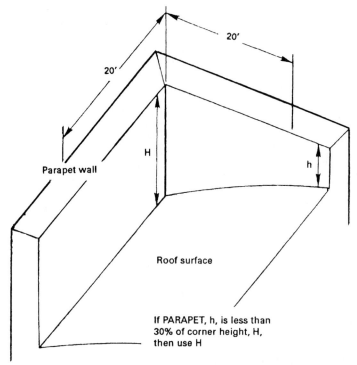

Figure 7.16 If the lowest parapet height h (outside corner areas) is *less* than 30 percent of the parapet height H (at corner), the designer uses this height, h, for the design. If h is *more* than 30 percent of H, the designer uses the minimum corner height H. (*Single-ply Roofing Institute.*)

- Eaves, overhangs, and canopies may increase wind-uplift loading, because of their influence on airflow up the wall. For roof systems with impervious decks—e.g., poured concrete and gypsum—consider the edge strip as the perimeter (see Fig. 7.19). For pervious decks—e.g., lightgage steel, precast concrete, etc.—treat the entire eave, overhang, or canopy like a corner (see Fig. 7.20). For a System 1 design with an eave, overhang, or canopy, upgrade the affected area to a System 2 corner design. Pervious decks also require extension of the corner and perimeter strips as shown in Fig. 7.20.

Basic ballasted wind-design systems

For conventional (membrane-atop-insulation) ballasted roof systems, there are three basic wind designs, System 1, 2, and 3, for increasing wind-loading severity. These three categories cover all except the extraordinarily severe wind loadings (see Table 7.6).

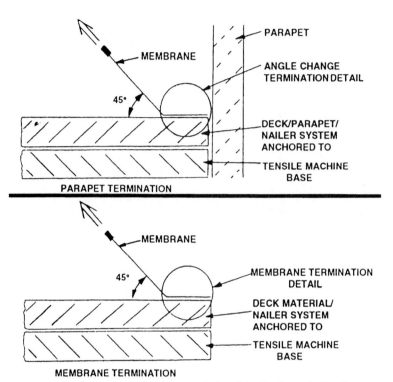

Figure 7.17 In roof-edge termination test, designed to simulate a ballooning membrane, a 12-in.-wide mockup of the edge anchorage detail, with parapet (top) and without parapet (bottom), is tested with a minimum 75-lb force. Failure can occur at the fastener or via membrane tearing. (*Single-ply Roofing Institute.*)

System 1, the least severe, requires a ballast of nominal $1\frac{1}{2}$-in. smooth, river-bottom stone spread at a minimum uniform weight of 10 psf over the entire roof. Concrete pavers can substitute for ballast—a minimum 15 psf for standard pavers, 10 psf for tongue-and-groove, interlocking lightweight pavers for System 1 design.

System 2 differentiates between corner areas and the field. It requires $2\frac{1}{2}$-in. smooth, river-bottom stone ballast at a minimum rate of 13 psf in corners, 10 psf for perimeters and the field. Again, as for System 1, pavers can substitute for stone ballast—22 psf minimum for standard pavers or "approved" interlocking, tongue-and-groove lightweight pavers with documented wind-performance data.

System 3 requirements, still more complex, limit ballast to the field of the roof, where nominal $2\frac{1}{2}$-in. ballast at a 13-psf rate is required. Corners and perimeter strips require an adhered or mechanically anchored membrane capable of resisting 90-psf uplift (per FM4470, UL580, etc.). System 3 also requires an extension of the general perimeter anchorage requirements for ballasted membranes. The

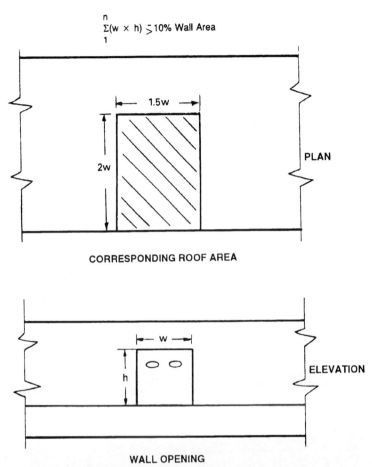

Figure 7.18 For buildings with large openings that could be left open in a storm, the roof section requiring additional wind-uplift resistance is a rectangle with dimensions of $1.5w$ and $2w$, as shown in plan (top) view (*Single-ply Roofing Institute.*)

junctures of loose-laid membrane with the corner or perimeter adhered or mechanically anchored sections require the same 75-psf resistance to diagonal uplift pressure as the perimeters for all systems. And these anchorage details must also provide a gravel stop of sufficient height to prevent ballast spillover.

As a fourth, special category for any building where the wind-loading exposure exceeds the wind speeds tabulated in Table 7.5, an extraordinary design produced jointly by the building owner, architect, and manufacturer will probably exceed the System 3 requirements.

Concrete pavers may be substituted for stone ballast. For nominal 1½-in. stone, you can substitute pavers weighing 15 psf or interlock-

TABLE 7.5 Maximum Wind Speeds for Varying Parapet Heights, Exposures, and Systems for Ballasted Single-ply Roof Assemblies (SPRI)*

A. From 2-in.-high gravel stop to 5.9-in.-high parapet

Maximum wind speed (mph)

Bldg. ht., ft	System 1		System 2		System 3	
	Exposure A+C	Exposure B	Exposure A+C	Exposure B	Exposure A+C	Exposure B
0–15	80	90	100	100	110	120
>15–30	80	90	100	100	110	120
>30–45	70	80	90	100	110	120
>45–60	70	80	90	100	110	120
>60–75	70	70	90	90	100	100
>75–90	No	No	90	90	100	100
>90–105	No	No	80	80	90	90
>105–120	No	No	80	80	90	90
>120–135	No	No	80	80	90	90
>135–150	No	No	80	80	90	90

TABLE 7.5 Maximum Wind Speeds for Varying Parapet Heights, Exposures, and Systems for Ballasted Single-ply Roof Assemblies (SPRI)*
(Continued)

D. For parapet heights from 18.0 to 23.9 in.

Maximum wind speed (mph)

Bldg. ht., ft	System 1		System 2		System 3	
	Exposure A+C	Exposure B	Exposure A+C	Exposure B	Exposure A+C	Exposure B
0–15	90	90	100	100	120	120
15–30	90	90	100	100	120	120
30–45	80	90	100	100	120	120
45–60	70	80	100	100	120	120
60–75	70	70	90	90	120	120
75–90	70	70	90	90	120	120
90–105	No	No	90	90	120	120
105–120	No	No	90	90	120	120
120–135	No	No	80	80	120	120
135–150	No	No	80	80	120	120

TABLE 7.5 Maximum Wind Speeds for Varying Parapet Heights, Exposures, and Systems for Ballasted Single-ply Roof Assemblies (SPRI)*
(*Continued*)

E. For parapet heights from 24.0 to 35.9 in.

Maximum wind speed (mph)

Bldg. ht., ft	System 1		System 2		System 3	
	Exposure A+C	Exposure B	Exposure A+C	Exposure B	Exposure A+C	Exposure B
0–15	90	90	100	100	120	120
>15–30	90	90	100	100	120	120
>30–45	80	90	100	100	120	120
>45–60	70	90	100	100	120	120
>60–75	70	70	100	100	120	120
>75–90	70	70	90	90	120	120
>90–105	70	No	90	90	120	120
>105–120	No	No	90	90	120	120
>120–135	No	No	90	90	120	120
>135–150	No	No	80	80	120	120

TABLE 7.5 Maximum Wind Speeds for Varying Parapet Heights, Exposures, and Systems for Ballasted Single-ply Roof Assemblies (SPRI)*
(*Continued*)

F. For parapet heights from 36.0 to 71.9 in.

Maximum wind speed (mph)

Bldg. ht., ft	System 1		System 2		System 3	
	Exposure A+C	Exposure B	Exposure A+C	Exposure B	Exposure A+C	Exposure B
0–15	90	90	100	100	120	120
>15–30	90	90	100	100	120	120
>30–45	80	90	100	100	120	120
>45–60	80	90	100	100	120	120
>60–75	70	70	100	100	120	120
>75–90	70	70	100	100	120	120
>90–105	70	70	90	90	120	120
>105–120	70	70	90	90	120	120
>120–135	70	70	90	90	110	110
>135–150	70	No	90	90	120	120

*Some types are omitted from above tabulation. See ANSI/RMA/SPRI RP-4, 1988, for missing categories (B, C, G).

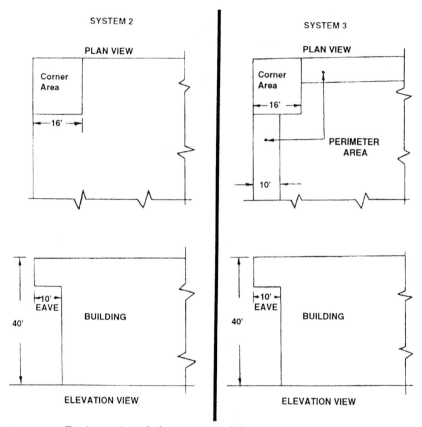

Figure 7.19 For impervious decks—e.g., monolithic structural concrete—overhangs require the corner and perimeter dimensions shown above for Systems 2 and 3. (*Single-ply Roofing Institute.*)

ing (tongue-and-groove) lightweight pavers, minimum 10 psf. For $2\frac{1}{2}$-in. stone, you can substitute standard pavers weighing 22 psf or "approved" interlocking lightweight pavers with demonstrated equivalent wind-uplift resistance.

PMR wind design

A protected membrane roof assembly (PMR) adds a further complication to the uplift problems of ballasted single-ply roofs. The insulation placed atop the membrane is another component requiring uplift resistance.

In general, the same ballasting and special anchorage requirements apply for PMRs as for conventional loose-laid ballasted roof assemblies. Additional requirements are as follows:

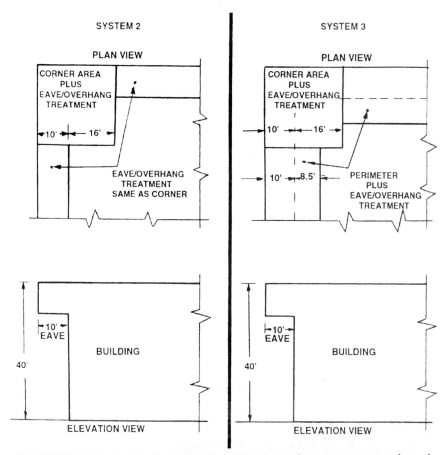

Figure 7.20 For pervious (i.e., jointed) decks—lightgage steel, precast concrete, plywood—corner and perimeter areas are extended.

A System 3 design requires a minimum 24-in. parapet height, plus 22-psf minimum weight standard pavers (or equivalent in uplift resistance) over the insulation in corner and perimeter strips.

Insulation with a top surface of integrally bonded cementitious facing shall have tongue-and-groove, interlocking joints and weigh a minimum of 4 psf. Requirements for the three basic designs follow:

- For System 1 PMR design, a 2-ft-wide perimeter strip shall be ballasted with minimum 22-psf standard pavers atop the cementitious-faced insulation units.
- For System 2, as in System 1, the entire membrane shall be covered with 4-psf cementitious-faced insulation panels. Corner areas, plus a 2-ft-wide perimeter strip, require an additional ballasting with 22-psf standard pavers.

TABLE 7.6 Ballast Weights for Single-ply Roof Assemblies

Conventional stone-ballasted assembly*		System 1	System 2	System 3
Ballast size (in.)†		1½	2½	2½
Ballast weight (psf)	Corners	10	13	No ballast
	Perimeter	10	10	(adhered or mechanically anchored)
	Field	10	10	13

*Crushed stone of similar gradation may substitute for smooth, river-bottom stone if placed on a shielding layer acceptable to membrane manufacturer.
†Ballast gradation size #4, or alternatively, #3, #24, #2, or #1 per ASTM D-448, "Standard Sizes of Coarse Aggregate."

Conventional paver-ballasted		System 1	System 2	System 3
Standard pavers weight (psf)	Corners	18	22	No ballast
	Perimeter	18	22	(adhered or mechanically anchored)
	Field	18	22	22
Interlocking lightweight pavers (psf)	Corners	10	"Approved"*	No ballast
	Perimeter	10	"	(adhered or mechanically anchored)
	Field	10	"	22

**"Approved" interlocking lightweight pavers with documented (or demonstrated) equivalent wind-uplift resistance.

Additional requirements for PMRs		System 1	System 2	System 3
Parapet ht (in.)		—	—	24
Standard pavers weight (psf)	Corners	—	—	22
	Perimeter	—	—	22
	Field	—	—	—
Cementitious-faced, interlocking insulation panels	Corners	—	22†	Special design required
	Perimeter*	22	22	
	Field	—	—	

*Perimeter strip for additional ballast is 2 ft wide; tabulated weight is for standard pavers.
†Tabulated weight is for standard pavers.

- A System 3 design with these cementitious-faced insulation units requires a special design to be reviewed by the owner and membrane manufacturer and approved by the building official.

Retroactive redesign

If wind scour occurs during the service life of a ballasted roof, the owner should reposition stones for an area up to approximately 12 ft^2. If the scoured area exceeds 12 ft^2, the entire system should be upgraded to the next level of ballast coverage—e.g., from System 1 to System 2.

Design example for ballasted, loose-laid system

Consider the same building used for the previous design example, a 40-ft-high office building with no parapets in Atlanta, for this example. (At publication time, SPRI was still following ANSI/ASCE 7-88.)

1. Review for maximum slope (= 2 in./ft, 17 percent).
2. Get same wind speed from ASCE 7-88 wind map = 72 mph.
3. From Table 7.5, for a building height of 40 ft, exposure C, maximum wind speed is 70 mph for a 2-in.-high gravel stop or a maximum 5.9-in.-high parapet.

Excess wind speed requires a choice between two alternatives:

1. Add parapets with a minimum 18-in. height and use System 1 ballast (1½-in. diameter, 10 psf) throughout roof.
2. Use System 2, with corners receiving ASTM No. 2 stone (2½-in. diameter) at 13 psf or 22-psf pavers (see Table 7.6). (At 40 percent of building height, corner dimensions = 0.4×40 = 16 ft square.)

Wind-Uplift Testing

One of the most difficult problems created by the multiplication of roof systems over the past two decades concerns wind-uplift testing. Simulation of actual wind-uplift pressure experienced by in-service roof systems is more easily achieved in the laboratory. Simulation can, however, be extremely difficult in the field. The trend toward mechanical fastening adds a tremendous complication to the task of field-testing conventional compact systems. Any randomly chosen location is virtually certain to produce a sample subject to eccentric loading, which overstresses some fasteners and understresses others.

Adhered test samples, on the contrary, pose no such problem, especially if the specified adhesive is uniform.

Laboratory tests

For conventional, compact roof systems, both FM and UL conduct laboratory wind-uplift tests and publish lists of approved roof-deck assemblies. FM publishes its approved list annually, in the *Approval Guide,* under "Building Materials and Construction." UL publishes a list of roof-deck assemblies rated Class 30, 60, or 90, depending on their successful resistance to 45-, 75-, or 105-psf total negative pressure in the UL uplift test. (See UL *Roofing Materials and Systems Directory,* "Roof Deck Constructions.")

The UL laboratory test (UL580) features a sophisticated apparatus designed to simulate actual wind loading on a roof system. This UL apparatus comprises three basic units: a lower chamber, where positive pressure is applied to the deck soffit; a 10×10 ft test specimen (deck-through-membrane); and an upper chamber, where negative pressure is applied to the roof membrane. Both upper and lower chambers contain glazed ports for viewing tests. A complete test cycle lasts 1 h and 20 min. The assembly passes if it withstands the full loading for its particular rating (i.e., Class 30, 60, or 90).[7]

FMRC 4470 features two pieces of test equipment with different sizes depending on fastener spacing. For systems with fastener spacing of 4 ft or less (or grid-spaced fasteners of 2×4 ft or less), tests are conducted in a steel pressure vessel with a hollow prism 9×5 ft in plan and 2 in. deep. For larger fastener spacing, up to 12 ft, there is a much larger, 12.5×24 ft frame.

Test procedure is the same for both frames. The test sample is clamped to the top angle of the vessel, with a rubber gasket controlling air leakage. Compressed air, entering through an opening in one side of the pressure vessel's frame, pumps up pressures in 15-psf increments from 30 to 180 psf, measured via a manometer connected to a ¼-in. opening on the opposite side. Each 15-psf pressure increment is held for 1 min. To qualify for a 1-60 rating, the roof-deck assembly must withstand a minimum 60 psf for 1 min without showing evidence of any bond failure between components or any delamination of insulation.

Ultimate load testing may be inadequate in many instances, as accumulating field experience points to repetitive loading at relatively low-velocity wind as the predominant failure mechanism. FMRC engineers are devising a test for fatigue failure of mechanically fastened single-ply systems (and perimeter flashing). Anchored by fasteners driven through batten bars (or individual stress-plate assemblies), these mechanically fastened single-ply systems can fail by fatigue stress associated with repetitive, wind-induced oscillations that produce ballooning of the membrane between fasteners at wind velocities well below maximum. The failure modes of these fatigue-stressed single-ply systems are normally either fastener backout or membrane tearing.

The fatigue-testing equipment applies pressure via three pneumatic pistons designed for easy pressure adjustment and quick cycling (see Fig. 7.21). The test program comprises a total of 1.5 million loadings in the 20- to 50- mph range, representing a 10-year service life (normal extent of a roof guaranty or warranty).

Tested systems include batten bars with continuous 1-in.-wide metal or plastic straps with prepunched holes. Stress plates for individual anchors are 2-in.-diameter metal and plastic disks (see Fig. 7.22). (On some stress plates, barbs penetrate into the membrane to resist fatigue-induced twisting of the fasteners and membrane abrasions from the frictional force of rotating stress plates.)

Among the first tested membranes were glass-reinforced PVC and unreinforced EPDM, respectively representing the high and low of membrane stress-strain modulus. As a result of their vastly different properties, the flexible, low-modulus EPDM membrane balloons with an amplitude more than 10 times that of the stiff, high-modulus, reinforced PVC (see Fig. 7.23). At the batten-bar spacing of 36 in. for the glass-reinforced PVC membrane, amplitude h was less than 1 in., whereas the unreinforced EPDM ballooned with an amplitude of nearly 10 in. at a 42-in. batten-bar spacing. Consistent with the theory that lower-modulus materials generally have higher strain capaci-

Figure 7.21 Fatigue-testing equipment features three pneumatic pistons designed for quick cycling for 1.5 million loadings in the 20- to 50-mph, equivalent-pressure range. (*Factory Mutual Research Corporation.*)

Figure 7.22 The reinforced PVC membrane tear shown above (arrow) appeared after 21,000-plus stress cycles. The screw had backed out one-half turn. Easily rotated, the screw retained no residual torque resistance. (*Factory Mutual Research Corporation.*)

ty, the unreinforced EPDM sheets proved superior in this respect to the reinforced PVC sheet in the FMRC test.

For PVC or CSPE single-ply membranes, eccentric loading of fasteners results from the normal method of field lap sealing (via heat or solvent welding). A typical field-welded lap seam along one side of the sheet adjacent to an individually anchored row of fasteners or a batten bar stresses the fasteners unsymmetrically, as shown in Fig. 7.24. This eccentrically loaded joint is feasible for PVC sheets because properly welded lap seams have peel and shear strength equal or superior to the membrane material's tensile strength. In EPDM membranes, however, lap seams made with contact adhesives or self-adhering tapes lack the peel strength to resist such eccentric loading. They must be adhered on both sides of the fastener to produce more or less concentric fastener loading (see Fig. 7.25).

When developed into its final form, the FMRC fatigue test will provide an essential ancillary laboratory test for evaluating the wind-uplift resistance of mechanically fastened single-ply membranes, as important as, if not more important than, conventional tests for ultimate wind-uplift resistance for single-ply membranes with welded joints.

Field wind-uplift tests

Field wind-uplift tests are generally limited to adhered and, to a much more limited degree, mechanically anchored roof systems, with

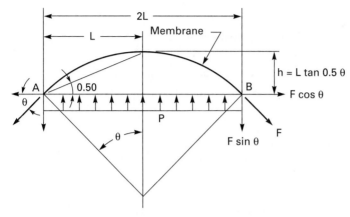

Figure 7.23 Cross section through the ballooning batten-bar-anchored single-ply membrane takes a circular membrane arc restrained by batten bars (points A and B, extended perpendicular to page), deflected upward by uniform uplift pressure. Membrane strain can be calculated via the following formula:

$$\epsilon = 0.602 \left(\frac{h}{L}\right)^{1.975}$$

Membrane stress $\sigma = C\epsilon^D$

where D and C are constants derived via tensile testing of the membrane. Under corner location test conditions (the most severe) for a glass-reinforced PVC membrane and an unreinforced EPDM membrane, the FMRC fatigue test yields an upward deflection h of ⅝ in. for the PVC and 9-plus in. for the EPDM. (The angle for the PVC is 2°; that for the EPDM, 26°). Fastener spacing was slightly higher (42 in.) for the EPDM than for the PVC (36 in.). (*Factory Mutual Research Corporation.*)

vacuum equipment or mechanically applied pressure acting vertically on the roof system to simulate the uplift pressure of wind blowing horizontally across the roof. Field testing of ballasted systems poses a much more difficult problem—i.e., direct simulation of wind blowing across the roof. This requires much heavier equipment: huge fans or jet engines capable of exerting the high wind speed required to displace the aggregate ballast. (Tests sponsored by Roofblok have subjected ballast aggregate to wind speeds of 123 mph.)

For new steel-deck roof assemblies, or those of doubtful wind-uplift resistance—e.g., unpeeled areas in a roof system that has already experienced a blowoff in another area—FM recommends one of two field uplift tests:

- The Negative Pressure Test
- The Pull Test

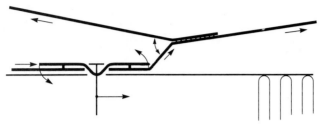

Figure 7.24 Typical lap-seam detail for PVC single-ply membrane results in unbalanced fastener stress, which can (a) loosen the fastener through prying action, or (b) abrade the sheet via lateral friction force between stress plate and membrane. Repetitive stress can produce fatigue failure. (*Factory Mutual Research Corporation.*)

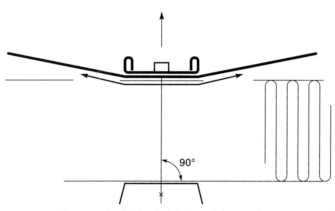

Figure 7.25 Symmetrically loaded joint, with membrane material patch over batten bar (or stress plate for individual fasteners) is mandatory for EPDM membranes, optional for PVC. (*Factory Mutual Research Corporation.*)

As its chief advantage, the Negative Pressure Test method is nondestructive (unless, of course, the roof assembly fails). The Negative Pressure Test apparatus (shown in Fig. 7.26) comprises three basic components:

- A 5-ft-square, transparent acrylic plastic dome, fabricated in four segments and field-bolted through flanges sealed with rubber gaskets and sealed at the roof surface with a flexible PVC foam strip
- An electrically powered vacuum pump, mounted on the vacuum chamber dome
- A manometer indicating vacuum pressure created within the chamber

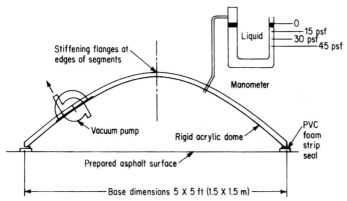

Figure 7.26 Portable vacuum equipment field-tests roof systems. (*Factory Mutual Research Corporation.*)

As a safety precaution for workers and observers of the test, locate test areas at least 10 ft from the perimeter (unless perimeter and corner areas are suspect). Follow FM-prescribed procedures for surface preparation of BUR adhered modified-bitumen and single-ply membranes to provide a smooth, level surface for sealing the vacuum chamber's gasketed bearing flange (a PVC foam strip seal). On gravel, slag, or granule-surfaced BUR or modified-bitumen membranes, sweep loose material from the test area and a 12-in. wide strip. Pouring bitumen over the perimeter to a maximum ½-in. thickness can provide a smooth, sealable surface that will prevent air leakage into the vacuum chamber and spoil the test. Smooth-surfaced BUR or modified-bitumen membranes may not require this peripheral bitumen. For the small proportion of single-ply membranes with surfacing sand or granules, these materials should be removed, along with any loose material within the vacuum test chamber.

Calculate the FM-required wind-uplift resistance in accordance with tabulated data in FM *Loss Prevention Data Sheet 1-28* and run the test as prescribed in FM *Loss Prevention Data Sheet 1-52*, which requires incremental pressure applications, holding the pressure for 1-min intervals, up to U_1 (required anchorage resistance). The field-tested sample fails if (*a*) the membrane suddenly balloons or (*b*) gage-measured deflection exceeds ¼ in. (6 mm).

Vacuum testing of mechanically fastened systems is best suited to single-ply membranes of limited fastener spacing. Testing is practicable for mechanically fastened single plies because their batten bars or individual fasteners (indicated by membrane patches above fastener disks) are visible. The severe spacing limitations of the vacuum test, however, disqualify the vast majority of mechanically fastened single-ply systems from the test. Uplift resistance can be measured only for a full membrane span on both sides of a fastener row. This limits fastener spacing to a maximum of 2.5 ft (0.76 m) for a valid test.

The older *Pull Test,* which relies on mechanically applied uplift force instead of a vacuum uplift, is limited to fully adhered BUR or modified-bitumen systems.

The destructive Pull Test apparatus comprises a tripod supporting a 200-lb-capacity spring scale connected with a hand-chain hoist, or block and tackle, to a plywood stiffener bonded to the membrane.

Portable pullout testers are another class of field-test equipment useful for field checks of mechanical fasteners. This equipment, furnished by the fastener manufacturers for testing their proprietary products, comes in compact units weighing as little as 7 lb and costing less than $1000. Some work on a mechanical torquing principle, calibrated to correlate the torque with pullout resistance up to a 600-lb maximum. Availability of these relatively inexpensive, lightweight testers gives roof designers added assurance of conformance with rigorous wind-uplift requirements.

Repair Procedures after Wind-Uplift Failure

Wind uplift damage occurs in one or more of the following modes:

- Loosened (or ruptured) edge flashings and gravel stops
- Membrane and insulation peeled from deck
- Membrane peeled from insulation
- Membrane and insulation broken loose from deck but finally dropped back in place

The last condition obviously constitutes a grave threat of future peeling. Unless eyewitnesses have seen the ballooning (sometimes observed as high as 3 ft above the deck), it may totally escape notice. If you have any suspicion that ballooning (or "floating") has occurred, you should cut samples to examine the deck-insulation and insulation-membrane adhesion. Also examine for backed-out fasteners or damaged stress plates.

Replace damaged flashing and gravel stops with new metal. For flashing that remains in place, additional nailing (if practicable) may suffice. If such repair is impracticable, consider replacing all flashing to conform with FM requirements in *Loss Prevention Data Sheet 1-49.*

Replace peeled insulation boards with new insulation and anchor it to the deck with mechanical fasteners unless adhesives are practicable. Be careful to examine insulation and reject damaged boards with crushed corners or edges or fracture lines resulting from the wind rolloff. Exercise the same care required for approval of new materials. Limit hot asphalt application to ambient temperatures above 50°F.

A Class 1 built-up or modified-bitumen roof with asphalt remaining on its deck is a special problem. If the old asphalt is not removed, its weight plus that of the new asphalt adhesive may exceed allowable weight for Class 1 rating. If it is impracticable to remove the old asphalt, anchor the insulation with FM-approved mechanical fasteners. (Note, however, that for a sprinklered building, retention of Class 1 rating may be unimportant.) Before readhering the insulation, remove any delaminated insulation layers stuck to the deck surface.

When the membrane is peeled from insulation that remains in place, test the insulation for solid anchorage before replacing the peeled membrane with a new membrane. Apply a good-quality membrane solidly adhered to the insulation, observing all other rules of good membrane application.

For ballooned insulation broken loose from the deck, first determine the extent of the affected area. Cut 12-in.-square test samples through the insulation, if necessary. Anchor the membrane and insulation with FM-approved mechanical fasteners. If the membrane has three or more plies, you can omit disk washers with the fasteners.

Apply a minimum of two shingled plies (solid mopping of felt with a solid mopping of felt surfacing) above the existing membrane.

Mechanical Fasteners

The proliferation of "high-tech" fasteners that began in the early 1980s (see Fig. 7.27) was a prescription for several serious ailments afflicting low-slope roof systems at that time. Defective hot-asphalt adhesion of insulation to steel decks not only produced the previously reported epidemic of roof blowoffs but also promoted BUR membrane splitting, a result of stress concentrations at opened joints between insulation boards sliding on their supporting steel decks. By the late 1970s, several major manufacturers had withdrawn approval of adhered insulation boards for guaranteed roof systems. FM followed with mandatory mechanical anchorage of insulation on steel decks in 1984. (Nailing of base sheets to wood decks is a traditional practice, with pullout resistance shown in Fig. 7.28.)

As usual, the solution of some problems creates a new set of problems, and the advent of high-tech fasteners is no exception to this general rule. Coming simultaneously with the trend toward re-covering of existing roof systems, combined with the trend toward thicker insulation, the new high-tech fasteners often required shanks several multiples longer than the earlier generation of mechanical fasteners used in new construction with insulation thicknesses of 1 in. or so. When these fasteners give signs of failing, even temporary remedial measures can be expensive (see Fig. 7.29).

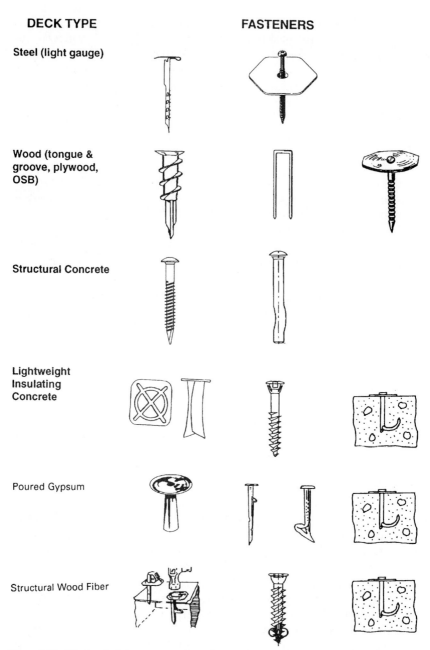

Figure 7.27 Mechanical fasteners and roofing nails come in varied shapes, sizes, and anchorage devices, designed for different substrates. (*Roofing Industry Educational Institute.*)

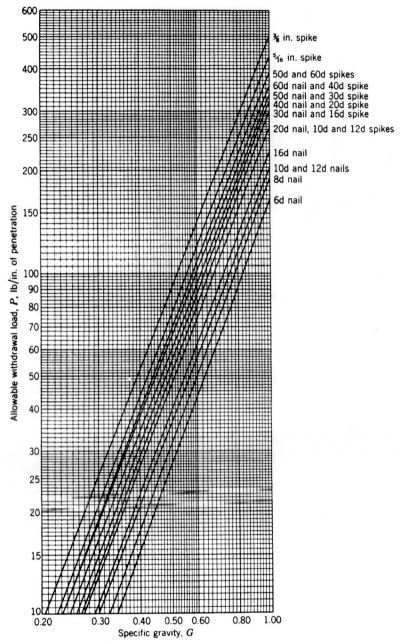

Figure 7.28 Allowable withdrawal loads for nails and spikes (lb/in. penetration) for one nail or spike installed in side grain under normal loading duration. Resistance per in. penetration = $1.380\, G^{2.5}D$, where (G = specific gravity and D = nail diameter. (*American Institute of Timber Construction.*)

Figure 7.29 At great expense and inconvenience, temporary bagged ballast prevents blowoff of this inadequately anchored, mechanically fastened EPDM membrane roof system on a building reroofed several years earlier with tapered, loose-laid EPS insulation boards. The steel deck lacks an air seal, and the five-story building is lightly pressurized. (*Roofing Industry Educational Institute.*)

Fasteners driven through existing roof systems with wet insulation require increased corrosion resistance. Moreover, the variety of deck materials supporting re-covered roof systems posed another challenge to the new high-tech fasteners: design for dependable anchorage in such inherently uncooperative materials as structural concrete, aged gypsum decks, and preformed mineralized wood fiber.

This extension of fastener anchorage of insulation into decks formerly adhered with hot-mopped asphalt also extends roof designers' problems. Like many other roof properties—e.g., membrane elasticity and tensile strength, insulation's thermal resistance—fasteners' pullout resistance declines with age. It varies with the deck material into which the fasteners are driven, according to consultant Colin Murphy.[8] Because of dynamic loading conditions observed in these different deck types in his survey of mechanically fastened thermoplastic roof systems, Murphy recommends the following reduction in initial pullout resistance:

Metal: 15 percent

Structural concrete: 25 percent

Lightweight insulating concrete: 30 percent

Corrosion resistance, considered by some fastener manufacturers as their major problem, has promoted a switch from cheaper carbon steel fasteners to stainless steel and corrosion-resistant platings and coatings. Horror stories abound, including accounts of wind blowoffs of re-covered roof systems with wet insulation in which the screw-type fasteners had corroded through the entire shank cross section.

Fluorocarbon and other polymeric coatings have become the predominant corrosion-prevention technique. (Stainless steel has a fatal handicap: It cannot be hardened into a satisfactory drill point.) FM's Kesternich test for corrosion resistance presents the standard industry hurdle. To evaluate abrasion of the fasteners' coating, fasteners are installed in a test specimen of the subject roof deck and placed in a cabinet at 100% RH and sulfur-dioxide atmosphere. The appearance of rust is evaluated against acid concentration and number of cycles.

Like other aspects of roof design, corrosion resistance can necessitate a balancing of contradictory performance requirements. The head of a popular fastener for lightweight insulating concrete is polymer-coated as required for Factory Mutual approval, to resist acidic water attack. The split shank, however, is not coated. Uncoated galvanized steel develops greater pullout resistance than polymer-coated, as the polymer is a partial lubricant. Moreover, lightweight insulating concrete is *alkaline,* not acidic.

Corrosion-resistant fasteners provide an excellent illustration of life-cycle costing benefits. Polymer coatings can drastically increase fastener cost. In the context of the entire roof system, however, fastener costs are a minor factor. And considered in the even broader context of incremental insurance, the additional capital cost for corrosion-resistant fasteners becomes trivial.

Fastener backout is a major mode of fatigue failure. (See the previous discussion of laboratory testing.) Backout can sometimes be confused with insulation compression around fasteners, a "tenting" phenomenon that can puncture the membrane. True backout, however, also produces tenting, plus the additional hazard of drastically weakened fastener anchorage. It is affected by the following fastener properties:

- Shank length
- Nature of drill point
- Thread design

- Stress plate and fastener head

Shank length can become a major factor, especially on a re-covering project where the vertical distance between the deck and the new membrane is 6 in. or more. Long fastener shanks provide longer levers to work fasteners loose at their fulcrum anchorage points. As a vertical cantilever, a fastener's flexural stiffness against lateral motion diminishes proportionately with the cube of its height (i.e., a fastener of a given length is eight times as stiff as a fastener of equal shank cross section that is twice as long). Lateral deflection of the fastener can accompany the wave motion associated with cyclical wind loading. Uplift pressure induces a wave motion in the membrane in the windward direction and then a reactive wave motion in the opposite direction. Movements induced by thermal contraction in the membrane can also stress fastener shanks. The combination of wind-induced oscillations, longer term thermal cycling, and vibrations from rooftop mechanical equipment can work a fastener loose from its anchorage at the deck. Rocking of the screw enlarges the threaded hole and loosens the anchorage.

Drill points of high-tech, screw-type fasteners are self-drilling (the only type of fastener that is both practical and economically competitive). They are classified as either *extruding* or *cutting*. Extruding drill points push displaced metal aside, whereas cutting points lift removed metal out of the drilled hole. Extruding points depend upon heat hardening for efficient drilling performance.

One proprietary design features a threaded conical point instead of a plain conical point and threaded cylindrical shank. This drill-point design produces better "bite," with a smaller hole, packed with laterally displaced drill-hole material between the hole and the screw shank. The result: better backout resistance as well as 50 percent higher ultimate pullout resistance in a 22-gage deck.

Thread design of mechanical fasteners has been improved by one major change: reduction of the top slope of the threads (i.e., their angle with the horizontal). Thread design with less top slope reduces the risk of backout by providing much higher stripping torque than initial tapping torque.

Still another development in high-tech fastener technology is the substitution of plastic stress plates for steel in some fasteners. With their rounded edges and greater thickness, plastic stress plates prevent the cutting punctures sometimes inflicted by steel stress plates in single-ply membranes. "Cupping" of steel stress plates from overdriving of fasteners turns sharp metal edges up, thereby creating a puncture hazard. The thicker cross-section dimensions practicable for plastic stress plates allow for a recessed screw head. This permits compression of insulation without projection of the screwhead, which can puncture the membrane if it is not kept below the stress-plate elevation.

As usual, however, there is no free lunch for the roof designer. Plastic stress plates may fail from fatigue or from excessive torque during application, as well as from poor basic design and inappropriate materials. Degradation from hot asphalt or torching also poses a threat to plastic stress plates.

Fastener technology thus challenges roof-system designers with a constantly evolving branch of roof technology. Performance criteria for fasteners—notably for dependable anchorage at the deck, lateral stability, and corrosion resistance—fly warning flags against reroofing over existing systems with wet insulation.

Alerts

General

1. Before specifying a roof assembly, consult with local building official and insurance agent or rating bureau to make sure that your proposed design satisfies local wind requirements.

2. Before specifying an anchoring technique:
   a. Request data on wind-uplift resistance from manufacturer.
   b. Check UL or FM listings for wind-uplift approval of proprietary products.

Design

1. Beware of substituting even a single component within a roof assembly rated for wind uplift. Such changes require restudy of the resulting new roof assembly. Wind-uplift ratings apply to an entire roof assembly, not to its individual components.

2. Specify mechanical fastening instead of cold-applied or hot-mopped adhesive whenever practicable, for any type of nailable deck.

3. Specify minimum 22-gage thickness for steel decks.

4. Specify minimum 18-gage for metal gravel stops or fascia strips.

5. Specify a continuous cleat or hook strip for stabilizing the bottom of fascia strips.

Field

1. Require weighting of all membrane edges left incomplete before splicing with other sections of membrane.

2. Exercise extreme caution about high winds and/or low temperature when hot-mopped adhesives are specified to anchor insulation or other roof components to a substrate.

References

1. W. A. Dalgliesh and W. R. Schriever, "Wind Pressures and Suctions on Roofs," *Can. Build, Dig.*, no. 68, May 1968, p. 68-4
2. H. J. Leutheusser, "The Effects of Wall Parapets on the Roof-Pressure-Coefficients of Block-Type and Cylindrical Structures," Department of Mechanical Engineering, University of Toronto, April 1964.
3. Phillip J. Smith, *Development and Analysis of a Method to Evaluate Cyclic Wind Loading of Mechanically Attached Single-Ply Membrane Roof Cover Systems,* Master's thesis, 1993, p. 114.
4. Thomas Smith, "Hurricane Hugo's Effect on Metal Edge Flashings," *International Journal of Roofing Technology,* vol. 2, 1990, pp. 65ff.
5. R. J. Kind and R. L. Wardlaw, *Design of Rooftops against Gravel Blowoff,* National Research Council of Canada, NRC no. 15544, September 1976.
6. R. J. Kind and R. L. Wardlaw, "Model Studies of the Wind Resistance of Two Loose-Laid Roof-Insulation Systems," *Nat. Res. Counc. Can., Lab. Tech. Rep.,* LTR-LA-234, May 1979.
7. See UL 580, "Tests for Wind-Uplift Resistance of Roof Assemblies."
8. Colin Murphy, "A Long Term Evaluation of Mechanical Attachment of Thermoplastic Systems," *RCI Interface,* November 1992, pp. 1ff.

Chapter

8

Fire Resistance

The roof designer must often satisfy two fire-resistive requirements: (1) building code criteria and (2) insurance company requirements for qualifying the building for fire coverage. Building code requirements are of course mandatory, and from a practical viewpoint, so are insurance company requirements. Use of a combustible instead of a fire-resistive roof assembly will usually raise the owner's total building cost by making a costly sprinkler system mandatory.

The most important standard-setting organizations for fire and wind-uplift resistance are the Underwriters' Laboratories, Inc. (UL) of Northbrook, Illinois, and Factory Mutual Research Corporation (FM), of Norwood, Massachusetts. Both organizations classify roof assemblies for fire and wind-uplift resistance for many of the nation's insurance companies. UL and FM maintain laboratories for testing and listing manufacturers' building products that satisfy their standards, and insurance companies use these construction listings to develop specifications and recommendations for their insureds. UL standards are also the basis for building code requirements for fire resistance of roof assemblies.

Nature of Fire Hazards

Fire hazards that concern building code officials and insurance companies are broadly classified as follows:

- *External, above-deck fire exposure* from flying brands or burning debris blown over from neighboring buildings on fire
- *Internal, below-deck fire exposure* from interior inventory or equipment fires

Criteria for testing and certifying these two classes of fire hazard fall into three categories. A set of UL criteria and tests is generally the basis for external fire resistance. Criteria for internal fire resistance comprise the following:

- Limitation of flame spread along the roof-assembly soffit
- Time-temperature rating, per ASTM E119 furnace test

Because it is simpler, the less serious external fire hazard, with its evaluating tests and standards, is discussed before the more serious internal fire hazard.

External Fire Resistance

External fire presents the risk of fire spreading from a burning nearby building. The membrane's chief fire-resistive function is to reduce this risk. Membranes are accordingly tested and rated for their resistance to external fire. A membrane should not spread flame rapidly, produce flying brands endangering adjacent buildings, or permit ignition of its supporting roof deck.

According to the industry standard for rating roof coverings, ASTM E-108, UL790 "Test Methods for Fire Resistance of Roof Covering Materials," classified roof coverings "are not readily flammable, do not slip from position, and possess no flying brand hazard." Their performance is rated for different fire intensities:

- Class A roof coverings "are effective against *severe* fire exposure."
- Class B roof coverings "are effective against *moderate* fire exposure."
- Class C roof coverings "are effective against *light* fire exposure."

Still another distinction among roof-deck coverings is the degree of protection they give the roof deck. Under "severe fire exposure," a Class A roof covering affords "a *fairly high degree* of fire protection"; under moderate fire exposure, a Class B roof covering affords "a *moderate degree* of protection"; and under light fire exposure, a Class C deck affords "a *measurable degree* of fire protection."

Aggregate-surfaced membranes generally qualify as Class A roof coverings, regardless of deck type, combustible or noncombustible. Mineral-surfaced cap sheets sometimes qualify as Class A, but smooth-surfaced membranes seldom rise above Class B rating.

To determine roof-covering fire classification, UL has three tests:

- Flame spread

- Flame exposure
- Burning brand

The testing apparatus consists of a 3-ft-long gas burner placed between a large air duct and the roof-covering specimen, which is mounted on a standard timber deck and set at the maximum slope for which the tested covering is recommended by the manufacturer (see Fig. 8.1). In all three tests, a 12-mph air current blows across the test specimen, simulating wind.

The *flame-spread test* is conducted in all roof-covering testing programs regardless of deck type (i.e., combustible or noncombustible). Flame spread depends chiefly on the following parameters:

- Roof slope
- Nature and quantity of surfacing
- Membrane composition

Roof slope affects roof-covering rating because hot, burning gases tend to rise from a horizontal surface, thus lengthening the flame-spread time. A vertical surface, representing the limiting case, obviously lengthens the dimension of flame spread up the surface. Depending on slope, a given roof covering can have several ratings. (The rating naturally declines with increasing slope.)

Both the nature and the quantity of surfacing and membrane composition affect surface combustibility. Clay-based asphalt emulsions have better ratings than asphalt cutbacks on smooth-surfaced roofs because the presence of noncombustible clay tends to retard flame spread.

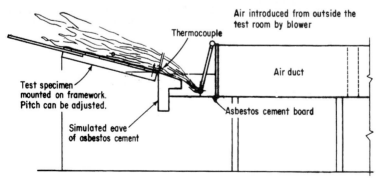

Figure 8.1 Testing apparatus for qualifying roof coverings through flame exposure, flame spread, and burning tests. (*Underwriters' Laboratories, Inc.*)

The flame-spread test subjects a 13-ft-long test specimen to flames around 1400°F for either 10 min (Classes A and B), 4 min (Class C), or until the flame recedes from the point of maximum spread. For Class A rating, the extent of burning, measured from the bottom of the deck, must not exceed 6 ft; for Class B, it must not exceed 8 ft; and for Class C, 13 ft.

Application of a maintenance coating later in the roof's service life could increase roof-surface combustibility and reduce its roof-covering classification. If the particular coating is not UL-listed for the tested membrane, its application could nullify UL classification.

The two flame-penetration tests—*flame exposure* and *burning brand*—are conducted only on roof assemblies with combustible decks (wood or plywood), not on noncombustible decks (concrete, gypsum, steel, preformed mineralized wood fiber).

The *flame-exposure test* subjects the specimens to intermittent flame exposures of 1400°F for Classes A and B, 1300°F for Class C. This test evaluates the roof covering's protection of the underlying combustible deck. Cracking or charring during this test's cooling cycles can reduce this protection.

The *burning-brand test* specifies different sizes of test brands simulating the firebrands hurled by wind or by uprushing gases of an actual fire onto the roof. Test-brand size varies from a 12-in.-square lumber lattice (4.4-lb weight) for Class A down to a $1\frac{1}{2}$-in.-square block (0.02-lb weight) for Class C. The brands are ignited in a gas flame at temperature over 1600°F for varying periods of time, depending on the sought classification, placed on the tested roof membrane, and allowed to burn until consumed.

To qualify for classification, the roof covering must withstand the three tests (flame exposure, flame spread, and burning brand) without (1) any portion of the roof covering material blowing or falling off in glowing brands; (2) exposing the roof deck by breaking, sliding, cracking, or warping; or (3) permitting ignition or collapse of any portion of the roof deck. (Other conditions of varying severity are established in "Test Methods for Fire Resistance of Roof Covering Materials," UL790 or ASTM E108.)

The classifications of roofing membranes are published in the *Roofing Materials and Systems Directory* (Underwriters' Laboratories, Inc.) under "Roof Covering Materials." Materials used in these UL-listed roofing membranes qualify for the UL label.

Internal Fire Hazards

Research on acceptable fire spread from internal (i.e., below-deck) fire has focused chiefly on steel-deck roof assemblies, which are especially important because of steel's predominance among deck materials. Because of its extremely high thermal conductivity, steel deck quickly transmits the heat energy of an interior fire to the above-deck roof components, which must be carefully selected to ensure adequate fire resistance.

Industry concern with steel-deck roof assemblies dates from the historic fire in 1953 at a huge General Motors plant in Livonia, Michigan. Representing the nation's greatest fire-insurance loss (until the 1966 McCormick Place fire), the Livonia fire gave the industry a stark lesson on the fire-feeding hazards of hot-mopped bitumen applied to a steel deck. Melted by the fire, the asphalt used to adhere the combustible insulation to the deck dripped through the steel-deck joints, drastically prolonging the fire and advancing it 100 ft ahead of the burning contents. As a consequence, less hazardous cold-applied adhesives and mechanical fasteners replaced the heavy bituminous moppings formerly applied directly to steel decks with combustible insulation. Note, however, that some of these fire-resistive improvements—especially cold adhesives and plastic sheet vapor retarders—have exacted sacrifices in other aspects of roof-system quality: notably wind-uplift resistance and vapor-retarding effectiveness.

Below-deck fire tests and standards

Following the Livonia fire, tests conducted in a 20×100 ft standard test building by FM and UL confirmed the hazards of placing large quantities of bitumen directly on a steel deck with combustible wood-fiber insulation. The bitumen and wood fibers spread test fires through the 100-ft-long building within 10 to 12 min, regardless of variations in the number of moppings (from one to three) between the deck and the insulation. The full-scale building tests conclusively demonstrated that no hot bituminous mopping of sufficient thickness to bond the insulation could be part of a fire-resistive steel-deck roof system with wood-fiber or similar insulation.

Noncombustible insulation—foamed glass, glass-fiber, perlite board—proved satisfactory with limited amounts of hot-mopped bitumen applied directly to the steel deck.

Flame-spread test

To satisfy building code and insurance requirements, a roof system must always resist internal fire (fire within the building). The chief safeguard required against internal fire is limitation of flame spread along the underside of the roof assembly.

As a result of the Livonia fire, a steel-deck roof assembly with 1-in., mechanically anchored, plain vegetable fiberboard insulation and a four-ply, aggregate-surfaced built-up membrane became the standard roof construction for both UL and FM, the criterion for evaluating other roof-deck assemblies. To qualify for UL listing as "acceptable," a roof-deck assembly must not spread the flame of a UL1256 test farther than the standard steel-deck assembly described above.

In this UL1256 standard, "Test Method of Roof Deck Construction," the test roof assembly forms the top of a test tunnel, with twin gas burners delivering flames against its soffit. Gas supply and other variables are adjusted until the furnace produces a flame-spread rate of $19\frac{1}{2}$ ft in $5\frac{1}{2}$ min on select-grade red-oak flooring. (In this flame-spread rating spectrum, asbestos = 0, red oak = 100.) To qualify as "acceptable" in this tunnel test, the roof assembly must not spread flame on the underside more than 10 ft during the first 10 min, 14 ft during the next 20 min (see Fig. 8.2).

In the more expensive 20×100 ft test building, an acceptable roof assembly must not spread flame more than 60 ft from the fire end of the test structure during the 30-min test.

FM fire classification

Factory Mutual's classification of resistance to interior fire divides roof assemblies into two basic categories: sprinklered and unsprinklered. (These classes refer to the roof construction for nonhazardous occupancies because some interiors containing combustible materials require sprinklers *regardless of roof-system fire resistance.*)

Figure 8.2 To qualify as a Class 1 steel-deck roof assembly, a test specimen, installed as the test tunnel roof, must not spread flame from twin gas burners more than 10 ft in the first 10 min, or 14 ft during the next 20 min. (*Underwriters' Laboratories, Inc.*)

Roof assemblies *not* requiring sprinklers include:

- Class 1 steel-deck assemblies
- Noncombustible decks—concrete, gypsum, asbestos cement, and preformed structural mineralized wood fiber
- Wood decks treated with fire-retardant inorganic salts limiting flame spread to 25 ft or less

Roof assemblies requiring sprinklers include:

- Class 2 steel-deck assemblies
- Combustible decks (untreated wood)

FM Research Corporation's calorimeter test for Class 1 steel-deck assemblies measures the fuel contributed to combustion by a 4×5-ft roof-assembly test specimen placed in the test-furnace roof. As the basis of comparison, a noncombustible panel undergoes the same 30-min fire exposure, with auxiliary fuel added to match the time-temperature curve recorded for the test specimen. The metered auxiliary fuel thus equals the fuel contributed by the test specimen. To qualify as a Class 1 assembly, the test specimen's average fuel contribution must not exceed certain tabulated values [in $Btu/(ft^2 \cdot min)$] for specified time intervals. Large-scale fire tests, conducted in a 100×20-ft building, have demonstrated the acceptability of assemblies that satisfy the tabulated limits.

The distinction between Class 1 and Class 2 steel-deck assemblies depends upon the heat release of the above-deck components.

The use of hot-mopped bitumen on a steel-deck surface to form a vapor seal or to bond combustible, organic insulation to a built-up membrane disqualifies a steel deck for Class 1 rating. One exception to this rule occurs when a deck soffit is sprayed with a noncombustible insulation. Otherwise, only FM-approved adhesives or approved mechanical fasteners can qualify a steel-deck assembly for a Class 1 rating. (For approved manufacturers' adhesives and fasteners, see Factory Mutual *Approval Guide,* latest edition.)

Application of foamed plastic insulation (polystyrene or urethane) directly to its top surface generally disqualifies a steel-deck roof assembly for Class 1 rating. Subjected to heat from an interior fire, most foam plastics disintegrate and expose the bitumen of the built-up membrane to heat conducted by the steel deck. Plastic foam thus creates the same fire-feeding hazard as a bituminous mopping.

To qualify as an FM Class 1 steel-deck roof assembly, an assembly using many plastic foams requires a lower layer of noncombustible insulation—e.g., glass-fiber or perlite board—between the steel deck and the foam. Nonetheless, some foamed plastics, notably isocyanurate boards, qualify as Class 1 without the interposition of a fire-barrier material.

Note the irony of the systems concept in roof design in regard to urethane. To improve the steel-deck roof system's fire resistance requires another insulation (e.g., perlite board) *below* the urethane. But to prevent urethane-promoted blistering may require another insulation board *above* the urethane.

Time-temperature ratings

Time-temperature rating is a second index of internal fire resistance, complementing the previously discussed index of flame spread along the deck soffit. Given in time units—hours and fractions—a roof assembly's time-temperature rating is established by its performance in a standard ASTM E119 furnace test, which subjects the tested assembly to a constantly rising temperature (see Fig. 8.3).

A time-temperature rating is required by many building codes. For example, the National Building Code, recommended by the American Insurance Association, requires a 2-h fire rating for roof assemblies in fire-resistive Type A construction. To qualify for this rating, a tested roof assembly (or any other tested system or component) must endure a test fire of progressively rising temperature—from 1000°F at 5 min to 1700°F at 1 h and 1850°F at 2 h. This standard test is promulgated by four major organizations: Underwriters' Laboratories (UL263), American Society for Testing and Materials (ASTM E119), National Fire Protection Association (NFPA 251), and American National Standards Institute (ANSI A2.1).

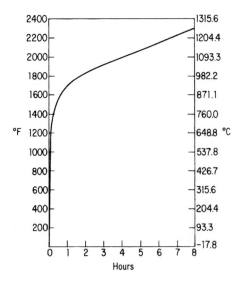

Figure 8.3 Standard time-temperature curve (ASTM E119) shows rising temperatures that construction assemblies must endure to qualify for fire-resistance ratings, in hours. (*American Society for Testing and Materials.*)

The fire-endurance test measures roof-assembly performance in carrying loads and confining fire. To qualify for a given fire rating, the tested assembly (minimum 180 ft^2 with minimum 12-ft lateral dimension) must (1) sustain the applied design load, (2) permit no passage of flame or gases hot enough to ignite cotton waste on the unexposed surface, and (3) limit the maximum temperature rise of the unexposed surface to an average of 250°F above its initial temperature or a 325°F rise at any one point.

(For steel assemblies with structural steel, prestressed, or reinforced concrete beams spaced more than 4 ft on centers, there are several other complex requirements, set forth in "Standard for Safety: Fire Tests of Building Construction and Materials," UL263, Underwriters' Laboratories, Inc., December 1976.)

Fire rating vs. insulating efficiency

With the advent of skyrocketing energy costs and the trend toward more thermally efficient roof insulation, a serious conflict has developed between fire resistance and energy conservation. Although some directory-listed, fire-rated roof assemblies have unlimited insulation thickness, many still contain only the minimal, usually 1-in.-thick insulations generally specified before the mid-1970s' impact of the energy crisis. These minimal insulations fall far below current standards of thermal-insulating efficiency. Thus the designer may be forced to choose between a listed fire-resistance rating and a thermally efficient roof system.

The problem arises because of the hazards of tampering with any component in a rated fire-resistive roof assembly. As noted in Chapter 2, "The Roof as a System," thickening the insulation of a fire-rated roof-ceiling assembly beneficially reduces deck-surface temperature. But in retarding heat loss *through* the roof assembly, this added insulation increases the ceiling space temperature, and this higher ceiling-space temperature could buckle steel joists or accelerate the burning of combustible structural members that would otherwise continue carrying their loads. Thus increased insulation thickness could reduce the roof assembly's time-temperature rating.

Resolution of this conflict between fire-resistive standards and roof-system thermal efficiency requires more research sponsored by manufacturers marketing efficient new insulating materials.

Building Code Provisions

Building code provisions for roofs, as well as other building components, almost universally follow FM or UL standards. All four model codes, the basis for most of the nation's local codes, classify buildings by degree of fire resistance and occupancy.

Under the BOCA National Building Code, promulgated by the Building Officials and Code Administrators International, Inc. Conference of America (BOCA), the degree of fire resistance is defined by five basic types of construction:

1. Noncombustible, protected
2. Noncombustible, protected and unprotected
3. Noncombustible exterior walls
4. Heavy timber
5. Combustible

Each type has subclasses; e.g., Type 1 includes Type 1A and 1B, and each subclass carries its own required fire rating for the roof assembly, for example, 2 h for Type 1A, $1\frac{1}{2}$ h for Type 1B, and 1 h for Type 2A.

For any given use, the larger the building, the more stringent the fire requirements. Conversely, the more fire-resistive the construction, the larger the permitted building. For example, any residential hotel more than nine stories high (10 stories if fully sprinklered) must have at least Type 1B roof assembly ($1\frac{1}{2}$-h fire rating) when the top floor ceiling dimension is 15 ft or less.

The BOCA National Building Code rates roof coverings in conformance with the UL classification: Class A, "effective against *severe* fire exposure"; Class B, "effective against *moderate* fire exposure"; and Class C, "effective against *light* fire exposure."

The BOCA code requires a Class A roof covering on every building except

- Buildings and structures of Type 5B construction with a *fire separation distance* of not less than 30 ft from the roof's leading edge
- Occupancies in Use Group R-3 located in detached buildings and accessory buildings with a *fire separation distance* of not less than 6 ft from the roof's leading edge

Alerts

1. Before specifying a roof assembly, consult with the local building official and insurance company representative to make sure that your design satisfies local fire requirements.

2. Beware of substituting even a single component, especially insulation, within a fire-rated roof-ceiling assembly. Such changes require restudy of the new roof assembly thus created. A fire rating applies to an entire roof-ceiling system, not to its individual components.

3. Make sure listed products are ordered with correct labels on bundles or rolls.

4. Verify that labels conform with project specifications.

Chapter 9

Historical Background of Contemporary Roof Systems

Industry experts' predictions cited in the second (1982) edition of this manual proved remarkably prescient. In 1980, single-ply and modified-bitumen membranes accounted for less than 10 percent of the low-slope roofing market. By the mid-1990s, they accounted for nearly 70 percent. Single-ply elastomers and plastic sheets claim more than one-third of the total, modified bitumens claim a quarter of the market, and conventional built-up systems still account for about one-third (see Fig. 9.1). Equally accurate were predictions that most of the hundred or so proprietary single-ply systems marketed in 1980 would disappear, leaving only a minority of these products still available in the mid-1990s.

Behind this roofing revolution is a combination of economic and technological factors:

- Skyrocketing petroleum prices during the energy crises of the 1970s closed the cost gap between more expensive synthetic materials and conventional built-up systems.
- Dissatisfaction with conventional built-up roofing provided a ready market of unhappy building owners and architects eager to try something new.
- The long term trend toward greater reliance on prefabrication and lesser reliance on heavy field labor favored the generally less labor-intensive field installation of the newer, lighter materials.
- Environmental regulations increasingly limited the heavy polluting installation techniques of conventional built-up roofing, with its smoke-producing hot bitumen kettles.

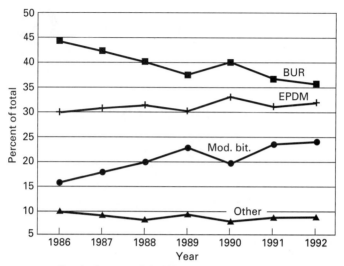

Figure 9.1 Graph shows modified-bitumen membranes gaining market share at expense of BUR in recent years, while EPDM, PVC, etc., remain essentially stable. (*National Roofing Contractors Association.*)

- The new single-ply systems are generally cleaner and more esthetically pleasing than built-up systems.
- They are also much better adapted to heat-reflective colors, an energy-conserving feature in warm climates.

Built-up Bituminous Membranes

Despite the limitations that have promoted the many alternatives to the traditional built-up membranes as a technique for waterproofing low-sloped roofs, today's built-up membrane is itself continually evolving, if more slowly than its single-ply competitors. As demonstrated by its retention of roughly one-third of the flat roof market, it remains a formidable competitor of the new systems. As its most notable continuing advantage, few, if any, competitive systems can match the toughness, impact resistance, and general durability of a well-constructed aggregate-surfaced built-up roof system.

Today's built-up membranes are the sesquicentennial descendants of the crude prototypical membranes invented in the 1840s, when square sheets of ship's sheathing paper, treated with a mixture of pine tar and pine pitch, served as the felts. The interply bitumen of these early membranes was coal tar, a waste product readily available from the production of coal gas for illumination at plants near the nation's cities. As the next advance, coal tar was substituted for pine tar as a more fluid saturant for the sheathing paper, but square sheets were still dipped manually into the melted saturant and the excess pressed out. Then came the substitution of paper or felt rolls, running through continuously operating saturators. The use of felt rolls accelerated the application process and promoted more uniform membrane quality. Substitution of distilled coal tar pitch for the more expensive pine pitch–coal tar mixture followed.

Asphalt became a competitor of coal tar pitch before the end of the nineteenth century, both as mopping bitumen and as felt saturator. Asphalt was not only more economical than coal tar pitch but also a more versatile material. (Its wide range of viscosities adapts asphalt to much steeper slopes than coal tar pitch.) As a consequence of these advantages, asphalt ultimately replaced coal tar pitch as the most popular bitumen in built-up systems.

More recently, built-up membrane felts have undergone even greater changes. Organic felts containing cellulosic fibers monopolized the market for decades until the introduction of asbestos felts in the 1920s. Today, however, fiberglass felts, introduced soon after World War II, have virtually wiped out the competition. The cellulosic fibers in organic and, to a lesser extent, asbestos felts make them vulnerable to fungus rot, whereas fiberglass felts are virtually immune to this roofing disease. The moisture-absorptive propensity of both organic and asbestos felts makes them vulnerable to blistering. This formerly major problem is virtually eliminated with the porous fiberglass felts. They vent the entrapped air–water vapor mixture that, upon heating, can cause blister growth in poorly applied organic- or asbestos-felted built-up membranes. Fiberglass felts also resist splitting better than organic and asbestos felts, which are drastically weakened by moisture absorption. Moreover, organic and asbestos felts are much weaker in the transverse direction than in the longitudinal direction. The greatly improved modern fiberglass felts have virtually isotropic (i.e., equal in both longitudinal and transverse directions) tensile strength.

In summary, the contemporary built-up membrane is potentially a vast improvement over its predecessors. It still, however, depends upon highly skilled application technique to realize this potential.

The First-Generation Single-ply Pioneers

In the late 1960s, during the writing of this manual's first edition, the new synthetic single-ply materials held a minor fraction of 1 percent of the low-slope roofing market. At that time, single-ply roofing was far more expensive than conventional built-up roofing. Accordingly, the new systems were generally restricted to complex roof structures—hyperbolic paraboloids, cable-supported inverted arches (like the Dulles Airport terminal), and the like. In some of the early single-ply failures, the complexity of roof structure was probably a greater factor than deficiencies in the membrane material.

Nonetheless, some materials introduced with the first generation of single-plies were defective. One pioneering entry in the single-ply sweepstakes comprised a 2-mil (0.05-mm) polyvinyl fluoride film bonded to a neoprene-bonded asbestos felt. Adhering this nonweldable film was extremely difficult. A self-adhering tape that covered adhesive-sealed laps and flashings lacked the required durability. Another membrane material, polyisobutylene (PIB), also fabricated with an asbestos-felt backer, had a short service life. In this instance, the weak link was the material itself, not the field seams. Incorporating PIB-based adhesives for the soluble PIB made the field seams fairly durable. But the PIB sheets themselves developed stress cracks under long term exposure, despite the application of ultraviolet-radiation-protective coatings.

Other first-generation prototypes in the 1960s included vulcanized butyl and neoprene rubber, compounded as black sheets for weather resistance. High material cost was compounded by high installation cost for the fully (or partially) adhered anchorage. Adding to this high cost were color coatings, generally based upon Hypalon (generically, chlorosulfonated polyethylene). This low-solids material requires multiple coatings. It also posed maintenance problems throughout the roof's service life. High initial cost plus high maintenance cost thus tended to price these early single-ply systems out of the market.

Liquid-applied elastomeric coatings also appeared among the first-generation single-ply membranes. Limited to taped plywood and poured concrete deck substrates (apart from sprayed polyurethane foam), these liquid-applied membranes generally comprised multiple coatings of black neoprene followed by two coats of Hypalon. Their chief disadvantage was an overwhelming dependence on high-quality fieldwork—notably, elaborate substrate preparation. Training roofing crews to seal minute concrete cracks is, to say the least, difficult. Worse yet, the need to maintain clean surfaces, an offense against the construction workers' practical religion, dooms liquid-applied coatings as a practical roofing system for general use. In summary, liquid-applied coatings require a degree of care and precision that is generally impracticable for construction work. As a consequence, they should be considered only as a last resort, when conventional systems prove impracticable.

The cold-process dead end

Contemporary with the first-generation single-ply roof systems was a less radical modification, but apparent dead end: so-called cold-process built-up membranes. Designed to eliminate the hazards and pollution attending hot-bitumen application, these systems featured (a) substitution of asphalt cutbacks for hot-mopped bitumen as the interply and surfacing bitumen, and (b) substitution of coated felts for the saturated felts used in conventional, hot-applied built-up membranes. Surfacing is usually an asphalt-based emulsion for smooth-surfaced membranes. The counterpart of conventional aggregate-surfaced built-up membranes is a sprayed surface of the same No. 11 mineral granules used to coat asphalt shingles. The granules are sprayed by compressed air into a fibrated asphalt cutback.

Cold-process membranes have, however, never given serious competition to hot-applied built-up systems. The advent of modified-bitumen membranes offered a generally better alternative for owners disenchanted with conventional built-up roofs, but not committed to the more radical change represented by elastomeric or plastic sheet systems. The cold-process technique is restricted to maintenance and repair of damaged built-up systems when it is uneconomical to bring in heavy equipment—kettles, asphalt dispensers, etc.

The Second-Generation Success

Big improvements in material quality marked the second generation of single-ply systems, introduced in the early 1970s. Polymer-coated fabrics, originally used as trench tarpaulins, were adapted as roof membranes. Pond and ditch liners made of reinforced polymeric sheets were also elevated to the status of roof membranes. Strong, durable sheets of plasticized polyvinyl chloride (PVC), butyl, and neoprene rubber augmented the supply of synthetic single-ply membrane materials. PVC had a major advantage over synthetic rubber material in its weldability, by either heat or solvent welding of field seams.

Physically, the new single-ply materials differ in their elastic properties. So-called elastomers are synthetic polymers with rubberlike properties, i.e., elastic materials that rapidly regain their original dimensions upon release of deforming stress. Synthetic rubber materials—neoprene, butyl, EPDM—satisfy this definition. But other materials used for single-ply membranes, e.g., plasticized polyvinyl chloride (PVC) or rubberized or modified asphalt, do not. These materials exhibit some permanent elongation after tensile stressing. Still other materials only partially satisfy the definition of an elastomer.

Chemically, the new synthetics are *polymers.* Their high tensile strength depends upon long-chain molecules built up from monomers (the basic molecular units). Polymerization increases monomeric molecular weights from the 30 to 150 range to so-called macromolecules, 100 to 10,000 times as large. Tensile strength depends on the degree of polymerization; the longer the molecular chain, the stronger the polymer.

A major chemical distinction differentiates *thermosetting* from *thermoplastic* polymers. Synthetic rubber materials (the elastomers) are thermosetting. They harden permanently when heated, like an egg. Thermoplastics soften when heated (like butter) and harden when cooled. With thermoplastic materials, thermal cycling can repeat indefinitely; viscosity changes accompany temperature changes.

This contrasting behavior stems from a basic difference in molecular structure. Thermosetting resins start as tiny threads. Heat promotes chemical reactions that cross-link these tiny molecular threads, creating a permanently rigid matrix. This molecularly cross-linked structure makes a thermosetting material more resistant to heat, solvents, general chemical attack, and creep (i.e., plastic elongation).

Thermoplastics comprise long, threadlike molecules, so intertwined at room temperature that they are hard to pull apart. However, when heated they slide past one another, like liquid molecules.

Never content with relative simplicity, the modern chemist has further complicated matters by creating intermediate polymers—neither exclusively thermosetting nor thermoplastic. Some thermoplastic materials are cross-linked for special uses, and some cross-linked resins are available in thermoplastic form. Examples used in roofing are chlorinated polyethylene (CPE), chlorosulfonated polyethylene (Hypalon), polyisobutylene (PIB), and polyvinyl fluoride.

As one tentative general conclusion about thermosetting vs. thermoplastic materials, you might consider thermosetting materials as potentially superior to thermoplastics on a direct material-vs.-material basis. But thermoplastic materials offer superior, easier field joint-sealing processes. Thermosetting synthetic rubbers require use of self-adhering tapes or contact adhesives, which require a wait before the field-seaming process can be completed. Thermoplastic sheets can be joined by heat-fused welds, performed with no delay.

As a practical consequence, you might favor a thermosetting material—EPDM, for example—on a roof with few openings and other penetrating elements because this roof requires fewer field seams. (The membrane could be prefabricated into large sheets; see Chapter 12, "Elastomeric Membrane.") A thermoplastic material—plasticized PVC, for example—might prove better for a roof with numerous penetrations and consequently numerous field seams. (For basic properties of different single-ply sheets see Table 9.1.)

Among the second-generation elastomers, ethylene propylene diene monomer (EPDM) has virtually eliminated such first-generation elas-

TABLE 9.1 Properties of Commonly Used Single-ply Roof Membrane Materials

Generic category	Principal polymer	Typical thickness, mils	Colors	Fire treatment	Reinforcement: N = none S = scrim F = fleece-backed	ASTM specs	Typical widths furnished
Elastomer (nonweldable)	EPDM	45, 60, 75, 90	B, W	Standard & retardant	N, S, F	D4637	10–50 ft
	Neoprene	55, 60	B	N/A	N	D4637	Calender width
	Polyepichloro-hydrin	48	B	—	N	—	To order
Weldable	CSPE (Hypalon)	36, 40, 45, 60	W	N/A	S, F	D5019	Meter to 7 ft
	CPE	38, 45	W	N/A	S, F	D5019	Calender width
Self-adhered	PIB	90	W	—	S, F	D5019	45 in.
Thermoplastic (weldable)	PVC	45, 48, 60	W, Grey	N/A	S, F	D4434	Calender width
	Copolymer	37, 45, 60	W	—	S, F	D4434	Calender width

SOURCE: Roofing Industry Educational Institute.

tomers as neoprene and butyl rubber as competitors. EPDM's excellent ozone and ultraviolet resistance, previously exploited by the automotive industry in tires, weather stripping, and trunk-lid gaskets, has now been exploited by roofing material manufacturers. EPDM's capacity for absorbing large quantities of cheaper reinforcing filler materials—notably carbon black—lowers its cost while increasing its tensile strength.

New Roof Systems

Complementing these improvements in material quality and variety, new roof-system concepts accompanied the introduction of the second-generation single plies. Just as the early moviemakers tended to perpetuate the tradition of stage plays by filming set scenes in drawing rooms, so the first-generation pioneers of the single-ply revolution followed the traditional concepts of the century-old built-up roof systems. These were limited to adhered assemblies. *Fully adhered* assemblies are the older, more conventional type, but *partially adhered* systems were later developed for substrates subject to shrinkage cracking that could spilt a built-up membrane, with its low breaking strain. Partially adhered systems for the new elastomers were designed for economy, not safety, as these new membrane materials, with their extremely high breaking strains (normally 100 times or more that of built-up membranes) posed no splitting hazard. Various mechanically anchored systems, with individual or bar-fastened assemblies, soon provided a variation on the partially adhered theme, offering more dependable anchorage for the new single-ply membranes.

A truly radical departure from the conventional built-up system concept came with the loose-laid, ballasted system, which the second-generation single-ply pioneers saw as a tremendous advance in installation speed and overall economy. The loose-laid, ballasted system offers several notable advantages. It exploits the tremendous advantage of the new membrane materials' flexibility. It also permits the use of efficient thermal insulating materials that are unsuitable for adhesion with hot-mopped bitumen. The combination of expanded polystyrene insulation and ballasted single-ply membranes generally results in the lowest-cost roof system (see Fig. 9.2).

In addition to its economy, the loose-laid, ballasted assembly eliminates (or at least greatly alleviates) several problems that plagued conventional built-up systems until the elimination of asbestos and organic felts. Anchored only at the perimeter and at openings, the loose-laid, ballasted system virtually eliminates the risk of splitting, by isolating the membrane from substrate movement. A loose-laid

Figure 9.2 Ballasted roof systems generally provide the lowest-cost low-slope roof system, with the additional advantage of eliminating several problems (e.g., blistering) that have plagued traditional adhered BUR systems. (*Roofing Industry Educational Institute.*)

membrane cannot blister, because there are no confined voids between membrane and substrate (or between membrane plies) for air–water-vapor pressure buildup. There are also fewer worries about moisture entrapped in the insulation. Evaporated by solar heat, this moisture can either vent downward through a permeable deck or diffuse upward through a relatively permeable membrane.

Loose-laid, ballasted assemblies are not only less expensive than adhered or mechanically fastened assemblies, but also much easier to install. Fast, economical installation of loose-laid systems requires no costly, time-consuming mopping or adhesive-application operation such as is required for built-up or adhered single-ply systems. (Loose-laid systems are anchored only at roof perimeters and openings.) Not only is the application process simplified, but it also becomes less dependent on good weather. There is no need to stop work for a rain shower, as the loose-laid system can generally tolerate some entrapped moisture. And work can proceed even at subfreezing temperatures, depending only on the ability to make the field sheet seams.

Unfortunately, structural requirements limit the United States market more than the European market for loose-laid, ballasted systems, which require additional structural capacity of 10 to 15 psf in

Figure 9.3 Polyester "stone mat," installed between ballast aggregate and a single-ply membrane, provides a shield against puncture from sharp-edged aggregate pieces. (Aggregate subject to freeze-thaw cycling can develop sharp edges through cleavage.) (*Roofing Industry Educational Institute.*)

additional dead load for loose aggregate and up to 25 psf for pavers. In Europe, where structural concrete is more popular for decks, loose-laid, ballasted systems are more popular. In the United States, the popularity of long-span steel joists with flexible, lightgage steel decks designed tightly for minimum load capacity limits the market for loose-laid, ballasted systems. (The drive for *structural* economy can reduce *overall* construction economy by eliminating a far greater economy. Savings of a few *pennies* psf in reduced structural roof-framing capacity can cost extra *dimes* psf for a more expensive lightweight adhered roof system.)

The new single-ply membranes thus give the roof designer a vastly expanded scope for cutting construction costs or solving other design problems. Elastomeric and thermoplastic sheets are the most versatile. They can be used in either a conventional loose-laid, ballasted system (i.e., membrane-atop-insulation) or a protected membrane roof (PMR, insulation-atop-membrane) system. And they can be mechanically anchored, either with individual fasteners or through fastener bars.

Modified Bitumens

Concurrent with the single-ply elastomers came a more direct attack on conventional built-up membranes: the modified bitumens, pioneered in Italy and other European countries. By reducing the traditional four-ply built-up membrane to two (underlayment plus single-ply sheet), modified-bitumen sheets fulfill the perennial construction goal of reducing heavy field labor. And with their tremendous improvement in material quality—improved flexibility, elasticity, and ductility at subfreezing temperatures—modified-bitumen systems preserve most of the advantages of conventional built-up membranes, their superior puncture resistance and general toughness. In a sense, modified bitumens can be considered more an evolutionary improvement of conventional built-up membranes than a revolutionary alternative.

Like thermosetting and thermoplastic sheets, modified bitumens are a triumph of polymer chemistry. They evolved as an incidental technological by-product of more directly sought advances. One major ingredient in many modified bitumens, atactic polypropylene (APP), is a by-product of isotactic polypropylene (IPP), familiar in tool or hair dryer cases as the integrally molded hinge capable of flexing thousands of times without fatigue failure. The process producing IPP resin yielded a soft, noncrystalline by-product, APP. Seeking markets for this material, manufacturers found that it plasticizes asphalt, endowing it with greatly enhanced physical properties—the previously noted flexibility, elasticity, and ductility down into the subfreezing range ($-15°F$, $-26°C$). APP-modified sheets were first used in Italy, where APP polymer was readily available. They were soon imported into the United States.

Because APP-modified sheets have elevated softening points, they proved unsuitable for adhering with hot-mopped asphalt. They required heat fusion by propane torch, operating at temperatures approaching 3000°F, not only to adhere the modified bitumen sheet to its substrate but also to seal field seams and form flashings.

Another plasticizer, styrene butadiene styrene (SBS) copolymer, produces a modified-bitumen membrane with greater elasticity and low-temperature flexibility than the thermoplastic APP modified-bitumen membranes. Introduced in France in the 1960s, SBS-modified bitumens exploit the unique ability of SBS to form a polymer dispersion within a mass of asphalt. Even though the polymer represents only 12 percent or so of the mix, the SBS molecules form a network. The SBS-modified asphalt behaves something like a water-soaked sponge: despite the much greater weight of the water, the sponge nonetheless behaves like a solid, not a liquid. SBS-modified bitumen thus exhibits truly elastic behavior, recovering its original shape upon removal of deforming stress.

This unique polymer-dispersing property gives SBS a major advantage over APP as a bitumen modifier. Unlike APP-modified-bitumen sheets, SBS-modified-bitumen sheets can be field-adhered with conventional hot-mopped asphalt, because the greatly reduced quantity of SBS polymer elevates the melt point much less than APP does (see Fig. 9.4).

Modified bitumens offer almost as great a range as elastomeric and thermoplastic sheets, although they are not suitable for loose-laid, ballasted systems or for bar-type mechanically anchored systems. Least adaptable is conventional built-up, which is unsuitable either for loose-laid, ballasted assembles or for bar-type, mechanically anchored systems.

The immediately following chapters will expand in greater detail on the advantages and offsetting disadvantages of the various membrane materials and systems.

Figure 9.4 SBS-modified-bitumen sheets are normally hot-mopped, like traditional BUR, though torching grades are available. (*Roofing Industry Educational Institute.*)

Chapter 10

Elements of Built-up Membranes

The roof membrane is the roof system's weatherproofing component. In built-up roof membranes, there are two basic elements—felts and bitumen—combined into a laminate that is normally shielded by a third element: a mineral aggregate surface embedded in a bituminous "flood" coat. Properly designed and applied, the felt-bitumen laminate forms a flexible roof cover with sufficient strength to resist normal expansion and contraction forces.

Bitumen is the adhesive and waterproofing agent and thus the most important membrane element. If it had sufficient fire resistance, strength, rigidity, and weathering durability, the membrane could theoretically (although not practicably) be fabricated entirely of bitumen.

Felts stabilize and reinforce the membrane, like steel reinforcement in a concrete slab, providing about 90 percent of its tensile strength. The felt fibers restrain the bitumen from flowing in hot weather and resist contraction stresses and cracking in winter (see Fig. 10.1). Felts also isolate the different layers of bituminous waterproofing, which helps the mopper apply the bitumen uniformly.

Aggregate surfacing protects the bitumen from damaging solar radiation. Through a combination of heat and photochemical oxidation, sun rays accelerate bitumen embrittlement and cracking. Mineral aggregate surfacing forms a fire-resistive skin that prevents flame spread and protects the membrane from abrasion caused by rain, wind, and foot traffic. It can help resist the corrosion from acid mists condensing on the roof in industrial areas. An aggregate surfacing also acts as ballast, offering some wind-uplift resistance, and as a shield against the impact of hailstones.

Figure 10.1 Porous glass-fiber ply felts (top) are often embedded into hot bitumen with a squeegee (bottom) rather than a broom, normally used to promote good mopping adhesion for organic felts. (*Roofing Industry Educational Institute.*)

Aggregate surfacing also makes feasible the pouring of a heavy surface flood coat of bitumen. The closely massed aggregate forms tiny dams that retard the lateral flow of heated bitumen and allow it to congeal, to a depth about three times that of a mopped interply layer, 60 lb/square (293 kg/m^2) of asphalt for the flood coat vs. 20 lb/square of interply mopping.

When properly designed and constructed, an *aggregate-surfaced membrane* is *water-resistant*; it can tolerate minor local ponding last-

TABLE 10.1 Minimum Recommended Roof Slopes

Membrane Type	Minimum slope, in./ft	Notes
Aggregate-surfaced	1/8	
Smooth-surfaced	1/4	
Mineral-surfaced roll roofing	1	With concealed nails, minimum 3-in. top lap, 19-in. selvage, double-coverage roll
Mineral-surfaced roll roofing	2	With exposed nails

ing 1 or 2 days. A *smooth-surfaced membrane* or *mineral-surfaced cap sheet* (or roll roofing) membrane is essentially *water-shedding*, requiring steeper slope for faster drainage (see Table 10.1). Lacking equivalently surfaced protection, these sloped roofs normally require recoating of their initial protective coating or replacement much sooner than aggregate-surfaced roofs. (For more detailed discussion, see the "Surfacing" section later in this chapter.)

As with surfacing, the number of felt plies affects membrane durability. Each felt ply adds a layer of waterproofing bitumen adhesive to the multi-ply membrane, providing an additional line of defense. A large number of felt plies also tends to equalize membrane properties throughout the membrane and increase tensile strength proportionately. The roofing industry has traditionally assigned 5 years' anticipated service life to each felt ply, that is, 20-year life for a four-ply membrane, 15-year life for a three-ply membrane, and so on.

This crude rule has some validity, judged by a statistical study conducted by Simpson, Gumpertz & Heger in 1977–1978 and reported by Cash.[1] Calculated life expectancies for four-ply membranes were double those calculated for two-ply membranes, with a three-ply membrane in between, according to an expert group of 104 respondents comprising roofers, consultants, materials manufacturers' representatives, government researchers, and so forth. (No type of built-up membrane was collectively rated at more than 50 percent probability of lasting 20 years.)

Membrane Materials

Roofing bitumens

Bitumens are basically heavy, black or very dark brown hydrocarbons, divided into three major classes: natural asphalt, petroleum asphalt, and coal tar pitch. Only petroleum asphalt and coal tar pitch concern the designer of built-up roof systems.

Despite several significant physical and chemical differences, both asphalt and coal tar pitch have the following physical properties:

- Excellent resistance to water penetration and extremely low water absorptivity
- Durability under prolonged exposure to weather
- Good internal cohesion and adhesion (e.g., to roofing felts and insulation)
- Thermoplasticity (i.e., reversible, temperature-produced changes from semielastic to viscous fluid)

These four properties contribute to the long histories of successful roofing performance of both roof bitumens—well over a century for coal tar pitch and nearly a century for petroleum asphalt.

The most basic functional distinction between the two roofing bitumens is that asphalt comes in a much greater range of viscosities, which make it suitable for slopes up to 6 in. (see Table 10.2). On the other hand, coal tar pitch comes in only two roofing grades (plus a waterproofing grade), with maximum viscosity about the same as that of so-called "dead-level" (ASTM Type I) asphalt. It is generally limited to $\frac{1}{4}$-in. maximum slope.

Asphalt is the dense, "bottom-of-the-barrel" residue left from petroleum distillation. The asphalt content of crude petroleum varies from zero to more than half. Moreover, asphalts from different sources may vary greatly, so an asphalt-producing refinery must carefully select its crude oils to ensure a sufficient quantity of satisfactory asphalt.

The heavy asphalt residue left after distillation of the crude's more volatile constituents requires further processing to qualify as roofing asphalt. The raw asphalt "flux" is heated to about 500°F. Air (or pure oxygen), blown through a perforated, hollow pipe pin wheel, bubbles through the hot liquid asphalt (see Fig. 10.2). The longer the blowing continues, the tougher and more viscous (i.e., the less fluid, or "steeper") the asphalt becomes, and the more suitable it is for roofs of steepening slope. The same flux can produce the whole range of roofing asphalts—from Type I (dead-level) to Type IV (special steep), a softening-point range of 135°F (minimum for dead-level asphalt) to 225°F (maximum for special steep).

Chemically, the accelerated oxidation process in the blowstill lengthens the long, chainlike hydrocarbon molecules by dehydrogenation, driving off hydrogen atoms in gaseous water molecules, lengthening the carbon chains, and also distilling some lighter hydrocarbon molecules (see Fig. 10.3). Continued blowing creates a gel structure in the asphalt, and the growing number of lengthened hydrocarbon mol-

TABLE 10.2 Required Roofing Bitumen Properties

Property	Asphalt*								Coal tar pitch†			
	Type I		Type II		Type III		Type IV		Type I		Type III	
	Min.	Max.	Min.	Max.	Min.	Max.	Min.	Max.	Min.	Max.	Min.	Max.
Roof slope (in./ft)	0	¼	¼	1	¼	3	¼	6	0	¼	0	¼
Softening point, °F (°C)	135(57)	151(66)	158(70)	176(80)	185(85)	205(96)	210(99)	225(107)	126(52)	140(60)	133(56)	147(64)
Flash point, °F (°C)	475(246)		475(246)		475(246)		475(246)		374(190)		401(205)	
Penetration (tenths of mm):												
32°F (0°C)	3		6		6		6					
77°F (25°C)	18	60	18	40	15	35	12	25				
115°F (46°C)	90	180		100		90		75				
Ductility, 77°F, cm	10		3		2.5		1.5					
Specific gravity,‡ 25/25°C									1.22	1.34	1.22	1.34

*Per ASTM D312-95.
†Per ASTM 450-91.
‡Asphalt specific gravity, although not limited by ASTM, generally runs around 1.03 for roofing asphalt.

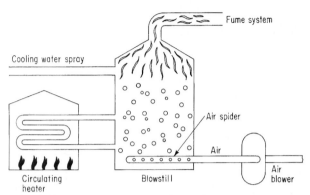

Figure 10.2 A single charge of asphalt in the blowstill can produce all four types of roofing asphalt, ASTM Types I to IV, with constantly rising viscosity and softening points, from 135°F (minimum for Type I) to 225°F (maximum for Type IV), determined by time in the blowstill at temperatures of 325°F and higher. Oxygen, bubbled through the liquid asphalt by the revolving air spider, reacts with hydrogen in the asphalt to form water vapor. This dehydrogenation process links smaller units into longer, carbon chain molecules. (*National Roofing Contractors Association.*)

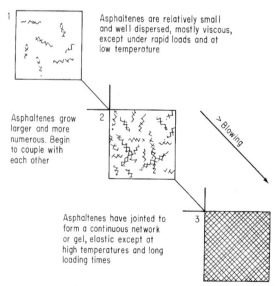

Figure 10.3 Hydrocarbon molecules lengthen into so-called asphaltenes as the blowing process continues, producing a more viscous, elastic asphalt. (*National Roofing Contractors Association.*)

ecules (asphaltenes) accompanying continued blowing makes the asphalt stronger, tougher, and more elastic, all desirable qualities in a roofing asphalt. Compared with the less oxidized asphalts with a lower softening point, the steeper-blown asphalts are less temperature-susceptible; it takes a greater temperature range to change them from brittle solids to viscous fluids, or vice versa.

But in roofing asphalts, the good qualities become inextricably associated with some bad. Compared with asphalt of a lower softening point, steep asphalt has the following liabilities:

- Less durability
- Less water repellence
- Higher probability of overweight interply moppings with blister-forming voids because of the greater difficulty in applying the moppings properly
- Higher probability of "fallback" when overheated, resulting in softer bitumen than specified

This greater difficulty in getting continuous, void-free interply moppings is the major drawback of steep (Type III) asphalt compared with dead-level (Type I) asphalt. Here is why steep asphalt poses a more difficult field-application problem. National Bureau of Standards (NBS) researchers have associated a 50- to 150-cSt viscosity range with proper mopping fluidity to produce a uniform, void-free interply asphalt film of 15 to 20 lb/square weight.[2] To maintain hot asphalt within this viscosity range requires a minimum 390°F *application* temperature for Type III asphalt, minimum 345°F for Type I asphalt. Because hot asphalt generally cools from 500 to 300°F in about half the time it takes to cool from 500 to 345°F, the timing of the asphalt-mopping, felt-laying operation is much more critical for Type III than for Type I asphalt. With a lower initial temperature—say, 450°F—the asphalt cools to 395°F in less than half the time it takes to cool to 345°F. Even if Type III asphalt is heated 50°F hotter than Type I asphalt, it normally cools to the equiviscous temperature (EVT—the temperature required to produce the theoretically ideal 125-cSt viscosity for mopping asphalt) in far less time than Type I asphalt because cooling rate increases exponentially with temperature difference between hot asphalt and ambient air. In 1988, the NRCA recommended use of two different EVTs, 75 cP for mechanical spreaders and 125 cP for hand mopping. Tolerances for EVT are generally 25°F (14°C).

Like minimum temperature for proper mopping viscosity, minimum temperature for good adhesion of felts is lower for Type I asphalt than for Type III, apparently by the same 45 to 50°F range that differentiates its softening point and EVT range from Type III asphalt's range.

Minimum surface temperature (as detected by a hand-held, infrared thermal detector) should not fall below 250°F for Type I asphalt or coal tar pitch or below 300°F for Type III asphalt, according to roofing specialist consultant Gerald B. Curtis, Camden, New Jersey. Based on extensive field experience, these temperatures are absolute minima. NBS is currently conducting field-based research to establish safe minimum temperatures for interply adhesion for different bitumens. Until such definitive research is available, the foregoing temperatures seem reasonable.

An even greater temperature difference than the 50°F figure just cited is indicated by Johns-Manville (now Schuller) research into the effect of temperature on the mopping characteristics of different asphalts. Testing six different Type I and III asphalts, the Schuller researchers required 450°F to get "good" mopping characteristics for all six Type III tested specimens, but only 375°F, 75°F less, to get similar good results with Type I asphalt. (The average weight of these good moppings of both asphalt types was 20 lb/square.) Moreover, Type I asphalt moppings at 350°F were rated generally better than Type III asphalt moppings at 425°F. The Schuller researchers also noted the greater safety factor, or spread, between the EVT and flash points for Type I asphalts compared with the narrower safety factor with Type III asphalts: an average 195°F spread for 12 tested Type I asphalts vs. 124°F for 13 tested Type III asphalts.

Overweight moppings with blister-originating voids and greater vulnerability to thermal-contraction stress are thus more likely to occur with Type III than with Type I asphalt. Corroborating field evidence from a Schuller research report indicates that 85 percent of the blistered low-slope roofs and 90 percent of the split low-slope roofs analyzed by laboratory test cuts had Type III asphalt interply moppings. The validity of this statistical evidence depends, of course, on the use of Type III asphalt in significantly lower percentages than the 85 to 90 percent range on the problem roofs. It appears to be a reasonable assumption.

Thus, as a general rule, specify the softest bitumen consistent with roof slope. The disadvantages of steeper asphalts normally outweigh their benefits. Only in special circumstances—for example, in a severely cold climate where asphalt embrittlement is a problem, or in an extremely hot climate like Phoenix's, where liquefication can disrupt roofing application—should you consider a steeper asphalt than required by slope. And when it is specified for interply moppings, steeper asphalt requires tighter field controls to prevent premature congealing before the felts are rolled in.

Type II asphalt has been proposed by Schuller researchers as a satisfactory compromise, multipurpose asphalt that solves the "two-asphalt" problem: the need for two kettles or tankers to supply Type III asphalt for adhering insulation, base sheet, and flashing and Type

I asphalt for membrane moppings and flood coat. Some roofers solve this practical problem by using Type III asphalt throughout the entire roof system, even for a dead-level roof's flood coat. As noted, such roofs run higher risks of failure.

Where this practical obstacle looms large, Type II asphalt can be used for everything except flood coats on poorly drained roofs and possibly for adhering flashing. With its 158 to 176°F softening-point range, Type II asphalt can serve as an insulation adhesive without Type I asphalt's threat of bleeding through deck joints. And with its lower EVT, Type II asphalt reduces the threat of interply voids and generally poor adhesion posed by the faster congealing Type III asphalt. These conclusions spring from field investigations reported by the Schuller researchers.[3]

The current dominance of glass-fiber felts, achieved since this Schuller research was conducted, complicates the recommendations. Because of "floating" or "sieving" attributable to the porous glass felts, Type III asphalt has been recommended for interply moppings, even on low-sloped roofs. The glass-fiber felts' porosity permits asphalt migration through the membrane cross section. This impairs the integrity of the laminated membrane.

Glass-fiber membrane failures have been attributed to this migration phenomenon. Since glass-felted membranes have less tendency to blister (owing to their greater porosity, which releases instead of entrapping vapor), the use of Type III asphalt for interply moppings is probably advisable. Nonetheless, because of its greater weather and water resistance, Type I asphalt would still be preferable for the flood coat. The great difficulty here, of course, springs from the two-asphalt problem—i.e., the inconvenience of monitoring tanker deliveries of two types of asphalt instead of one.

Coal-tar pitch is a jet black substance, denser and more uniform than asphalt. Normally recommended only for slopes of ½ in or less, coal tar pitch comes in an acceptable softening-point range of 126 to 147°F, compared with asphalt's 135 to 225°F (minimum for Type I to maximum for Type IV). Recommended EVTs for coal tar pitch, designed to produce 25-cP viscosity, are 355°F for Type I, 375°F for Type II.

Coal tar pitch is a by-product of the so-called "destructive distillation" of bituminous coal in the manufacture of coke, the carbon used to alloy steel. Subjected to temperatures well over 2000°F in the coke ovens, the coal is heated without air, and this destructive distillation process drives off gases and vapors, leaving the coke as a more or less pure carbon for charging the blast furnace. After the evaporated coal gases are cooled and condensed, crude tar is separated from the other compounds in the condensate. Coal tar pitch is one of some 200,000 products made from this crude coal tar.

Coal tar pitch is a good waterproofing agent because of a physical property known as *cold flow*. Even at moderate ambient temperatures—say, around 60°F—coal tar pitch slowly heals cracks formed at lower temperatures. This cold-flow property stems from a chemical peculiarity of coal tar pitch. Although its molecules have very strong *intra*molecular attractions between linked benzene rings, they have relatively weak *inter*molecular bonds; the molecules readily flow over one another, in slow response to gravity. This desirable waterproofing property is obviously what limits coal tar pitch's use to low-slope roofs. Used even on a moderately sloped roof, coal tar pitch, over years, migrates from higher to lower areas, leaving the roof's high points with a thinned layer of waterproofing bitumen.

Type I (dead-level) asphalt also has this cold-flow property, but usually to a lesser degree than coal tar pitch. More viscous asphalts, Types II to IV, lose this property in inverse proportion to their viscosities.

ANSI-ASTM Standard D450 was revised in 1978 to add a new "low-fuming" coal tar pitch (Type III) to the already specified Type I "old-style pitch."

Incompatibility of asphalt and coal tar pitch

Mixing asphalt and coal tar pitch violates a well-publicized roofing industry taboo because the two bitumens are chemically incompatible. The worst mixing occurs when asphalt melts in a kettle coated with the remnants of coal tar pitch (or vice versa). Such mixing can produce a mongrel bitumen with unpredictable properties.

Applying one bitumen over the other can produce similarly deleterious consequences:

- Asphalt applied over coal tar pitch may soften and flow off, leaving exposed coal tar pitch, which weathers rapidly.
- Coal tar pitch applied over asphalt may harden and crack.[4]

In mixtures of the two materials, coal tar pitch, the donor or "exudative" bitumen, becomes hardened and embrittled through loss of its lighter or more actively solvating molecules. Asphalt, the recipient or "insudative" bitumen, becomes softened and more fluid.

Tests conducted at Owens-Corning Fiberglas Corporation's Granville, Ohio, laboratory indicate that the incompatibility between asphalt and coal tar pitch is greatest when the two bitumens have similar softening points, i.e., when Type I asphalt is mixed with roofing-grade coal tar pitch. But when the asphalt has a high softening point—220 to 260°F for the coating-grade film, which is further stabilized by addition of mineral fillers—the mixing hazard abates. This principle explains the successful combination of flashings mopped with steep asphalt on coal-tar-pitch roofs and even the compatibility of coal-tar-pitch flood coats on asphalt-coated-felt membranes or glass-fiber felts, which are coated with a similarly hard asphalt.

Coal tar pitch vs. asphalt

Commercial passions flare when the subject of coal tar pitch vs. asphalt arises; discussing this topic is much like refereeing the Hatfield-McCoy feud. To establish as cool an atmosphere as possible, we shall preface this discussion by reiterating a previously stated thesis: Both coal tar pitch and asphalt are intrinsically excellent roofing materials, with long records of proven performance and thousands of roofs still performing satisfactorily, some after service lives of 30 to 50 years. For the majority of built-up roof projects—perhaps the vast majority—both asphalt and coal-tar-pitch roof systems will perform satisfactorily, *if properly designed and constructed.*

Nonetheless, there are significant differences between the two materials. As noted, asphalt is a more varied, more versatile bitumen than coal tar pitch. Properly processed asphalt is suitable for slopes up to 50%, whereas coal tar pitch requires special precautions on slopes over 2 percent. Under the right conditions, asphalt can be left exposed on the roof surface, but coal tar pitch always requires a protective aggregate surfacing because of its excessive fluidity at high temperature. Asphalt also has markedly less temperature susceptibility than coal tar pitch; i.e., it takes a greater temperature drop to change asphalt from a viscous fluid to a brittle solid. Coal tar pitch makes this transition over a narrower temperature range, whereas asphalt retains its plasticity at considerably lower temperature than coal tar pitch and at higher temperature retains higher viscosity.

As a countervailing advantage, coal tar pitch has a generally better cold-flow, self-healing quality and somewhat less vulnerability to deteriorative oxidation than asphalt. This self-healing property, shared in a less dependable degree by Type I asphalt, is required to protect the underlying felts from destructive moisture and ultraviolet radiation.

Felts

Roofing felts are nonwoven fabrics classified as either *organic* (paper or synthetic nonwovens) or *inorganic* (glass fiber). (This classification holds despite the fact that the binder on glass-fiber mats is an organic resin.) The paper-manufacturing process used to produce organic felts orients the fibers in their longitudinal (machine or roll) direction, making them roughly twice as strong in that direction as in the transverse (cross-machine) direction. Some glass-fiber and synthetic-fiber felts, however, are virtually isotropic (i.e., they have nearly equivalent longitudinal and transverse tensile strengths).

Today's roofing felts have evolved from square sheets of ship's sheathing paper for the prototypical built-up roofs to the organic-fiber felts. Coal tar was next substituted for pine tar, as a more fluid saturant. Then came the substitution of paper or felt rolls, running through continuously operating saturators.[5] Asphalt-saturated felts appeared before the end of the century.

Organic felts contain cellulose fibers—shredded wood and felted papers. (Although sometimes erroneously called rag felts, they now have little rag content.) Organic felts are impregnated with coal tar pitch or very soft asphalt called "flux."

Glass-fiber mats entered the United States market in the late 1940s. Glass-fiber filaments are drawn from molten-glass streams through tiny orifices made of precious metals. The molten glass comes from batches of sand, limestone, and soda ash, continuously fed into a furnace fired above 2500°F. The long fiberglass filaments, 12 to 15 per inch cross section, are usually bound with a thermosetting binder, phenol-formaldehyde, urea-formaldehyde, or acrylic resin. Glass-fiber mats may have cross-directional tensile strengths exceeding 60 lb/in. (0.5 kN/m), four times the strength of organic felts and five times that of the now obsolete asbestos felts (see Table 10.3).

Glass-fiber mats also differ from organic felts in not having a low-viscosity impregnating asphalt. Because glass fibers do not absorb asphalt, a much harder coating-grade asphalt is used. Synthetic polymeric mats of polyester fiber are also used as BUR reinforcement. These may be spun-bonded, using organic resins as binders. They are sometimes needle-punched to entangle the fiber layers and make them more resistant to delamination. These synthetic-polymer mats have been used, both unsaturated and bitumen-treated, in built-up membranes.

Felts are manufactured into three kinds of sheets:

- Saturated felts
- Coated felts
- Mineral-surfaced sheets

Each type represents a progressive stage in the manufacturing process.

Saturated felts—organic, polyester, or glass-fiber—are saturated with bitumen.

Coated felts are saturated felts that have been subjected to an additional manufacturing stage: coating with blown asphalt (generally

TABLE 10.3 Typical Glass-Fiber and Organic Roofing Felt Properties

MAT MILL	COATER			FINISHED PRODUCT					
Dry Wt. lb/100 ft²	Coating Wt. lb/100 ft²	Surfacing Wt. lb/100 ft²	Product Wt. lb/100 ft²	Description	ASTM Number	Strength lb/in.*		Moisture Percent	Ash Percent
						MD	CMD		
1.5 Min.	6.3 Min. (unfilled)	22 Max.	8.4 Min.	Standard Ply Sheet	D2178 Type III	22 Min.	22 Min.	1 Max.	70-88
1.7 Min.	3.0 Min. (unfilled)	32 Max.	6.0 Min.	Heavy-Duty Ply Sheet	D2178 Type IV	44 Min.	44 Min.	1 Max.	70-88
1.9 Min.	3.0 Min. (unfilled)	60 Min.	6.0 Min.	Ply Sheet	D2178 Type VI	60 Min.	60 Min.	1 Max.	70-88
1.7 Min.	3.0 Min. (coal tar)	50 Max.	6.0 Min.	Coal Tar Glass Felt	D4990 Type I	44 Min.	44 Min.	1 Max.	70-88
				Premium Tar Glass	---	60 Min.	60 Min.	---	---
1.7 Min.	--- (Max. 55% Filler)	24.0 Min. (Granules)	63.2 Min.	Mineral Surfaced Glass Fiber Cap Sheet	D-3909	---	---	1 Max.	70-88
1.5 Min.	20.0 Min. (Max. 60% Filler)	8.0 Min. (Granules)	50 Min.	Asphalt Coated & Mineral Surfaced Glass Fiber	D4897 Type I	22 Min.	22 Min.	1 Max.	70-88
1.7 Min.	22.0 Min. (Max. 60% Filler)	8.0 Min. (Granules)	55 Min.	Venting Base Felt	Type II	44 Min.	44 Min.	1 Max.	70-88
1.4 Min.	5.6 (Asphalt)	---	12.8 Min.	Asphalt Coated Glass Fiber Base Sheet	D4601 Type I	22 Min.	22 Min.	1 Max.	70-88
1.7 Min.	7.0 (Asphalt)	---	14.5 Min.		D4601 Type II	44 Min.	44 Min.	1 Max.	70-88

*Multiply lb f/in. by 0.175 to obtain N/mm.

TABLE 10.3 Typical Glass-Fiber and Organic Roofing Felt Properties (*Continued*)

FELT MILL		SATURATOR			COATER		FINISHED PRODUCT				
Dry Wt. lb/100 ft²	Gage mils	Sat. %	Sat. Wt. lb/100 ft²	Prod. Wt. lb/100ft²	Coating Wt & Surf. lb/100 ft²	Product Wt lb/100 ft²	Description	ASTM Number	STRENGTH lb/in. MD	CMD	ASH PERCENT
5.2 Min.	27	120 Min.	6.2 Min.	11.5 Min.	----	----	No. 15 Asphalt Saturated Roofing Felt	D226	30 Min. Type I	15 Min.	10 Max.
10 Min.	50	150 Min.	15 Min.	26 Min.	----	----	No. 30 Asphalt Saturated Roofing Felt	D226	40 Min. Type II	20 Min.	10 Max.
5.2 Min.	27	140 Min.	7.3 Min.	13 Min.	----	----	No. 15 Tar Saturated Roofing Felt	D227	30 Min.	15 Min.	----
5.2 Min.	25	140 Min.	7.2 Min.	----	18 Min.	37 Min.	No. 40 Asphalt Coated Base Sheet	D2626	35 Min.	20 Min.	----
10 Min.	50	160 Min.	16 Min.	----	18 Min.	54.6 Min.	No. 55 Asphalt Coated Roll Roofing	D224	45 Min. Type II	----	----
10 Min.	50	150 Min.	15 Min.	----	Granule Surfaced	74 Min.	No. 90 Asphalt Mineral Surfaced Roll Roofing	D249	----	----	10 Max.

around 220°F softening point, but ranging up to 260°F) and stabilizing with finely ground minerals (silica, slate dust, talc, dolomite, trap rock, or mica) to improve their durability and resist cracking in cold weather. Finely ground minerals—commonly talc, mica, or silica—are also dusted on the inside surface of coated felts as a releasing, or parting, agent to prevent contacting surfaces from adhering when the felt is unrolled. Coating a saturated felt greatly improves its moisture resistance.

Mineral-surfaced sheets are coated felts taken still another step in the manufacturing process. Mineral granules (colored slate, rock granules) are embedded in the weather-exposed surface. The heaviest mineral-surfaced roll roofing with organic felt weighs a nominal 90 lb/square. (It is consequently known as a 90-lb mineral cap sheet.) Glass-fiber cap sheets typically weigh 20 percent less, or 72 lb/square (32 kg/m^2).

Saturated felts generally serve as ply felts; coated felts serve as base sheets, although lightly coated fiberglass mats are widely used as ply felts; mineral-surfaced sheets serve as surfacing (cap) sheets. Grid-grooved mineral-surfaced sheets serve as venting base sheets. Both saturated and coated felts are also used in vapor retarders, and mineral-surfaced sheets are sometimes used as exposed base flashing.

Organic felts absorb water, with threatening consequences to membrane performance, because water absorption can promote such major premature failures as blistering, ridging, and splitting.

Surfacing

Hot-mopped, built-up roof membranes have four basic types of surfacing:

- Mineral aggregate (embedded in hot bituminous flood coat)
- Asphalt (hot- or cold-applied)
- Mineral-surfaced cap sheet
- Heat-reflective coatings

Embedded aggregate surfacing has this basic advantage: It makes possible the pouring of a heavyweight flood coat, an average of $\frac{1}{8}$ in. thick, three times the thickness of a standard interply mopping. Individual aggregate pieces, applied simultaneously with the hot, fluid bitumen, dam the flood coat to its thickened depth; this thickened flood coat enhances the membrane's waterproofing quality.

Aggregate surfacing provides other important benefits:

- Shielding from damaging solar radiation, which accelerates photochemical oxidation of bitumen by a factor of 200 and ultimately threatens the felts.

- Improved resistance to bitumen erosion from scouring action by wind and rain.
- Improved impact resistance (from hailstones, foot traffic, dropped tools, and falling tree limbs).
- Improved fire resistance (preventing flame spread or ignition by burning brands).
- Superior wind-uplift resistance. (A rough surface disrupts the laminar airflow patterns associated with uplift pressure, the 3- to 4-psf aggregate weight acts as partial ballast, and the thick flood coat stiffens the membrane, making it less vulnerable to "ballooning" and undulation.)
- Reduced roof-surface temperature compared with a black, smooth-surfaced membrane.

The last benefit has several advantages. Most obvious, it reduces cooling-energy consumption (or produces cooler temperatures in unairconditioned buildings). It also tends to reduce the deteriorative oxidation that embrittles bitumen. According to figures cited by R. L. Fricklas, gravel surfacing can reduce peak roof-surface temperature by 20°F or more.[6] A rise of 18°F roughly doubles an organic material's oxidation rate.[7] Reduction of roof-surface temperature can be especially important in climates characterized by prolonged heat, humidity, and rainfall because water dissolves oxidized asphaltic compounds, exposing fresh asphalt to the photo-oxidative attack of sunlight and direct attack by atmospheric oxygen. Aggregate surfacing thus shields the bituminous flood coat from harmful ultraviolet radiation and excessive heat, agents that combine to greatly accelerate the chemical deterioration of the bituminous flood coat.

Aggregate surfacing does, however, have some disadvantages compared with smooth-surfaced roofs. Because it obscures the underlying felts, aggregate surfacing makes it more difficult to spot membrane defects (for example, small, growing blisters; ridges; fishmouths; curled felts; even splits). Repairing an aggregate-surfaced membrane is also more difficult than repairing a smooth-surfaced membrane, partially offsetting its advantage in generally requiring fewer repairs. And if an aggregate-surfaced membrane requires tearoff-replacement, it presents a much tougher problem than a smooth-surfaced roof. Disposing of the old aggregate can be extremely difficult and expensive. Slag or gravel sticks to workers' bitumen-covered shoes and equipment wheels; carried back to the new work area, it can get trapped between new felts, forming voids (the origin of blisters) and puncturing the felts. Entrapped aggregate within a new membrane almost certainly portends shortened service life, increased problems, and repair bills.

The maximum 3-in. slope limit generally recommended by roofing manufacturers for aggregate-surfaced membranes is designed to prevent both the heavy flood coat and the aggregate from sliding down the slope during application or during hot weather.

In addition to strength, hardness, and durability, aggregates require the following two basic properties:

- Opacity
- Proper sizing and grading

Opacity is an important property of surfacing aggregates. Translucent aggregates that permit passage of solar radiation lack a major feature of good surfacing aggregate. Carl G. Cash's research on roofing aggregates demonstrates the potential hazards of translucent aggregate. Comparing a highly opaque aggregate with a highly translucent quartz aggregate in test panels exposed for 3 months (September through November 1978), Cash discovered 100 percent light-stained area on felts covered by the translucent aggregate, compared with only 3 percent light-stained area for the felts surfaced with opaque aggregate.[8] After 3 months' exposure, felts surfaced with translucent quartz aggregate exhibited the same stain intensity as unsurfaced, totally exposed "control" felts after 1 month's exposure. As a shield against solar radiation, translucent aggregates thus appear virtually useless.

Proper sizing and grading of aggregates are needed to ensure proper nesting, which is vital to continuity of surface protection and continuous flood coat. ASTM Standard D1863, "Standard Specification for Mineral Aggregate Used on Built-Up Roofs," sets a maximum size of $3/4$ in., with a desired minimum of $3/16$ in., although permitting a small percentage (by weight) to pass through a No. 8, 2.36-mm ($3/32$-in.) sieve.

Undersizing of aggregate is more serious than a comparable degree of oversizing. Excess fines in undersized aggregate spread into hot, fluid bitumen sink into the flood coat. Submerged, these fine aggregate particles perform no protective function whatever, only the negative function of interrupting the continuous waterproofing film. On the surface, loose, undersized aggregate is less stable than larger, properly graded aggregate under the action of wind, flowing water, and roof traffic. This reduced stability also detracts from the aggregate's protective function as a shield against ultraviolet radiation.

Undersizing also reduces aggregate embedment, a fact established by Cash's cited research. The quantity of adhered aggregate for laboratory test samples was a direct linear function of mean aggregate diameter between the limiting sizes of 2 mm (0.08 in.) and 12 mm ($1/2$ in.). Adhered aggregate dropped from 300 lb/square embedment for the 12-mm mean aggregate diameter to 150 lb/square for the 2-mm aggregate. Moreover, because field-test samples generally had less adhered aggregate than laboratory test samples (up to 60 lb/square flood-coat weight), field-applied aggregate may suffer even greater embedment loss from undersizing than laboratory-prepared specimens. (ARMA requires a minimum adhered weight of 200 lb/square for gravel, 150 lb/square for slag.)

Oversized aggregate, a less common, less severe problem than undersized aggregate, can nonetheless cause trouble. Oversized interstices between pieces break the continuity of the surface protection. In the vertical plane, these larger interstices may leave voids that can entrap silt and promote plant growth.

The most common roof aggregates are river-washed gravel, crushed stone, and blast-furnace slag, a fused, porous substance separated in the reduction of iron ore, chemically comprising silicates and aluminosilicates of lime.

Slag is an excellent aggregate. Its opacity provides superb shielding from solar radiation. Its pitted, porous surface provides excellent adhesion with the cooled bitumen, which, as a hot fluid, flows into the slag's surface cavities and forms keys that lock the aggregate in place. The slag's alkaline surfaces also improve adhesion because of their greater affinity for bitumen than for water. Although gravel is a good aggregate, its smooth, often rounded surfaces provide less surface area for bonding, thus making it inferior to slag.

Roof designers should approach roofing aggregates other than gravel, slag, and crushed stone with intense suspicion. Georgia roofing-grade white marble chips may satisfy ASTM requirements, but a host of occasionally used aggregate materials do not. Dolomite (marble chips) can provide a good heat-reflective surface, but it has several flaws. Small translucent chips may admit damaging solar radiation that can degrade the flood-coat bitumen. The resulting embrittlement can loosen the aggregate, exposing more bitumen to photo-oxidative reactions and accelerating the membrane's deterioration. Moreover, dolomite chips are often coated with dust, which may weaken the aggregate bond to the bitumen. Excessive quantities of salt or free alkaline material also disqualify aggregates as suitable membrane surfacings.

Other occasionally used, but highly dubious, surfacing aggregates include limestone, volcanic rock, crushed oyster and clam shells, crushed brick, tile, or cinders. The crushed coral rock used extensively in southern Florida is a similarly dubious surfacing aggregate, consistently undersized. Canadian researchers D. C. Tibbetts and M. C. Baker warn against the temptation to use what is *available* instead of what is *suitable*.[9] The first-cost saving may be repaid many times over in reduced membrane service life.

Full-weight bitumen flood coat is required to bond an adequate quantity of aggregate to the membrane surface. Cash's cited research, paralleling earlier research by Jim Walter researchers, demonstrates this requirement. For asphalt field samples, adhered aggregate ranged linearly downward from about 270 lb/square embedded aggregate for a 60-lb asphalt flood coat to less than 90 lb/square embedded aggregate for a 20-lb flood coat (see Fig. 10.4).

Skimping on the flood coat thus compounds the problem of constructing a good, weathertight membrane. Not only does it reduce the thickness of the water-resistant bitumen film, it also impairs the shielding of this film by sufficient aggregate coverage, exposing the bitumen to damaging solar radiation and general weather erosion.

A double-pour aggregate surfacing, featuring two flood-coat applications, can solve several problems with aggregate surfacing. Blast-furnace slag as the lower surfacing exploits the superior embedment and opacity attainable through using this porous material, thus providing basic protection. In the top flood coat, where exposed bitumen is a lesser threat to the membrane's integrity than in the bottom flood coat, white marble chips, or similarly light-colored aggregate, can provide superior heat reflectance. This could more than pay for its initial cost in cooling-energy savings (from reduced surface temperature) and superior weatherproofing. (Double-pour membranes are required for deliberately ponded, water-cooled built-up membranes.)

Double-pour aggregate surfacing is credited as the chief reason for the excellent performance of built-up roof membranes at the U.S. Naval Base in Guam, where these specially designed roof systems have given 20 years or more of excellent service. Coral-gravel aggregate is washed with water, dried, and then sprayed lightly with diesel fuel (as a primer) before it is applied. This unique policy was instituted after typhoon winds had persistently swept aggregate from single-pour aggregate-surfaced roofs. The unwashed coral gravel used in these single-pour roofs is dusty and thus difficult to embed solidly in a bituminous flood coat.

An insufficient quantity of embedded aggregate, below 50 percent of weight total, often results from tardy spreading of aggregate—after the flood-coat bitumen has congealed, when it has lost the required fluidity for adhesion with the aggregate. Excessive moisture and dust can also reduce the quantity of embedded aggregate. Acceptable limits for these aggregate contaminants are not, however, as severe as formerly believed, according to research by Cash.[10] Up to 2 percent dust (defined

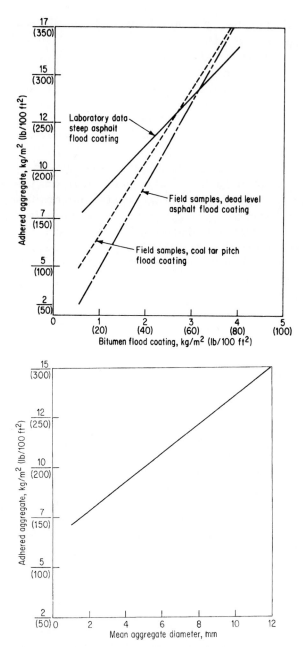

Figure 10.4 Tests reported by Carl G. Cash indicate linear relationship between flood-coat thickness and weight of adhered aggregate (top). Adhered aggregate also varies linearly with mean aggregate diameter (bottom). (*From Carl G. Cash, "On Builtup Roofing Aggregates," presented to ASTM D8 Main Committee, Dec. 6, 1978.*)

as aggregate passing a No. 200 mesh sieve) and 2 percent moisture caused no detectable loss of aggregate adhesion on the laboratory test samples. Mean adhered aggregate was roughly the same for 0, 0.5, 1, and 2 percent dust or water content. Furthermore, 2 percent water content makes the aggregate surface palpably wet. ASTM Standard D1863-80 accordingly permits maximum 2 percent moisture for crushed stone and gravel, 5 percent for slag, and 2 percent dust content.

Is the flood coat of an aggregate-surfaced built-up membrane the membrane's key waterproofing element? "No," said E. H. Rissmiller, former Senior Research Associate at the Celotex Corporation, as his comment on this old roofing-industry controversy. According to Rissmiller, this flood coat's major purpose is to bond the aggregate to the membrane. It does provide some additional waterproofing, but it cannot function as the principal waterproofing agent because its film is broken by the following factors:

- Embedded aggregate projecting through the flood-coat film
- Air and moisture bubbles entrapped during application of the aggregate
- Alligatoring and cracking due to weathering and temperature changes
- Thin spots where the bitumen flows off ridges
- Roof traffic and impact from falling objects

This theory highlights the importance of uniform, continuous interply moppings which, by general consensus, form the foundation of a good built-up roof membrane. The poured topcoat can be corrected if it is deficient. But repairing deficient interply moppings, which can also cause premature membrane failure, is much more difficult.

Regarding whether the interply moppings or the flood coat is *the* waterproofing element, the safest answer is to assume the indispensability of both. Built-up membranes require good flood coats *and* good interply moppings to ensure their watertight integrity.

Smooth-surfaced membrane

A *smooth-surfaced membrane* is generally topped with 15 to 25 lb/square of hot steep asphalt or cold-applied cutback or emulsion applied to an asphalt-saturated inorganic felt at a spread of 25 to 50 ft^2/gal (0.6 to 1.2 m^2/L). A smooth-surfaced roof offers several positive advantages:

- Easier inspection, maintenance, and repair than an aggregate-surfaced membrane

- Easier installation of new openings [for vents; exhaust fans; heating, ventilating, airconditioning (HVAC) units; and so on] in buildings subject to changing uses
- Easier cleaning than aggregate-surfaced roofs (important for paper mills and food-processing, fertilizer, and other industrial plants whose stack exhausts settle on the roof)
- Easier reroofing or replacement at the end of a roof's service life
- Slight reduction in dead load (by 3 to 4 lb/ft^2 aggregate weight)

As discussed, however, the disadvantages usually outweigh the advantages. Foremost among these disadvantages, a smooth-surfaced roof is a *water-shedding* rather than a *water-resistant* roof. All membranes, even those designed for deliberate ponding of water, should drain. But whereas aggregate-surfaced roofs can tolerate some intermittent ponding, a smooth-surfaced roof should be designed for rapid runoff—generally with a slope of at least ¼ in./ft, but preferably ½ in./ft.

Because of their extreme vulnerability to moisture, organic felts are unsuitable for smooth-surfaced roofs. Glass-fiber felts make the best smooth-surfaced membranes.

Mineral-surfaced roll roofing

Mineral-surfaced cap sheets (roll roofing) are also suitable for roofs with minimum ½-in. slope. A minimum 1-in. slope requirement should be considered an absolute for organic roll roofing felts. In the military services' "roofing bible," former NBS researcher W. C. Cullen recommends a minimum 3-in. slope for asphalt roll roofing (application parallel to eave, with concealed nailing), with a 2-in. minimum for emergency construction.

Heat-reflective surfacings

Heat-reflective surfacings perform three vital functions:

- Reduce cooling-energy consumption (especially important in warm climates with long cooling seasons)[11]
- Protect membrane bitumens from deteriorating photo-oxidative chemical reactions that accelerate exponentially with increasing temperature
- Reduce the rate of blister growth

This last benefit depends on two factors associated with reduced membrane temperature: (1) reduced internal pressure inside the blister void and (2) increased bitumen cohesive strength and cleavage resistance at lower temperature. Rising temperature produces both an exponential growth of pressure within a vapor-saturated blister chamber and an exponential decline in bitumen strength. Thus it is conceivable that heat-reflective coatings can even prevent blister formation in some critical instances; i.e., the reduced internal pressure attributable to the heat-reflective coating may be too slight to pry the adhered felts apart around the perimeter of a potential blister-forming void.

Color is the key to heat-reflective cooling. Under a sunny, summer sky, with air temperature 90°F, a black roof surface might climb to 170°F, compared with 135°F for an aluminum-coated surface, and still lower for a white surface.

Heat-reflective surfacings are available for all three types of built-up surfacing. Heat-reflective aggregates include white marble chips and other light-colored aggregates. Gray slag cuts the peak roof temperature by about 15°F from a black-surfaced membrane's temperature. Heat-reflective cap sheets are surfaced with light mineral granules. For smooth-surfaced roofs, aluminum or other light, pigmented coatings can be brushed, rolled, or sprayed.

Built-up Membrane Specifications

Today's variety of membrane specifications evolved from two basic ancestors:

- The older five-ply wood-deck specification, with the two lower plies nailed to the deck and the top three mopped
- The four-ply concrete-deck specification, with the base ply mopped to the deck and all four plies mopped

Subsequent modifications ultimately reduced the number of felt plies from four or five saturated felts to two coated felts, in some instances. With the widespread failure of the two-ply membranes, the trend has swung back toward the more traditional three- and four-ply systems.

The typical modern built-up roof specification over rigid insulation board calls for all shingled plies—preferably a minimum of four (three in mild climates). Over a lightweight concrete insulating fill or a gypsum deck, the typical specification calls for a nailed coated base sheet plus a minimum of three shingled plies.

Application of the felt starts at an edge, or low point (so that the shingled joints will not "buck" water), with a felt cut into strips whose widths are an even factor of the 36-in. felt width divided by the number of plied felts. Thus for three-plied shingled felts (a four-ply membrane, including base sheet), the edge strips are 12, 24, and 36 in. wide.

Next come the regular 36-in.-wide shingled felts. The overlap dimension is computed by dividing the felt width minus 2 by the number of shingled plies. Thus for three shingled piles, the overlap = $(36-2)/3 = 11\frac{1}{3}$ in., which is also the distance of the first sheet from the edge.

This overlap ensures that any vertical cross section always has at least the minimum number of required plies.

Shingling of felts

Shingling of felts is a universal practice in built-up roofing membranes. It originated as a means of facilitating and accelerating application of the membrane. Overlapping the felts enables the roofer to complete the membrane as the roofing crew proceeds across the roof, instead of applying alternating layers of felt and bitumen across large areas, or even over the entire roof, in separate stages (see Fig. 10.5).

Shingling of felts may reduce the membrane's resistance to hydrostatic pressure. Waterproofing membranes for foundations, walls, and concrete-topped basement roofs are often applied with alternating laminations, with an entire ply placed and mopped before the next ply is applied. These parallel planes of mopped interply bitumen probably provide better waterproofing defense than the moppings between shingled-felt plies. Shingled plies can slowly wick moisture from an exposed felt edge diagonally down through the membrane to the base sheet—or, in the case of a totally shingled built-up membrane, to the insulation substrate. Defective application, e.g., a fishmouth at a felt lap, provides a direct path to the substrate. Waterproofing membranes laid ply-on-ply are not vulnerable to this diagonal moisture penetration through the entire membrane cross section.

Accompanying this singular liability of shingling are two important advantages:

- Improved slippage resistance
- Reduced risk of poor interply mopping adhesion

Improved slippage resistance results from the shingled felts' structurally integrated construction. Phased, or laminar, construction promotes slippage because it provides a continuous, unbroken film of bitumen between felts. When a phased built-up membrane slips, the slippage invariably occurs at the phased plane. Shingling provides slightly more frictional resistance (equal to the slippage force times

Elements of Built-up Membranes 229

Figure 10.5 Felt-laying (top); flood-coat application (bottom). (*Roofing Industry Educational Institute.*)

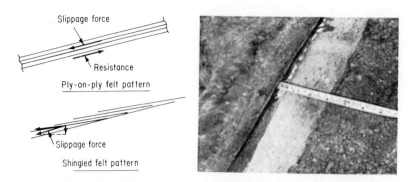

Figure 10.6 With an unshingled, ply-on-ply felt pattern (top drawing), slippage force parallel to the slope is resisted only by the horizontal shearing strength of the asphalt mopping film (unless the felts are nailed). A shingled felt pattern (bottom drawing) provides additional friction resistance, represented by the coefficient of friction multiplied by the slippage force and the sine of the tiny acute angle between the shingled felts and the roof-deck plane. A 3-in. slippage occurred on the 1-in. sloped, smooth-surfaced roof (photo) because the roofer omitted the specified nails to anchor the felts. Shrinkage of wet felts and overweight moppings evidently aggravated the problem.

the sine of the acute angle between the shingled felts and the membrane substrate plane) in addition to the pure parallel shearing resistance (see Fig. 10.6). Because the bitumen's tensile resistance exceeds its horizontal shearing resistance, a shingled-felt membrane resists slippage better than a totally laminated ply membrane or even a membrane with one laminated ply formed by phased application, usually a base plus two or three shingled plies.

Reduced risk of poor interply mopping adhesion results from the slower cooling rate of several simultaneously applied bituminous moppings compared with the cooling rate of individually applied moppings. With three or four shingled felts, there is three or four times as much hot bitumen as with application of merely a single felt, and this greater mass of hot bitumen cools at a significantly slower rate than a single mopping.[12]

Coated base sheet

Using a coated base sheet as the membrane's bottom felt ply became an industry standard for organic and asbestos felt membranes in response to widespread ridging, or "picture-framing," which afflicted the industry in the 1960s. Water vapor migrating upward through board-insulation joints can penetrate into saturated felts, condense, and cause ridging. Coated felts, with their added waterproofing film, can better resist this moisture penetration than saturated felts. As a highly impermeable asphalt film on the membrane's underside, the base sheet is applied as a single ply, not shingled with the ply felts. (Shingling the base sheet with the other felt plies would obviously defeat its purpose because it would no longer form a continuous coated-felt bottom surface.)

For some years the coated base sheet remained an industry standard for asphalt organic and asbestos felt membranes (but not for glass-fiber-felt membranes, which are less vulnerable to ridging). In the basic nailable wood-deck specification, one coated base sheet replaces two nailed saturated felts.

In recent years the trend has swung away from use of the coated base sheet toward the old specification, with all saturated, shingled felts, simultaneously applied. Because of their lower permeability and their superior water-absorptive resistance, coated base sheets still predominate in membranes applied over wet decks—lightweight insulating concrete, poured gypsum, or structural concrete—and also over nailable wood decks, where they have better nail-holding strength. But many specifiers have returned to the old saturated, all-shingled ply system for membranes over board insulation, for several reasons:

- Coated felts applied in a single operation (i.e., unshingled) are subjected to more rapid cooling of hot-mopped asphalt, with consequently higher risk of membrane thermal-contraction splitting, slippage, or blistering resulting from mopping voids.

- The high softening point of the coated-film asphalt (up to 260°F compared with the 100 to 160°F softening-point saturant in saturated felts) may shorten the time allowed for bonding the felt to the substrate with the hot asphalt. This shortened adhesion time also heightens the probability of mopping voids and consequent blistering.

- Phased application, in which the various membrane felt plies are installed in separate operations (often on different days) as opposed to the simultaneous shingling operation, heightens the risks of slippage and blistering. (See the following section, "Phased Application," for further discussion of this problem.)

Phased application

In phased application, the felts are applied in two (or more) operations, with a delay between operations. At the felting break plane, the felts are necessarily unshingled. This break plane usually occurs between application of a base felt (usually coated, as indicated in the preceding section) and the top three plies of a four-ply membrane.

To some roofing experts, phased application refers to *any* delay in the field fabrication of the membrane—including the application of aggregate surfacing and a delay in placing the top felt plies. But to others, phased application refers only to a break in the felt application. In this manual, "phased application" (or "phased construction") refers to a break in felt application; "delayed surfacing" refers to a delay in flood-coat and graveling-in operations.

Phased application offers convenience to a roofing contractor; while the coated base sheet serves as a temporary roof, the contractor can spread the workers around many roofing projects on good working days and get buildings "in the dry." But if the delay in completing the felt application lasts even overnight, it threatens the membrane's integrity, and the longer the felts remain exposed, the greater the risk. Although less for coated felts than for saturated felts, this risk nonetheless applies to all felts.

Phased felt application poses the following hazards:

- Heightened risk of slippage, which, in a phased membrane, always occurs at the bitumen plane between base sheet and shingled felts

- Heightened risk of interply blistering at the base sheet's top surface, from mopping voids caused by condensation (dew) or rain-deposited water on this exposed felt surface (particularly at a fishmouthed lap joint, where water can wick)

Some roofers have a cavalier attitude toward this latter hazard, arguing that the hot-mopped bitumen will evaporate all surface moisture, or even felt-absorbed moisture, when it hits the felt. Like most wishful thinking, this belief is totally false. You can count on a practical maximum of 50 percent moisture evaporation, according to research by T. A. Schwartz and Carl G. Cash, of Simpson, Gumpertz & Heger.[13] Because the Schwartz and Cash tests were performed in a laboratory at room temperature, the percentage of dissipated moisture doubtless declines in winter, when hot-mopped bitumen cools rapidly. Entrapment of undissipated moisture in moppings' voids creates the classic condition for blister formation and subsequent growth (see later discussion of blistering mechanics in this chapter).

Other hazards from phased application include the following:

- Exposing organic felts to condensation, rain, or even high relative humidity can cause the felts to expand from moisture. Moisture absorption can curl the felts' edges as the exposed side dries and contracts (see Fig. 10.7). These curled edges form open conduits for water entry by wicking action. They sometimes protrude through the flood coat and aggregate surfacing. Although saturated felts

Elements of Built-up Membranes 233

Figure 10.7 Classic shark fin (at end of tape measure) started as a curled organic felt, which forms a conduit wicking water into the BUR membrane. (*Roofing Industry Educational Institute.*)

are most vulnerable, even coated felts eventually absorb moisture if exposed long enough.
- Uncompleted membranes are more vulnerable to traffic damage than completed, surfaced membranes.
- Felts left unsurfaced for phased application accumulate dust that can weaken adhesion of the next interply mopping.

To avert the foregoing problems with phased application, the roofer often applies a glaze coat, a light bituminous mopping designed to protect the felts until they get their final surfacing. This expedient helps, but the glaze coat presents an easily damaged traffic surface, tacky in summer and brittle in winter.

This glaze coat, whether applied to the base sheet or to the top felt surface awaiting flood coat and surfacing aggregate, should be kept as thin as practicable, to prevent alligatoring. One major manufacturer recommends squeegeeing the glaze-coat surfacing to maintain it at 10 lb/square, roughly 0.02 in. thick.

In contrast with their unanimity in disapproving of phased felt application (with a day or more delay in membrane completion), roofing experts are divided on their opinion of phasing flood coat and aggregate surfacing. Fiberglass membranes can safely remain unsurfaced for up to 6 months, according to David Richards formerly of Owens-Corning Fiberglas Corporation. Even for organic membranes, there are benefits in delaying flood-coat surfacing operations until the completion of felt-laying instead of completing each membrane segment—from felt-laying through graveling-in—in the same day's work. According to architect Justin Henshell, delaying the graveling-in operations has two important advantages:

- Less chance of tracking aggregate into new areas being felted, thereby avoiding the blistering hazard from voids created by entrapped aggregate pieces
- Less chance (or temptation) to use the wrong flood-coat asphalt (Type III, instead of the normally correct Type I for low-slope roofs) because the roofer can then order the correct Type I asphalt for one continuous flood-coating operation

A compromise, allowing 4 days' maximum delay between felt application and flood-coat surfacing application, is recommended by R. J. Moore of ARMM Consultants, Gloucester City, New Jersey. This policy avoids several hazards of same-day completion and delaying flood-coat-surfacing until the total completion of the felt application.

Like so many other aspects of built-up roofing design and application, this problem requires compromises and trade-offs, regardless of which policy is chosen. There are many matters on which there is an overwhelming consensus. But when to schedule graveling-in operations is definitely not one of them. As a key requirement whenever surfacing aggregate application is delayed, the exposed felts must be glazed for temporary protection against moisture invasion. This glaze coat should be limited to 10 lb/square.

Temporary roofs

A temporary roof, comprising built-up layers of felt or a heavy, coated base sheet, is advisable when the job is hampered by any of the following conditions:

- Prolonged rainy, snowy, or cold weather
- Necessity of storing building materials on the roof deck
- Mandatory in-the-dry work within the building before weather permits safe application of permanent roofing system
- A large volume of work on the roof deck by tradespeople other than the roofer

A minimum thickness of insulation must be applied over fluted steel decks as a substrate for the temporary roof.

Temporary materials, including insulation, should generally be removed before the permanent roofing is installed. The risk of damaging the temporary roof, subjected to a heavy volume of construction traffic, is too great to warrant its use as a vapor retarder or bottom plies of the permanent roof. Owners who save the costs of replacing temporary roofing materials are gambling small current gains against potentially large future losses.

Some experienced roofing contractors, however—notably J. Roy Martin of Columbia, South Carolina—report satisfactory results repairing a felt-asphalt vapor retarder and leaving it in place after it served its stint as a temporary roof. Provision for drainage at the vapor-retarder plane both drains the temporary roof and then, after insulation and membrane are installed, disposes of water that invades the roofing sandwich.

Some manufacturers similarly allow for repair of temporary roofs left in place as vapor retarders. To justify this practice, however, requires *extraordinary*—in the literal sense of *more-than-ordinary*—care in the repair operations, with rigorous inspection and identification of punctures and other defects, removal of dust and general cleanup, and scrupulous follow-through on repair work.

Joints in built-up membranes

Roof expansion-contraction joints accommodate movement from thermal expansion or contraction, from drying shrinkage of poured decks, or from structural movement—e.g., deck bearing-seat rotation—without overstressing and cracking, buckling, or otherwise impairing the roof system.

Expansion-contraction joints should allow for movement in three directions: perpendicular and parallel to the joint in the horizontal roof plane, and perpendicular to the roof in the vertical plane. Roof expansion-contraction joints are recommended at the following locations:

- Junctures between changes in deck material (e.g., from steel to concrete)
- Junctures between changes in span direction of the same deck material
- Junctures between an existing building and a later addition
- Deck intersections with nonbearing walls or other surfaces where the deck can move relative to the abutting wall, curb, or other building component
- Maximum distances of about 200 ft

- Junctures in an H-, L-, E-, U-, or T-shaped building

Expansion-contraction joints are mandatory at changes in deck material or span direction because of the need to avoid relative movement in both the vertical and horizontal planes, where movement can occur parallel as well as perpendicular to the joint line. An expansion-contraction joint at these critical lines protects the membrane from a complex combination of concentrated stresses: splitting or cracking from joint rotation at deck or joist bearings, from differential vertical deflection (i.e., shearing forces), from axial separating forces, or from horizontal shearing forces that may develop along the intersection of two different deck materials or along the line where the same deck material changes span direction.

In lightweight insulating concrete decks designed for composite structural action between corrugated steel-deck centering and concrete fill, a line of changing deck direction presents a drastically weakened cross section. A 26-gage (0.02-in.) corrugated high-strength steel sheet provides nearly 100 times as much tensile strength (roughly 2000 lb/in.) as 3-in.-thick lightweight insulating concrete. A line where this steel is omitted is consequently highly vulnerable to concrete tensile cracking. And wherever the concrete cracks, the membrane is also likely to crack along this line of stress concentration.

The expansion-contraction joints just discussed are building, not merely roof, expansion-contraction joints. They generally continue through the entire building superstructure, from rooftop to foundation, with a minimum 1-in. width. Double, parallel columns and beams are required to ensure independent movement of the integral structural segments on each side of the expansion joint.

The building designer should consider the roof when designing building expansion-contraction joints. Most recommendations for joint spacing set a higher figure than the approximately 200 ft recommended, but these recommendations generally ignore the special requirements of the roof. Contraction, not expansion, is the major problem of the roof system. (For this reason, expansion joints are appropriately termed "expansion-contraction" joints.) Closer spacing, especially in colder climates, is warranted by the roof's extreme exposure. In insulated airconditioned buildings, the structure experiences perhaps 30°F seasonal and 10°F maximum daily temperature differential, but the roof membrane may experience nearly 200°F seasonal and 100°F daily temperature differential.

However, the chances of membrane-stress concentration increase with the distance between expansion-contraction joints. At 0°F, a built-up membrane's ultimate strain drops to about 1 percent (except for some glass-fiber-mat membranes), and its coefficient of thermal expansion-contraction increases rapidly in the $+30°$ to $-30°F$ temperature range. With increasing distance between joints comes a corresponding expanse of roof for any loose boards to move. This greater potential for lateral movement increases the risk of stress concentration—for example, where a curbed opening restrains an insulation board from accommodating itself to movement in other nearby boards, thus widening a joint and producing a membrane-stress concentration (see p .257).

Rotation of roof beams from live-load deflection can split the built-up membrane if the members are relatively deep. For a uniformly loaded span with $L/360$ live-load deflection, the end rotation at a simple span (i.e., noncontinuous) support is 0.51°. If the beam—say, a prestressed T—is 18 in. deep, that rotation offsets the top of the beam by tan $0.51° \times 18 =$ 0.16 in. If another similarly loaded beam bears on the opposite side of the support, the total movement equals $5/16$ in. In cold weather, that movement could split most built-up membranes, with their ultimate elongations of less than 1 percent. As a consequence of end rotation, a roof-system expansion-contraction joint may be required along a line where simple-span roof beams sit on a bearing wall or girder.

Membrane control joints are not recommended. A control joint is essentially a roof expansion-contraction joint over a continuous structural deck, i.e., a location where no building expansion-contraction joint occurs. The entire concept of a control joint is based on an erroneous theory of membrane behavior: that the built-up bituminous membrane behaves elastically, like steel, with completely reversible thermal movement, expansion as well as contraction.

A built-up membrane is not elastic, but *viscoelastic*. As time passes, tensile stresses at service loadings disappear at constant strain. Cold-weather membrane contraction has no counterpart hot-weather expansion, except in minor thermal ridging of the flexible membrane. Consequently, an aging membrane tends to experience cumulative shrinkage.[14] Because of the membrane's long-term shrinkage, a control joint's ultimate position is open. It is obviously highly vulnerable to leakage because it is in the roof plane.

In place of a control joint, use a *roof area divider* or *relief joint*. Unlike a control joint, a roof area divider anchors the membrane. It is especially useful for eliminating stress concentration at reentrant (270°) corners. Locate these joints at high points, with drainage away from the joint in both directions.[15]

Membrane Performance Standards

The 1974 publication, "Preliminary Performance Criteria for Bituminous Membrane Roofing" (*Build. Sci. Ser.* 55, NBS), briefly described in Chapter 2, promulgates 20 performance "attributes" required by built-up membranes:

- Tensile strength
- Notch tensile strength
- Tensile fatigue strength
- Limited creep
- Flexural strength
- Flexural fatigue strength
- Pliability
- Shear strength
- Impact resistance
- Ply adhesion
- Wind-uplift resistance
- Limited thermal expansion-contraction
- Limited moisture expansion
- Limited moisture effects on strength
- Abrasion resistance
- Tear resistance
- Impermeability
- Weather resistance
- Fire resistance
- Fungus-attack resistance

From this list of required attributes, Cullen and Mathey developed a "performance format" of 10 criteria, each with a stated requirement, criterion, test method, and commentary. A major goal of these preliminary performance criteria is to aid manufacturers developing new built-up membranes by providing scientific means of evaluating the membrane's prospective performance during its service life, without waiting 20 years for retrospective evaluation. But inevitably these criteria, preliminary though they are, have been used by consultants and some designers, on the sound premise that something, although admittedly imperfect, is still better than nothing but past experience, the traditional guide for roofing technology.

Let us consider several criteria to demonstrate the NBS modus operandi. First, take the simple tensile-strength requirement of 200 lb/in. in the membrane's weakest (i.e., transverse) direction. From three different sources—field observations and large- and small-scale tests of many commercially marketed membranes—Cullen and Mathey established 200 lb/in. as the minimum tensile strength required by a *new* membrane to resist external and internal stresses imposed during a typical membrane's service life. For this particular criterion there was a standard ASTM test available: ASTM D2523, for testing built-up membranes' load-strain properties.

For the related requirement of "tensile fatigue strength," however, there was no standard ASTM test, and so the NBS researchers devised their own criteria (100,000 cycles of repeated 20 lb/in. at 73°F and 100,000 cycles of repeated 100 lb/in. force at 0°F). Hailstone-impact resistance is another performance requirement for which NBS developed its own standard test, reported in *Hail-Resistance Test, Build. Sci. Ser.* 23, NBS.

The performance criterion generating the most confusion and controversy is doubtless thermal shock factor (TSF), discussed in several other sections of this manual. TSF is a *calculated* criterion, set at 100°F, for the temperature drop that must be theoretically resisted by a membrane test sample clamped at constant length. The formula for calculating TSF is

$$\text{TSF} = \frac{P}{M\alpha}$$

where TSF = temperature drop (°F)
P = tensile strength (lb/in., 0°F)
M = load/strain modulus (lb/in., 0°F)
α = coefficient of thermal expansion-contraction from 0 to $-30°F$ [in./(in. · °F)]

No one claims that this TSF precisely parallels the complex thermal stresses experienced by a roof membrane in service. The field membrane is normally restrained by its deck or insulation substrate. Moreover, a field membrane is seldom, if ever, *totally* restrained during a large temperature drop.

Objections that the TSF concept fails to simulate the field membrane's thermal behavior miss the point. There are numerous situations on actual roofs when the membrane's substrate moves. Additional contraction stress may act in combination with thermal stress. The TSF is merely an index for comparing different membranes. With other factors assumed equal, a membrane with a higher TSF resists thermal-contractive stress better than a membrane with a lower TSF.

There is another point to be made about TSF. To objections that a 100°F temperature drop is too severe, remember that this criterion applies to a *new* membrane. NBS studies of existing membranes, discussed elsewhere in this manual, indicate that aging deterioration of the membrane—strength loss and increase in thermal coefficient and stiffness as the bitumen age hardens—reduces even a good membrane's TSF by as much as 70 percent. The performance criterion for roofing, like a load factor for a structural beam, column, or truss, contains a safety factor. Moreover, for mild climates, free of temperature drops into the subfreezing or subzero range, the 100°F TSF criterion can be waived, according to Cullen and Mathey.

Flexural and tensile fatigue testing of typical four-ply built-up membranes, reported in a subsequent NBS research publication, is designed to demonstrate the membrane's durability under repeated cyclical loading. Flexural fatigue loading spans a spectrum from low-cycle, large-amplitude flexing from foot traffic to wind-induced, high-frequency oscillations of low amplitude. Thermal cycling exemplifies one kind of tensile fatigue, but there are others: e.g., long term tensile cycling from alternating contraction and expansion of membrane felts, a process that tends, however, toward permanent shrinkage.

BUR Failure Modes

The gradual replacement of organic (and the now extinct asbestos) felts with glass-fiber felts has changed the failure modes of BUR membranes. Up until the mid-1980s, blistering was by far the biggest problem with BUR membranes, followed by splitting and ridging, respectively. Because of their porosity, glass-fiber felts are less likely to retain the air–water-vapor mixture that, under thermally induced pressure, creates blisters. Far less permeable than glass-fiber felts, organic (and asbestos) felts retain pressure within interply mopping voids, the source of blisters. Glass-fiber felts also make a built-up membrane less liable to splitting. (Their greater tensile strength, especially in the transverse direction, and their resistance to the moisture-induced weakening that afflicts organic felts give glass-fiber felts this superior splitting resistance.) Ridging, which results from the moisture-induced expansion of organic (or asbestos) felts, is also less common in BUR membranes with glass-fiber felts.

Surveys of failure modes indicate these changes, but less decisively than might be anticipated. In the last year for which NRCA Project Pinpoint data are available (1990), blistering and splitting account for virtually the same proportion (22.8 percent and 22.3 percent, respectively) of BUR problems.[16] Blistering, however, has declined from its former predominance. Back in the late 1970s, blistering accounted for roughly twice as many problems as splitting, according to an NRCA survey.[17]

The convergence of blistering and splitting incidence in little more than a decade can be readily explained. The benefits of glass-fiber felts in reducing BUR blistering are considerably greater than their undeniable, yet more limited, benefits in reducing splitting. (Splitting results primarily from stress concentrations produced by insulation movement, which can exert stresses that cannot be resisted by any

BUR membrane.) The relatively slow improvement in the performance of BUR membranes as a whole is even more readily explained. Though glass-fiber felts were dominant well before the end of the 1980s, the prevalence of organic- and even asbestos-felted BUR membranes in the total stock of BUR membranes in service perpetuates the maladies of these membranes and thus distorts the statistics by *understating* the beneficial effects of glass-fiber felts. The *full* beneficial effects will not be known until the old organic-felted and asbestos-felted BUR membranes are removed from service. BUR membrane failure modes listed for 1990 (exclusive of flashing failures, which accounted for roughly 20 percent of BUR system problems) come in the following order:

- Blistering
- Splitting
- Ridging/wrinkling
- Slippage

The following pages repeat much of the second edition's discussion of various BUR maladies, since they still occur on old BUR membranes with organic and asbestos felts. Readers who need a fuller discussion should refer to this earlier edition, *The Manual of Built-up Roof Systems,* 2d ed., McGraw-Hill, 1982.

Blisters

Blisters can range in size from barely detectable spongy spots to bloated humps 6 in. high and 50 ft^2 in area (see Fig. 10.8). A few minor blisters are tolerable. If the blisters remain small and few in number, the BUR membrane can readily live out its full service life, although it may require minor repair.

For a BUR membrane with numerous large blisters, however, the prognosis is shortened service life. Blisters drastically increase the membrane's vulnerability to physical and chemical degradation. A large blister is easily punctured by foot traffic, dropped tools, and wheeled equipment loads. A blister's sloping sides make gravity a relentless agent of deterioration. Erosion of aggregate surfacing exposes the bitumen flood coat to ever-renewed photochemical oxidation, which accelerates to a rate 200 times faster in sunlight than in the dark.[18] Deterioration of the exposed bitumen, which develops cracks caused by accelerated embrittlement, exposes the membrane to direct invasion of moisture. When the combined chemical degradation and erosion of the bitumen expose bare felt, the membrane's deterioration enters an accelerated stage, with felts open to direct attack by water and atmospheric ultraviolet radiation.

Figure 10.8 Blisters occur with especial frequency in Florida and other southern states, where the semitropical climate provides the heat and humidity conducive to blister formation and growth. This roof was a poorly designed and constructed project in South Florida. The organic felts in this BUR membrane promoted the severe blistering. Glass-fiber felts would seldom, if ever, display blistering to this extent.

Blistering mechanics. *All blisters originate with the formation of a void, or unadhered area, in the mopping bitumen, either between the felt plies or between the substrate and the membrane* (see Fig. 10.9). Regardless of other disagreements in their theories of blistering mechanics, roofing experts generally agree on this point: A void in either the interply mopping or at the insulation-membrane interface is essential to the development of a blister in a built-up membrane. And this void generally, *but not necessarily,* dates from the application

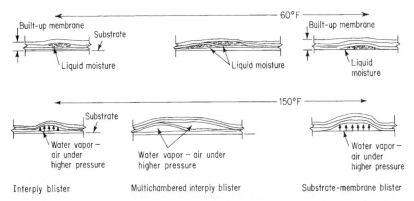

Figure 10.9 Blister configurations range from single-chambered interply or substrate-membrane blisters to multiple, interply blisters.

of the hot bitumen. That is why continuous, void-free interply moppings are so important: Mopping voids create the potential for a crop of future blisters.

Voids can result from a host of causes. Architect Justin Henshell, AIA, provides the following list:

- Moisture in or on the felts (either top or bottom ply being mopped)
- Use of felt rolls crushed into oval shape by storage in a horizontal instead of a vertical position
- Uncoated felt surfaces resulting from mop skips or clogged holes in the bitumen-dispensing machine
- Failure to broom out entrapped air
- Distorted insulation boards (warped or misaligned)
- Moppings of improper viscosity (usually too viscous, but sometimes too fluid)
- Trapping of foreign material: gravel, matchbooks, broken insulation, paper wrappings, and so on
- Tenting or ridging caused by expanding insulation or expanding blisters in plies below
- Unfilled voids in the insulation substrate surface—e.g., broken corners of insulation boards
- Improperly set, upturned metal flanges of curbs and so forth
- Unfilled edges (toes) of cants, tapered edge strips, or blocking
- Coated felts' side or end laps remaining uncoated when mop skims by
- Fishmouth or wrinkle (particularly in coated felts)

Most blisters grow from the evaporation of liquid moisture and the expansion of the resulting water vapor contained within the void. Because of liquid moisture's tremendous expansion when it evaporates, its presence within a void is the readiest agent promoting the spectacular growth of blisters often observed on built-up membranes. In late spring or summer, under midday and early afternoon sun, membrane temperatures can easily rise from 70 to 150°F, with a tremendous increase in pressure (see Fig. 10.10).

Under standard atmospheric pressure, this 80°F temperature rise expands water about 1500 times its original liquid volume. However, if it were confined within a constant volume, the water-vapor pressure would rise by more than 4 psi, about 600 psf. In an actual blister, the result lies between these two physical extremes: i.e., the membrane resists expansion of the evaporated moisture, and so the blister's volume is far less than the volume associated with unrestrained

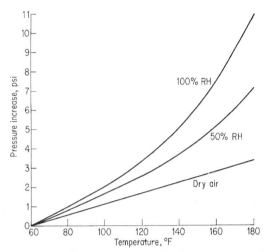

Figure 10.10 Vapor-pressure graph shows exponential increase in water-vapor pressure vs. linear increase in dry air pressure with constant volume and rising temperature. *Note:* Graph ignores effect of increased pressure (less than one standard atmosphere at 180°F) on the humidity ratio W = lb water vapor per lb dry air, which would slightly reduce above values.

expansion at atmospheric pressure. Nonetheless, the blister volume greatly expands, relieving the pressure that would occur within an unyielding, constant-volume void space.

Although liquid moisture accelerates blister growth most rapidly, it is not indispensable to blister formation. Trapped air–water-vapor mixture can also promote blister growth. Here is how it happens. When the hot-mopped bitumen cools and hardens, it seals in the trapped air–water-vapor mixture at a temperature probably exceeding 100°F, depending on ambient as well as bitumen temperature. Cooling of the hot-mopped bitumen creates negative pressure within the void, and the still-flexible felt plies tend to press together tightly, so that they appear to be bonded. Unless additional air and/or moisture enter the void, subsequent heating of the membrane cannot produce positive pressure within the void until the temperature of the trapped air–water-vapor mixture exceeds the temperature at which it was originally entrapped.

But the negative pressure naturally occurring within a mopping void promotes entry of additional air and water vapor. This entry can occur via two mechanisms: (1) filtration through tiny membrane cracks and (2) simple gaseous diffusion.

In either case, the void behaves like a bellows. It expands and contracts vertically, but it cannot expand horizontally until the total internal force is great enough to break the bond between felt and bitumen at the void perimeter. Obviously, the smaller the void, the less probability that it will grow into a blister. Mathematical equations demonstrate this fact (see Fig. 10.11). You can calculate the force F (lb) tending to break the peripheral bond around a circular void and enlarge the blister from the following formula:

$$F = p\frac{\pi d^2}{4} \qquad (10.1)$$

where F = force tending to enlarge blister (lb)
p = total internal (air-vapor) pressure (psi)
d = void diameter (in.)

The force resisting this expansion, R (lb), equals the product of the mopped bitumen's peripheral bond strength by the void circumference:

$$R = s_p \pi d \qquad (10.2)$$

where R = resisting force (lb)

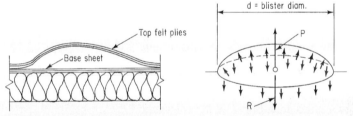

Figure 10.11 *Blister growth mechanics.* For the interply circular blister shown, blister growth requires a cyclic prying action, in which the total interply pressure P exceeds the prying resistance at the blister's perimeter. Because the internal pressure is uniform in all directions, its total vertical component P must exceed R, the peripheral resistance, for the blister to expand its base area. Then

$$P > R$$

$$\frac{p\pi d^2}{4} > s_p \pi d$$

$$p > \frac{4s_p}{d}$$

where p = internal gas pressure (psi)
s_p = prying resistance (lb/in.)
d = blister diameter (in.)

For a circular blister, the pressure required to expand its area (and volume) varies inversely with the diameter. This same rule holds for an elongated oval (e.g., an ellipse), so long as the blister retains its basic shape (i.e., the same length/width ratio).

s_p = peripheral bond strength at void perimeter (lb/in.)

Equating these expressions, $F = R$, yields the following:

$$p = \frac{4s_p}{d} \quad (10.3)$$

Equation (10.3) explains why a small void is less likely to grow into a blister than a larger void. The internal void pressure required to expand the blister varies inversely with void diameter. Thus, for example, it takes five times as much pressure to expand a 1-in.-diameter void as it does to expand a 5-in.-diameter void. Note also that the precise blister shape—whether it is a circle or, like most blisters, an elongated oval—does not affect the basic theory. So long as the blister retains its basic shape (i.e., its length-width ratio remains constant), the pressure required to expand it decreases in inverse proportion to its width or length. Experiments established peripheral bond strength at a constant 1.0 lb/in. (for steep asphalt) at 160°F.

Blister growth encounters several other limitations indicated by the formulas. In a small blister, membrane weight is insignificant compared with internal pressure, which may rise to 2 psi or so, 50 times the weight of an aggregate-surfaced membrane. Moreover, the buckled membrane can act as a dome or arch for spans measured in inches. But reduced pressure and longer dome spans accompanying blister growth make membrane weight an increasingly important factor limiting further growth. In fact, for an aggregate-surfaced membrane weighing 6 psf, the theoretical limit for a circular blister is 8 ft because that is the point at which the membrane weight exceeds the theoretical pressure required to overcome peripheral bond resistance.*

As still another practical limitation on blister growth, a blister's increasing area exponentially increases the probability that a small membrane puncture will vent the blister and thus relieve internal pressure. (When you apply foot pressure to a spongy blister, you can sometimes hear the expelled gases hiss as they escape through tiny fissures.) The theoretical approach thus helps establish the void theory of blister origin.

The theory of blister development yields useful, practical conclusions about void size and the role of liquid moisture in originating a blister. The most basic point is: *It takes at least a dime-sized (roughly $3/4$-in.-diameter) void to originate a blister.*

---

*For a membrane weighing 6 psf (= 0.042 psi), the blister diameter required to balance membrane weight = $d = 4s_p/p = 4 \times 1/0.042 = 96$ in. (8 ft). Thus, for growth beyond 8-ft diameter, an aggregate-surfaced membrane's weight becomes the theoretically limiting factor rather than a 1-lb/in. peripheral bond resistance.

A $¾$-in.-diameter blister requires about 5 psi to enlarge it, which is roughly the pressure gain within a saturated (100% RH) constant-volume space for a temperature rise from 60 to 140°F. Because any enlargement of a blister relieves pressure (by increasing the volume), it is highly unlikely that a blister could originate with less than a $¾$-in.-diameter void.

A second basic point is: *The void needed to originate a blister can be smaller if it contains liquid moisture than if it contains only an atmospheric air–water-vapor mixture.*

This inference follows from Eq. (10.3), $p = 4s_p/d$. Because the internal pressure required to expand the void varies inversely with void diameter, it logically follows that the higher pressure resulting from a vapor-saturated condition produced by liquid moisture can expand a smaller void than the lower pressure resulting from an unsaturated air-vapor mixture.

The hazard of wet felts as blister-initiating agents is highlighted in a provocative research paper by T. A. Schwartz and C. G. Cash of Simpson, Gumpertz & Heger, Cambridge, Massachusetts, consultants.[19] This paper convincingly refutes some roofers' assurances that hot mopping bitumen evaporates *all* moisture contained in the felts. Hot bitumen removes, at best, 50 to 60 percent of the moisture from felts containing excessive moisture. And some of this liberated moisture may remain trapped within the interply adhesive layer.[20]

What keeps a blister growing is a daily cyclic pumping action, with the daily volume of air inhaled into the blister chamber exceeding the daily volume of exhaled air. To see how this cyclic pumping mechanism works, consider the daily pressure changes within a membrane mopping void. This mopping void originally sealed as the bitumen cooled between mopping temperature (say, 350°F) and ambient temperature (say, 70°F). Let us assume that the sealing temperature is 120°F, slightly below the 135°F softening point of Type I asphalt. (The precise assumed temperature is unimportant to this argument.) Now, if this sealed mopping void—a potential blister chamber—was at atmospheric pressure at 120°F, it will be at negative pressure whenever membrane temperature drops below 120°F and at positive pressure when membrane temperature exceeds 120°F. As shown in Fig. 10.12, even on a hot, sunny day, membrane temperature exceeds 120°F only for 6 h or so (roughly 12 noon to 6 p.m.) and falls below 120°F during the remaining 18 h. Because of these changing pressure differentials, air and water vapor tend to diffuse *out* of the sealed blister chamber during a short part of the day (6 h at most) and *into* the blister chamber during most of the day. On cloudy and rainy days, suction into the blister will probably continue throughout the day because membrane temperature probably will never reach the 120°F assumed equilibrium sealing temperature.

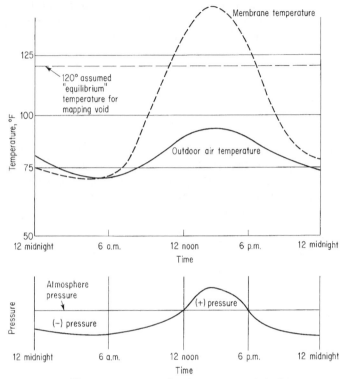

Figure 10.12 How a vapor retarder (or nonvented deck) promotes growth of blisters at the insulation-membrane interface. (1) During the 12 to 16 h with negative air pressure within the roof sandwich, air can enter the blister void via diffusion and infiltration through tiny cracks in the membrane. These cracks form because the membrane is only partially elastic—i.e., it does not contract as much for an 80°F temperature drop as it expands for an 80°F temperature rise. Moreover, below 80°F the bitumen has a higher coefficient of thermal expansion-contraction than the reinforcing felts. Thus it contracts more than the felts at low temperature, and this additional contraction would open minute cracks. If the insulation contains a normal 2 percent or so moisture (by weight), there is enough moisture to keep the trapped air–water-vapor pressure saturated at temperatures above 160°F. Thus the major gain during the 16-h negative pressure period is air, not water vapor. Daily additions of dry air can maintain blister growth through cyclic pumping action until the natural limiting factors stop it. (2) During spring or summer daytime hours, solar radiation heats the membrane to 150°F or so, raising pressure within the void and within the insulation. A vapor retarder raises this internal pressure because it obstructs underside venting of the roof. This internal positive pressure is maintained during the 8- to 12-h period when the membrane is warmer than the outside air because the heated, plastic bitumen expands and flows into the microscopic cracks, sealing them against escape of the internal air–water-vapor mixture.

Excess daily air intake into the blister chamber expands the blister's volume. On hot, sunny days, when the membrane is heated to its peak 130 to 155°F temperature, the cumulative, increasing pressure breaks the peripheral bond at the blister boundary until the expanding volume relieves the pressure and restores a new equilibrium of forces within the blister. Thus, by a continual, rachetlike pumping action, a blister expands by tiny daily increments, possibly $\frac{1}{8}$ to $\frac{1}{4}$ in. diameter a day during the hottest weather, with each day's prying action expanding it a little farther.

The conditions promoting blister growth are really worse than those postulated for the simplified model. When membrane temperature rises to 150°F or so, the softened, semifluid bitumen seals microscopic membrane cracks and reduces membrane permeability. By reducing the membrane's exhalation, this process increases internal blister pressure. When the membrane cools to 65 or 70°F, contraction of the stiffened bitumen reopens the microscopic cracks, thus increasing membrane permeability and consequent air intake. So the blister tends to suck in more air than it expels under daytime pressure.*

Clues to blister origins. A blister's interior, cut open in the field or laboratory, usually contains telltale clues pointing to its origin, notably:

- Tiny foaming craters in interply moppings
- Smooth, shiny bitumen surfaces
- "Legs" (sometimes called "stalactites" or "stalagmites")
- Bare, uncoated felt spots
- Foreign substances—e.g., aggregate—entrapped between plies

Foaming craters ($\frac{1}{8}$ in. or so in diameter) provide circumstantial evidence that liquid moisture was present during the felt-laying operation.

---

*This inference, that the longer "inhalation" time implies a greater volume of "inhaled" than "exhaled" air from the blister chamber, assumes roughly comparable negative and positive pressures. Note by the previous argument that internal blister pressure decreases as the blister grows. (Pressure varies inversely with blister diameter.) Because internal blister pressure decreases with increasing blister size, further growth by cyclic pumping action, resulting from a daily excess of inhaled over exhaled air, grows easier with increasing size once the blister's internal pressure starts it growing from a small originating void. We thus have a rational evolutionary explanation of blister growth: It requires high internal pressure to start blister growth. Then, once the blister starts growing and high internal pressure is no longer necessary, reduced internal pressure promotes greater inhalation of air into the blister, perpetuating the cyclic pumping action that keeps the blister growing.

To boil water at standard atmospheric pressure requires a minimum temperature of 212°F, 150°F or more below the temperature of properly heated bitumen and well above the maximum 190°F or so that the hottest black-surfaced, sun-baked roof will experience (short of fire).

A *shiny bitumen surface* may (but does not necessarily) indicate a lack of interply mopping cohesion dating from the original application (see Fig. 10.13). These shiny surfaces often contain foaming craters, which indicate the sudden boiling of liquid moisture. From these two clues, you can infer that the sudden expansion of boiling liquid moisture, augmented by the expansion of heated, saturated air, produced the shiny-surfaced blister cavity.

Shiny bitumen does not, in itself, necessarily prove lack of original adhesion, *unless* there is a dry top felt (thus indicating that the bitumen cooled before the felt was laid and thus failed to adhere). The interply bitumen may develop a shiny surface merely from flow induced by solar heating.

So-called legs, or stalactites and stalagmites, usually form in an alligatored pattern when the adjacent felts are pulled apart. These legs reveal the heated bitumen's resistance to the internal blister pressure prying adjacent felts apart (or prying a bottom felt off its substrate). Rising membrane temperature *increases* the blister's internal pressure and simultaneously *reduces* the bitumen's tensile strength. As the temperature approaches the bitumen's softening point, the increasingly fluid material flows like taffy or chewing gum, with little or no increase in stress. Legs are thus lines of contact between the blister's separated plies inside the blister's periphery, where prying action has succeeded in breaking contact between plies. The legs break only when the blister's vertical growth exceeds the bitumen's ductility. (Ductility diminishes with falling temperatures that may not significantly contract the blister.)

A *bare, uncoated felt spot* often marks the original blister void. A radial, starlike uncoated felt pattern emanating from the original void indicates the sudden boiling of liquid moisture in the bare spot when it was hit by 300 to 400°F asphalt (see Fig. 10.13).

Foreign substances—aggregate, insulation fragments, broken matchbooks—are sometimes found in the original void. These are the most direct circumstantial evidence for a blister's origin.

Other hypotheses—notably incompatibility between mopping bitumen and felt coating or saturant—are sometimes offered as explanations of blister formation. By a fairly general expert consensus, the incompatibility theory is highly dubious.

Though it cannot compare with medical malpractice litigation as a source of junk science in the courtroom, roof litigation is by no means

Elements of Built-up Membranes 251

Figure 10.13 Exposed blister chamber (top) exhibits two telltale clues to the blister's origin: (1) Countless foaming craters in the shiny bottom asphalt surface indicate the presence of liquid moisture on the base felt or on the felt above when the hot asphalt was mopped; (2) bare stretches of uncoated base felt ($3/4$ in. wide, at least 6 in. long) indicate the origin of mopping voids, probably resulting from stopped holes in the asphalt dispenser. Laboratory test samples also revealed overweight interply moppings, a probable consequence of the application of overcooled asphalt. Blister in bottom photo occurred on the same project. It shows how blisters can promote ponding even on a well-sloped roof: in this instance, a 1-in./ft slope.

unrepresented. The most ridiculous hypothesis presented in courtrooms as an explanation for extensive BUR membrane blistering was the carbon dioxide "theory." This theory attributed interply blisters in two-ply coated-felt membranes to the manufacturer's use of crushed seashells as the stabilizer in the coated felts, which allegedly released carbon dioxide in the presence of moisture. The reported discovery of three times the normal percentage of ambient carbon dioxide inside some blister chambers purportedly accounted for the blisters.

In the legal cases built to validate this theory, the proponent chemists repealed Dalton's law of partial pressures, which has stood unshaken for nearly two centuries. According to Dalton's law, each individual gas in a mixture exerts its share of total pressure in direct proportion to its *volume*. Normal carbon dioxide atmospheric content is about 0.03 percent by volume. The excess carbon dioxide inside the blisters thus accounted for $(3-1) \times 0.03 = 0.06$ percent of total internal pressure. How 0.06 percent, or even 100 times that amount, could be positively identified as the incremental pressure that caused the blisters remained an unexplained mystery. Yet at least one scientifically illiterate judge in the Florida court system found this argument convincing and ruled accordingly.

Another myth about blistering mechanics merits rebuttal. Some roofing experts have argued in the past that blisters can heal themselves by resealing at high temperatures when the bitumen becomes more fluid. The chief objection to this hypothesis, sufficient in itself as a rebuttal, is the coincidence of a blister's maximum *positive* internal pressure with the bitumen's maximum fluidity. Resealing would require the opposite situation: i.e., maximum *negative* internal pressure, tending to compress the blister chamber and reseal the separated felt plies. Moreover, dependable adhesion requires temperatures that are around 100°F (56°C) higher than the hottest temperatures attained in service by a BUR membrane. And as a third forensic nail in the coffin of the resealing hypothesis, bitumen—especially asphalt—progressively gains viscosity as it ages.

Splitting

Apart from wind blowoffs, splitting is obviously the most serious failure mode for conventional roof systems. Splits allow immediate infiltration of leakwater into the roof system and ultimately into the building.

Typically, splitting makes its dismal presence known late in a northern winter, during the first thaw, when melting snow and ice join rain as abundant sources of leakwater. Late February/early March is a typical split discovery date—a bad time for reroofing or even repair.

What makes built-up BUR membranes so vulnerable to splitting is their brittle, glasslike nature at low temperatures. At $-30°F$ ($-34°C$), the breaking strain of the best BUR membrane is still less than 2 percent. By comparison, synthetic rubber membranes have breaking strains around 200 percent and modified bitumens range up to 100 percent and higher. To make matters even worse, BUR membranes have extremely high coefficients of thermal expansion-contraction in cold weather. (For a temperature drop in the subfreezing zone, a length of BUR membrane contracts about five times as much as the same length of steel.)

Built-up membranes' viscoelastic nature—brittle and elastic at subfreezing temperatures, plastic and semifluid at moderate and hot temperatures—explains the higher incidence of membrane splitting in cold climates than in warm climates. In Key West, Florida, for example, where the winter design temperature is 55°F (only 22 h at or below 55°F in an average winter), a BUR membrane acts totally differently from the way it acts in Duluth, Minnesota, where the winter design temperature is −21°F.

Stress concentration, a result of poor design or application, is the basic cause of BUR membrane splitting. Several research projects have demonstrated beyond all reasonable doubt that thermal contraction stress alone cannot split a properly applied BUR membrane.[21] Stress concentrations are created by the following:

- Insulation movement
- Shrinkage cracking of cast-in-place substrates (e.g., concrete or gypsum)
- Relative movement at joints where decks change span direction, where bearing ends rotate from live-load deflection, or where deck material changes
- Metal contraction concentrated at gravel-stop flashing joints

Movement of unanchored insulation boards is, by far, the most common cause of BUR membrane splitting. One of the most correctly belabored rules of BUR (or any other conventional) roof systems concerns the vital importance of anchoring insulation boards to the deck. Unaided by any other factor, unstable insulation boards can split a BUR membrane. A major function of insulation sandwiched between deck and membrane is to restrain insulation movement, especially thermal movement imposed by the membrane above.

Loose insulation promotes a progressive contraction that ultimately splits the membrane (see Fig. 10.14). Under cold-weather temperature cycling, especially when the temperature drops below 0°F (−18°C), the contracting membrane drags loose insulation boards toward the securely anchored sections. During warmer daytime hours, the flexible membrane can only buckle, usually over insulation joints, where the insulation-membrane bond is weakest. Cumulative contraction-buckling cycles, with loose insulation moving across the deck and some boards restrained by curbs and other obstructions, inevitably open some insulation joints. Membrane stress concentration is the result.

The common occurrence of membrane splits over insulation joints is readily explained. Consider a widening insulation-board joint, with insulation unanchored on at least one side of the joint. At any vertical cross section where the membrane is anchored to the insulation, tem-

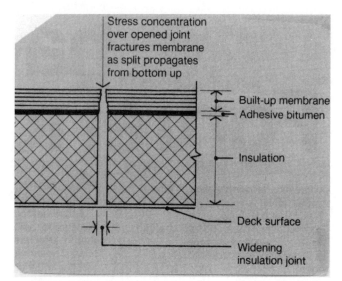

Figure 10.14 Stress concentration above widening insulation board joint splits membrane as loose insulation boards are moved by contracting membrane. Membrane breaking strain declines as temperature drops, to a 1 percent range at subzero temperatures. Despite some stress relief from shearing deformation in the adhesive bitumen at the membrane-insulation interface, stress concentrates over the widening joint. Thermal cycling reduces membrane tensile strength. And if an incipient split occurs in a water-ponding area, water intrusion drastically weakens organic felts. These hazards highlight the vital importance of two basic rules of good roof design and construction: Anchor insulation solidly to the deck; slope the roof for positive drainage.

perature contraction of the insulation relieves tensile stress on the membrane. The membrane is restrained by its adhesion to the insulation from contracting into a stress-free state. But at a vertical cross section through the membrane where there is movement widening the joint between adjacent insulation boards, membrane tensile stress is increased, not reduced, by the insulation's contraction. The opened joint between unanchored boards, or between an anchored and an unanchored board, becomes a line of stress-strain concentration in the BUR membrane (see Fig. 10.14).

Secure anchorage of the insulation with fasteners rather than unreliable adhesives—the now unapproved cold adhesives or hot-applied asphalt—is thus a preventive of BUR membrane splitting as well as a safeguard against wind blowoff.

Membrane-splitting risk increases exponentially when insulation boards cantilever over metal deck flutes. (They should have minimum $1\frac{1}{2}$-in. bearing.) When insulation boards are cantilevered, the contractive tensile stress along the stress-concentration line above a con-

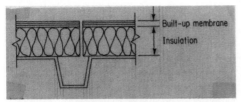

Figure 10.15 Where insulation boards cantilever over a steel-deck flute, the membrane splitting hazard from inadequately anchored insulation multiplies exponentially. In addition to contractive tensile stress, this condition can induce flexural stress from insulation-board warping and even vertical shearing stress from differential movement at the ends of the cantilevered boards. Minimum bearing of insulation boards on steel deck flange is $1\frac{1}{2}$ in.

tinuous joint is compounded by other stresses: flexural stress from the warping of the insulation board and, perhaps even more hazardous, shearing stress from the relative vertical movement of the insulation boards (see Fig. 10.15).

Discovery of cantilevered boards constitutes *prima facie* evidence of negligent field application. This condition can be readily prevented merely by field trimming of insulation boards if cumulative dimensional errors reduce bearing on the deck flanges below the $1\frac{1}{2}$-in. minimum.

Stress concentration capable of splitting a membrane also occurs with shrinkage cracking in poured-in-place decks that serve as a membrane substrate (structural concrete, gypsum, or lightweight insulating concrete fill). Shrinkage cracking can easily split a BUR membrane, unaided by any other factor. Solidly mopping a membrane onto a poured, monolithic substrate exposes the membrane to an extremely high splitting risk, averted only if the substrate material is considerate enough not to crack after the membrane is applied.

Relative movement at bearing ends of structural members—e.g., precast concrete tee sections—or where deck spans change direction or material can easily split BUR membranes. These conditions require expansion joints to avert membrane splits (see Fig. 10.16).

Among the secondary factors contributing to BUR splitting failures are:

- Repetitive thermal stress cycling (which reduces tensile strength)
- Overweight interply moppings (which raise the membrane thermal coefficient of expansion)
- Water weakening of organic felts (which drastically reduces membrane tensile strength)

Figure 10.16 Failure to specify an expansion joint at the line where the steel deck material changes direction produced the slightly offset crack (left of measuring tape tip) in lightweight insulating concrete. The membrane split directly above the line (see pencil point). Steel deck corrugations are vertically oriented to right of gap, horizontally on left of gap. A vertical cross-sectional plane of weakness is created where a lightweight insulating concrete is poured continuously over a break in the supporting high-strength steel deck that changes span direction. That cross section loses about 98 percent of its tensile strength.

Junk science rears its ugly head in splitting analysis even more than in blistering analysis. Among the most common scapegoat explanations of BUR membrane splitting are the following:

- Thermal shock
- Excessive expansion joint spacing
- Structural movement

Thermal shock is a phrase that is widely parroted, but usually misunderstood. By itself, it refers to sudden membrane cooling—for example, when a rain shower suddenly cools a sun-baked membrane.

The phrase *thermal shock factor,* however, refers to something entirely different. It is a calculated property, representing the temperature drop required to break a membrane that is totally restrained from contracting (see p. 239).

Expansion joint spacing in excess of the 200-ft limit often recommended is sometimes cited in litigation as an absolute taboo. Violating this taboo is tantamount to running a grave splitting risk, according to this benighted theory.

For lawyers and "experts" seeking easy scapegoats, this is a handy tactic for befuddling judges and jurors. But roof designers should know better. For a roof system with perfect attachment at the deck-insulation interface, there would be no splitting hazard for a healthy BUR membrane regardless of expansion-joint spacing. With uniform stress distribution, membrane stress for a 10-ft and a 1,000-ft expansion-joint spacing would be essentially the same.

What the 200-ft limit between expansion joints recognizes is merely the greater probability of insulation movement or another stress-concentrating mechanism splitting the membrane on a large roof area unrelieved by expansion joints. The 200-ft maximum is merely an empirical rule that has been found satisfactory. There is nothing sacred about it.

Expansion-joint spacing exceeding 500 ft is approved for some air-conditioned, steel-framed buildings in New York City. The same building in Florida can have 600-ft expansion-joint spacing, according to a report by the old Building Research Advisory Board of the National Academy of Sciences.[22]

Excessive structural deflection is another popular scapegoat for BUR membrane splitting. Deck deflection, especially in lightgage steel, can break the adhesive bond at the deck-insulation interface. By loosening the insulation, it can act as an indirect cause of membrane splitting. But structural deflection within normal ranges cannot come close to splitting a BUR membrane.

The most obvious example of structural movement directly causing splitting involves the rotation of roof-beam bearing ends from live-load deflection (see p. 237).

Their vulnerability to splitting is a large factor in the decline in the use of BUR membranes over the past several decades. Modified bitumens, elastomerics, and weldable thermoplastics all offer considerable advantages over BUR in split resistance.

Ridging

Ridging, sometimes called wrinkling, is another failure mode that should be following the declining path of blistering, but instead

Figure 10.17 Ridging normally occurs over the joints of dimensionally unstable insulation boards. The staggered grid pattern is sometimes called "picture framing," a phenomenon generally limited to smooth-surfaced membranes with asbestos felts.

appears to have accounted for a stable share of BUR problems over the past decade.[23]

Ridges were a major BUR membrane problem several decades back, when organic and asbestos felts were predominant. Ridges would appear in parallel lines above continuous longitudinal insulation-board joints. When they also appear above the transverse insulation-board joints, the pattern is called "picture framing" (see Fig. 10.17).

In a less orderly pattern, ridges may resemble long blisters. But the mechanics and causes of ridging differ from the causes of blistering. A blister depends upon the volume expansion of a heated air-vapor mixture entrapped within a closed, sealed chamber (generally an interply mopping void); a ridge normally depends on the moisture-absorbing expansion of organic (or asbestos) felts.

Because they are far less susceptible to moisture absorption, BUR membranes containing glass-fiber felts should be virtually invulnerable to conventional ridging. They are not, however, immune to random wrinkling, which results from other causes.

Classic ridging, of the type plaguing the industry back in the 1960s, formed in the following way. Hot asphalt applied over the insulation boards leaks into the joints between boards, and because of this bitumen leakage, the base felt is only lightly coated, or even bare, along the insulation-joint line. When water vapor from the building interior flows upward through the open insulation joint, it condenses within the base felt. With cumulative moisture absorption, organic or asbestos felts swell and buckle, forming ridges directly above the insulation joints (see Fig. 10.18).

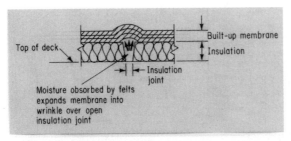

Figure 10.18 Ridging mechanics.

Ridging failure can result from membrane flexural fatigue. Repeated flexing, caused by cyclic elongation and contraction accompanying wetting-drying of the felts, ultimately cracks the membrane.

Moisture appears to be virtually the sole cause of ridging expansion of the felts, with little or no contribution from temperature cycling. In an experiment demonstrating ridging mechanics, Professor E. C. Schuman, formerly of Penn State University, found no ridges whatever formed on a test sample made with dry material under daily temperature cycling from −20 to 160°F continued over a 6-month period.[24] Expansion of a thoroughly wetted organic-felt membrane is 1.2 percent, compared with a computed expansion for a 180°F temperature rise of about 0.2 percent.

Schuman's studies also explain the persistence of ridging even after a wetted felt dries. Plotting curves with water absorption (lb/ft^3) as abscissa vs. linear change (percent) as ordinate, Schuman found that felt expands rapidly with wetting but contracts very slowly with drying. Physically, the explanation for this slow drying contraction is the presence of *inter*cellular water preventing the removal of *intra*cellular water. Intracellular expansion accounts for the major part of a wet organic-felt membrane's expansion.

Use of coated felts, with their increased resistance to moisture absorption, helped reduce the problem of ridging in the past. The current persistence of this problem indicated in Project Pinpoint may indicate confusion concerning the definition of ridging. Random wrinkling is sometimes called "ridging."

Membrane slippage

Slippage is a minor problem in terms of incidence (less so in its consequences) that appears quite persistent, according to NRCA data. Slippage can be defined as relative lateral movement between felt plies. A membrane plagued by slippage often assumes a randomly wrinkled appearance, like a badly laid carpet (see Fig. 10.19).

Figure 10.19 Membrane slippage, a problem on sloped roofs, is caused by one or more of a complex of factors. (*National Institute of Standards and Technology.*)

Because it can expose the base sheet, slippage often reduces the membrane from a multi-ply to a single-ply covering, exposed to weather and other destructive forces. Although relatively uncommon, slippage is a costly mode of failure, extremely difficult to rectify.

Slippage failures usually occur within the first year or two of the roof's service life, with most failures on roofs from ½- to 1-in. slope. Slippage seldom occurs on roofs of less than ¼-in. slope, but roofs of ¼- to ½-in. slope are more vulnerable to slippage than is generally recognized. Membranes with slopes greater than 1 in. seldom slip because the roof designer, recognizing the risks of slippage on steeper slopes, usually takes the precaution of specifying backnailing or another positive anchorage technique.

Slippage-promoting factors. These findings are from an intense NBS laboratory investigation featuring slip-and-sag tests. This investigation followed field studies that identified slippage problems and parameters. Former NBS materials expert W. C. Cullen identified six factors involved in slippage:

1. Slope
2. Bitumen
3. Felts
4. Climate
5. Substrate heat capacity
6. Surfacing

Slope is the prime factor in slippage (see Fig. 10.20). As a general rule, preventive backnailing of felts is recommended for Type III asphalt membranes over $1\frac{1}{2}$-in. slope and for Type I asphalt or coal-tar pitch roofs of $1\frac{1}{2}$-in. or greater slope. Even lesser slopes may require backnailing under unusual circumstances.

Low-softening-point bitumen is often blamed for slippage. And low softening point is often attributed to softening-point fallback from overheating steep asphalt. Slippage from low-softening-point asphalt has prompted several major roofing manufacturers to recommend steep asphalt (ASTM D312, Type III) for interply mopping on all roofs, regardless of slope.

But this remedy has its own special hazards. In many important respects, steep asphalt is inferior to dead-level asphalt as a plying cement. In fact, as pointed out in the discussion of equiviscous temperature (EVT), softening point, penetration, and ductility may not give a true picture of temperature susceptibility, or even flow characteristics. Viscosity is the most important index of flow. Heating

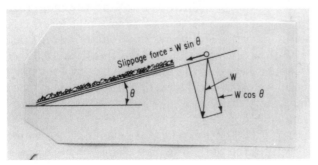

Figure 10.20 Vector diagram shows how slippage force ($W \sin \theta$) increases with increasing slope and membrane weight.

asphalt above its blowing temperature in the absence of air produces a more fluid material at high roof-surface temperatures. For steep asphalt, softening-point fallback may reach 20 to 25°F. Steep asphalt is much more susceptible to softening-point fallback than dead-level asphalt.

Phased construction, when a base sheet is glazed and the shingled felts applied on a later date, with another interply mopping added to the glaze coat and thickening the asphalt film, is a prime cause of slippage. In phased membranes, slippage, if it does occur, is invariably at the continuous interply mopping plane at the base sheet's top surface. Eliminating phased construction, or shingling all felts, obviously eliminates this source of slippage.

Another flaw in felt application can cause a "wheeling effect" that promotes slippage. Where the membrane's felt plies are not properly staggered but instead terminate at the same vertical line, which may extend from eave to ridge, slippage potential may be especially high, with slippage exposure increasing as it nears this vertical line. The explanation for this wheeling effect, so-called because of its apparent rotation about two hubs located some distance back from the vertical line, is as follows. Slippage is aggravated along this line because of unbalanced, one-sided restraint (i.e., back toward the hub), whereas at other segments of the staggered felt lines there is generally two-sided restraint to slippage (i.e., tensile restraint from the felts on both sides of a given point from areas where the asphalt might be more stable, thus providing better anchorage for the felts). As another analogy, the line of spliced felt ends is like the ends of cantilevered beams, which have greater freedom to deflect because of their one-sided support. In fact, the slippage line may resemble the deflection curve of a continuous row of beams with cantilevered end spans.

Built-up membranes constructed of glass-fiber or perforated felts are most resistant to slippage. They let the interply bitumen flow through and form shear keys with the other interply moppings.[25]

Slippage risk increases with the thickness of interply moppings between felts. Adhesives should be as thin as practicable to provide waterproofing while still ensuring continuous, unbroken coverage and total separation of the felts. Increased adhesive thickness reduces horizontal shear resistance between the felt plies. Overweight moppings are likely to occur in phased construction because the phased construction plane may get two moppings: a glaze coat on top of the base ply and a final interply mopping to bond the top plies.

Climate and roof orientation affect slippage because they determine membrane temperature and the daily duration of high temperature. Obviously, the hotter the climate, the greater the slippage hazard. Of even greater importance is exposure to solar radiation, which explains the prevalence of slippage failures on south- and west-facing roof slopes.

Substrate heat capacity is another factor affecting the slippage risk, which varies *inversely* with substrate heat capacity. A structural concrete deck substrate (150 lb/ft^3) continually moderates membrane surface temperature—lowering it during a hot, sunny day, raising it during a cool night—through its high thermal conductivity and its high heat-storage capacity. In contrast, highly efficient lightweight foamed insulation weighing 2 lb/ft^3 isolates the membrane and raises surface temperature on a sunny day because it retards heat transmission through the roof sandwich.

Membrane surfacing affects the slippage hazard in two ways: color and surfacing aggregate weight. As discussed in Chapter 5, "Thermal Insulation," surface color has a tremendous effect on membrane temperature. A black surface increases the slippage risk because it promotes much higher temperatures than a light-colored roof surface.

Aggregate surfacing weight increases slippage risk by increasing the gravitational-force component acting parallel to the roof plane (see Fig. 10.20). Do not, however, try to alleviate this problem by eliminating aggregate surfacing in favor of a smooth-surfaced roof; this cure normally entails excessive sacrifice of other desirable membrane qualities.

Cullen has combined the previously described six factors into an equation relating slippage distance and time:

$$\frac{s}{t} = \frac{1}{\eta} \times \frac{FD}{A} \tag{10.4}$$

where s = slippage distance
 t = time
 F = force
 D = interply bitumen mopping thickness
 A = area
 η = bitumen viscosity

Multiplying through by t yields

$$s = \frac{1}{\eta} \times \frac{FDt}{A} \tag{10.5}$$

Examination of Eq. (10.5) shows that slippage distance varies *directly* with force (which depends on slope and surfacing weight), interply mopping thickness, and time (which depends on duration of high roof temperatures). Slippage distance varies *inversely* with viscosity. (This equation does not apply to membranes with backnailed felts.)

Poor field-application practices promote slippage, notably through overweight interply bitumen moppings and overheating of hot-mopped asphalt, which can result in softening-point fallback and consequent reduced bitumen viscosity. Improperly installing mechanically anchored base sheets obviously increases the membrane-slippage risk. Here are Cullen's recommendations for reducing membrane-slippage risk:

1. Set minimum viscosity limits for various asphalts at 140°F as well as 450°F. Cullen suggests a minimum viscosity of 5×10^6 P at 140°F as a guide value to reduce slippage potential.
2. Limit asphalt interply moppings to a minimum 15 lb/square, maximum 20 lb/square.
3. Limit coal-tar-pitch interply moppings to 20 lb/square minimum, 25 lb/square maximum.
4. Limit coal tar pitch to $\frac{1}{4}$-in. maximum slope without mechanically fastening the base sheet.
5. Avoid phased construction in climates subject to high slippage risk; substitute four- or five-ply shingled felt membranes for coated base sheet plus three or four shingled plies.
6. Design light, heat-reflective roof surfaces.

Backnailing is required generally for slopes of $1\frac{1}{2}$ in. or greater for Type III asphalt, $\frac{1}{2}$ in. for Type I asphalt (see Fig. 10.21).

For the owner who needs a cure for a currently developing slippage problem, Cullen advises sprinkling water on the roof during hot portions of the day or painting a smooth-surfaced roof with whitewash.

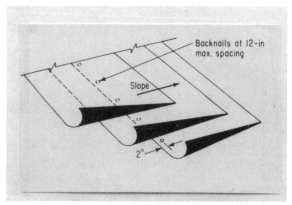

Figure 10.21 Backnailing, to prevent slippage, is covered by overlapping felts.

Because slippage usually occurs within the first year or two (before natural aging has significantly increased bitumen viscosity), these inexpensive remedies may be enough to hold off the problem until the membrane's aging, hardening asphalt develops its own slippage resistance.

For a more drastic, and expensive, remedy, consider mechanical fastening to the substrate. Fastener heads must be waterproofed with plastic cement (or substitute) and at least two layers of reinforcing felt or fabric.

Alerts

General

1. Check with the manufacturers of built-up roofing and composition flashing to make sure that the proposed roof system and materials are compatible with the roof deck, vapor retarder, and insulation.

2. On major roofing projects, require either inspection by the manufacturer or inspection service provided by the owner.

Design

1. Use a nailed, asphalt-coated base sheet for the first ply in built-up roofing membranes over "wet" decks—e.g., poured gypsum, lightweight insulating concrete fill.

2. Do not specify coal tar pitch on roofs with slope exceeding $\frac{1}{4}$ in.

3. Specify asphalt with the lowest practicable softening point suited to slope and climate as follows:

Slope (in /ft)*	ASTM designation	Softening point, °F
$\frac{1}{4}$ in. or less	D312, Type I	135–151
$\frac{1}{4}$–$1\frac{1}{2}$	D312, Type II	158–176
$\frac{1}{4}$–3	D312, Type III	185–205
$\frac{1}{4}$–6	D312, Type IV	205–225

*Warm climates may limit the slopes for asphalts of lower softening point to less than the maximum slopes specified above.

4. When hot-mopped bitumen is used to bond a base sheet to insulation, specify Type III (or in some instances Type IV) asphalt.

5. Provide trafficways or complete traffic surfacing for roofs subjected to more than occasional foot traffic (see Fig. 10.22).

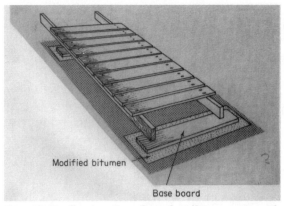

Figure 10.22 Construction of wood walkway starts with embedment of mineral-surfaced modified-bitumen strips in hot bitumen. Aggregate surfacing follows between and around the walkway bases.

6. In roofs over cold-storage space, do not place membrane in direct contact with cold-storage insulation. (For discussion of cold-storage insulation, see Chapter 5, "Thermal Insulation.")

7. Avoid contact with built-up membrane and metal flashings and other metal accessories, especially below the level where water collects.

Field

1. Prohibit storage of felts over new concrete floors. Require pallets covered with kraft paper. Stack felts on end. Avoid prolonged storage of felts on site. Organic felts should be stored in an environment with RH ≤ 40%.

2. Require a smooth, dry, clean substrate, free of projections that might puncture the felts. Take precautions on wood and precast decks to prevent bitumen drippage [see Chapter 4, "Structural Deck" (Design Alert No. 4)].

3. Make sure insulation is firmly attached to substrate.

4. Prohibit use of heavy mechanical roof-construction equipment that may puncture the membrane or deflect the deck excessively.

5. Prohibit use of asphalt-saturated felt with coal tar pitch and vice versa.

6. Require a visible thermometer and thermostatic controls on all kettles, set to manufacturer's recommended limits. Require rejection of bitumen heated above the specified maximum and reheating of bitumen too viscous for mopping. Avoid prolonged storage of bitumen above the *finished blowing temperature* (FBT) (approximately 490°F). (It can lower the softening point by 10 to 20°F.)

7. When EVT is available, require hot bitumen temperature at application within ± 25°F of the EVT.

8. When an asphalt tanker is on the site, maintain asphalt temperature below the finished blowing temperature and feed asphalt into a kettle for further heating before pumping it to the roof.

9. In cold weather, require double or insulated lines and insulated asphalt carriers and mop buckets. (Also store felts in a warm enclosure.)

10. Check for continuous, uniform interply hot moppings, with no contact between adhered felts, with average interply mopping-weight tolerance = ± 15 percent.

11. Require manual brooming following 6-ft maximum distance behind unrolling felt.

12. Require immediate repair of felt-laying defects: fishmouths, blisters, ridges, splits.

13. Terminate day's work with complete glaze-coated seal stripping.

14. Aggregate-surfacing membrane may be glaze-coated with flood coat and aggregate surfacing delayed until later, if necessary. Limit delay between felt application and flood-coat aggregate surfacing to a maximum 4 days.

15. Prohibit phased application, i.e., leaving felts exposed overnight or longer before top felts are applied.

References

1. Carl G. Cash, "Durability of Bituminous Builtup Membranes," *Durability of Building Materials and Components*, ASTM STP691, 1980, pp. 741-755.
2. W. J. Rossiter and R. G. Mathey, "The Viscosition of Roofing Asphalts at Application Temperatures," *Build Sci. Ser.* 92, NBS, December 1976, p. 14.
3. C. Slahetka and J. McCorkle, "Type II Asphalt: The Way to Go on Low-Slope Roofs," *RSI Magazine*, October 1979, pp. 82ff.
4. See P. M. Jones, "Bituminous Materials," *Can. Build. Dig.*, no. 38, May 1968, p. 38-3.
5. Herbert Abraham, *Asphalts and Allied Substances: Their Occurrence, Modes of Production, Uses in the Arts, and Methods of Testing*, 6th ed., Van Nostrand Reinhold Company, New York, 1966, p. 48.
6. R. L. Fricklas, "Technical Aspects of Retrofitting," *Proc. 5th Conf. Roofing Technol.*, NBS-NRCA, April 1979, p. 32.
7. P. G. Campbell, J. R. Wright, and P. B. Bowman, "The Effect of Temperature and Humidity on the Oxidation of Air-Blown Asphalts," *Mater. Res. Stand.*, ASTM, vol. 2, no. 12, 1962, p. 988.
8. Carl G. Cash, "On Builtup Roofing Aggregates," presented to ASTM D8 Main Committee, Dec. 6, 1978, p. 5.
9. D. C. Tibbetts and M. C. Baker, "Mineral Aggregate Roof Surfacing," *Can. Build. Dig.*, no. 65, May 1968.
10. Cash, "On Builtup Roofing Aggregates," op. cit.

11. See C. W. Griffin, "Cost Savings with Heat-Reflective Roof Coatings," *Plant Eng.*, July 10, 1980.
12. See R. M. Dupuis, J. W. Lee, and J. E. Johnson, "Field Measurement of Asphalt Temperatures during Cold Weather Construction of BUR Systems," *Proc. Symp. Roofing Technol.*, NBS-NRCA, September 1977, pp. 272ff.
13. T. A. Schwartz and Carl G. Cash, "Equilibrium Moisture Content of Roofing and Roof Insulation Materials and the Effect of Moisture on the Tensile Strength of Roofing Felts," *Proc. Symp. Roofing Technol.*, NBS-NRCA, September 1977, p. 242.
14. For an excellent discussion of this phenomenon, see R. G. Turenne, "Shrinkage of Bituminous Roofing Membranes," *Can. Build. Dig.*, no. 181, April 1979.
15. For another excellent discussion on this general topic, see R. G. Turenne, "Joints in Conventional Bituminous Roofing Systems," *Can. Build. Dig.*, no. 202, January 1979.
16. W. C. Cullen, *Project Pinpoint Analysis: Ten-Year Performance Experience of Commercial Roofing 1983–1992*, NRCA, 1993, Table 3.
17. "The Shape of Roof Construction," *The Roofing Spec*, November 1979, p. 40.
18. K. G. Martin, "Evaluation of the Durability of Roofing Bitumen," *J. Appl. Chem.* (Australia), vol. 14, 1964, p. 427.
19. T. A. Schwartz and Carl G. Cash, "Equilibrium Moisture Content of Roofing and Roof Insulation Materials and the Effect of Moisture on the Tensile Strength of Roofing Felts," *Proc. Symp. Roofing Technol.*, NBS-NRCA, September 1977, pp. 238–243.
20. Ibid., p. 242.
21. W. C. Cullen, *Effects of Thermal Shrinkage of Builtup Roofing*, NBS Monog. 89, March 1965. J. W. Lee, R. M. Dupuis, and J. E. Johnson, "Experimental Determination of Temperature-Induced Loads in BUR Systems," *Proc. Symp. Roofing Technol.*, NBS-NRCA, September 1977, p. 43. Joel P. Porcher and Herbert W. Busching, "A Study of Thermal Splitting of Roofing Membranes," Department of Civil Engineering, Clemson University, July 1979, p. 21.
22. *Expansion Joints in Buildings*, Technical Report No. 65, Building Research Advisory Board, National Academy of Sciences, Washington, DC, 1974, p. 5.
23. Cullen, op. cit., Table 3.
24. E. C. Schuman, "Moisture-Thermal Effects Produce Erratic Motions in Builtup Roofing," *Engineering Properties of Roofing Systems*, ASTM STP409, 1967, p. 64.
25. W. C. Cullen, *Slippage of Builtup Roof Membranes—Causes and Prevention*, NBS Rep. 10950, p. 45.

Chapter

11

Modified-Bitumen Membranes

Modified-bitumen membranes represent a great forward leap in the perennial goal of building construction: to produce a superior, factory-made component that reduces field work, where quality control is most difficult. Considered as a stage in the evolution of the traditional built-up membrane, modified-bitumen membranes reduce the four-ply, field-fabricated membrane to two plies. The slightly higher material cost of modified bitumens—normally 10 percent or so—is generally offset by the greater labor costs of installing built-up membranes.

Modified bitumens replace traditional built-up membranes with a more flexible, ductile sheet. Modified bitumens can be hot-mopped, like built-up membranes, or they can be torched, with welded side and end lap seams. Their vastly increased flexibility gives them a tremendous advantage over the stiff fiberglass flashings otherwise required for built-up assemblies, now that asbestos-felt flashings have been long banished from respectable roofs. With its thick, tough cap sheet, modified bitumen offers a membrane almost equal to an aggregate-surfaced built-up membrane (and superior to a smooth-surfaced or cap-sheeted membrane) in puncture and impact resistance. Unlike the more puncture-prone elastomeric or thermoplastic single-plies, modified bitumens thus enable the roof designer to exploit the labor savings of a new factory-fabricated component without sacrificing the toughness of traditional built-up roofing.

For conservative building owners who want to exploit the benefits of technological progress without being quite so radical as to adopt single-ply synthetic materials, modified bitumen thus offers a comfortable compromise. Thicknesses of 120 to 180 mils, three times as thick as the 45 to 60-mil sheet thickness of elastomeric or thermoplastic single-ply sheets, make modified bitumens more resistant to puncture, impact forces, and fastener backout. On the other hand, compared with traditional built-up roof assemblies, modified bitumen offers advantages in roofs with many penetrations and flashings, the trend in modern buildings. Modified bitumen is especially adaptable to reroofing projects. Unlike elastomeric and thermoplastic single-plies, it is totally compatible with asphalt built-up construction. In instances where it is permissible—i.e., in existing roof systems not plagued with wet insulation—a single-ply modified-bitumen sheet can sometimes be applied directly to an old smooth-surfaced membrane. Such adaptability is a tremendous convenience on projects requiring only spot repair or partial reroofing—for example, on sections where the membrane surface has been severely eroded, but where large portions of the roof have substantial remaining service life.

In summary, the many advantages of modified-bitumen systems add up to a unique versatility. Of all roof systems available for low-slope roof construction, modified bitumens are the most generally adaptable. They can be considered for the widest range of building uses, for new construction as well as reroofing. Neither major competitor, conventional BUR or single-ply, offers such a wide range.

Historical Background

Despite their great waterproofing property, conventional bitumens, asphalt and coal tar pitch, have undesirable temperature-dependent variations in their physical properties. At subfreezing temperatures, they become brittle and glasslike. [Around 25°F (-3°C), a chunk of coal tar pitch thrown against a hard surface will shatter into sharp-edged fragments.] At elevated summer temperatures, conventional roofing bitumens become viscous fluids; in the intermediate, moderate temperature range, they are viscoelastic solids. Tensile strength, breaking strain, and elasticity all vary widely—even wildly—through the normal cycle of temperatures experienced by roofs in most parts of the continental United States. The same built-up membrane has radically different properties (e.g., breaking strain, tensile strength, thermal expansion coefficient) at 0°F (-18°C) and at 100°F (38°C).

Because of asphalt's undesirable temperature susceptibility, the idea of alloying asphalt with rubber to give it greater uniformity over a wide temperature range and to make it more elastic has been pursued since the mid-nineteenth century. British patents date back to 1843. A century later, there were more than 100 references and articles on the subject. Early applications included roofing mastics, protective coatings, caulking, highway joint sealants, and bituminous paving.

Modified-bitumen roof membranes were introduced in Italy in 1967, via an atactic polypropylene (APP) polymer-asphalt mix produced by Romolo Gorgati. By 1969, modified bitumens were available in Germany; they were available in Holland and Great Britain by the mid-1970s, and in the United States by the mid-1970s. By the mid-1980s, they had completed their conquest of Europe, claiming nearly 90 percent of the market in Italy and 60 percent in France and Norway. Though the United States has lagged behind Europe, modified bitumen's share of the market has grown from 10 to 15 percent in the mid-1980s to roughly 35 percent in the mid-1990s. This phenomenal growth shows no sign of decelerating.

Skyrocketing petroleum prices in the early 1970s spurred the development of modified-bitumen membranes in the United States. Prior to that enormous price rise, the unit cost of rubberizing polymers had been up to 20 times that of asphalt. This cost differential imposed an economic limit of about 5 percent polymer, which doubled the cost of the asphalt-polymer blend. When this extraordinarily high price ratio dropped rapidly in the early 1970s, polymer-asphalt ratios as high as 35 percent became economical. Rapid growth of modified-bitumen membranes followed.

Accompanying this economic development were two polymer production breakthroughs:

- The introduction of a plastic material, isotactic polypropylene (IPP), leading to the production of atactic polypropylene (APP)
- The development of elastomeric block copolymers, sequenced butidiene-styrene (SBS) and styrene (or "sequenced") ethylene-butylene-styrene (SEBS).

IPP has remarkable fatigue resistance, a property making it useful for applications involving integrally molded plastic hinges—e.g., tool boxes and computer cases. Polymerization of this isotactic plastic, via solution and slurry processes, yields APP as a waste by-product. The search for productive uses for this waste by-product was successful. APP blends readily with petroleum asphalt, and, unlike more crystalline plastics, remains well dispersed in asphalt mixtures.

Homogeneity is obviously essential to the long term performance of a modified-bitumen roofing material. Phase separation would create randomly distributed defects that would inevitably produce premature failure of the material. Discovery of this inexpensive elastomeric polymer thus toppled the economic barrier to rubberized asphalt.

Though no longer a by-product of IPP production, which has shifted from the less efficient solution-and-slurry process to the more efficient gas-phase process, APP remains relatively inexpensive. APP is now produced via a melt-phase process for amorphous polyolefins (APOs), one type of which is amorphous polypropylene (APP), the other amorphous propylene ethylene copolymer (APE). These polymers can be custom-blended for a prescribed set of physical properties.

With the second breakthrough, the development of SBS and SEBS elastomeric copolymers, came unique macromolecules ideally suited to impart rubberlike properties to asphalt blends. Though APP-modified bitumens were the early leader, SBS-modifieds had pulled roughly equal in market share by the mid-1990s. Reinforcement, predominantly glass fiber and polyester, sometimes used in combination, adds further variety to modified bitumens, creating a broad spectrum of strain-energy properties, from high strength–low strain to low strength–high strain. And in the most recent advance, the mopping asphalt itself is modified with APP or SBS polymers, thereby further expanding the benefits of polymer-modified-bitumen roof systems.

Modified-Bitumen Materials

Rubberizing polymers

With modified bitumens, polymer chemistry replaces the airblowing process that produces the asphalts used in traditional BUR membranes. Airblowing is a dehydrogenation process, removing lighter oils from the asphalt as water molecules are formed and evaporated from the heated, oxygenated asphalt. For modified bitumens, airblowing is reduced or totally eliminated. This reduced airblowing preserves the lighter oils, thereby improving flexibility and weatherability. The added polymer increases asphalt viscosity. APP polymer raises the softening point of raw asphalt from 90 to 300°F (32 to 150°C), nearly 100°F (56°C) above the softening point of ASTM Type III (steep) asphalt.

The two major polymers used in modified bitumens, APP and SBS, differ fundamentally in their chemical nature. APP is a *plastomer,* whereas SBS is an *elastomer.* This chemical difference manifests itself physically in much greater elasticity for SBS-based modified bitumens, with more nearly uniform properties through a wide temperature range—e.g., greater flexibility at low temperature. APP-modified bitumens are generally stronger and stiffer than SBS-modifieds. They also have greater resistance to high temperatures.

APP thermoplastic polymer forms a uniform matrix in the blended asphalt. It increases the blown asphalt's ultraviolet resistance, flexibility at high and low temperatures (though not as much as SBS), resistance to flow at high temperatures, breaking strain, and even waterproofing quality.

SBS (and SEBS) work via a more complex chemical process, and as a consequence, the chemical constitution of the blending asphalt is far more critical than for APP blends. With SBS-modified bitumens, it may be necessary to adjust the aromatic or aliphatic content of the bitumen by adding selected petroleum oils to the mix.

In its results, the asphalt-rubberizing process with SBS polymer parallels the elasticizing process that converts brittle iron into ductile steel through the addition of carbon, manganese, and other elements. SBS and SEBS are copolymerized blocks of thermoplastic crystalline material, polystyrene, with blocks of rubbery butadiene. Incompatible with the bitumen, the glassy polystyrene end blocks of adjacent molecules tend to associate into small, rigid domains. They form cross links, locking the long polybutadiene chains in a three-dimensional rubber network (see Fig. 11.1). The rubbery butadiene chains (copolymer midblock segments) provide high elasticity and fatigue resistance. The polystyrene, which constitutes the copolymer end blocks, adds low-temperature flexibility, via its extremely low glass-transition temperature [−121°F (−85°C); see Fig. 11.2].

A sponge metaphor makes it simpler to visualize. At concentrations of only 10 to 15 percent polymer, the SBS absorbs the 85 to 90 percent

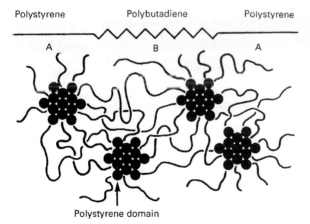

Figure 11.1 In SBS-rubberized asphalt, glassy polystyrene end blocks form cross links, locking long polybutadiene chains into a three-dimensional rubber network. (*Shell Chemical Co.*)

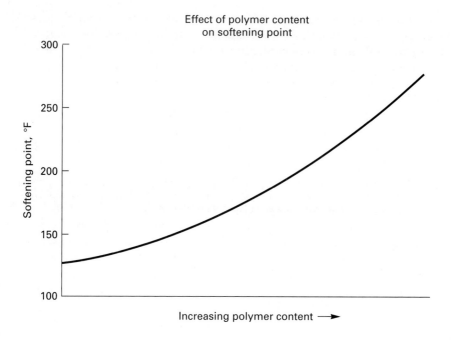

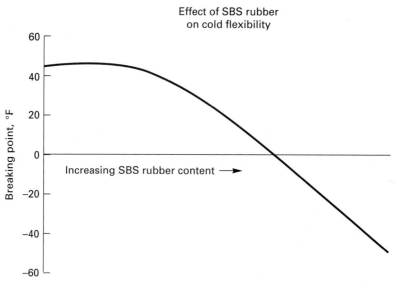

Figure 11.2 Top graph shows rising softening point of rubberized asphalt with increasing SBS polymer content. Bottom graph shows increasing SBS content increasing low-temperature flexibility. (*Tamko.*)

bitumen and becomes chemically the predominant ingredient. This process resembles the action of a sponge, which can soak up many multiples of its weight in water. Despite the greater weight of the water, the result is a wet sponge, not spongy water. Just as the water dominates in the sponge-water mix, the SBS polymer dominates the polymer-bitumen blend in an SBS-modified bitumen.

APP- and SBS-modified bitumens differ not only in softening-point temperature, but in their response to this temperature. APP's crystalline structure gives it a definite melting point, with rapid conversion from solid to liquid at 300°F (149°C). In contrast, SBS-modified bitumen gradually melts in a 210 to 250°F (approximately 100 to 120°C) temperature range. This behavior is characteristic of a rubbery, cross-linked molecular material.

SBS-modified membranes offer greater versatility in application techniques than APP-modified membranes (see Table 11.1). APP-modifieds, with their high polymer content, can be melted only via propane torching (see Fig. 11.3). With their much lower polymer content, SBS-modified bitumens can be hot-mopped at application temperatures around 425°F (220°C). Special torching-grade SBS-modifieds are manufactured with thicker material on the bottom. (The added bitumen compensates for the lack of the bitumen supplied by hot mopping.)

Reinforcement

Located within the modified-bitumen sheet cross section, reinforcement serves several purposes:

- As a carrier during factory production
- As a means of increasing tensile strength and puncture resistance
- As a fire-protection enhancement
- As a structural element bridging substrate gaps

The particular properties imparted by reinforcement depend upon two factors: the type of fabric and its material. Reinforcing fabrics come in three types:

- Scrims
- Nonwoven (i.e., mats)
- Composites (i.e., scrim and nonwoven)

TABLE 11.1 Properties of Basic MB Sheets, Atactic Polypropylene (APP) and Sequenced Butadiene-Styrene (SBS)

Generic category	Typical thickness, mils (1 mm = 39 mils)	Type of reinforcement G = glass fiber E = polyester	Weight of polyester reinforcement, g/m^2	Factory surfacing N = none G = granules F = foil	ASTM specs	Application method
APP	150, 160, 180, 200	G, E	160, 170, 250	N, G, F	No	Torch
SBS	120, 150, 160, 180	G, E	160, 170, 250	N, G, F	No	Mop, torch, cold-adhered

Figure 11.3 Propane torching of APP-modified-bitumen sheets is mandatory because APP-modified asphalt cannot be melted with hot-mopped asphalt. (*Tamko.*)

Scrims are open fabrics comprising two sets of yarns perpendicular to each other and held together by weaving or adhesive bonding. Scrims offer high tensile and tear strength in both machine (longitudinal) and cross-machine (transverse) directions.

Nonwoven fabrics comprise randomly distributed fibers bonded thermally, chemically, or mechanically (or some combination of these three techniques). Their dense overlapping arrangement of fibers provides good puncture resistance. Breaking strain depends upon the fiber material, its orientation, and its bonding strength.

Composites combine scrim and nonwoven fabrics, chemically or mechanically bonded. Materials are chosen for special properties. Composites and nonwovens are common as modified-bitumen reinforcement. Because they must function as carriers during the manufacturing process, when they must retain the polymer-bitumen mix, scrims are generally unsuitable for modified-bitumen reinforcement, except in combination with nonwovens.

The predominant materials used as modified-bitumen reinforcement are *glass fibers* and *polyester*. Glass fibers provide better dimensional stability, fire resistance, and ultraviolet resistance. Polyester provides greater strain energy. (At low temperature, it has nearly 50 percent breaking strain vs. 10 percent for glass-fiber mats.) Polyester also has greater flexibility and fatigue and puncture resistance.

A composite reinforcement might be designed as a polyester nonwoven, chosen for superior puncture resistance and strain energy, adhered to a glass-fiber scrim, chosen for dimensional stability.

The no-free-lunch rule asserts itself in these material and fabric selections. Polyester's relatively low ultraviolet and fire resistance detracts from its puncture resistance. Glass fiber's greater dimensional stability comes with a countervailing liability: the notorious glass-fiber "memory" (i.e., its tendency to retain kinks and other irregularities instead of relaxing like more flexible materials). Woven glass cloth reinforcement exhibits less memory than nonwoven glass-fiber mats. It makes a good reinforcement for modified flashing sheets, where memory poses a special hazard. (Flashing material must be moldable, to conform with its backing surfaces.) But even in membrane sheets, memory poses a problem, causing sheets to bridge over irregularities in a substrate or to leave end lap voids that can grow into blisters.

In an effort to exploit the complementary properties of polyester and glass fiber, manufacturers sometimes use two layers of reinforcement instead of a single composite layer. This increases the risk of delamination, especially if the sheet is heated excessively on the top side, where the reinforcement is close to the surface. Another approach features use of a glass-reinforced base sheet combined with a polyester-reinforced cap sheet.

Surfacing

Surfacing for modified-bitumen membranes provides heat resistance (via solar reflectance), ultraviolet resistance, and fire resistance. SBS-modified sheets require surfacing because of their slight ozone and ultraviolet resistance. Because of their greater weatherability, APP-modified sheets may be left unsurfaced. This is, however, a bad practice, often resulting in surface checking that shortens membrane service life. All modified-bitumen membranes should have some kind of surfacing.

Surfacing comes in three basic types:

- Mineral granules
- Metallic foil
- Field-applied protective coatings

The most common surfacing for SBS-modified sheets is the same mineral-granule surfacing used on asphalt shingles and the mineral-surfaced coated cap sheets used in traditional BUR membranes. Granules rolled into the sheet's top surface are retained longer than those on sheets lacking mechanical embedment. If these shielding granules are lost, resurfacing, generally with a liquid coating, will be required later in the membrane's service life. Granule loss often indicates manufacturing defects.

Metallic foil (copper, aluminum, or stainless steel) is another type of surfacing that requires special care in the manufacturing process. An earlier version of proprietary modified bitumen surfaced with aluminum foil was withdrawn from the market after widespread evidence of dimensional instability. With its original embossed foil surfacing, this product generally performed satisfactorily in its early days. Later modifications in the embossed pattern resulted in a tendency toward delamination, leaving the exposed polymer-modified asphalt vulnerable to ultraviolet degradation. Successful bonding of aluminum foil to polymer-modified asphalt requires a pattern of tiny expansion joints in the foil, to accommodate differential rates of thermal expansion-contraction.

One major manufacturer coats just the underside vertices of the Vs in the embossed foil with a soft bitumen. When the foil is laminated to the modified-bitumen sheet, the vertices of the embossed Vs can release when the heated soft bitumen releases its tenuous bond with the underlying bitumen, and the Vs themselves can open and close without breaking the bond between foil and bitumen. There is a thermal stability test that can accurately predict field performance. Loss of the foil facing poses a serious problem for SBS-modified bitumens. They deteriorate rapidly when exposed to ultraviolet radiation.

Protective coatings for modified-bitumen sheets are essentially the same as those used for smooth-surfaced built-up roof membranes: fibrated aluminum, asphalt emulsions, acrylics, and the like. However, modified-bitumen membranes are more difficult to coat than smooth-surfaced BUR membranes. Some exude a waxy substance that must "cure" before application of the coating. Adhesion is more difficult to achieve, because of the modified bitumen's plastic nature.

Guidelines for specifying modified-bitumen coatings appear in the Roof Coatings Manufacturers Association (RCMA) publication, *Methods for the Preparation of MB Membranes for the Application of Surface Coatings*. These guidelines include procedures for brooming, scrubbing, cleaning, drying, power washing, aging, and priming. Reinforcing the cardinal rule of roof design, these guidelines include a recommendation for positive drainage, defined as enough slope to assure drying of the roof within 48 h after a rainfall.

In addition to the three basic surfacings, modified-bitumen membranes are sometimes covered with loose aggregate, like single-ply ballasted systems, or embedded aggregate, like conventional built-up membranes. For the latter condition, the same 60 lb/square flood coat is required. In its adaptability to varied surfacings, modified bitumen is the most versatile of all roof systems.

Modified-Bitumen Specifications

As modified-bitumen specifications have evolved, two-ply membranes, and occasionally three-ply, have become standard. When first introduced in the United States in the late 1970s, modified bitumens were classified as single-ply, along with elastomeric and thermoplastic sheets. Single-ply modified bitumens are now restricted to such limited uses as re-covering old smooth-surfaced BUR membranes. For new construction, experience has demonstrated the prudence of adding a base sheet.

The most common modified-bitumen specification comprises a base sheet and a cap sheet, and calls for phased construction (i.e., the base applied in one operation, followed by the cap sheet in a separate operation, rather than simultaneous shingled application (see Fig. 11.4). Three-ply hybrid membranes, comprising two hot-mopped, shingled 15-lb glass-fiber felts, followed by a modified bitumen cap sheet, have also become more popular in the 1990s.

(As discussed later in this chapter, phased construction poses a blistering hazard for modified-bitumen membranes as well as for BUR membranes. The cap sheet should be applied on the same day as the base sheet.)

Base sheets for the predominant two-ply, phased modified-bitumen membrane can vary widely. Listed in order of expense and quality, they include the following:

- Another modified-bitumen sheet
- Coated glass-fiber felt
- Standard glass-fiber felt
- Coated *organic* felt

Adding a base sheet to a single-ply modified-bitumen membrane offers several notable advantages. A base sheet protects vulnerable plastic foam insulation from damage by torch heat (up to 3000°F, 1650°C) or being melted by hot-mopped asphalt. A base sheet integrated with a modified-bitumen cap sheet into a two-ply membrane also adds tensile strength and impact resistance. The combination of a polyester-reinforced APP-modified-bitumen sheet with a polyester base sheet can virtually double membrane tensile strength and strain energy.

A comparison of different two-ply combinations of modified-bitumen membranes by French researcher Alan Chaize further demonstrates the

Modified-Bitumen Membranes   281

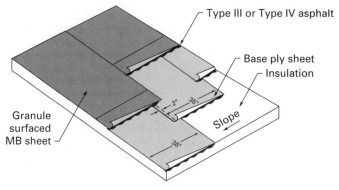

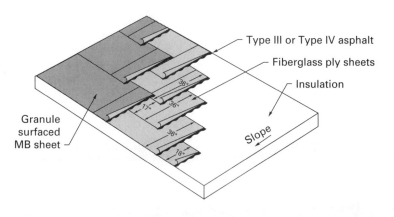

Figure 11.4 The most common modified-bitumen membrane comprises a two-ply membrane, with base sheet and granule-surfaced cap sheet applied in two separate operations (top). Three-ply membrane, with two shingled glass-fiber felts topped with granule-surfaced cap sheet, has become popular in the 1990s. (*Tamko.*)

advantages of two-ply over single-ply. When the base sheet for a polyester-reinforced SBS-modified-bitumen cap sheet is another such sheet, allowable movement at 14°F (−10°C) is nearly 4 percent. This qualifies it for an F5 (the highest) rating in the FIT (fatigue-indentation-temperature) roof classification. (For more detailed discussion of FIT, see Chapter 2, "The Roof as a System.") This allowable movement greatly exceeds that recorded for the same polyester-reinforced SBS cap sheet combined with a woven glass-fiber felt or even a glass-fiber-reinforced SBS sheet as the base sheet. (These data also demonstrate the greater membrane breaking strain achievable with polyester reinforcement.[1])

Hot mopping vs. torching is another decision to be made by the specifier. With APP-modified bitumen, there is obviously *no* choice: The high melt point of polymerized asphalt in these membranes requires torching. Where there is a choice, with SBS-modified bitumens, the designer should consider several advantages of hot mopping:

- Hot mopping generally produces a superior lap seal.
- Hot mopping is generally safer than torching.
- It is generally faster than torching.
- If hot asphalt is required for other roofing operations—to adhere insulation or base sheet—it is more economical to hot-mop the cap sheet as well. (See Fig. 9.4 in Chap. 9.)

Since defective lap seams constitute the most common failure mode of modified-bitumen membranes, the first-cited advantage of hot mopping is especially important. With torching, there is an optimum heat, and so more skill is required to produce a strong, dependable lap seam than with hot mopping. Overheating with the torch can destroy the joint material. Underheating produces a weak joint or "pseudoseal" that later separates. Many torch-grade sheets have a bottomside, "burn-away" plastic film facer (a releasing agent for the sheet roll). If torching heat is inadequate, this plastic facer film may merely shrivel instead of disintegrating in the torching heat. This shriveled plastic prevents the complete fusion of modified bitumen required for a strong lap joint.

Torching has another disadvantage compared with hot mopping. Along the overlap line, at the boundary of the lap-seam transition from two layers to one, if the torch operator fails to trowel the cap sheet down into the base sheet, an elongated blister void may occur. (Such a void can occur at the base sheet's lap seam or in the cap sheet's lap seam.) At a hot-mopped lap joint, such a void is less likely to form, because fluid bitumen can fill the cross-sectional area.

Hot mopping has several disadvantages, notably:

- The need for ancillary operations—preheating in a kettle, transport of hot asphalt to the application point—not required for torching.
- Greater slippage hazard, attributable to use of a softer, more fluid bitumen, producing thicker layers. (To combat this hazard, some manufacturers require ASTM Type IV mopping asphalt.)

Modified-Bitumen Flashings

Flashings for modified-bitumen membranes are normally torched rather than hot-mopped. Like hot-mopped built-up flashings, hot-mopped modified-bitumen sheets are difficult to adhere to their backing. Hot-mopped Type III or IV asphalt requires a minimum 400°F (205°C) temperature for dependable adhesion. Under normal field conditions, especially when there is appreciable wind, rapid cooling of the hot asphalt gives the workers only 5 s or so to embed the flashing sheets. Torched application, in contrast, allows a few additional seconds to adhere the sheets. Cold adhesives, still another choice, have not displaced torching as the most dependable and popular method.

Torched APP-modified-bitumen flashing sheets should always be used with APP-modified-bitumen membrane. Torch-applied SBS is more appropriate for SBS-modified bitumens. (See Fig. 11.5.)

"Torch and flop," the practice of heating the flashing sheets on their bottom side and then "flopping" them against their backing surface, is, at best, a hazardous procedure. Unless it is performed fast, it will produce poor (or nonexistent) adhesion of the modified-bitumen flashing sheet to its backing, as the liquid mix rapidly congeals (in accordance with Newton's exponential cooling law). It is sometimes necessary to prevent the torching flame from getting under the

Figure 11.5 Modified-bitumen flashing sheet is torched to lead flashing flange of vent-pipe flashing. (*Roofing Industry Educational Institute.*)

counterflashing and igniting combustibles. At best, however, torch and flop is the lesser of two evils.

"Mop and flop," sometimes used to keep the mopping asphalt adhering cant strips from running into the building, poses an even greater risk of poor adhesion than torch and flop. (Heated to a far lower temperature than torched modified-bitumen flashing sheets, hot-mopped asphalt cools to below its minimum dependable adhesion temperature even faster.) Mop and flop techniques should always be approved reluctantly and restricted to instances where they pose the lesser of two risks. And when these techniques are approved, the specifier should insist on rigorous inspection, to assure good adhesion of flashing to its backing. (See Fig. 9.4 in Chap. 9.)

Avoid the practice of preheating of cold flashing surfaces. Preheating is sometimes advocated as a precaution to assure good adhesion, since cold and moist surfaces accelerate the cooling of the heated polymer-asphalt mix and threaten its adhesion to its supports. In cold weather, keep modified-bitumen rolls in a heated storage space. If that is insufficient, suspend operations until the advent of warmer weather.

Like BUR base flashings, modified-bitumen flashings should be mechanically anchored along their top edge, with maximum spacing of 8 in. (200 mm). Hot-mopped flashing installations are most vulnerable to sagging. But even torched installations may not retain adequate adhesion, especially where moisture accumulates behind the flashing.

Performance Criteria for Modified Bitumens

Paralleling the pioneering work of NIST researchers W. C. Cullen and R. G. Mathey in "Preliminary Performance Criteria for Bituminous Membrane Roofing," published in 1974, NIST researchers W. J. Rossiter and J. F. Seiler produced "Interim Criteria for Polymer-Modified Bituminous Membrane Materials" (*Build. Sci. Ser.* 167) in 1989. In the mid-1990s, this publication remained the latest and most complete survey of performance criteria for modified-bitumen membranes. (An ASTM standard, under consideration for nearly a decade, remained incomplete as this third edition of the roof manual went to press.)

Once a generally recognized standard is promulgated, the designers' task in selecting modified-bitumen membranes will be greatly facilitated, to say the least. It will make comparison of different manufacturers' products much simpler. With industry recommendations for a complete array of tested properties available, the task of selecting a modified-bitumen specification will be far less lonely than it is now, with no industry-recognized format for reporting performance data. In the meantime, designers must make the best of an unresolved situation, using the best available data.

Build. Sci. Ser. 167 provides guidelines for a combination of performance and prescriptive criteria. They were promulgated for Department of Defense (DoD) construction agencies until voluntary consensus standards are available from ASTM. (Table 11.2 summarizes these criteria.) In addition to these suggested criteria for modified-bitumen membranes, *Build. Sci. Ser.* 167 also cited other criteria needed for the future (see Table 11.2, bottom section). Lack of an acceptable data base prompted the NIST researchers to exclude these from the suggested criteria. Certainly the most important of these omitted criteria is seam strength, the most common problem for modified bitumens.

The attributes that most definitively exhibit modified bitumens' superiority over conventional BUR are the prescriptive criteria for *low-temperature flexibility* and *strain energy*.

Low-temperature flexibility is the lowest temperature at which a membrane sample can be bent 180° around a 1-in. (25-mm) mandrel in approximately 2 ± 1 s without exhibiting *any* visual signs of flexural cracking. There are two criteria for the tested membrane specimens: (*a*) an absolute maximum of 25°F (-4°C), and (*b*) not more than 5°F (3°C) greater than the nominal value published by the manufacturer. The ASTM draft has an elaborate description of required testing procedure. It involves numerous trials at various temperatures, dropping the test temperature in 5°F (3°C) increments for samples that pass, and raising the test temperature by the same increments for samples that fail (i.e., samples that crack). The process continues until discovery of the limiting temperature.

Strain energy measures the membrane's ability to absorb the kinetic energy transmitted via the *relative* movement between substrate and membrane without rupture. This relative movement can result from thermal contraction of the membrane with respect to the substrate. It can also result from substrate movement, which concentrates tensile stress at shrinkage cracks and insulation joints. By elementary physics, energy is the capacity for doing work. For a unit cross-sectional area of membrane, it is measured in unit stress times unit strain (elongation)— i.e., lbf · in./in.^{2} or kN · m/m^2. Strain energy is thus the area of the stress-strain curve, with measurements carried out to the breaking strain (see Fig. 11.6).

Modified bitumens display a tremendous range in breaking strain (i.e., maximum elongation under tension, expressed as a percentage of unstressed original test-specimen length). Breaking strain of commercially marketed membranes tested by NIST researchers for *Build*

TABLE 11.2 Summary of Suggested Prescriptive and Performance Criteria for Modified Bitumens

Requirement	Test method	Criterion
Thickness	ASTM draft, Sec. 5	Min. 95% of nominal, 40 mils (1 mm) min.
Load, max.	ASTM draft, Sec. 6	Min. 85% of nominal
Elongation (max. load)	ASTM draft, Sec. 6	Min. 80% of nominal
Elongation (break)	ASTM draft, Sec. 6	Min. 80% of nominal
Low-temperature flexibility	ASTM draft, Sec. 11	Max. 5°F (3°C) above nominal, max. 25°F (−4°C)
Tear resistance	ASTM draft, Sec. 7	Min. 80% nominal, 30 lbf (130 N) min.
Dimensional stability	ASTM draft, Sec. 10	Max. change, ± 1%
Fire resistance	ASTM, UL, or FM test	Conform to applicable code
Flow resistance	UEAtc No. 27 Sec. 5.1.7	No slippage
Hail impact	NBS BSS 55	1.5-in. (38-mm) hailstone at 112 ft/s (34 m/s) without water penetration
	or ASTM D 3746	22 lbf · ft (30 J) without water penetration
Moisture absorption	ASTM draft, Sec. 9	1 g (per specimen), max.
Moisture content	ASTM draft, Sec. 8	0.5% by mass, max.
Pliability	a) ASTM draft, Sec. 11*	No cracking at temperatures of application
	b) UEAtc No. 27 Sec. 5.4.3	No cracking or tearing when unrolled at 32°F (0°C)
Strain energy	ASTM draft, Sec. 6	Not less than 3 lbf · in./in.^{2} (0.5 kN · m/m^2)
Uplift	ASTM, UL, or FM test	Conform to applicable code
Weather resist. heat exposure	ASTM draft, Sec. 12	15% max. loss of load elongation; low-temperature flexibility not to exceed 32°F (0°C); strain energy not less than 3 lbf · in./in.^{2} (0.5 kN · m/m^2)

Summary of Modified Bitumens for Which Criteria Have Not Been Proposed

Requirement	Test method	Criterion
Cyclic movement†	UEAtc No. 27	No tears, cracks, or wrinkles; no total loss of adhesion
Durability, granules	ASTM draft, Sec. 15	Not established
Durability, foils	CGSB, Sec. 7.2.6	Not established
Puncture (static)	UEAtc No. 27, Sec. 5.1.9	Not established
Puncture (dynamic)	French Std., NF P 84-353	Not established
Seam strength	Not established	Not established
Slippage of base flashing	ASTM draft, Sec. 15	Not established
Tear resistance	Not established	Not established
Water transmission	Not established	The membrane shall be watertight as tested
Weathering (UV)	Not established	Not established
Weathering (moisture)	Not established	Not established
Weathering (chemical)	Not established	Not established
Weathering (fungus)	Not established	Not established

*One could use either the method described in a or in b.

†Although a criterion is proposed for this requirement, a data base is lacking at this time for implementation of the suggested criterion.

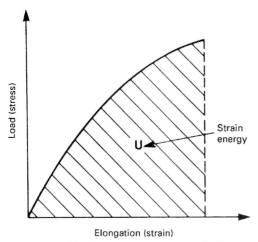

Figure 11.6 Strain-energy curve area indicates splitting resistance of modified-bitumen (or any other type) membrane. (*Roofing Industry Educational Institute.*)

Sci. Ser. 167 ranged from 2 to 150 percent. With high breaking strain, a membrane can have low tensile strength and nonetheless have satisfactory strain energy. With a low breaking strain, on the other hand, the membrane obviously requires high tensile strength. For a useful analogy, flexible earthquake-resistant structures can be designed to yield to ground movement, thus requiring less strength than more rigid structures, which experience greater shock with the transfer of dynamic stress. (As an even simpler analogy, you can absorb more kinetic energy when you jump from a height by flexing your knees during impact, thus slowing the deceleration rate, which is directly proportional to the impact stress exerted on your body.)

In addition to a *strain-energy* criterion as an index of the membrane's ability to absorb substrate movement, there is an associated concept that might be termed *usable strain energy*. Excessive membrane elongation short of breaking the membrane may cause cracking of the bitumen, with consequent leakage. A watertightness test is thus conducted at an elongation corresponding to this minimum required *usable strain energy*. This minimum may, in some instances, approach total strain energy—i.e., the strain energy at breaking strain. A minimum strain energy of 3 lbf · in./in.^{2} (0.5 kN · m/m^2) has been proposed for modified-bitumen membranes. At this strain energy, the tested specimen must pass the watertightness test.

Modified-Bitumen Failure Modes

Ranked in rough order of their incidence, the failure modes of modified-bitumen membranes are as follows:

- Defective lap seams
- Shrinkage
- Checking
- Blistering
- Delamination
- Slippage
- Splitting

Defective lap seams

Lap-seam failures, by far the most common failure mode for modified bitumens, account for more than three times the reported occurrence of the next most prevalent problem, according to one survey.[2]

Failure of SBS-modified-bitumen hot-mopped seams often results from underheated asphalt—i.e., application temperature below 400°F (204°C). Even with SBS-modified bitumen, the polymer-asphalt blend requires a hotter mopping asphalt temperature than felts in a conventional built-up membrane for solid fusion of the overlapped sheets. Solid fusion is essential for watertight side and end laps.

In contrast with hot-mopped lap seams, where failure almost always indicates too little heat, torched lap seams can be spoiled by either too much or too little heat. Proper torch lap sealing requires an applicator who is sensitive to balanced requirements, a worker imbued with Aristotle's doctrine of the golden mean. Underheating fails to fuse the asphalt mix, with consequent production of a weak joint. Too much heat, on the other hand, can cook the materials, also producing a weak joint. Overheating can sometimes be detected by a visible reinforcing pattern on the top sheet, indicating excessive loss of material.

Proper application of torch heat to lap seams depends on two factors: flame temperature and duration of torch application. Application time can be controlled with greater precision than flame temperature. A well-trained applicator knows the proper timing to produce a well-fused lap joint.

Lap seams should be field-checked as application proceeds. In properly constructed field joints, either hot-mopped or torched, fluid asphalt mix should bleed out in a continuous line about $\frac{1}{4}$ in. (6 mm) wide beyond the edge of the top sheet. This conventional roofing wisdom is not, however, an infallible index of a strong lap seam. NIST researchers investigating 53 modified-bitumen roofs found no definite correlation between "bleed-out" and strong joints.[3] On balance, however, "bleed-out" appears to be a positive indicator of a good joint, though not sufficient proof in itself of the joint's adequacy. (At the least, absence of bleed-out furnishes grounds for suspicion.)

Troweling of torched lap seams can help to assure the seam's fusion. (This practice is required by some manufacturers.) An on-site inspector should probe the field seams with a trowel shortly after their application (see Fig. 11.7). Unadhered areas require reheating or patching, 4 in. (10 cm) on each side.

Lap seams can sometimes separate in mineral-surfaced SBS-modified-bitumen sheets. Mineral-surfaced sheets should have selvage edges. But at end laps the granules can interfere with fusion of the two modified-bitumen surfaces. To avoid this problem, the granules can be heated and pressed into the coating. As an alternative, the lap area can be primed with asphalt primer. This practice promotes adhesion between the two polymerized asphalt surfaces.

Figure 11.7 Inspector probes modified-bitumen field-fabricated lap seam with trowel point shortly after application. Defectively sealed spots require reheating or patching. (*Roofing Industry Educational Institute.*)

Figure 11.8 Cracking of an 8-year-old APP-modified-bitumen membrane was worse in ponded areas than in dry areas. (*National Roofing Contractors Association.*)

Checking

Checking in modified bitumens resembles alligatoring in smooth-surfaced BUR membranes (see Fig. 11.8). Unsurfaced APP-modifieds offer the most common examples. Despite their superior resistance to ultraviolet radiation, APP-modified-bitumen membranes nonetheless succumb under long term exposure, especially in warm climates with high levels of solar radiation. The destructive photo-oxidation attack energized by the sun's ultraviolet radiation accelerates at rising temperatures, roughly doubling with each 18°F (10°C) increase in temperature. This exponentially increasing hazard from high surface temperature highlights the importance of heat reflectance of the membrane surface. Black APP-modified-bitumen-membrane surface temperature can reach 175°F (80°C) in Phoenix, where the *normal* summer high is 105°F (41°C), and extreme solar radiation levels increase the thermal load. Use of a light-colored coating capable of reducing surface temperature by only 36°F (20°C) can thus reduce the embrittlement reaction by 75 percent during peak-temperature hours, which are abundant in Phoenix's long, hot, sunny summers. Ponding appears to aggravate checking.

Shrinkage

Shrinkage appears to be a manufacturing problem, manifested in a readily observable shortening of modified-bitumen cap sheets. Shortening of about ½ in. (12 mm) exposure at end laps sometimes occurs within a 24-h period after installation. This contraction may result from the relaxation of tensile stress created during the manufacturing process, when the cooled material tends to contract but is restrained by frictional forces tending to retain elongated shape within the roll. One manufacturer has apparently solved the problem by machine-calendering the polyester reinforcement.

The deleterious consequences of shrinkage include removal of ultraviolet-shielding granules and creation of voids at tee joints (where the underlying sheet contracts).

Blistering

Compared with the epidemic scope of blistering in the bad old days of conventional built-up roof systems with organic and asbestos felts, blistering of modified-bitumen membranes is a relatively minor problem. Not only is it less common, but it is a less serious hazard when it does occur. Blisters in an aggregate-surfaced built-up membrane can set in motion a vicious downhill process—flood-coat flow down the blister's sides, eventual exposure of the vulnerable felts to ultraviolet degradation, fungus rot (of organic and asbestos felts), and eventual cracking and puncturing, admitting water into, and possibly through, the membrane. Blistered modified-bitumen membranes are less vulnerable to this accelerating degradation process. Despite this favorable comparison with built-up membranes, however, blistering in modified-bitumen membranes nonetheless ranks as a serious problem, making the membrane highly vulnerable to puncture from foot traffic, dropped workers' tools, hailstones, and the miscellaneous missiles that roofs seem to attract (see Fig. 11.9).

The triumph of glass-fiber felts in built-up membranes has drastically reduced blistering as the scourge of built-up roof systems. Glass-fiber felts have two advantages over their vanquished competitors, organic and asbestos felts. The greater porosity of glass-fiber felts prevents the containment of excess pressure within the blister chamber (i.e., gage pressure higher than atmospheric pressure) that is necessary to propel the growth of blisters from small voids. Glass-fiber felts also lack the high water absorptivity of the organic fibers used in both organic and asbestos felts. Release of this moisture into mopping voids and its evaporative expansion under solar heating promote blister growth in organic- and asbestos-felted built-up membranes.

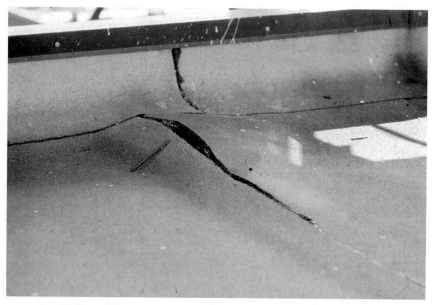

Figure 11.9 This modified-bitumen membrane blister was found on a 5-year-old SBS-modified-bitumen membrane. Note gaping seam split in cap sheet. This blister required repair, desperately. (*National Roofing Contractors Association.*)

Like the substitution of nonhygroscopic glass-fiber felts for water-absorbing felts in built-up membranes, use of polymer-modified asphalt sheets reinforced with glass fiber and polyester similarly reduces the hazard of blistering in modified-bitumen systems. It removes a major source of blistering: the release of moisture from highly water-absorptive materials into voids that later expand under solar heating.

Despite the elimination of perhaps its major cause in built-up membranes, however, blistering does occasionally occur in modified-bitumen membranes. Torched applications have less tendency to blister than hot-mopped applications. Torching at temperatures approaching 3000°F (1700°C) is more likely to evaporate moisture than hot-mopped asphalt, with its maximum application temperature of 450°F (232°C) or so. Hot-mopped SBS-modified membranes have accordingly demonstrated a greater propensity for blistering than torched APP membranes. But torched APP membranes have nonetheless blistered—notably over structural concrete decks to which they were directly applied without interposition of insulation.

Blisters in modified-bitumen membranes are more likely to be *interfacial* (i.e., between substrate and membrane base sheet) than *interply* (i.e., between base sheet and cap sheet). This distribution differs from that of BUR blisters, which are overwhelmingly interply. Nonetheless, interply blisters do occur in modified-bitumen membranes, and for the same reasons that they occur in BUR membranes. Phased application, with delays measured in days (or weeks) between application of base sheet and cap sheet, is cited as the primary cause of modified-bitumen blistering by a major manufacturer following an extensive investigation.[4] Moisture and/or contaminants (dirt, dust, debris) prevent adhesion and form voids that later expand into blisters through the cycling pattern described later.

Blisters originate with the formation of a void in the adhesive bitumen, either between the plies or between the membrane and its substrate. By general consensus, liquid moisture entrapped within the void promotes blister growth. Trapped within a small void, liquid moisture exerts a tremendous, exponentially increasing pressure, with consequent expansion of the void, under solar heating. In summer, midday and early afternoon membrane temperature can easily rise from 70°F (21°C) to 150°F (66°C). Under standard atmospheric pressure, this 80°F (45°C) temperature rise would expand the liquid moisture to about 1500 times its liquid volume. If the resulting water vapor were confined within a constant volume, the water-vapor pressure would increase by more than 4 psi (600 psf). In an actual blister, neither the extreme-case volume nor the extreme-case pressure occurs as the temperature rises. Because the membrane resists expansion of the evaporated moisture, the blister chamber's volume is far less than the volume of freely expanding atmospheric vapor. Nonetheless, the blister volume does expand, greatly relieving the pressure that would exist within an unyielding, constant-volume blister chamber.

When the pressure within a void exerts enough force to break the resisting bonding forces at the void perimeter, the void starts its cyclical growth into a blister that may ultimately measure several thousand times the volume of the original void. (See Chapter 10 for a more detailed description of blistering mechanics.)

Blister prevention for modified-bitumen membranes requires essentially the same precautions as for built-up roof membranes. Here are the most important rules:

- Store modified-bitumen rolls in dry locations.
- Protect the substrate from accumulating moisture. (Check dew-point temperatures in humid climates.)
- Over moisture-retaining substrates—e.g., lightweight insulating concrete—specify a venting base sheet or layer of porous roof insulation—e.g., glass-fiber insulation—capable of diffusing moisture as it emerges from a wet substrate material.

- Complete membrane application in a single day, applying the cap sheet to the base without delay.

Delamination

Unlike most roof-failure modes, which can usually be blamed on defective application or design, delamination generally indicates a manufacturing defect. The chief cause is use of two reinforcement layers by the manufacturer to exploit the complementary qualities of glass fiber and polyester. Squeezing two layers of reinforcement into the thin cross section of a modified-bitumen sheet creates several hazards. Placing the two layers of reinforcement too close together creates a plane of weakness through the central part of the modified-bitumen sheet's cross section. To prevent this tendency to delaminate requires (a) impregnation of both reinforcement layers with hot asphalt-polymer mix and (b) sufficient asphalt-polymer mix between them. Impregnation of reinforcement is required to develop adhesion between the modified bitumen and its reinforcement. Sufficient modified bitumen is required between the two reinforcement layers to avoid creation of a weak plane, vulnerable to horizontal shearing or peeling forces that can delaminate the sheet. Delamination almost always occurs at a plane between two reinforcement layers.

Delamination has an alternative explanation. If polyester reinforcement is placed too close to either surface, torching heat can either melt or shrink it. The safest location for polyester is thus at the center of the modified-bitumen cross section, with glass-fiber reinforcement located near the top, where its heat resistance enables it to resist ultraviolet degradation and fire exposure.

Slippage

Membrane slippage, relative lateral movement between membrane plies or between base sheet and substrate, can reduce a two-ply membrane to a single ply, with consequent severe loss in service life. It is also difficult and costly to rectify.

A relatively uncommon problem in BUR membranes, slippage is even less common in modified-bitumen membranes. Other factors being equal, it depends on the viscosity of the adhesive bitumen and the highest roof temperature.

Slippage force depends on three basic factors:

- Slope
- Membrane weight
- Bitumen viscosity

Slope is the basic determinant of slippage force. It is equal to $W \sin \theta$, where W is the weight of the membrane above the slippage plane and θ is the slope angle (see Fig. 11.10). Slippage force thus increases almost linearly with increasing slope—i.e., on a slope of 1 in./ft, slippage is almost four times the force on a ¼-in. slope.

Membrane weight, as the basic formula indicates, is also a linear variable affecting slippage.

Bitumen viscosity (and consequent resistance to flow) depends on two basic factors: the nature of the asphalt-polymer mix and roof temperature. Roof temperature, in turn, depends on several variables:

- Air temperature
- Solar radiation intensity
- Roof-surface reflectance
- Substrate thermal resistance
- Substrate heat capacity

Air temperature and *solar radiation,* in combination, impose a heat load on the roof. In a city like Phoenix, with frequent summer highs of 110°F (41°C) and intense, unrelieved solar radiation, a black roof can reach temperatures of 175°F (80°C). A light heat-reflective metal foil or organic coating can reduce this temperature by 50°F (28°C) or so. Roof-surface temperature also depends on the underlying substrate's thermal resistance and heat capacity. (A light, highly efficient foam insulation raises surface temperature, because it simultaneously retards heat flow from the surface and provides little capacity for absorbing heat. On the other hand, a structural concrete substrate, with its low thermal resistance and high heat capacity, promotes low roof-surface temperature.)

As you would anticipate, SBS-modified-bitumen membranes are more vulnerable to slippage than APP-modified bitumens, because hot-mopped asphalt has lower viscosity (i.e., lower flow resistance) than the harder, fused asphalt polymer resulting from torching of

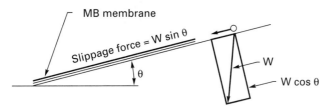

Figure 11.10 Vector diagram shows how slippage force ($W \sin \theta$) increases with increasing slope and membrane weight.

APP-modified sheets. (For a more detailed discussion of slippage mechanics, see Chapter 10.)

Splitting

Splitting, a major problem in conventional BUR membranes (especially in cold climates), has been greatly reduced in modified-bitumen membranes because of their vastly increased strain energy. Stress concentrations producing strains that exceed the roughly 1 percent low-temperature breaking strain of typical glass-fiber BUR membranes can split them. SBS-modified-bitumen membranes, in contrast, have much larger breaking strains (see Fig. 11.11).

Membrane splitting is more common in APP-modifieds than in SBS-modifieds, with their generally greater breaking strain and strain energy. As in BUR membranes, these splits are most common at lines of stress concentrations—notably at insulation joints, where thermal contraction can widen the gap and create intense stress concentration (see Fig. 11.12).

Reinforcement weakened during the manufacturing process may contribute to some modified-bitumen splitting, according to consultant/contractor Dick Baxter.[5]

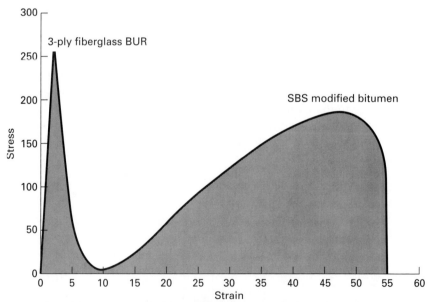

Figure 11.11 Stress-strain curves for three-ply, glass-felted BUR membrane (left) and SBS-modified-bitumen membrane (right) demonstrate the latter's vastly superior split resistance. (*Tamko.*)

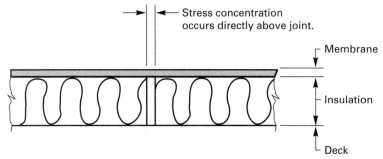

Figure 11.12 Thermal contraction of poorly anchored insulation boards can open joint gaps, concentrating tensile stress in membrane directly above the joint.

Alerts

General

1. Use these alerts in conjunction with manufacturer's recommendations.

2. Require certification of the roofer as a qualified modified-bitumen applicator.

3. Consider on-site inspection of roof application as part of the designer's function.

Design

1. Design a minimum $\frac{1}{4}$ in./ft (2 percent) slope for all modified-bitumen roofs.

2. Limit SBS-modified-bitumen membranes to a maximum 3-in./ft (25 percent) slope.

3. Specify a minimum two-ply membrane (base sheet plus modified-bitumen cap sheet).

4. Specify surfacing for all modified-bitumen membranes. (Consult the RCMA publication, *Methods for the Preparation of Modified Bitumen Membranes for Application of Surface Coatings.*)

5. Require selvage edges on mineral-surfaced modified-bitumen sheets (plus special field precautions to "sink" granules in mopping or torching of end lap seams).

6. Specify cant strips at 90° flashing corners (perlite cant strips for torched flashing application).

7. Specify two-ply, not single-ply, base flashings.

8. Provide an additional (third) flashing sheet at inside (270°) and outside (90°) corners.

9. Require termination bars where counterflashing is omitted.

10. Require taping of protective cover boards installed over EPS insulation. If cover boards are mechanically anchored, specify a barrier sheet (e.g., red rosin sheathing paper) to prevent melting of the EPS.

Field

1. Limit APP-modified-bitumen application to 40°F (4°C) minimum temperature.

2. Require a minimum temperature of 400°F (205°C) for hot-mopped application at point of application.

3. Trowel lap seams for torched applications to assure their integrity.

References

1. Alan Chaize, Centre Scientifique et Technique du Bâtiment, Paris, France, *Intern. Symp. Roofing Technol.*, NRCA, 1991.
2. W. C. Cullen, "Project Pinpoint's Database Continues to Grow," *Professional Roofing*, April 1990, pp. 28–31.
3. W. J. Rossiter and R. D. Denshfield, "A Field Study of Polymer-Modified Bituminous Roofing," *Tenth Conf. Roofing Technol.*, NRCA, April 1993.
4. Schuller, Manville Roofing Systems, "Blistering of Modified Bitumen Roofing Systems," *Technical Topics*, Feb. 11, 1994.
5. R. P. Baxter, "Field Performance of Polymer-Modified Bituminous Roofing Membranes," *Proc. Eighth Conf. Roofing Technol.*, NIST-NRCA, 1987.

Chapter

12

Elastomeric Membranes

Synthetic rubbers, especially ethylene propylene diene terpolymer (EPDM), dominate the single-ply market. Formed into wide sheets and applied as single-layer systems, these elastomeric membranes possess many desirable qualities for roof applications. Elastomers can elongate and recover their original shape repeatedly, easily accommodating stress concentrations that can split built-up bituminous membranes. Unlike built-up membranes, elastomers do not require uniform attachment to the substrate. Their breaking strains range upward from 200 percent, compared with 2 percent or even less for built-up membranes. Elastomerics' flexibility allows conformance with odd shapes and difficult flashing conditions. They are adaptable to a wide variety of roof systems—adhered, mechanically fastened, and ballasted.

Basic Material

Carbon black is the key element in the formulation of EPDM; it provides ultraviolet resistance and greatly enhances tensile and tear strength. Extending oils lower compound costs by allowing the polymer to accept high inert filler loadings. A curative system is designed to cross-link the EPDM polymer after initial thermoforming is completed. Additional ingredients may include processing aids, inert low-cost fillers, fire retardants, antioxidants, antiozonants, and antistick dust. White EPDM, desired for its heat reflectivity, has been compounded, but at a sacrifice in the strength and durability provided by the reinforcing carbon black.

In a typical EPDM manufacturing process, the ingredients are intensely mixed in a high-shear "Banbury" mixer, then fed to a multiroll calender machine where the adjacent rolls rotate at slightly different speeds, so that a thin ribbon of uncured membrane is formed. This uncured, tender material is then vulcanized by the application of heat and pressure into a strong, inert finished membrane. Typical sheet thickness is 0.045 to 0.060 in. (1.1 to 1.5 mm), although 90-mil (2.3-mm) sheets are available when additional puncture resistance is required. Table 12.1 shows the physical requirements of both EPDM and chloroprene-based (CR) roof membranes as prescribed by ASTM standard D4637.

Vulcanized rubber is insoluble and nonweldable. These facts make field seaming of elastomers more complex than field seaming of thermoplastic sheets, which can be heat- or solvent-welded. Factory-producing large roof EPDM sheets from narrow calender-width sheets prior to vulcanization reduces the number of field seams (see Fig. 12.1). These prefabricated sheets range from 10 ft (3 m) for fully adhered systems to 50 ft (15 m) for loose-laid, ballasted systems.

Elastomeric sheets are usually produced without reinforcement, since the sheet itself is quite strong. Recently, however, to satisfy the more rigorous wind loadings imposed on mechanically anchored systems, EPDM sheets with embedded woven scrim reinforcement have been produced. Though the reinforcement fabric restricts elongation, it more than compensates for this disadvantage by distributing local tensile and shear stresses.

Despite its excellent weather resistance, EPDM is vulnerable to chemical attack from oils and fats, which weaken and swell the membrane. One EPDM manufacturer recommends substitution of polyepichlorohydrin membranes for EPDM in areas where contamination from fats or oils is unavoidable. If the contaminated area can be isolated by curbs and separately drained, the rest of the roof can use standard EPDM.

Compared with PVC, CPE, and CSPE, EPDM polymer exhibits another shortcoming: It lacks halogen atoms, which are excellent fire retardants. Early EPDM membranes required an application of Hypalon and sand to achieve acceptable fire ratings. Contemporary EPDM ballasted roofs can use non-fire-retardant EPDM, relying on the aggregate ballast to provide fire resistance. For exposed applications, however, a specially compounded fire-resistant (FR) sheet is available.

Because of its relative chemical inertness, EPDM can be installed over virtually all substrates except those containing soft asphalt. In *adhered systems,* the integrity of the substrate is critical. Faced isocyanurate insulations are an acceptable substrate if the foam facer is well bonded to the foam. Delamination of isoboard facers has occurred on some adhered EPDM systems. Membrane manufacturers have

TABLE 12.1 Physical Requirements for Vulcanized Rubber Sheet (EPDM and Polychloroprene)

Type:	I (EPDM)		II (CR)	
Grade:	1,2	1	1	1
Class:	U	SR	U	SR
Thickness, min., mm (in.):				
Sheet overall	1.0 (0.039)	1.0 (0.039)	1.0 (0.039)	0.76 (0.030)
Coating over scrim	—	0.4 (0.015)	—	0.2 (0.008)
Breaking strength, min., N (lbf)	9.0 (1305)	400 (90)	11.0 (1595)	355 (80)
Tensile strength, min., MPa (psi)	300	—	250	—
Elongation, ultimate, min., %	300	250*	250	200*
Tensile set, max., %	10	—	10	—
Tear resistance, min., kN/m (lbf/in.)	26 (150)	—	26 (150)	—
Tearing strength, min., N (lbf)	—	22 (5)	—	45 (10)
Brittleness point, max., °C (°F)	−45 (−49)	−45 (−49)	−35 (−31)	−35 (−31)
Ozone resistance, no cracks	Pass	Pass†	Pass	Pass†
Heat aging:				
Breaking strength, min., N (lbf)	—	355 (80)	—	335 (75)
Tensile strength, min., MPa (psi)	8.3 (1205)	—	9.3 (1350)	—
Elongation, ultimate, min., %	200	200*	150	150*
Tear resistance, min. kN/m (lbf/in.)	22 (125)	—	22 (125)	—
Linear dimensional change, max., %	±2	±2	±2	±0.5
Water absorption, max, mass %	+8, −2	+8, −2	±10	±10
Factory seam strength, min., kN/m (lbf/in.)	9 (51) or sheet failure	9 (51) or sheet failure	9 (51) or sheet failure	9 (51) or sheet failure
Weather resistance, no cracks or crazing	Pass	Pass	Pass	Pass

*Coating failure—reinforcement fails before coating.
†Test performed on coating rubber only.
SOURCE: ASTM D4637.

Figure 12.1 Large EPDM sheets, with factory-prefabricated seams, reduce the need for vulnerable field lap seams. Prior to application, sheets are unrolled and allowed to "relax" to relieve roll-winding tensile stresses before field seaming and edge anchorage. (*Roofing Industry Educational Institute.*)

consequently restricted isoboard by publishing lists of approved insulation products.

Fleece backers, separating the polymeric sheet from potentially incompatible substrates—e.g., fresh asphaltic mastic or Type I asphalt—have been applied to EPDM sheets. Some manufacturers of fleece-backed EPDM recommend that it be laid in hot asphalt. (Field seams must still be taped or adhesively formed.)

Ballast aggregate for EPDM membranes requires careful selection, as the thin EPDM sheets are vulnerable to puncture by sharp edges. Typical specifications call for clean, rounded stones with diameters in the $1\frac{1}{4}$-in. (32-mm) range (ASTM D448 No. 4 stone). In higher-wind areas, larger No. 2 or 3 stones with diameters greater than 2 in. (50 mm) may be necessary (see Fig. 12.2). If the aggregate is considered too sharp, a sheet of nonwoven polyester "stonemat" or fleece may be inserted between membrane and aggregate.

Field Seaming

Field seaming of the EPDM sheets is the critical part of their application. Early systems used a solvent-based primer applied to both mating surfaces. A gum tape was then applied to one surface, the other

Figure 12.2 Ballasted EPDM roof system receives rounded river gravel spread at 10 to 12 psf weight (top). Larger aggregate, No. 2 stones (2-in.-plus diameter), is applied at 20 psf in roof corners and perimeters (bottom). (*Roofing Industry Educational Institute.*)

surface embedded, and the seam fully and thoroughly rolled to assure complete contact. This practice gradually gave way to application of solvent-based adhesives. Cleaning of talc and other deleterious material prior to application of the solvent-based neoprene contact adhesive was critical. After evaporation of the solvent from the adhesive, the two surfaces were mated and rolled, and the exposed lap edge was caulked with an elastomeric sealant (see Fig. 12.3).

Judged by extensive field experience, these neoprene-based lap adhesives lacked durability. In the middle and late 1980s, butyl-based adhesives were substituted. They have proven more water-resistant and durable. Recent air-quality regulations restricting the release of volatile organic chemicals (VOCs), such as the solvents used in butyl-based adhesives, have limited their use.

As a consequence of these air-quality regulations, pressure-sensitive tapes have made a comeback, as an alternative to solvent systems and associated lap sealants (see Fig. 12.4). Surface cleanliness is probably more critical with tapes than with solvent-based adhesives. Since the tapes are thicker and less conformable than liquid-applied adhesives, some manufacturers require a cover patch at all tee joints to ensure watertightness.

Flashings

The "memory" (i.e., a material's tendency to regain its original configuration) of fully vulcanized rubber makes it difficult for the roofer to apply rubber flashing solidly to irregular, complex wall shapes, even right-angled corners. The answer to this problem was use of a semicured, gumlike sheet of greater plasticity, which allowed the flashing to be molded to its backing surface. Early applications used gum neoprene (chloroprene). (Uncured EPDM had not yet been developed.)

These neoprene flashings often failed years before the EPDM membrane, requiring replacement of the flashings to save the entire system. They succumbed to overcure and embrittlement after long term exposure to rooftop heat and ultraviolet radiation (see Fig. 12.5). Semicured EPDM has replaced uncured neoprene, but is used less extensively in contemporary EPDM systems. With the aid of screw-anchored restraining strips of EPDM at edges and curbs, EPDM membranes can be extended up the wall, eliminating the need for separate flashing sheets (see Fig. 12.6). Shrinkage is controlled by bonding the bottom side of the membrane sheet to the solidly fastened restraining sheet.

As a major advantage over BUR systems, EPDM single-ply exploits the convenience of molded boots and jacks to eliminate the pitch pockets used to flash pipe penetrations in BUR systems. Formed into coni-

Elastomeric Membranes 305

Figure 12.3 The field seam for this EPDM membrane comprises an "in-seam sealant" in addition to lap sealant to provide an extra line of waterproofing in case the lap seal peels in service or opens because of defective application. (*Roofing Industry Educational Institute.*)

Figure 12.4 Pressure-sensitive tapes have replaced the multistage field-seaming process with solvent-based contact adhesives for EPDM roof membranes. Lap surfaces are primed, tape installed, and roller pressure applied. The tape edge projects slightly beyond the lap edge. Unlike field seams made with solvent-based adhesives, taped joints require no separate lap sealant. (*Roofing Industry Educational Institute.*)

Figure 12.5 Uncured neoprene flashing displays cracking from heat and ultraviolet radiation. Uncured EPDM, a much more durable material, has replaced uncured neoprene as base flashing for EPDM membranes. (*Roofing Industry Educational Institute.*)

Figure 12.6 Curb is flashed by (*a*) turning up EPDM membrane, (*b*) anchoring it with mechanically fastened restraining bar, and (*c*) covering curb, bar, and membrane edge with uncured EPDM flashing sheet. Corners are overlapped for double protection. (*Roofing Industry Educational Institute.*)

Elastomeric Membranes 307

Figure 12.7 Premolded pipe boots simplify flashing. Top of boot is tapered, providing varied-diameter O-ring seals. Boot is pulled down and adhered to EPDM membrane. Sealant applied to pipe under O ring provides extra moisture protection. Clamp holds boot snugly to pipe. (*Roofing Industry Educational Institute.*)

cal shapes with integral O rings, boots are pulled down onto the penetrating pipes with the O ring, providing a snug, watertight top seal (see Fig. 12.7).

A typical installation proceeds as follows:

- Place sealant around the pipe in the vicinity of the O ring.
- Clean the membrane at the base of the boot.
- Apply splicing adhesive or sealing tape to both the membrane and the underside of the pipe boot flange.
- Embed the flange.
- Install a pipe clamp to secure the top of the boot to the pipe.

In contrast to a pitch pocket, which must be periodically refilled to remain watertight, these EPDM flashing boots require no field maintenance.

Mechanically Fastened Systems

Among the fastest-growing of the steadily evolving elastomeric systems is the mechanically fastened assembly. As one advantage, these assemblies eliminate the need for the hot asphalt or bonding adhesives required for adhered systems. The most popular contemporary mechanically fastened systems feature batten bars or individually through-fastened discs, weather-protected by membrane cover strips or patches (see Figs. 12.8 and 12.9). Spacing is determined by design wind loading, with additional fasteners required in corners and perimeters (see Chapter 7, "Wind Uplift"). When wide sheets are used, fastener spacing may be 5 to 10 ft, dimensions that can result in some insulation boards lacking any fasteners and thus shifting position when the flexible membrane balloons under wind-uplift pressure. Every insulation board should thus be anchored with at least two "preliminary" fasteners.

Figure 12.8 Mechanically fastened EPDM system has plastic batten bar screwed through membrane and waterproofed with adhered cover strips. (*Roofing Industry Educational Institute.*)

Figure 12.9 Mechanically fastened EPDM membrane has metal batten bars installed in seam area and weatherproofed by overlapping sheet. (*Roofing Industry Educational Institute.*)

The new mechanically fastened EPDM systems supplant an older generation of mechanically fastened systems featuring "nonpenetrating" fasteners. In these systems, base plates or tracks were first installed on the substrate and anchored to it. Following installation of the EPDM membrane, caps or insert bars were installed, clamping the membrane to the base. These systems were abandoned, however, when the field labor demands for assuring dependable cap connections proved too arduous.

Elastomeric Failure Modes

Surveys of EPDM failure modes invariably highlight lap-seam failure as the biggest problem. Lap-seam failures account for 37 percent of EPDM system problems, according to a recent NRCA survey.[1]

Following lap-seam failure in incidence were the following:

- Flashing problems (21 percent)
- Punctures/tears (14 percent)
- Shrinkage (8 percent)

- Wind-uplift problems (8 percent)

Minor problems, each accounting for 3 percent or less of the total, included fastening, blistering, and embrittlement.

Lap-seam failures stem from the nature of EPDM itself. Its chemical inertness, which enables it to resist heat and ultraviolet radiation, makes EPDM a difficult surface to wet with an adhesive and thus makes a strong, durable field seam difficult to form. (The no-free-lunch law strikes again!) As previously indicated in this chapter, the care required in adhering EPDM lap seams, especially the care required to clean the adhered surfaces for both solvent-based seams and pressure-sensitive tapes, makes field seaming the critical operation in installing EPDM membranes. For an improperly applied lap seam, it is downhill all the way, with chemical aging and thermal-cycling stresses combining to weaken the seam and open it to leakage.

Flashing and *shrinkage* problems, though listed separately in the NRCA survey, often signal a combined problem, aggravated by poor field anchorage at the perimeter. Flashing problems can be aggravated, or even caused, by long term EPDM membrane shrinkage, which, according to one manufacturer, may ultimately reach 2 percent in its service life.

EPDM shrinkage evidently results from a combination of factors:

- A slow loss of processing oil used in the manufacturing process
- Residual tensile stress (developed during the manufacturing process when the molecular cross-linking occurs)
- Insufficient relaxation of the EPDM sheet just prior to installation

One reason for relaxation time (minimum 30 min) is to smooth out wrinkles and creases. But it also helps to dissipate roll-winding tensile stress and reduce future shrinkage. Membrane shrinkage from residual tension resembles long term creep shrinkage in prestressed concrete structural members, which gradually shorten under pretensioning stress.

The flashing-shrinkage problem is manifested in a "bridging, tenting, contraction, or shrinkage," according to a Midwest Roofing Contractors Association (MRCA) bulletin on repair methods.[2] Repair is considered necessary under either of the following two conditions:

- Loose perimeter flashing pulled away 2 in. or more from wall (or vertical flashing surface pulled into a 45°-plus angle)
- Flashing anchorage defective over a minimum 3-ft-long section

Elastomeric Membranes 311

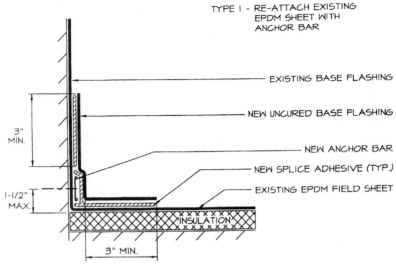

Figure 12.10 Repair detail when EPDM flashing has pulled away from its backing but is loose enough to be pressed back into correct position. (*Midwest Roofing Contractors Association.*)

Recommendations for repair are as follows:

1. If the flashing/membrane can be pressed back into the base of the wall or angle change, then reattach the detail according to the Type 1 detail (Fig. 12.10).
2. If the flashing/membrane cannot be easily pulled back in, or the tenting is too widespread, then repair the wall detail with either plain EPDM or reinforced EPDM.
  a. Prepare for the repair as shown in Fig. 12.11.
  b. Use anchor bars and fasteners with plain EPDM according to the Type 2 detail (Fig. 12.12).
  c. For reinforced EPDM sheets and fasteners with stress plates, use the Type 3 detail (Fig. 12.13).

Ridges and *buckles* are common and are best left alone unless the membrane is cracking or under obvious strain.[3]

Open laps or *splices* should be pulled back to a sound part of the splice or seam and any debris removed. If water infiltration is suspected, open the membrane and inspect for damage. The open seam should be glued together with seam adhesive, followed by a patch.

312   Chapter Twelve

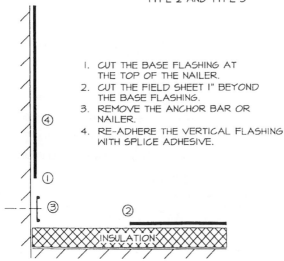

Figure 12.11 First step in the repair of unattached EPDM flashing that cannot be readily pulled back into position. (*Midwestern Roofing Contractors Association.*)

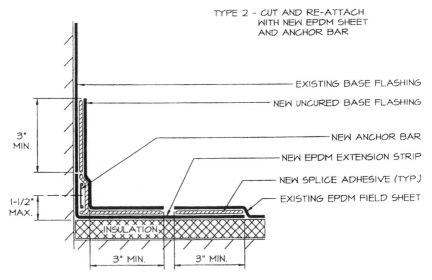

NOTES:
1. DRAWINGS NTS
2. SEAM SEALANT NOT SHOWN FOR CLARITY
3. WHEN FASTENING ANCHOR AT HORIZONTAL POSITION, FASTENERS SHALL BE NO MORE THAN 6" MAX. O.C. FROM WALL

Figure 12.12 EPDM flashing repair for *unreinforced* EPDM. (*Midwest Roofing Contractors Association.*)

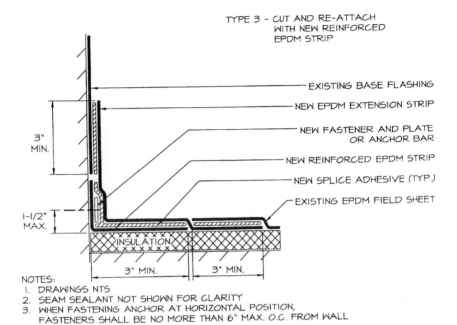

Figure 12.13 EPDM flashing repair for *reinforced* EPDM. (*Midwest Roofing Contractors Association.*)

The occasional *loose, backed-out,* or *popped-up fastener* can be repaired by removing it and installing a new fastener and stress plate in an adjacent area. The existing hole should not be reused. A patch is installed extending 3 in. (75 mm) beyond the plate edge.

Ballast scour, a common problem with loose-laid, ballasted single-ply systems, requires redistribution of ballast, following the ANSI/RMA/SPRI RP-4 guidelines for ballasted single-ply roof systems. Remedial action may require use of larger aggregate, heavier application, and/or use of pavers. Whenever dead load is increased, a review of the structural framing—including deck, purlins, beams, girders, and even columns—may be required.

References

1. W. C. Cullen, *Project Pinpoint Analysis: Ten-Year Performance Experience of Commercial Roofing, 1983–1992,* NRCA, 1993, Table 7.
2. "Repair Methods for Re-Attaching EPDM Membrane and Flashing Experiencing Shrinkage," MRCA, October 1994.
3. *Repair Manual for Existing Roofing Systems,* NRCA/SPRI draft.

Chapter 13

Weldable Thermoplastics

Weldable thermoplastic single-ply membranes have a notable advantage over their chief competitor, nonweldable elastomerics. This is the relative ease with which dependable, heat-fused lap seams, as strong as the basic sheet itself, can be field-fabricated (see Fig. 13.1). This feature makes weldable thermoplastics especially advantageous on projects with many roof penetrations, with a high ratio of lap-seam length to roof area. Thus, on cluttered rooftops, with an abundance of rooftop equipment, stacks, vents, and other penetrations, weldables (along with modified bitumens) may be prime candidates. Elastomeric single-plies, with their more difficult, less dependable adhered lap seams, are a better bet on roofs with large, unobstructed areas, with a low lap-seam-to-roof-area ratio.

In contrast with EPDM, where problems are concentrated on the lap seams, the major problem with weldable thermoplastics concerns the basic sheet material itself. A recent NRCA study shows the stark contrast between the two materials. For polyvinyl chloride (PVC), the major weldable roof membrane, NRCA recorded seven times as many material failures (shrinkage and embrittlement) as lap-seam failures (see Fig. 13.2). For EPDM, the ratio was reversed, with nearly five times as many reported lap-seam failures as material failures (in the form of shrinkage).[1]

Materials

Weldable thermoplastics have polymer chains that are not cross-linked (or vulcanized) like those in elastomeric sheets. In contrast with the thermosetting elastomerics, thermoplastics melt and flow when heated. They are also capable of solvating when solvents are

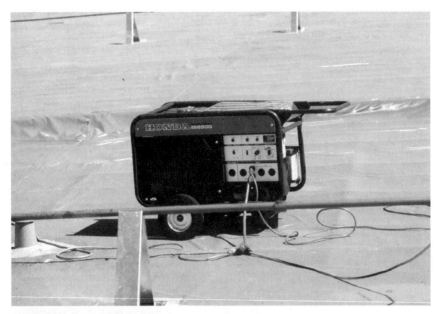

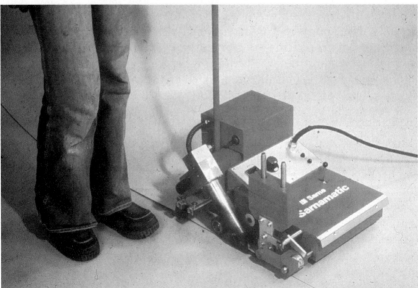

Figure 13.1 Weldable thermoplastic membrane requires a uniform power supply (top) to produce side lap joints heat-fused by self-propelled hot-air welder (bottom). (*Roofing Industry Educational Institute.*)

Figure 13.2 Weldable PVC membrane is covered with a polyester stone shield before application of paver walkways and stone ballast. (*Roofing Industry Educational Institute.*)

applied to their surfaces. Under stress, polymer chains can slip past one another, a process called "cold flow" or "creep." Most thermoplastic membranes are scrim-reinforced to control creep (or relaxation).

Chemically, weldable thermoplastics belong to the polyolefin group, which includes polyethylene, polypropylene, polybutenes, polyisoprene, and their copolymers. All have been used in waterproofing, vapor retarders, and occasional roof-membrane application. A narrower group, the chlorinated polymers—CPE, CSPE, and especially PVC—became the mainstays of the thermoplastic roof membranes. The presence of the chlorine atom makes these chlorinated olefins fire retardant, and also makes them "polar"—i.e., it results in a separation of positive and negative charges on the plastic molecule. Polar molecules can be easily dissolved in polar solvents to facilitate the solvent welding or gluing of membranes. The polarity also permits the formation of hydrogen bonds, or weak van der Waals forces. These bonds resemble the cross-linking of elastomers, resulting in more rigidity and resistance to creep. At elevated temperatures, these weak attractions are overcome, and the products flow the same way as their nonchlorinated cousins.

Polyvinyl chloride (PVC), the most popular weldable roof membrane, owes its flexibility to a plasticizer, which softens the otherwise rigid PVC. The choice of an appropriate plasticizer for roofing use is critical. Some membranes with easily extractible plasticizers quickly embrittled and/or shrank. This jeopardized the reputation of all PVC membranes. Other systems, using nonmigratory plasticizers or blending PVC with polymeric plasticizers ("copolymer blends" or "copolymer alloys"), have performed better. Plasticizer migration has also been controlled by use of separation sheets installed over plasticized sheet.

Heat welding with hot compressed air at temperatures ranging from 500 to 600°F (260 to 315°C) has proved more dependable than solvent welding, formerly used as an alternative welding technique for weldable thermoplastics (see Fig. 13.3). Properly performed heat welding produces a true fusion of joined materials, analogous to a structural steel weld, dependably watertight.

Historical Background

The development of weldable thermoplastic roof membranes has a history of three decades of trial and rejection and constant reformulation. One chlorinated polymer, polyvinylidene (Saran), had a brief use in roofing (Saraloy 400). This was soon replaced by chlorinated polyethylene (CPE), a product with less chlorine and a lower crystal structure. This made a fairly successful flashing (Saraloy 640R) and was eventually introduced as an entire roofing system. The first version was a reinforced CPE, laminated to a $\frac{1}{8}$-in.- (3.2-mm-) thick layer of flexible polyurethane foam (Chemply). The concept was good, in that no plasticizer was needed in CPE, and the resistance to some chemicals was superior to that of PVC. However, as mentioned in Chapter 9, this product may have been ahead of its time. In the 1980s, CPEs were still available (Cooltop 40), but the 1990s saw the last of them. They were replaced with a PVC terpolymer blend (C-3 and Ultraply 78 plus). These terpolymer blends are reportedly easier to weld and may have other advantages.

Still another chlorinated polymer, chlorosulfonated polyethylene (CSPE, best known as Hypalon), has had great success in roofing. First as a liquid elastomeric coating and then as a reinforced polymeric sheet, Hypalon is one of the few polymers that has outstanding weather resistance even in a white color. (Elastomerics such as EPDM rely heavily on the carbon-black filler for ultraviolet protection, and the white EPDMs are considered inferior to the black sheets.) Some early Hypalon sheets were laminated to neoprene-

Figure 13.3 Mechanically fastened batten-bar strip (top) is covered with heat-welded strip (bottom), simultaneously welded on both sides by compressed air at 500°F to fuse joint material. (*Roofing Industry Educational Institute.*)

bonded asbestos backing paper for adhesion and fire-resistive properties. These evolved to fleece-backed and nonbacked products.

Hypalon is unique among plastics in that it can be manufactured and installed in an uncured state, but upon exposure to heat and moisture will convert from a plastic to an elastomer. This allows for welded laps, with their inherent advantages, but by cross-linking later on, Hypalon picks up the creep resistance and inertness of cured elastomers. On the other hand, patching the cured material is more complicated than patching true thermoplastics, which can be welded at any time during their life.

Hypalon sheets have at times suffered from excess chalking. In this phenomenon, the pigment gradually powders and washes away. Some chalking in exterior white paints is desirable, as the surface is self-cleaning. However, if the chalking is so severe that water flowing from downspouts whitens asphalt paving, this can be objectionable. In addition, if the erosion is so severe that mil thickness is rapidly lost, membrane service life is shortened. Another reported problem with some Hypalon sheets was algae attack. While this has also been noted in other membranes, the large volume of white CSPE in warm, humid climates may have resulted in a disproportionate number of reported algae failures.

Both chalking and algae attack indicate *formulation* problems. These problems may be solvable through use of pigment, biocides, and other ingredients. They do not appear to indicate an inherent failure of the polymer itself.

Thermoplastics without chlorine, such as plain polyethylene, have disadvantages in fire resistance and bondability. Being nonpolar, polyethylene is very difficult to adhere to. Weatherability and acceptance of fillers are also poor. However, there are opportunities for other non-chlorine-containing weldables. The blending of copolymers, terpolymers, interpolymers, and even weldable EPDMs or EP rubbers is being intensely explored. Some of these materials have more than 15 years of field exposure, while others are new to the marketplace.

The low puncture resistance of weldable polymeric sheets makes walkway overlay sheets a necessity, to shield the underlying membrane from puncture and abrasion attending rooftop traffic (see Fig. 13.4). Weldable thermoplastic sheets tend to be thinner than elastomerics. Most rely on the scrim reinforcement for puncture resistance and tensile strength. The polymeric coatings need to be thick enough merely to waterproof the membrane and to provide a weldable coating.

Plasticized sheets tend to be thicker, as plasticizer is theoretically lost from the surface, thus making a thicker sheet more durable.

Figure 13.4 Polymeric walkway, a necessity on single-ply membranes because of their inferior puncture resistance, is edge-welded to a thermoplastic single-ply membrane. (*Roofing Industry Educational Institute.*)

ASTM D4434 requires a minimum 40-mil thickness. This prescriptive requirement may, however, be waived for a thinner sheet that satisfies specified performance requirements.

Flashings

Flashings and edgings of PVC and other copolymer systems come in two basic types: polymer-clad lightgage steel and reinforced polymer sheets. Polymer-clad steel flashings are shop- or field-formed into curbs, wall flashings, and gravel-stop fascias. Prior to application, they are anchored to nailers, decks, or walls. The membrane sheet is brought up onto the clad steel and welded directly to the flashing (see Fig. 13.5). This unique metal-based flashing system serves several purposes. First, the steel serves as a separator, keeping substrate contaminants from attacking the plasticizer. Second, the steel distributes stresses, so that the sheeting does not thin or creep at the locations of maximum loading. The steel can be through-fastened without fear that the fasteners will cut through the membrane. Screws with elastomer washers are frequently installed on vertical surfaces without the need for further weatherproofing.

Figure 13.5 Weldable single-ply base flashing features polymer-coated metal. A joint cover has been welded to joint adjacent sections to provide a flexible connection. Metal cap flashing will cover the base flashing. (*Roofing Industry Educational Institute.*)

Failure Modes

The major failure mode reported for PVC membranes is embrittlement, sometimes leading to irremediable shattering.

The mysterious discovery that ballasted PVC membranes tended to embrittle faster than weather-exposed membranes led to a new theory of PVC degradation. Ultraviolet radiation has traditionally been the prime suspect in degradation of polymeric materials. But the discovery that ballast-shielded PVC membranes can become embrittled faster than sun-baked PVC membranes forced a reconsideration of this theory. Worldwide confirmation of this phenomenon has forced industry experts to find another explanation for PVC embrittlement. In one now rejected hypothesis, the problem was attributed to "dirty" ballast, with adhered clay blamed for extracting the plasticizer from the PVC. Now, however, it appears that some other process—hydrolysis of the plasticizer—must be the culprit, since ballasted roofs remain wet far longer than exposed surfaces.

An even more spectacular failure mode, called "the shattering of aged unreinforced PVC roof membranes" in a warning bulletin issued in 1990 by NRCA and SPRI, constitutes an advanced stage of embrittlement that culminates in the PVC sheet's irremediable shattering. These PVC-shattering failures share several common factors:

- Lack of reinforcement
- Ballasted systems
- Substantial age (8½ years median)
- Cold-weather occurrence

This last-cited factor prompted NRCA/SPRI and FM (in *Loss Prevention Data Technical Advisory Bulletin 1-29, PVC Roof Coverings,* June 1992) to ban foot traffic on older PVC membranes at low temperatures—NRCA/SPRI used 50°F (10°C) and FM used an even more conservative 60°F (16°C).

This shattering problem should be confined to older, thinner, unreinforced PVC membranes. Virtually all contemporary weldable thermoplastic sheets are reinforced. And today's thicker PVC sheets—45, 55, and even 60 mils—should resist plasticizer loss far better than the older unreinforced sheets, which were sometimes only 32 mils (0.8 mm) thick. ASTM D4434 requires a minimum 1.14-mm (45-mil) thickness. Plasticizer loss can occur more readily in a thin sheet, with reduced distance from core to surface for plasticizer migration.

Alerts

Follow these ASTM specifications for weldable membranes:

D4434, "Specification for Polyvinyl Chloride Sheet Roofing"—Covers nonreinforced and reinforced sheets

D5019, "Specification for Reinforced Non-Vulcanized Polymeric Sheet Used in Roofing Membrane"—Covers Hypalon, CPE, and PIB (polyisobutylene)

D5036, "Standard Practice for Application of Fully Adhered Single Ply Poly(vinyl Chloride) Roof Sheeting"

Reference

1. W. C. Cullen, *Project Pinpoint Analysis: Ten-Year Performance Experience of Commercial Roofing, 1983–1992,* NRCA, 1993, Tables 7 and 15.

Chapter 14

Flashings

Flashings, the most common source of roof problems, are also the most often slighted aspect of roof design and normally the most difficult. Many flashing failures, probably the majority, stem from the designer's failure to provide adequate flashing details on the architectural drawings. Depending upon contract and fee structure, the designer may not even have the responsibility for coordinating roof design with installed equipment—e.g., HVAC. As a consequence, the roof becomes an elevated dump and work platform for equipment that the designer cannot comfortably include in the building interior.

A modern building's roof is penetrated by a host of components: electrical conduits, piping, ducts, vents, stacks, antennae, and hatches. Roof-mounted HVAC units, industrial process equipment, skylights, ladders, stairs, signs, and structural supports all require flashing. And so do such ubiquitous components as penthouses, walls, parapets, and edge fascias. A competent architect not only will provide standard details for all flashed components but will also furnish detailed drawings for the unusual, often unique, conditions that unerringly seem to turn up even on apparently prosaic buildings. At the least, a responsible architect will specify standard details—e.g., NRCA, SMACNA, etc.—though this practice is far inferior to providing details on the project drawings.

Some architects, however, defer to the roofing contractor for the provision of shop drawings, though it should remain the designer's responsibility. Since roofers have a financial motive to provide the cheapest acceptable details, the result is often a leaking system of defective flashings. Roofers sometimes attempt to justify the use of cheap, defective flashings by citing "local practices," which often contradict the flashing principles of NRCA, SMACNA, and other standards-promulgating organizations. (Cheap roof-level expansion-contraction joints instead of more expensive curbed joints are one example of substandard local practices.)

Designer irresponsibility is, of course, only one of several possible causes of flashing failure. Poor field application and defective materials cause flashing failures as well as membrane failures. These causes can occur in combination: Defective design is often aggravated by defective field application. Defective materials are less likely to be a contributing cause than defective design and application. New materials lacking the requisite performance requirements (listed later) have nonetheless contributed to flashing failures in the past. And as a final cause of flashing failure, there is abuse: maltreatment by rooftop workers, often compounded by neglected maintenance, which permits accelerated deterioration of holes or tears in flashing material.

What makes flashing design so difficult (and doubly difficult on reroofing projects) is the demand for coordination with all other aspects of design—notably the drainage plan, but also mechanical, electrical, and plumbing design. The designer should control both the quantity and the location of roof penetrations and rooftop equipment. Every rooftop component presents an opportunity for a flashing failure, especially if it is located in a low spot, subject to invasion by ponded water. And, of course, in addition to these overall requirements, each project requires the designer's attention to unusual conditions for which no published standard detail may be directly applicable. For this reason, the designer must be well grounded in the principles of flashing design. He or she must be capable of adapting generally recommended details to specific conditions—e.g., the intersection of a sloping asphalt-shingled roof with a conventional built-up roof or a modified-bitumen roof.

Flashing Functions and Requirements

Flashings seal the joints at gravel stops, curbs, chimneys, vents, parapets, walls, expansion joints, skylights, scuttles, drains, built-in gutters, and other places where the membrane is interrupted or terminated. *Base flashings* are essentially a continuation of the roof membrane at the upturned edges of the watertight tray. They are normally made of bituminous, plastic, or other nonmetallic materials applied in an operation separate from the application of the membrane itself. *Counterflashings* (or *cap flashings*), usually made

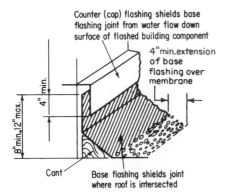

Figure 14.1 Base flashing and counterflashing.

of sheet metal, shield or seal the exposed joints of base flashings (see Fig. 14.1).

There are two basic types of base and counterflashing combinations:

- Vertical terminations (at walls, parapets, or curbs)
- Horizontal terminations (e.g., at roof edges, drains, or vents)

To serve its varying functions, flashing requires the following qualities:

- Impermeability to water penetration
- Flexibility, for molding to supports and accommodating thermal, wind, and structural movement
- Compatibility with the roof membrane and other adjoining surfaces, notably in coefficient of thermal expansion-contraction and chemical constitution
- Stability (resistance to slipping and sagging)
- Durability, notably weather and corrosion resistance (flashings should last at least as long as the roof membrane)

Materials

Flashing materials span a broader spectrum than built-up membrane materials. They include a wide variety of metals plus modified bitumens, bituminous laminations used in fabricating built-up membranes, and new plastic or elastomeric sheets. Because repair of leaking and deteriorating flashings is so common and costly, first-cost economy should rank last among the factors governing material selection and design. Quality and durability are paramount.

TABLE 14.1 Thermal Expansion of Counterflashing Metals

Metal	Thermal coefficient × 10^{-6} in./(in.) · (°F)	Expansion of 10-ft length for 100°F temperature rise, in 64ths in.
Galvanized steel	6.7	5
Monel	7.8	6
Copper	9.4	8
Stainless steel (300 series)	9.6	8
Aluminum	12.9	10
Lead	15.0	12
Zinc, rolled	17.4	13

Using nonmetallic materials for base flashings and metals for counterflashings exploits the best qualities of each. Because it has a similar coefficient of thermal expansion, modified bitumen works best with the built-up membrane as base flashing. And because of their good weather and corrosion resistance, metal counterflashings provide superior surface protection. As long as the base and counterflashings are designed to allow relative movement, they can function together despite their different thermal-expansion coefficients (see Table 14.1).

Cold-applied and bituminous plastic cements constitute an important class of flashing materials. A crucial distinction among these materials is between ordinary asphalt (*plastic*) roofing cement and the stiffer *flashing* cement. Both are basically so-called asphaltic cutbacks—Type I asphalt (Type II for some manufacturers) dissolved in mineral spirits of 300 to 400°F boiling range to produce 55 to 65 percent asphalt-solution plastic at ordinary temperatures. They can be troweled at temperatures down to 20°F.

Flashing cement differs from plastic cement in both the quantity and the quality of the mineral fibers added to the cutback to give it special characteristics. Plastic cement contains 15 to 45 percent fiber and filler and is designed to maintain some plasticity in the ordinary range of roof-surface temperatures after the solvent thinner evaporates. Flashing cement, however, typically contains longer and/or a greater quantity of fibers and mineral filler in order to produce a stiffer, more sag-resistant cement.

These differing properties set their proper uses:

- Specify flashing cement for vertically flashed and canted surfaces where sag resistance is important.

- Specify plastic cement for horizontal or slightly sloped flashing joints that must accommodate relative movement (e.g., the joint between a gravel-stop flange and the membrane felts underneath).

Heat-reflective, aluminum-fibrated coatings normally provide a better surface coating for base flashings than flashing cement, especially in hot climates and locations exposed to direct sunlight. Compared with a black coating, an aluminum coating can drop flashing surface temperature by 20°F or more, even after it ages. This temperature reduction not only reduces the risk of sagging, it enhances the asphalt's durability because its chemical degradation rises exponentially with increasing temperature.

Base flashing materials

Before the disappearance of asbestos felts as a viable material for built-up membranes in the early 1980s, base flashing materials for built-up membranes were generally either fabric-based bituminous sheets or plied felts similar to the felts in the built-up membrane itself. The current domination of built-up membranes by fiberglass felts has changed the flashing materials for built-up membranes. With their "memory" (i.e., the tendency to recover their original shape) and their general lack of flexibility, glass-fiber felts have inherent disadvantages as base flashings, which require flexibility, and even some plasticity, to fit the contours of supporting surfaces. Modified bitumens, tough and flexible despite their greater thickness, have become the preferred base flashing material for built-up membranes as well as for modified-bitumen membranes.

Base flashings for single-ply membranes are generally the same as (or closely related to) the membrane material—PVC, PIB, EPDM, etc. Uncured EPDM has replaced uncured neoprene as a versatile, moldable flashing material. Its superior plasticity enables it to fit tightly against its backing even at awkward angles. For straight flashing runs, cured membrane is generally specified. Flexible, premolded boots for inside and outside corners greatly simplify the flashing of such components as pipe penetrations (see Fig. 14.2). In a built-up roof system, this detail requires either a counterflashing skirt on the roof-penetrating pipe or an expensive lead flashing (see Fig. 14.3).

For wall and parapet base flashings in built-up membrane systems, the choice between hot-mopped asphalt, torched modified-bitumen sheets, and cold-applied flashing cement as the bonding and interply adhesive may be dictated by anchorage provisions. Because hot-mopped flashings must be nailed, a precast concrete parapet or similarly nonnailable wall section requires cold-applied flashing cement. With nailable walls—e.g., concrete with wood nailing strips—the choice involves important tradeoffs. Hot-mopped steep asphalt offers speed, convenience, and economy. With noncombustible walls, a

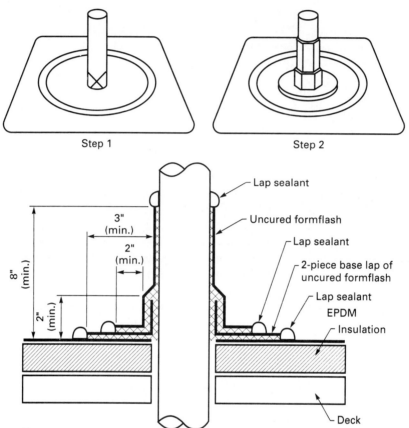

Notes:
1. Pipe batten required when this detail is used on mechanically anchored systems.

2. Remove all existing flashings, leads, etc. Pipe surface must be free of all rust, grease, etc.

3. Pipe must be anchored to bottom side of deck to assure stability.

4. Do not install clamping ring around top of flashing

5. Pressure treated wood nailer required when outside diameter of pipe exceeds 18".

Figure 14.2 Pipe flashing detail for EPDM roof system shows how field-fabricated flashing boots simplify flashing of single-ply systems. (*Firestone Building Products Co.*)

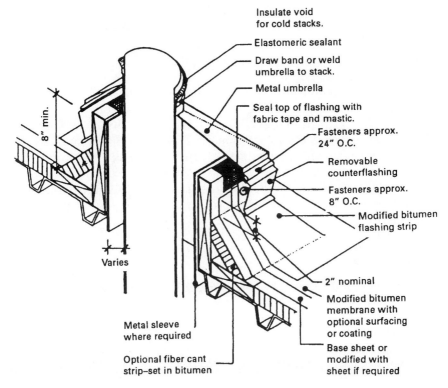

Figure 14.3 Skirted counterflashing for pipe penetration through built-up membrane shows the added complexity (or added expense of lead flashing) of flashing a conventional BUR roof system. (*National Roofing Contractors Association.*)

torched modified-bitumen flashing is often selected. The roofer can restrict the heat to where it is needed, ensuring conformability and adhesion.

Metal is generally unsuitable as base flashing material. It normally lacks the required flexibility for molding to supports, it is difficult to connect to the built-up membrane, and metals have incompatible coefficients of expansion-contraction. At hot summer temperatures, when metal expands, bitumen-saturated felts experience little or no expansion. But at the coldest winter temperatures, bitumen-saturated felts contract at rates varying from about 30 to 500 percent greater than those of the various metals used as counterflashings.

Metal base flashings are occasionally detailed at skylights, ventilators, HVAC equipment, scuttles, and similar roof-penetrating components. These are generally poor details. The roof-penetrating components should be set on curbs flashed with bituminous materials, as shown in NRCA standard details.

A notable exception to the rule against metal base flashings is at roof drains, where the need for stability and a good bolted connection to the metal drain frame favors lead or copper. A drain flashing seldom measures more than 2 ft, at most. Thermal contraction of a lead flashing of that width amounts to $1/32$ in. maximum; the bituminous materials bonded to the lead flashing can readily accommodate movements of that tiny magnitude. Polymer-coated metals have been successful, as details are designed to create "expansion joints," exploiting the polymeric sheets' flexibility.

Uncured EPDM offers a superior combination of qualities for flashing: ease of installation (because it readily conforms to corners and other flashed surfaces), excellent ozone and ultraviolet resistance, and flexibility at extreme low temperature. As a recent innovation, polyester "fleece" is bonded to the back of EPDM and plastic membrane sheets. This permits adhesion with either hot-mopped asphalt or solvent-based adhesive. The fleece serves as a separator, limiting incompatibility reactions between the asphalt and the polymer.

Before specifying new materials, designers should investigate their past performance, demanding technical data from laboratory and field tests. Costly failures have resulted from using unsuitable flashing materials.

Counterflashing materials

Counterflashings (or cap flashings) shield the exposed joints of base flashings and shed water from vertical surfaces onto the roof. Because of their exposed locations, counterflashings must be rigid and durable; thus metal generally proves the best material. The metals used for counterflashings include galvanized steel, copper, aluminum, and stainless steel.

Where no counterflashing is provided, the top edge of the base flashing is sometimes terminated with a mechanically anchored pressure bar. This, in turn, is sealed to the wall with caulking (see Fig. 14.4).

Whenever there is a possibility of delay by the sheet metal contractor installing a metal counterflashing, the top edge of all base flashing requires at least a temporary seal to protect the vulnerable base flashing joint. It can consist of a three-course seal of felt or mesh and roof mastic.

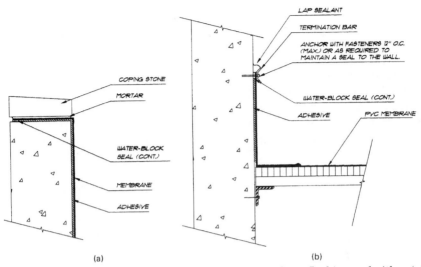

Figure 14.4 PVC membrane is carried up the wall as base flashing and either (a) anchored at the top of the wall or (b) anchored on the side of the wall with a termination bar. (*Lawrence Roof Consultants.*)

Principles of Flashing Design

Flashing design requires observance of these basic principles:

- Eliminate as many roof penetrations as practicable.
- Consolidate as many roof openings as possible into a smaller number of larger openings.
- Locate flashed joints above the highest water level anticipated on the roof and provide positive drainage away from the flashed joints.
- Allow for differential movement between base and cap flashings.
- Contour flashed surfaces to avoid sharp bends (45° maximum) in bituminous base flashings.
- Anchor flashings securely to supports.

Eliminating penetrations

Good flashing design starts with a preliminary effort to minimize roof penetrations. There are obvious first-cost advantages in, for example, locating HVAC equipment on the roof. But the perennial operating and maintenance costs are often overlooked. Rooftop equipment not only increases the number of flashed components to be maintained and repaired, but also means increased wear on the roof, with hazards to membrane and flashing from dropped tools and pedestrian traffic.

The hazardous conditions created by rooftop penetrations have been documented by infrared moisture detection surveys by Wayne Tobiasson, research engineer with the U.S. Army Cold Regions Research and Engineering Laboratory (CRREL). Tobiasson's surveys detected wet areas around the overwhelming majority of "breather vents" on the surveyed roofs. These wet areas evidently resulted almost exclusively from exterior water entry at the vents' flashed joints, which usually consisted of metal flanges set atop the membrane and stripped in with felt strips and plastic cement.

These topside vents were apparently worthless in their design role of releasing accumulated moisture or preventing moisture buildup in the insulation. Though long sanctioned by the roofing industry's conventional wisdom, breather vents are obviously a positive liability. They are the first penetrations that should be eliminated from the roof, but not necessarily the last.

Reducing openings

The benefits of consolidating openings into a smaller number of larger openings are best illustrated by rooftop HVAC units. Via long-sanctioned, but erroneous, practice, rooftop HVAC units are often supported on wood sleepers. Holes to accommodate associated electrical and refrigerant lines are simply poked through the membrane, with each penetration getting its own miniature pitch pan. (Even this second-rate protection against leakage is sometimes omitted by the most benighted, irresponsible roofers.)

Flashing elevation

As the most basic principle of flashing design, often ignored, locate flashed joints above the roof water line. Like the preceding two principles, this one demands preliminary coordination of drainage and flashing design. Only through such planning can the designer avoid slovenly and costly errors. Locate edges, wall intersections, skylights, equipment bases, and, above all, expansion-contraction joints at high points.

When it is impracticable to follow this rule, at least keep flashings out of low areas where ponding might occur. Even well-flashed joints are more vulnerable than the membrane to penetration by standing water, and poorly flashed joints are extremely vulnerable. Failure to keep flashed joints out of standing water is one of the most common and costly errors in roof design. It is also one of the most difficult and expensive to correct.

There should be no compromise in expansion-contraction-joint design. These joints should be located only at high points, at the terminating boundaries of drain planes. Expansion-contraction joints should incorporate minimum 8-in.-high curbs and counterflashing cover, with the roof surface sloped away in opposite directions.

The rule that flashings should be elevated obviously works better with interior draining than with perimeter drainage. With interior drainage, the designer can elevate perimeter edges or parapets, which form high, level lines.

Peripheral scupper drains require awkward crickets or saddles. These often leave small, local ponds around the scupper drains, which often freeze up and malfunction during cold weather.

If at all practicable, elevate penthouses, with drainage directed away from the walls in all directions. Adding a few extra drains is a trivial price to pay for the safety of elevated flashings, secure from ponded water. At the very least, keep flashings out of low areas where water may pond. Where a curbed opening occurs in a drain plane, design crickets to divert the flow around the curb.

Tapered fiberboard (or other material) edge strips can keep vulnerable gravel stops reasonably high and dry.

Base flashings at vertical surfaces should extend at least 8 in. above the highest anticipated waterline (see Fig. 14.1) for two reasons. First, there is the necessity of protecting the flashed joint from being penetrated by rain driven against the vertical surface or snow piled against it. The second reason concerns the field-application procedure. A vertical dimension of 8 in. (absolute minimum of 6 in.) provides working room for application of the base flashing, i.e., for nailing or otherwise anchoring the flashing felts above the cant strip. Maximum elevation for bituminous base flashing on a vertical surface is 12 in. above membrane grade line. This maximum-height limit is a safeguard against sagging, which becomes more probable as the weight of the flashing increases with its vertical dimension. A vertical flashing must resist the force of gravity throughout every second of its service life. Even perfectly formulated flashing cement or steep asphalt tends to flow a little at the highest range of roof temperatures reached even in the northern United States and Canada, and this plastic flow is always downward. Limiting the height of the base flashing above the membrane level, especially hot-mopped flashing, thus limits the sagging stress applied to nails, reglet friction wedges, or other flashing anchorage devices. Limiting flashing height also reduces the vulnerability to differential movement at the intersection of vertical and horizontal surfaces.

An exception to the rule limiting flashing height occurs with lightweight polymeric single-ply flashings. These flashings (or the membrane sheet itself) are frequently extended to the top of parapet walls and turned under the coping (see Fig. 14.4). Restraint is required at the base of the wall to resist membrane contractive stresses, which can pull flashing away from the wall.

Differential movement

Flashing details must provide for differential movement among the different parts of the building. Anchor base flashings to the structural roof deck, free of the walls or other intersecting elements, and anchor the counterflashing to the wall, column, pipe, or other flashed building component.

This need to accommodate relative movement between structurally independent building elements is often overlooked—notably in flashing details that incorrectly connect base flashing directly to walls. Where steel joists or other independent structural members span parallel to curtain (i.e., nonbearing) walls, relative movement normally occurs between the two structurally independent components. Such relative movement poses obvious hazards—notably, diagonal wrinkling—to flashing anchored to both components.

Flashing contours and cants

A third basic rule of flashing design is to avoid right-angle corner bends in bituminous base flashing. To reduce the risk of cracking the felts, use cants of 45° slope.

Wood cants (southern yellow pine, Douglas fir, or equivalent), pressure-treated with water-base preservative, are generally preferable to fiberboard cants where they abut vertical nailers at equipment curbs. Wood cants help brace these vertical nailers, whereas fiberboard nailers perform no structural function. Wood cants are also more durable. When flashing leaks wet a fiberboard cant, it tends to deteriorate and soften. However, a treated wood cant continues to provide firm, solid support for the flashing. Wood cants should be securely anchored, to prevent warping.

Flashings should be supported continuously, firmly adhered to their backing, and not left to bridge openings.

Flashing connections

Anchorage for flashings is generally the same as for adhered membranes and subject to the same limitations. On vertical surfaces, flashings exert a constant gravity force against flashing nails. Flashing nail spacing should not exceed 8 in.

Roof-edge failures initiate many wind-uplift roofing failures, according to *Factory Mutual*. Flashing nails driven into wood (or lag screws used for the same purpose) should satisfy the following FM requirements:

- Twisted or threaded shanks
- Corrosion-resistant
- 1½-in. minimum penetration into wood
- 100-lb withdrawal for each nail holding flashing or cleats, 150-lb withdrawal resistance for nails anchoring wood cants, top nailers, and fascias
- 3-in. maximum spacing for lighter metal flashings (24-gage steel, 0.032-in. aluminum, or 20-oz copper)
- Staggered nailing patterns for roof-edge nailers, with spacing *in any one row* not exceeding the foregoing 3-in. requirement

Steel nails driven through dissimilar flashing metals—e.g., aluminum or copper—can cause galvanic corrosion if water provides an electrolytic circuit linking the two metals on exposed metal—e.g., termination bars. Neoprene or EPDM washers help isolate incompatible materials from corrosion-generating contact. Metal edging embedded in roofing mastic is also isolated and water-protected.

For nailing into masonry mortar joints, specify 1½-in.-long barbed, hardened roofing nails; for concrete anchorage, specify case-hardened, large-headed steel nails or metal expansion plugs placed in predrilled holes.

Specific Flashing Conditions

Flashings are required for the following conditions:

1. Edge details, e.g., gravel stops, eaves
2. Walls, parapets, and other vertical surfaces
3. Roof-penetration connections—for vents, skylights, roof drains, scuttles, airconditioning equipment, columns, and so forth
4. Expansion-contraction joints
5. Water conductors, e.g., built-in gutters, valleys, scuppers

Each kind of flashing shares in the general problems previously discussed and also has its own peculiar problems, discussed in the following text.

Edge details

Gravel stops perform four functions:

- As a barrier preventing loose aggregate from rolling off the roof
- As edge termination for the membrane
- As a rain shield
- As top surface for anchorage of a fascia strip

To serve these functions, gravel stops require a rigid material: metal or premolded plastic. End laps, expansion joints, and membrane connections must all be watertight yet flexible enough to accommodate thermal movement and differential movement of dissimilar materials. Flange widths should be a minimum of 4 in., but not greater than the horizontal width of the underlying nailer. The flange should be primed before installation.

Gravel stops, like other flashed joints, must be raised above the general roof elevation. This can be accomplished by simply sloping the roof down from the gravel stop or via a tapered edge strip (see NRCA details). On roofs with perimeter drainage systems, the tapered edge strips should form a cricket surface.

Edge strips are subjected to especially high wind-uplift forces; they form the leading edge of the roof airfoil. Most blowoffs start with wind loosening the roof-edge detail, bending and twisting the metal fascia strip, then peeling the membrane off the insulation. To ensure adequate stiffness in the fascia, the roof designer should consult Factory Mutual's *Loss Prevention Data Sheet 1-49* for recommended metal thickness and anchorage details. For metal-deck roofs, edge flashing details require proper blocking as a shield against wind access to flutes under the insulation at the upper deck surface.

To accommodate thermal expansion and contraction stresses in a gravel stop, the roof designer has two alternative strategies:

1. Closely spaced (3-in.) nailing, designed to restrain the gravel stop from expansion and contraction
2. Sleeved joints for 10-ft lengths of fascia units, designed to accommodate linear expansion and contraction

The first strategy, nailing, is suitable for thin-gage metal, in which expansion-contraction forces are light enough to be resisted by the closely spaced nails (see NRCA details). The second strategy is for thicker sections—e.g., extruded aluminum sections—whose thermal forces might distort closely spaced nails. Here the 10-ft sections of fascia are anchored at midlength, with ends left free to expand and contract, inserted into sleeves anchored to the cant (NRCA Heavy Metal Roof Edge, p. 361).

For a stark example of how *not* to detail a gravel stop, see Fig. 14.5.

Flashings 339

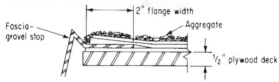

Figure 14.5 Gravel stop violates several basic principles of flashing design. (1) The horizontal flange is inserted between the membrane base sheet and two shingled plies, instead of being placed on top of the completed membrane and then stripped in. (2) It violates the rule against using metal base flashing. See NRCA and SMACNA details for proper detailing.

This detail violates the two most basic rules of flashing design:

- It is *not* located above the highest water level anticipated on the roof. (It is in fact used on a dead-level roof, on which ponding occurred over some segments of flashed joints.)
- It does *not* allow for differential movement.

Where metal gravel stops are stripped in with the membrane, splits often occur, extending from the edge perpendicular to the perimeter line toward the roof interior. Such splits will most likely occur at subzero temperatures. In the 0 to −30°F temperature range, bitumen is a brittle solid, and the membrane coefficient of expansion-contraction becomes extremely high (three to five times that of the metal gravel stop). In climates subject to such extremely cold temperatures, notably the northern and midwestern states, it is especially important to provide a layer of plastic cement to absorb some of the thermal contraction stress in horizontal shear deformation.

Because of flashed joints' vulnerability to ponded or running water, edge cant strips are preferable to gravel-stop–fascia details. (See RCABC detail, p. 360.)

Vertical flashings

Flashings at masonry walls or parapets come in two types, depending on the need to accommodate differential movement. Where the roof deck and the wall are structurally connected—e.g., where the wall is supported on a monolithically cast concrete frame that incorporates the roof deck as a concrete slab, or where roof beams bear on the wall—the flashing detail may not need to allow for differential movement (see NRCA details p. 362). Where the roof structure is not tied to the wall—e.g., where the roof's structural members span parallel to the wall, or where the wall is a nonbearing curtain wall—an expansion joint is required to accommodate differential movement. For this detail, a vertical nailer behind the cant forms a blocking backer for the base flashing. Like the cant, this vertical nailer must be anchored to the structural deck. (Even when an expansion joint is not required, a vertical backer board against the wall may be generally advisable.)

Compressible insulation inserted in the 1-in. gap between the back of the vertical nailer and the wall blocks convection currents that waste heating energy and carry moisture to the underside of the counterflashing through joints in the wood blocking. Because the conveyed air would come from the warm interior, water vapor would condense on the cold surfaces and leak back into the building. In cold climates, freezing of this condensed migrating water vapor can build up ice layers, possibly damaging the flashing. And upon thawing, this condensed moisture could leak back into the building.

A convenient insulation for sealing off this convective vapor flow is 1-in.-thick fiberglass batt insulation with a vapor retarder (two layers of kraft paper adhered with bitumen) on one face. The roofer folds this insulation double, with the vapor retarder on the outside, squeezes the envelope (nailing tabs up) into the gap behind the vertical nailer, bends the tabs over, and nails or staples them to the nailer's top surface. NRCA (p. 362) details show a flexible vapor retarder, required to obstruct vapor flow up through the insulation filler.

Two-piece, through-wall metal counterflashing has several advantages:

- Delayed installation of the vertical shielding segment permits installation of the base flashing without bending up the projecting metal and possibly deforming it from proper alignment when it is rebent. Turning up a one-piece metal counterflashing to repair or replace base flashings work-hardens the metal and often forms a concave lip where the metal enters the wall. Water collecting on this lip usually works its way into the wall, and in time this invading water causes spalling.

- The horizontal receiver segment shields against water filtering through the core of the masonry wall and then laterally into the rear of the base flashing.
- Metal flashings can sometimes be salvaged when reroofing is required.

Note that cleats are generally required to brace the bottom of the exposed counterflashing strip.

Because of the threat of water penetrating into the back of the base flashing, through-wall flashing rather than a reglet should be used. Avoid using a prefabricated cant and reglet block system into which the base flashing is inserted. Gravity stress on the base flashing and cracking and loosening of the reglet caulking make this an extremely vulnerable joint requiring constant surveillance. However, if for one reason or another through-wall flashing is impracticable—e.g., in a concrete wall—use a prefabricated cap-flashing system and vertical waterproofing detail. (See Fig. 14.6.) The best elastomeric sealants are silicone or polyurethane. Flashing cement and oil-based caulking are short-lived.

Roof penetrations

Curbs are required around most kinds of roof openings. It is necessary to allow for differential movement by isolating the base flashing from the counterflashing and the roof-penetrating component from the membrane.

It is also necessary to provide structural framing, or stiffeners, around openings, especially in metal or plywood decks. Flexible decks can be wracked by wind or other lateral forces on the roof-penetrating component, and this wracking can damage the flashing and promote leakage at the vulnerable flashed joint.

Some roof-penetrating components—notably skylights, scuttles, and HVAC rooftop units—come with metal curbs and flanges that act as base flashings, with direct connection to the membrane (see Fig. 14.7). These details violate the general flashing rule to isolate bituminous materials from metal. In cold weather, the membrane contracts much more than the metal, promoting splits and delamination at points of stress concentration.

Rooftop airconditioning units require a decision as to what kind of curbed opening to detail: support on the roof or an opening through the roof (see NRCA details).

The National Roofing Contractors Association (NRCA) has published a helpful leaflet, "NRCA Roof Curbing Criteria," with a listing of NRCA-approved curb details furnished by manufacturers of rooftop equipment. Here are the major requirements:

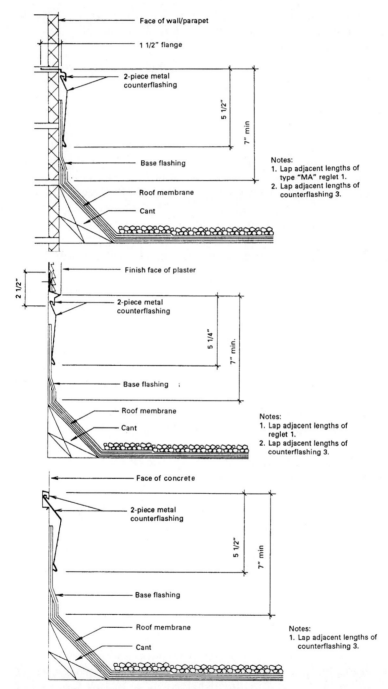

Figure 14.6 Stainless-steel clip-spring assemblies provide superior counterflashing shielding the wall base flashing for masonry (top), stucco (middle), and concrete (bottom). (*Fry Reglet Corporation.*)

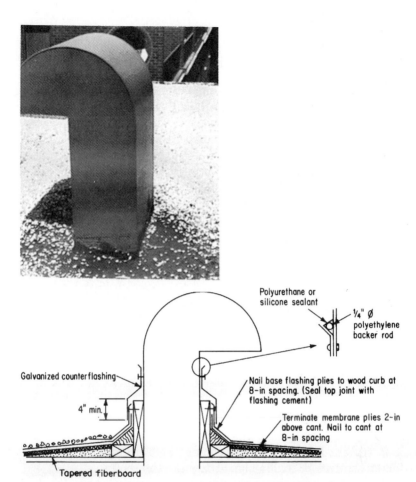

Figure 14.7 This improperly flashed vent (top) leaked around the opening perimeter. It was corrected with the sketched detail (bottom), which (1) raised the flashing above the previously ponded roof elevation and (2) substituted a properly counterflashed bituminous base flashing for the defective metal flashing in the original detail.

- A continuous metal-framed curb (16-gage minimum) furnished by the manufacturer of the roof-penetrating equipment, with a 2-in. wood nailer mounted at the curb top and minimum 8-in. vertical clearance (to top of nailer) above the finished roof surface
- No penetration of the curb for drains, electrical conduit, etc.
- Provision for metal counterflashing by sheet metal contractor
- 45° cant at the base of the curb
- Installation instructions requiring the supplier of rooftop equipment to provide a weathertight seal between unit and curb top

- Flashing of all piping and plumbing with roof flange-and-sleeve assembly extending a minimum 8 in. above the roof surface and the sleeve edge turned down inside the vent stack

To avoid the cost of providing separate support for base flashings at small pipes, vents, and similar small roof penetrations, designers usually omit these base flashing supports. Instead, the flanges or metal or plastic pitch-pan sleeves sit directly on the roof membrane and the horizontal flanged sheet metal support is stripped in like a gravel stop. Such a detail is justified around a small roof-penetrating element because it is subjected to far less differential movement than a larger roof-penetrating element. But these details nonetheless require more frequent inspection and maintenance than basic flashing details with curbed flashings.

Tubular penetrations can be effectively counterflashed with sheet metal umbrellas.

Pitch pans are the solution of last resort—for TV antennae, sign posts, and other components with irregular cross sections or diagonal entry into the roof (see Fig. 14.8). When forced to accept a pitch pan as a necessary evil, follow these rules:

- Use metal pans, 4 in. high with a 4-in. flange.
- Pack any gap between the roof-penetrating element and the deck with fiberglass insulation, mineral wool, or other compressible material to prevent drippage.
- Fill the bottom stratum of the pan with cement grout or plastic cement stiffened with portland cement before applying the final surfacing stratum of sealer, sloped down to the roof.
- Use pourable sealer rather than plastic cement (which embrittles and shrinks faster).
- Anchor a treated wood nailer (the same thickness as the insulation) around the pan opening, for nailing the flange. This is essential in nonadhered single-ply systems.
- For BUR or modified-bitumen systems, set the flange in plastic cement on top of the membrane plies (bottom ply enveloped).
- For weldable thermoplastic systems, polymer-coated metal can be formed into a pitch pan placed directly on the wood nailer and nailed in place. The membrane sheet is welded to the polymer coating on the pan. Corners can be sealed with preformed polymer fittings.
- For EPDM, follow a procedure similar to that for flange setting for BUR, except that EPDM flashing can be carried up the vertical pitch-pan dam and tucked into the interior of the pan itself.

Figure 14.8 Fabricating a sheet metal box, detailed for lateral entry of pipes or conduit (top left), is a far better practice than allowing vertical entry into a pitch pocket (top right). The sleeper pitch-pan detail shown at the rooftop airconditioning units (middle and bottom) is a cheap, inferior substitute for the more expensive curbed or column-supported details approved by the NRCA.

- If the roof system has a vapor retarder, flash the opening at the retarder elevation to preserve continuity.

Drain flashings present a unique combination of hazards. Clogged drains may impound water for long intervals. Drains are also subjected to differential vertical movement between drainpipe and roof. A structural roof deck supported on shrinking 2×12 in. wood joists could lower the deck elevation ½ in., breaking or distorting the drain flashing at its connection. (An expansion joint or 45° offset in the drainpipe can accommodate most differential vertical movement.)

Lead or copper sheets generally provide the best connection to drain frames; they also provide the stability needed to maintain a watertight joint. As noted earlier, drain flashings constitute an exception to the general rule against metal base flashings because of severe erosive and corrosive forces at the drain. Differential movement of the metal drain flashing is a minor problem because of its small plan dimensions (see NRCA details).

Expansion-contraction joints

Expansion joints, generally spaced about 200 ft and at changes in building plan shape, relieve stresses that otherwise accompany thermal expansion and contraction and other building movement accompanying material shrinkage or foundation settlement. They should be provided at changes in span direction in the structural deck or framing, at changes in deck material, and at expansion joints extending through the structure. (See Chapter 10, "Elements of Built-Up Membranes," for a further discussion of expansion-joint design.)

An expansion joint must accommodate tensile and contractive movement and shear movement in the roof plane and shearing movement perpendicular to the roof plane. It requires a curb for base flashing, and the membrane should slope down from this curb (see page 366).

Like the gap between a vertical nailer and a wall, an expansion joint should be filled with compressible material to prevent condensation and icing from vapor migrating upward from the warm interior. Condensation on the underside of a metal bellows may drip back into the interior if the expansion-joint gap is not filled. (Some prefabricated expansion joints come with an insulated bellows.)

Uncurbed, unelevated expansion-joint details in built-up membranes are a reckless attempt to economize. These very bad practices take several forms. Expansion joints featuring a fold in the membrane may behave like a roof ridge, with bitumen flowing down the sloping sides. Inverted V galvanized lightgage steel or other metal expansion joints in the membrane plane violate the rules against metal base flashing and elevated flashings (see Fig. 14.9). A designer who locates a joint designed for movement down at the general membrane level is inviting leakage. Even for single-ply elastomeric systems, roof-level expansion-contraction joints are hazardous.

Some designers betray a total misunderstanding of the function of expansion-contraction joints, and their ignorance (or carelessness) produces some flashing conditions that are not merely defective, but ludicrous. In some cases, these joints are simply drawn on roof plans as an afterthought, without the slightest conception of their purpose. Expansion-contraction joints across a roof sometimes run directly into a parapet, where they are abruptly terminated (see Fig. 14.10). This practice obviously nullifies the entire concept of an expansion-contraction joint. It is supposed to accommodate relative movement between two structurally separate areas. But by terminating the joint at a parapet, the designer is asking the structure to refrain from relative movement at the parapet intersection, although no such request is made of the roof areas on opposite sides of the joint.

To function properly, expansion-contraction joints must be carried continuously down to the foundation tops. As shown in details pro-

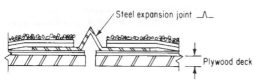

Figure 14.9 The cheap, defective expansion-joint detail shown flouts the rules of good flashing design with a vengeance. This inverted V, galvanized steel expansion joint violates the rule against metal base flashing and the rule to elevate flashing, especially important for an expansion joint. Compounding these serious flaws, it aggravates the already serious leak hazard by inserting the metal flanges between membrane plies instead of placing them on top of the felts and stripping them in with two plies of felt.

Figure 14.10 The photograph shows several flashing defects. The roof-level expansion joint terminates at a parapet corner, thus nullifying the entire concept of a joint permitting relative movement between two independent structures. It should have been located coincident with the *outside* face of the parallel parapet, with the joint continued through the parapet. Compounding the design felony, the expansion joint in the photograph is a cheap, substandard galvanized sheet metal detail with an elastomeric bellows instead of a curbed detail as recommended by NRCA or SMACNA. Note also (in the top left corner of the photograph) a defective, caulked horizontal joint in the parapet cap flashing.

mulgated by NRCA and SMACNA, the counterflashing for such joints is designed to shield the expansion-contraction joint from water intrusion. Wherever it intersects a parapet, the joint must be carried through the parapet and down the wall. Another instance where this elementary rule was violated is shown in Fig. 14.11, where a cheap, defective expansion joint was capped with a wood 2×4. Some parapets adjoining expansion-contraction joints behave as one would expect, with tensile cracks from contraction stresses in the masonry.

Water conductors

Built-in gutters are subject to surface abrasion. They are also difficult to connect to the membrane. For relatively flat gutters, which may retain water for long intervals, leakage in sliding expansion joints is also a problem.

Metal offers superior durability, but largely because of its differential thermal coefficient of expansion, it is difficult to connect permanently to the roof membrane. Gutters made of bituminous materials

Figure 14.11 This figure, like Fig. 14.10, shows a defectively designed and constructed expansion joint detail terminated at a parapet instead of continuing through it. The inadequately elevated expansion joint has a 2×4 board cover plate. How it was supposed to function was an unresolved mystery, evidently beyond the interest of all concerned.

readily accommodate thermal movement, but they deteriorate so rapidly under wetting-drying cycles and the abrasive flow of water that they are limited to short valleys. NRCA details show a satisfactory method of flashing scuppers. Interior drainage is nonetheless preferable to perimeter drainage, for these reasons:

1. Interior drains are less vulnerable to ice-dam blockage than exterior drains.
2. Interior drains are generally less vulnerable to wear and damage (e.g., distortion from freeze-thaw cycling) than scuppers.

Flashing Failures

Flashing failures, ultimately resulting in leaks, occur in several modes, and each failure mode has several possible causes, acting singly or in combination. Here are the basic flashing failure modes:

- Sagging
- Ponding leakage
- Leakage *around* the flashing
- Leakage directly *through* the flashing

- Separation of flashing materials
- Diagonal wrinkling (resulting in splits)
- Damage after construction (abuse)

Sagging

Sagging is a vertical (or steeply sloped) flashing's tendency to slide downward, ultimately tending to pull out from reglets, to tear, or to split. Sagging has numerous causes. It can result from omission or excessive spacing of flashing nails, an overly fluid adhesive—hot-applied asphalt with too low a softening point—or use of plastic cement instead of the more viscous flashing cement. At summer roof temperatures, which may exceed 150°F, plastic bitumen gradually flows downward, stressing the flashing fabric and possibly tearing it. Flashings extended more than 12 in. up a vertical surface are also subject to sagging. The added weight of the flashing materials aggravates the general problem of perpetual gravity force. Membrane shrinkage, caused by inadequate insulation anchorage to the deck, can also exert sagging stress on base flashings.

Still another cause of sagging is omission of a primer on masonry walls. A primer of solvent-thinned (cutback) asphalt is required to prepare a masonry surface for either hot-applied steep asphalt, cold-applied flashing cement, or a torched-on flashing system. If the masonry surface is not primed, mortar flakings, efflorescent precipitate, and other foreign particles may prevent good adhesion between the masonry surface and the bitumen.

Ponding leakage

Ponding is a greater threat to flashings than to the membrane. Flashing occurs at joints between different building components. In addition to the increased probability of leakage at any joint between materials, there is usually much greater probability of relative movement at a flashing than in the membrane. Thus arises the rule to keep flashed joints at high points, if practicable. If this is not practicable, at least keep flashing out of low points, with rapid drainage away from them. This policy requires active attention; the designer cannot simply let the elevation chips fall where they may. Yet many roof designers do just that.

Leakage can occur *around* poorly designed flashings. When vertically flashed joints are too low—i.e., less than 8 in. high, with less than a 4-in. cap-flashing overlap of base flashing—wind-driven rain, snow, or even deep ponded water can blow or rebound up and under the cap flashing, penetrating the masonry wall above the base flashing and descending inside the wall to invade the built-up roof system. If water ponds at such an inadequately detailed flashing, it is even more vulnerable to leakage from wind-driven rain and capillary rise when snow is banked against flashings. Even with a well-designed flashing of the correct height (between 8 to 12 in.), water can get behind the base flashing if the masonry is porous.

Leakage directly through flashing is usually the result of improper application—e.g., omission of a backing ply from the system, or inadvertent thinning of an uncured elastomeric flashing while stretching it to conform to the curb or wall. A loose flashing felt is especially vulnerable to puncture and consequent admission of water. Hot-mopped flashings are especially vulnerable to loose, faulty backing.

Separation of flashing materials often results from details that ignore metallic and bituminous flashings' different coefficients of expansion-contraction. As noted earlier, these failures often occur at gravel stops. They also occur at curbs around roof-penetrating components—scuttles, airconditioning equipment, and so forth. Such failures can be avoided by heeding the design principles set forth elsewhere in this chapter.

Diagonal wrinkling

Diagonal wrinkling in base flashing along parapet walls follows a consistent pattern, with the wrinkles sloping down from the top toward the center of the wall (see Fig. 14.12). The most severe wrinkling occurs at building corners, diminishing in proportion to the distance from the corners and proximity to the center. Diagonal wrinkles obviously break the flashing felts' bond with their backing and thus make the flashing highly vulnerable to punctures and mechanical damage. And like membrane ridging, diagonal wrinkling of base flashing often sets in motion a degenerative process that ends in cracking and consequent leaking.

The worst diagonal wrinkling from membrane shrinkage usually occurs along a parapet perpendicular to the felt direction because built-up membranes generally shrink more in the transverse than in the longitudinal direction.

Another cause of diagonal wrinkling in base flashing, sometimes in combination with membrane shrinkage, is differential thermal expansion of the masonry parapet walls. Long term wall growth can produce the relative lateral movement that produces diagonal wrinkling in base flashing. This long term expansion results from the largely irreversible elongation caused by thermal cycling and moisture absorption in the inelastic masonry. In the northern hemisphere, sun-baked parapets may experience daily temperature differentials of

Figure 14.12 Diagonal wrinkling in parapet flashing can result from (1) relative movement of membrane parallel to the wall from insecurely anchored insulation boards or (2) long term masonry wall expansion.

100°F and extreme annual differentials approaching 200°F, compared with 15°F daily and 25°F seasonal extreme temperature differentials for the roof deck and structural framing.

On parapets with stone or concrete copings, freeze-thaw cycles can also promote diagonal wrinkling of the base flashings. Thermal cycling cracks the mortar joints, letting water enter. Subsequent freezing expands the water, prying apart the coping stones and possibly dragging the top edge of the base flashing along with them.

This parapet-promoted diagonal wrinkling in base flashing can be avoided by isolating the base flashing from the parapet, as shown in NRCA details.

Post-construction damage

Post-construction damage accounts for a significant proportion of flashing failures. Equipment and scuttle locations are the busiest locations on the roof, and a dropped tool or a carelessly swung worker's boot can puncture an inadequately supported flashing. Ladder locations, where a roof visitor climbs over a parapet or up a wall, with the lowest ladder rung starting over a foot above roof level, are likely spots for a careless kick that may puncture base flashing that is not solidly bonded to a firm backing. Maintenance crews working on ventilating or airconditioning equipment may even fail to reinstall counterflashing removed during repair operations.

Alerts

General

1. Check with roofing materials manufacturer for approval of all flashing details and materials.

2. Check "NRCA Roof Curb Criteria" list for rooftop equipment to ensure that roof-penetrating components conform with good flashing practice.

3. Complement specifications with large-scale flashing details on drawings.

Design

1. Locate flashed joints at a roof's high points, if possible; in any event, locate them above anticipated water level.

2. Keep HVAC units off the roof, if economically feasible. (This policy reduces the number of vulnerable flashed joints, keeps hazardous workers off the roof, and lengthens HVAC equipment service life.)

3. Minimize the number of penetrations through the roof. Group piping and roof-mounted equipment requiring flashing with pads, curbs, and so on into a smaller number of larger areas to reduce the total quantity of flashing (see standard details by NRCA, SMACNA, etc.).

4. Locate stacks and other small roof-penetrating components at least 2 ft from walls or other vertical surfaces. (They are vulnerable to damage if placed too close to curbs requiring periodic maintenance and repair.)

5. Avoid right angle bends in base flashing; use cant strips to provide an approximate 45° bend.

6. Favor treated, solid wood cant strips over fiberboard where curb support is needed.

7. Specify water-based, salt-type preservative treatment, compatible with bitumen, for wood cant strips, nailers, and blocking used with roof assembly.

8. Never anchor base flashing directly to walls, structural members, pipes, or other building elements independent of the roof assembly. Always provide for differential movement.

9. Where metal cap flashing overlaps base flashing and at horizontal joints between cap-flashing segments, detail the joint for differential movement.

10. Extend base flashing at vertical surfaces to a minimum 8-in. vertical dimension, maximum 12 in. above membrane elevation.

11. Avoid pitch pockets whenever practicable. (Use pedestals or curbed openings.)

12. Check Factory Mutual minimum gages for gravel stops (consult Factory Mutual's *Loss Prevention Data Sheet 1-49*).

13. Specify an expansion joint between drain and leader pipes, or offset the leader pipe 45°. (When these units are rigidly attached to columns, movement can break flashing connections and permit water entry into the membrane.)

14. For single-ply membranes, check that nailers have been installed to resist shrinkage stresses—75 lbf/ft for elastomerics, 175 lbf/ft for weldable thermoplastics.

15. Extend single-ply membranes under metal copings and gravel stops.

Field

1. When hot-mopped, felted flashings are specified, check to ensure use of steep asphalt (to prevent eventual sagging of base flashings adhered with low-softening-point asphalt mistakenly applied). Back mopping of flashing felts is often preferable to application of hot-mopped asphalt on vertical surfaces.

2. Check for proper priming of walls and gravel stops' metal flanges before applying base flashing or stripping felts.

3. Modified-bitumen flashings should be top-anchored to resist slippage. Corners may be reinforced with "butterfly" caps or sealed with asphalt mastic and mesh.

4. Check for use of specified nails.

5. Check temperature for application of plastic or flashing cement (25°F minimum).

6. Install flashing, including counterflashing, as roof application progresses. If delay in installing counterflashing is unavoidable, apply a three-course stripping seal (one ply of fabric flashing between two trowelings of flashing cement) to seal the joint and prevent water from entering behind the base flashing until counterflashing goes in place.

7. Check for use of sealer strips at vertical flashing joints.

8. Establish a maintenance program with semiannual inspection followed by resurfacing of deteriorating surfaces, restripping of opened joints, renewal of all joint and reglet caulking, and repair of rips, tears, and other defects.

9. Verify compatibility of elastomer with substrate. Avoid asphalt contact by interposition of separator sheets, plywood, or fleece-backed membrane.

Standard Details

In the following pages appear standard details promulgated by NRCA, CRCA, and RCABC. These details demonstrate the principles promoted in this chapter. Preceding these standard details are others (Figs. 14.13 through 14.16) depicting violations of industry-approved flashing principles. The standard details begin on p. 359.

Figure 14.13 This negligently detailed parapet cap flashing has two futile nails visible at a 1 in. vertical gap, with its open invitation to water intrusion. Compounding the error is a horizontal lap between the two spliced segments of 1 in. or less. One satisfactory horizontal splice features a vertical leg on one segment overlapped by an inverted U on the adjacent segment.

Figure 14.14 The lead-flashing vent pipe intersecting parapet base flashing violates the NRCA-recommended minimum distance of 12 in., measured from the toe of the cant.

Figure 14.15 This flashing atrocity shows rooftop electrical conduit penetrating a parapet counterflashing. Note also the exposed stucco ledge and vertical nailer flange.

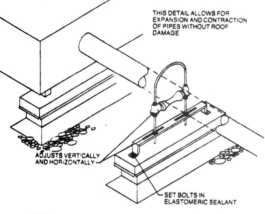

Figure 14.16 Rooftop piping on the project depicted lacks even pitch pans shielding the strut penetrations through the membrane (to a concrete deck). Recommended NRCA detail features roller-supported steel frames designed to accommodate lateral pipe movement and set on flashed and counterflashed curbs. (*National Roofing Contractors Association.*)

Standard details promulgated by NRCA, CRCA, and RCABC begin here and continue through p. 371.

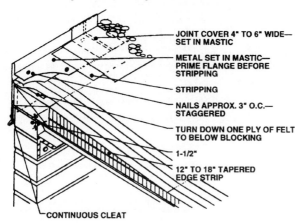

Light Metal Roof Edge — NRCA DETAIL

- JOINT COVER 4" TO 6" WIDE— SET IN MASTIC
- METAL SET IN MASTIC— PRIME FLANGE BEFORE STRIPPING
- STRIPPING
- NAILS APPROX. 3" O.C.— STAGGERED
- TURN DOWN ONE PLY OF FELT TO BELOW BLOCKING
- 1-1/2"
- 12" TO 18" TAPERED EDGE STRIP
- CONTINUOUS CLEAT

NOTES:

ENVELOPE SHOWN FOR COAL-TAR PITCH AND LOW-SLOPE ASPHALT.

ATTACH NAILER TO MASONRY WALL.

THIS DETAIL SHOULD BE USED ONLY WHERE DECK IS SUPPORTED BY THE OUTSIDE WALL.

THIS DETAIL SHOULD BE USED WITH LIGHT-GAUGE METALS, SUCH AS 16-OZ. COPPER 24-GAUGE GALVANIZED METAL OR 0.040" ALUMINUM. A TAPERED EDGE STRIP IS USED TO RAISE THE GRAVEL STOP. FREQUENT NAILING IS NECESSARY TO CONTROL THERMAL MOVEMENT.

WOOD BLOCKING MAY BE SLOTTED FOR VENTING WHERE REQUIRED.

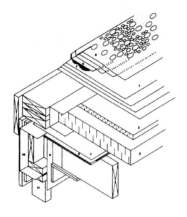

FL. 4. Roof Edge for Roofs Drained to Exterior — Isometric (CRCA)

1. Roofing membrane and membrane protection.
2. Secondary insulation, usually fibreboard
3. Primary Insulation
4. Vapor retarder
5. Acceptable structural deck and support for vapor retarder, by others.
6. Metal gravel stop
7. Roofing mastic
8. Three plies of stripping, one of which is mesh or cotton
9. Vapor retarder inside building by others
10. Air barrier, by others
11. Wall construction by others

Embed prime lengths of gravel stop in mastic. Nail back edge of flange at 100mm staggered centers.

Because of the expansion and contraction of the metal flashing, this detail will require more than average maintenance.

Eaves trough may be installed.

Roof Edges and Fascias: Gravel Stop (RCABC 1989)

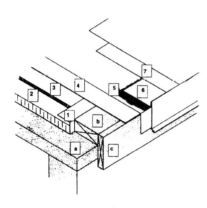

WORK INCLUDED (2 year guarantee only on built-up roofing and 5 year guarantee for flexible membranes where approved by manufacturer.)

1. Vapor retarder: the requirement for a vapor retarder is determined by the design authority.
2. Insulation
3. Insulation overlay: where required
4. Primary membrane and membrane protection: mechanically fasten to outside face of fascia (C), top ply only in BUR. Do not mop outside edge.
5. Mastic: trowel coating of compatible mastic
6. Metal Gravel Stop Flashing: minimum 100mm (4") flange onto roof membrane, gravel stop maximum 25mm (1") high; embed in mastic and mechanically fasten at 100mm (4") o.c.; lap flashing 25mm (1") minimum
7. Membrane flashing

Related work by others: (a) Acceptable deck; (b) Wood blocking; and (c) Wood fascia.

Note: Avoid this detail; wherever possible use "Roof Edge Cant Strip."

Roof Edges and Fascias: Roof Edge Cant Strip (RCABC 1989)

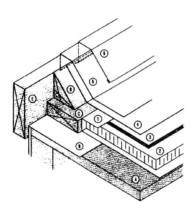

WORK INCLUDED

1. Vapor retarder: the requirement for a vapor retarder is determined by the design authority.
2. Insulation
3. Insulation overlay: where required
4. Primary membrane and membrane protection
5. Membrane flashing: mechanically fasten to outside face of fascia (E). Do not mop outside edge.
6. Metal flashing

Related Work by others: (a) Acceptable deck; (b) Air/vapor seal; (c) Wood blocking; (d) Wood cant strip; and (e) Wood fascia.

This raised detail is preferred — metal is not stripped into membrane — allowed to move. Drainage to interior of building.

HEAVY-METAL ROOF EDGE — NRCA DETAIL

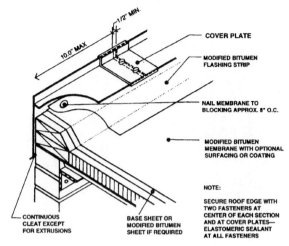

NOTES:

THIS DETAIL SHOULD BE USED ONLY WHERE THE DECK IS SUPPORTED BY THE OUTSIDE WALL.

METALS OF 22-GAUGE STEEL, 0.050" ALUMINUM, 24-GAUGE STAINLESS STEEL OR HEAVIER ARE APPROPRIATE FOR THIS DETAIL. METALS OF THIS WEIGHT ARE VERY RIGID WHEN FORMED, AND FASTENING AT THE CENTER-LINE AND JOINT COVER WILL ALLOW EXPANSION AND CONTRACTION WITHOUT DAMAGING THE BASE FLASHING MATERIAL.

ATTACH NAILER TO MASONRY WALL. REFER TO FACTORY MUTUAL DATA SHEET 1-49.

WOOD BLOCKING MAY BE SLOTTED FOR VENTING WHERE REQUIRED.

SCUPPER THROUGH ROOF EDGE — NRCA DETAIL

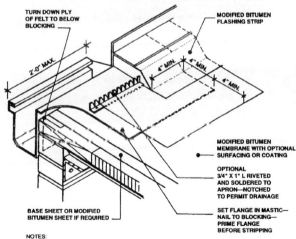

NOTES:

THIS DETAIL SHOULD BE USED ONLY WHERE THE DECK IS SUPPORTED BY THE OUTSIDE WALL.

THIS DETAIL CAN BE ADAPTED TO ROOF EDGES AS SHOWN IN OTHER DETAILS AND IS EASY TO INSTALL AFTER THE BUILDING IS COMPLETED. THIS DETAIL IS USED TO RELIEVE STANDING WATER IN AREAS ALONG THE ROOF EDGE. ALL ROOF SURFACES SHOULD BE SLOPED TO DRAIN.

ATTACH NAILER TO MASONRY WALL.

WOOD BLOCKING MAY BE SLOTTED FOR VENTING WHERE APPROPRIATE.

BASE FLASHING FOR WALL-SUPPORTED DECK — NRCA DETAIL

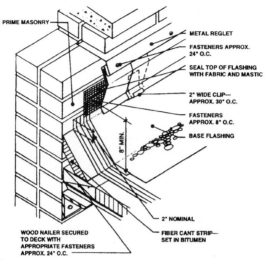

NOTES:

THIS DETAIL SHOULD BE USED ONLY WHERE THE DECK IS SUPPORTED BY THE WALL.

THE JOINTS IN THE TWO PIECES OF FLASHING SHOULD NOT BE SOLDERED. BREAKS IN SOLDERED JOINTS COULD CHANNEL WATER BEHIND THE FLASHING. CLIPS AT THE BOTTOM OF THE FLASHING ARE NOT NECESSARY ON FLASHINGS OF 6" OR LESS.

BASE FLASHING FOR NON-WALL-SUPPORTING DECK — NRCA DETAIL

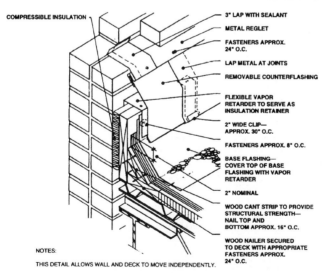

NOTES:

THIS DETAIL ALLOWS WALL AND DECK TO MOVE INDEPENDENTLY.

THIS DETAIL SHOULD BE USED WHERE THERE IS ANY POSSIBILITY THAT DIFFERENTIAL MOVEMENT WILL OCCUR BETWEEN THE DECK AND A VERTICAL SURFACE, SUCH AS AT A PENTHOUSE WALL. THE VERTICAL WOOD MEMBER SHOULD BE FASTENED TO THE DECK ONLY. THIS IS ONE SATISFACTORY METHOD OF JOINING THE TWO PIECE FLASHING SYSTEM. OTHER METHODS MAY BE USED.

Membrane Flashing: 4 Ply Built-Up Roof (RCABC 10/88)

Wall Flashing — Use cant strip to support membrane and flashing
— Terminate membrane at top of cant.

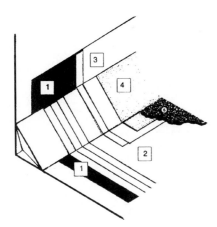

WORK INCLUDED (2 year guarantee only)

1. Primer: apply primer to wood and concrete surfaces directly adhered with hot asphalt.
2. Four Ply Built-Up Roof Membrane: four plies No. 15 asphalt saturated perforated felt in moppings of hot asphalt; carry to top of cants.
3. Two Ply Membrane Flashing: two plies of asphalt saturated, perforated felts in moppings of hot asphalt; with the first ply extending 50mm (2") from toe of cant and the second ply 100mm (4") from toe of cant and both plies carried up vertical surface a minimum of 200mm (8"). Mechanically fasten top ply along top of outside edge as indicated on flashing details.
4. One Ply Mineral Surfaced Membrane Flashing: either 40 kg (90 lb) mineral surfaced roofing or granulated 2mm reinforced SBS modified bituminous membrane; adhered in a full mopping of hot asphalt; extend from 50mm (2") on roof deck to top of cants.
5. Pour Coat and Gravel: minimum single pour coat and gravel is required.

Note: Cant strips (wood or fiberboard) are required for all built-up roof systems.
Refer to Minimum Standards for additional requirements.

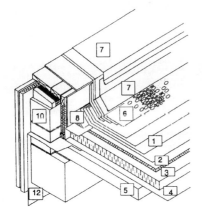

FL.1. Low Parapet Wall — Isometric (CRCA)

1. Roofing membrane and membrane protection
2. Secondary insulation, usually fiberboard
3. Primary insulation
4. Vapor retarder
5. Acceptable structural deck and support for vapor retarder, by others
6. Membrane base flashing
7. Metal flashing
8. Fiber cant. Secondary and primary insulation continued under cant
10. Wall construction by others
12. Air barrier, by others

LIGHT-METAL PARAPET CAP

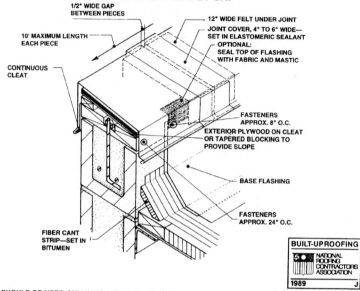

NOTES:
THIS DETAIL SHOULD BE USED ONLY WHEN THE DECK IS SUPPORTED BY THE WALL. AN EXPANSION JOINT DETAIL SIMILAR TO DETAIL E SHOULD BE USED FOR NON-WALL SUPPORTED DECK.

REFER TO BUILT-UP ROOFING DETAIL Y FOR METAL THICKNESS AND CLEAT REQUIREMENT.

GAUGE OR THICKNESS GUIDE
FOR METAL FASCIA EXPOSED TO VIEW

COPING   CAP FLASHING AND FASCIA   VARIATIONS

RECOMMENDED MINIMUM GAUGES FOR FASCIA SHOWN ABOVE

EXPOSED FACE WITHOUT BRAKES "A" DIMENSION	CLEAT REQUIRED	GALVANIZED IRON	COLD ROLLED COPPER	ALUMINUM 3003-H14
UP TO 4" FACE	NO	26 GA.	16 OZ.	.032" (20 GA.)
UP TO 6" FACE	YES	26 GA.	16 OZ.	.040" (18 GA.)
6" TO 8" FACE	YES	24 GA.	16 OZ.	.050" (16 GA.)
8" TO 10" FACE	YES	22 GA.	20 OZ.	.064" (14 GA.)
10" TO 15" FACE	YES	20 GA.	ADD BRAKES TO STIFFEN	.080" (12 GA.)

NOTE: WHEN USING THE ABOVE TABLE, OTHER ITEMS SHOULD BE CONSIDERED, SUCH AS FASTENING PATTERN. FOR INSTANCE, IF THE METAL CAN ONLY BE FASTENED AT 10 FOOT INTERVALS, A HEAVIER GAUGE METAL WOULD BE REQUIRED. ALL CLEATS SHALL BE CONTINUOUS AND OF SAME MATERIAL OF EQUAL OR GREATER THICKNESS THAN THE FASCIA METAL USED.

CURB DETAIL FOR ROOFTOP AIR HANDLING UNITS — NRCA DETAIL (EPDM)

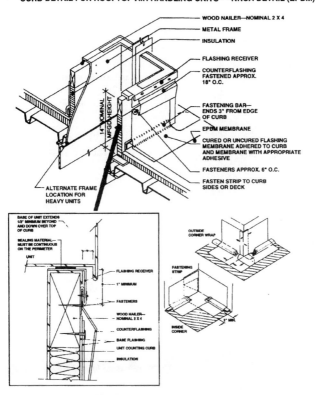

FL.23. Typical Curbed Opening (CRCA 11/90)

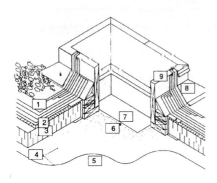

1. Roofing membrane and membrane protection
2. Secondary insulation, usually fiberboard
3. Primary insulation
4. Vapor retarder
5. Acceptable structural deck and support for vapor retarder, by others
6. Air barrier, not shown — by others. Design flexible membrane to accommodate deck deflection. Flexible membrane must connect air barrier to ductwork or other equipment installed inside curbed opening.
7. Mineral wool insulation, by others, not shown. Install insulation in close contact with the air barrier. Entirely fill space between curbed opening and installed equipment.
8. Metal flashing
9. Construct so that top of curbed opening is minimum 200mm above membrane.

Left Hand Side shows: one piece metal flashing
Right Hand Side shows: two piece metal flashing

FLASHING STRUCTURAL MEMBER THROUGH ROOF DECK — NRCA DETAIL

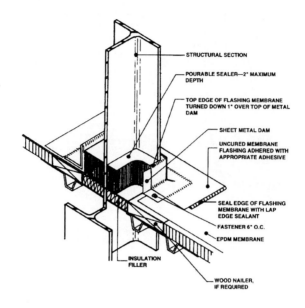

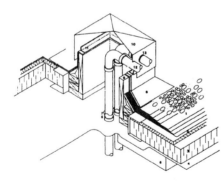

FL.24. Grouping Pipes Through Roof Membranes (CRCA)

1. Roofing membrane and membrane protection
2. Secondary insulation, usually fiberboard
3. Primary insulation
4. Vapor retarder
5. Acceptable structural deck and support for vapor retarder, by others
6. Air barrier, not shown — by others. Design flexible membrane to accommodate deck deflection. Flexible membrane must connect air barrier to ductwork or other equipment installed inside curbed opening.
7. Mineral wool insulation, by others, not shown. Install insulation in close contact with the air barrier. Entirely fill space between curbed opening and installed equipment.
8. Metal flashing
9. Construct so that top of curbed opening is minimum 200mm above membrane.
10. Metal hood with sloped top by others. Design and install hood so that it will overlap curb.
11. Hood support angle with slotted bolt holes to permit vertical movement.
12. Slope pipes away from hood, by others.
13. Caulk joints between pipes and hood, by others.

9

STACK FLASHING — NRCA DETAIL

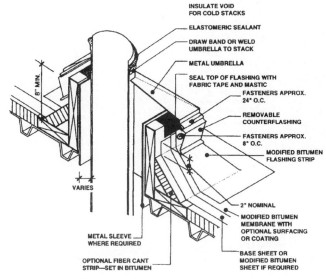

NOTE:

THIS DETAIL ALLOWS THE OPENING TO BE COMPLETED BEFORE THE STACK IS PLACED. THE METAL SLEEVE AND THE CLEARANCE NECESSARY WILL DEPEND ON THE TEMPERATURE OF THE MATERIAL HANDLED BY THE STACK.

EXPANSION JOINT — NRCA DETAIL

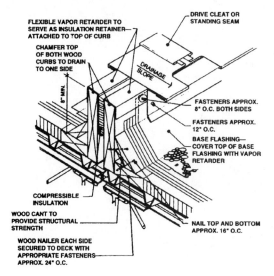

NOTE:

THIS DETAIL ALLOWS FOR BUILDING MOVEMENT IN BOTH DIRECTIONS. IT HAS PROVEN SUCCESSFUL WITH MANY CONTRACTORS FOR MANY YEARS.

CRCA SPECIFICATIONS

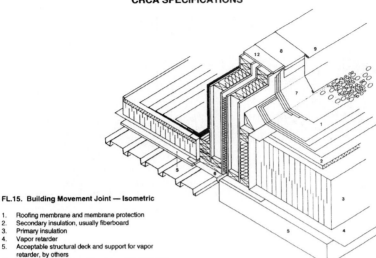

FL.15. Building Movement Joint — Isometric

1. Roofing membrane and membrane protection
2. Secondary insulation, usually fiberboard
3. Primary insulation
4. Vapor retarder
5. Acceptable structural deck and support for vapor retarder, by others
6. Air barrier. Flexible membrane to accommodate a movement of at least 25mm vertically and horizontally, by others.
7. Membrane base flashing
8. Flexible membrane flashing
9. Metal flashing. Design flashing for 25mm movement in both directions.
12. Nail plywood support on one side leaving other side free

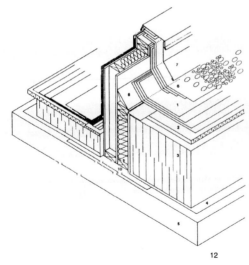

FL.13. Roof Area Divider For Thick Insulation — Isometric

1. Roofing membrane and membrane protection
2. Secondary insulation, usually fiberboard
3. Primary insulation
4. Vapor retarder
5. Acceptable structural deck and support for vapor retarder, by others.
6. Membrane base flashing
7. Metal flashing
8. Wood Cant
9. Air barrier, by others

ROOF DRAINS — NRCA DETAIL

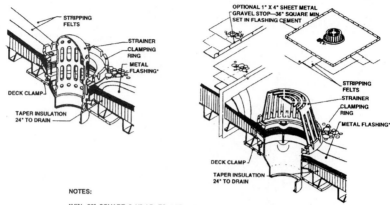

NOTES:

*MIN. 30" SQUARE 2-1/2 LB. TO 4-LB. LEAD OR 16 OZ. SOFT COPPER FLASHING SET ON FINISHED ROOFING FELTS IN MASTIC. PRIME TOP SURFACE BEFORE STRIPPING.

MEMBRANE PLIES, METAL FLASHING AND FLASH-IN PLIES EXTEND UNDER CLAMPING RING.

STRIPPING FELTS EXTEND 4" AND 6" BEYOND EDGE OF FLASHING SHEET, BUT NOT BEYOND EDGE OF SUMP.

THE USE OF METAL DECK SUMP PANS IS NOT RECOMMENDED.

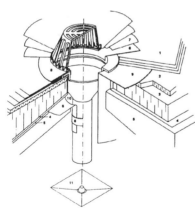

FL.21. Roof Drain (CRCA)

1. Roofing membrane, and membrane protection.
2. Secondary insulation, usually fiberboard.
3. Primary insulation.
4. Vapor retarder.
5. Acceptable structural deck and support for vapor retarder, by others.
6. Air barrier, by others. Design flexible membrane and clamping ring of air barrier to accommodate a deflection of at least 25 mm.
7. Three ply felt stripping.
8. Lead sleeve by others.
9. Optional metal sleeve by others.
10. Insulation to prevent condensation on vapor retarder, not shown — by others.
11. 1 200mm x 1 200mm sump with 25mm slope to drain.

Right Hand Side shows:
•optional metal sleeve (9) by others.

13

Flashings 371

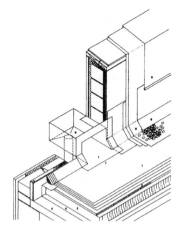

FL.6. Scupper For High Parapet Wall (CRCA)

1. Roofing membrane, and membrane protection. Aggregate is not continued into scupper opening. As an option, a metal gravel stop may be soldered in place across opening.
2. Secondary insulation, usually fiberboard.
3. Primary insulation.
4. Vapor retarder.
5. Acceptable structural deck and support for vapor retarder by others.
6. Membrane base flashing.
7. Three ply membrane flashing over metal scupper (not indicated). One ply cotton, two plies felt.
8. Metal flashing.

NRCA DETAILS

PIPE ROLLER SUPPORT

THIS DETAIL ALLOWS FOR EXPANSION AND CONTRACTION OF PIPES WITHOUT ROOF DAMAGE

ADJUSTS VERTICALLY AND HORIZONTALLY

SET BOLTS IN ELASTOMERIC SEALANT

NOTE:
NRCA REAFFIRMS ITS OPPOSITION TO PIPES AND CONDUITS BEING PLACED ON ROOFS, HOWEVER, WHERE THEY ARE NECESSARY THIS TYPE OF PIPE ROLLER SUPPORT IS RECOMMENDED. THE DETAIL SHOWN IS GENERIC ONLY.

CLEARANCES FOR MULTIPLE PIPES — BETWEEN PIPES AND FROM WALLS AND CURBS

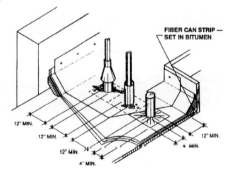

FIBER CAN STRIP — SET IN BITUMEN

12" MIN.
12" MIN.
12" MIN.
12" MIN.
4" MIN.
4" MIN.

Material attributed to SMACNA in this booklet was used with permission from sources copyrighted by SMACNA.

R-30 9/95 © The Roofing Industry Educational Institute

Chapter

15

Protected Membrane Roofs and Waterproofed Decks

The protected membrane roof (PMR), also known as the insulated roof membrane assembly (IRMA), inverted roof assembly (IRA), or simply upside-down (USD) roof, reverses the positions of membrane and insulation in the conventional insulated roof assembly. Instead of being in the conventional exposed position on top of the insulation, a PMR membrane is sandwiched between the insulation above and the deck below (see Fig. 15.1).

Historically, the prototypical PMR is the sod roof, a centuries-old roof system in northern countries. Birch bark provides a shingled drainage surface off a timber-framed roof structure, and an earth-sod cover is insulation over the shingles.

A still later version of PMR is represented by a United States manufacturer's now-discontinued specification for a "garden" roof, installed directly on a concrete deck, comprising a five-ply, aggregate-surfaced, coal-tar-pitch membrane topped with loose gravel aggregate, earth fill, and sod. Before the flood coat was applied, this roof was test-flooded with 2 in. of water to ensure its watertight integrity before it was covered with earth fill.

Over the past several decades, the PMR has commanded steadily increasing international attention as a possible solution to problems encountered with the conventional roof system. By 1995, there were more than 50,000 PMRs in service in the United States and Canada, with thousands more in Europe, Asia, and the Middle East, performing in the torrid climate of Saudi Arabia and in the Arctic climate of Alaska.

Since Canada has been the world's pioneer in PMR systems, the Canadians' favorable experience is especially significant because of its

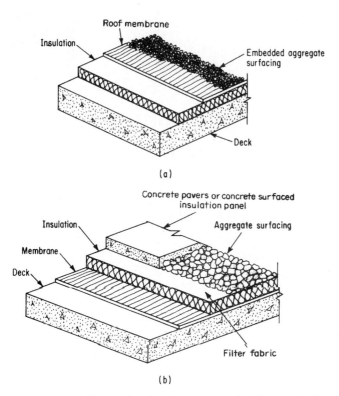

Figure 15.1 (a) Conventional roof system sandwiches insulation between deck and membrane; (b) protected membrane roof (PMR) places the membrane directly on the deck (or on leveling board on steel decks), with insulation above the membrane, and with concrete pavers, concrete-surfaced insulation panels, filter fabric, and aggregate atop the insulation.

extent and Canada's rigorous climate. According to a survey by the Division of Building Research, National Research Council of Canada, which began promoting the PMR concept back in 1965, PMRs are recommended by the overwhelming majority of architects who have specified and roofers who have built them. They also rate PMRs as superior to conventional roof systems for roof gardens and plazas.[1]

Much United States' experience with PMRs has been similarly favorable. Since 1969, the University of Alaska has used PMRs as its primary specification for new construction for its Fairbanks and Anchorage campuses. University planners estimate an additional 25 percent first cost for their version of PMR. But a 70 to 80 percent estimated reduction in operating and maintenance cost, plus longer anticipated service lives, more than pays back the added first cost, making PMR a highly profitable long term investment.[2]

A major city's Board of Education has had similarly favorable experience with PMRs. Over the past few years the Board has installed 30 or more PMRs—some reroofing projects, others new. According to the Board's chief specification writer, these roofs have generally outperformed the conventional roof systems they have replaced. Countless problems with conventional built-up roof systems prompted the switch to PMRs.

The PMR is not the panacea long sought by harassed architects and building owners. However, it is a new technique that already has established its claim to a steadily growing share of the roofing market, with unquestionable strengths as well as drawbacks. If not *the* wave of the future, it is nonetheless *a* wave of the future.

Why the Protected Membrane Roof?

The conventional roof-system arrangement is inverted to protect the membrane from the many hazards of the conventional roof's exposed membrane location, notably:

- Accelerated oxidation and evaporation of volatile oils, leading to premature embrittlement and cracking of bitumen
- Ultraviolet degradation of organic polymeric materials
- Low-temperature contraction, causing splitting of bituminous membranes and lap-shear stress on adhesively lapped single-ply systems
- Roof traffic and hailstone impact
- Warping or delamination from ice contraction
- Blistering and ridging
- Stress concentrations over insulation joints

The PMR concept also avoids the vapor-entrapment problem of conventional roof assemblies, with insulation sandwiched between vapor retarder and membrane. It combines the vapor retarder and membrane into one unit.

The life-shortening hazards to the membrane are drastically reduced in a PMR. Whereas a conventionally exposed membrane may experience a range of 100°F daily and 200°F annual temperature change, a properly designed PMR should normally experience less than 10°F daily and 30°F annual temperature change. Insulation shields the membrane from photo-oxidative reactions that exponentially accelerate the chemical deterioration of exposed areas of bitumen, especially in the higher temperature ranges.

Illustrating the benefits of this solar shielding, bitumen specimens taken from a conventional 10-year-old asphalt built-up membrane reportedly suffered a roughly 200°F rise in softening point, from 190°F (Type III asphalt) to over 400°F. In contrast, the Type III asphalt from several PMRs of equal age gained less than 30°F in softening point.

A severe increase in softening point indicates drastic deterioration of a bitumen's waterproofing quality. The affected bitumen becomes more brittle and permeable. Cracks admit water, and the felts wick it deeper into the membrane. Heat load on polymer-modified bitumens causes phasing and embrittlement (see Chapter 11). Protection from this combined physical and chemical degradation of the waterproofing bitumen constitutes a major benefit of the PMR concept, along with protection from physical abuse.

Another major benefit of the PMR concept is its virtual elimination of thermal stress in the membrane. Subfreezing temperatures, especially subzero (0°F) ones, produce the greatest tensile splitting hazard because of the cold, brittle bitumen's vastly increased coefficient of thermal expansion-contraction at these low temperature ranges. (In the −30 to 0°F temperature range, the coefficient of expansion-contraction for various built-up membranes ranges from about 5 to 10 times as much as the same coefficient in the +30 to +73°F temperature range.)

A PMR membrane apparently remains above freezing temperature even in the coldest weather. Laboratory research and field experience both corroborate this phenomenon. First consider the problem theoretically. As the example in Fig. 15.2 indicates, under the extreme condition of a 55°F setback for interior temperature and just 2 in. of extruded polystyrene (XEPS), membrane surface temperature is 44°F when outside temperature is −30°F. And although common sense suggests that open joints between insulation boards might act as thermal bridges, with substantially reduced membrane temperatures at those points, laboratory tests and field observations indicate otherwise. In laboratory tests designed to test the effects of open (¼-in.) insulation joints, with simulated outside temperature at −30°F and inside temperature at 80°F, temperatures measured at the membrane directly under the open joints were only 1 to 3°F colder under dry conditions and 3 to 7°F colder under wet conditions than membrane temperatures under the insulation.[3] Insulation in these tests was a maximum 1¼ in. thick. Thicker insulation, with narrower joints, should reduce even the slight temperature differences measured in the laboratory test.

Two layers of insulation, with staggered joints, should eliminate the slight thermal bridges created by joints in single-layered insulation.

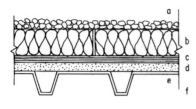

Element	R
a Outside air	0.25
b 2" polystyrene	8.50
c BUR membrane	0.33
d ½" gyp. board	0.45
e Steel deck	0.00
f Inside air	0.61
R_t =	10.14

Figure 15.2 Under extreme thermal steady-state condition, with outside temperature T_o at $-30°F$ and indoor (setback) temperature T_i at $55°F$, the top of the membrane shown would drop only to $44°F$.

$$T_{membrane} = T_i - \frac{\Sigma R_x}{R_t}(T_i - T_o) = 55 - \frac{0.61 + 0.45 + 0.33}{10.14}(55 + 30)$$

$$= 55 - 11 = 44°F$$

Laboratory tests indicate that this temperature might drop as much as 3°F lower at a wide joint. In any event, the membrane would remain well above freezing temperature, so long as it remained dry.

Even in Fairbanks, Alaska, where the 99 percent design temperature is −53°F, properly designed PMRs are free of ice buildup from freezing water. In such a severe climate it is necessary to cover the drains to insulate them (see Fig. 15.3).

The PMR arrangement offers another possible advantage in doubling the membrane's function. To its basic function as the roof system's waterproofing component, the PMR concept adds that of vapor retarder, properly located on the warm side of the roof assembly. In its conventional position above the insulation, the membrane retards the escape of migrating water vapor moving up from the warm interior toward the cold exterior. An exposed membrane thus tends to trap moisture (entering in liquid form through faults in the membrane, or as upwardly migrating vapor) within the system. Especially when there is a vapor retarder under the insulation, the conventionally located membrane helps trap moisture and prevent its escape once it invades the insulation.

The PMR concept has another practical advantage, resulting from the insulation's vulnerable but easily accessible location. If the insulation deteriorates from exposure to sunlight or physical abuse, it can be readily removed and replaced with no disruption in the building's operations because a properly applied protected membrane should live a long service life. This is much easier and more economical than a tearoff-reroofing job, or even a membrane replacement or reroofing over an old roof. Reroofing with a conventional system entails much

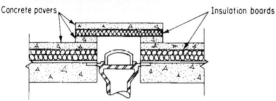

Figure 15.3 In severely cold climates, covering the drains with pavers protects them from ice clogging.

greater inconvenience and expense than simple replacement of ultraviolet-deteriorated insulation on a PMR.

Still another advantage is the easy dismantling of a PMR. On a PMR surfaced with loose aggregate or pavers, you can add insulation by simply removing the surfacing and placing another layer. This makes the PMR an excellent choice for a building designed for vertical expansion. It is also an easy way to add insulating capacity if energy costs escalate.

Conventional roof systems have one notable advantage over PMRs in reroofing. With a conventional roof system you can slope a flat or irregular roof surface that ponds water with tapered board insulation or sloped insulating concrete fill. This remedy is limited in a PMR, since at least half of the total thermal resistance is normally required above the membrane to keep the membrane temperature above the dew-point temperature.

Another noteworthy disadvantage concerns the potential vegetation and fungal growth promoted by the warm, humid environment at a

PMR membrane surface. Vegetation may grow through paver joints or loose aggregate ballast; such growth may promote rot and even root penetration of the membrane. (Filter fabric helps to reduce this threat.)

How PMR Changes Component Requirements

As indicated in the previous discussion, rearrangement of the components in a PMR drastically alters their performance criteria:

- The *insulation* becomes the critical component. Besides its primary function of conserving heating and cooling energy, it must protect the membrane from temperature cycling, physical damage, and solar radiation. It must maintain long term insulating efficiency despite perennial cyclical exposure to water, freeze-thaw cycles, and water-vapor pressure.
- The *membrane* function is reduced to waterproofing (and possibly vapor retarder). Unlike the membrane in a conventional roof system, it no longer has to withstand extensive temperature change, foot traffic, solar heat, and ultraviolet attack.
- The *surfacing* (aggregate or pavers) must provide ballast against wind uplift, and counterbalance flotation of loose-laid insulation. These functions are added to the surfacing's functions in the conventional roof concept—i.e., shielding against ultraviolet radiation, hailstone impact, foot traffic, and small fire sources.

These changing performance requirements have promoted significant changes in materials. The PMR insulation's exposed location eliminates many insulation materials suitable for the protected sandwich-filler location of the conventional roof system. These more rigorous requirements—notably resistance to moisture absorption—have thus far limited PMR insulation to extruded polystyrene.

Conversely, the less rigorous demands on the PMR membrane have expanded the range of suitable materials. A PMR is similar to plaza and below-grade waterproofing, and many products formerly limited to waterproofing can now be used on roofs. In the United States, a glass-fiber felt with water-resistant binder is generally considered the best felt type for a PMR built-up membrane. In Canada, however, a properly glazed organic felt membrane is generally used. Because of the PMR membrane's continual exposure to moisture, single-ply membranes formulated from polymers with low water absorption and lap-sealed with water-resistive adhesives are preferred.

PMR Design and Construction

The PMR concept is compatible with any type of membrane: conventional hot-applied bituminous, fluid-applied or synthetic single-ply sheet, loose or adhered. All systems currently marketed feature insulation considered, in design concept, loosely laid on the membrane.

The major commercially marketed PMR system is applied in a four-stage construction sequence:

- Membrane and flashings
- Insulation in one or two layers, with butted joints
- Filter fabric
- Stone ballast at 10 psf in the field of the roof, 20 psf at the perimeter

In stage 1, a minimum three-ply shingled built-up membrane, modified-bitumen membrane, or polymeric single-ply membrane is installed and flashed. Next comes extruded polystyrene insulation, with precautions taken to prevent it from adhering to the membrane. Two insulation layers with offset joints keep more water above the insulation layer.

To qualify as a Factory Mutual Class 1 steel-deck roof assembly, this roof system requires installation of a $\frac{1}{2}$-in. gypsum leveling board (Type X core) anchored by approved mechanical fasteners before the membrane is applied.

Filter fabric serves several purposes: to keep aggregate from being trapped in insulation joints, where it can promote board buckling under thermal expansion on hot, sunny days, and to prevent buildup of sediment, which can clog drains with fine material that can build up in insulation joints or on the membrane.

In conditions of extreme heat and interior humidity—e.g., paper mills or laundries—a single membrane may not provide optimum resistance to condensation. It may be necessary to install a vapor retarder at deck level, followed by *some* thermal insulation. The roof membrane is then installed and covered with extruded polystyrene as in a standard PMR design. This strategy may be necessary when the primary membrane is temporarily cooled during periods of cold rain. Cold water percolating past the extruded boards chills the primary membrane, and vapor may condense on the underside of the membrane or on the deck. The lower insulation layer atop the vapor retarder maintains the vapor retarder's temperature above the dewpoint temperature.

Drain location

Interior drain location is preferable to exterior, scupper-style drains on any low-slope roof system, but especially on PMR systems in cold climates. Ice can block perimeter drains. Expanding, freezing water trapped at the perimeter can also displace insulation boards. Interior drains, conducting heat from the building interior, keep drains open and drain water flowing in even the coldest weather.

PMR Flashings

Flashings for a PMR are governed by the same design principles as conventional flashings (see Chapter 14). However, there are several differences:

1. Base flashings should extend at least 6 in. above the top of the PMR surfacing. Thus, on a PMR with 3-in.-thick insulation and 2-in. pavers, minimum base-flashing height above the membrane should be 11 in., not the 8-in. minimum generally required above the membrane of a conventional built-up membrane.
2. On a PMR, base flashings should be installed *before* the insulation and surfacing are applied.

PMR flashings are vulnerable to damage from paver edges adjacent to cant strips. Lateral displacement of the paver can puncture the flashing. It is better to hold pavers back from flashing and fill the spaces with round stone aggregate. The pavers should weigh a minimum 22 psf at the perimeter to prevent floating and allowing stone migration under the insulation boards.

PMR Insulation Performance Requirements

According to Dow Chemical Company researchers K. A. Epstein and L. E. Putnam, the ideal PMR insulation would have (in addition to the obviously required thermal-insulating quality) the following seven properties:

1. Impermeability to water
2. Resistance to freeze-thaw cycling
3. High compressive strength
4. Dimensional stability
5. Resistance to ultraviolet radiation

6. Nonbuoyance
7. Incombustibility

No commercially available insulation comes close to meeting these ideal requirements, but the PMR concept makes an ideal insulation unnecessary. Protective surfacing can shield the insulation from ultraviolet radiation and provide ballast against flotation and a noncombustible shield against fire brands. Moisture-absorptive resistance, tolerance of freeze-thaw cycling, and compressive strength are indispensable; of these requirements, moisture-absorptive resistance is paramount. Extruded polystyrene (XEPS) remains the only insulating material that has proven successful in PMR assemblies.

Moisture trapped between the membrane and the insulation can create a potential moisture accumulation problem with paver-ballasted PMR assemblies. In cold weather, when the bottom insulation surface is warmer than the top surface, with the vapor-pressure gradient paralleling the temperature gradient, water vapor moves up from the membrane-insulation interface into the insulation. This vapor may condense within the insulation. Alternatively, because the insulation is not impermeable, the water vapor may escape from the top surface, if that surface is properly vented—i.e., with pavers and pedestals, ribbed insulation, 3-in. aggregate, or a sand layer in poured concrete plaza applications. Concrete pavers laid directly on the insulation's top surface can inhibit vapor escape.

PMRs with loose pavers, pedestals, or aggregate surfacing rely totally on the system's self-drying characteristics to limit the insulation's moisture content. To promote self-drying, you can use one or more of the following strategies:

- Slope the deck to provide drainage.
- Design the system to ventilate the insulation.

Laboratory tests confirm the efficacy of both methods.

Drainage is probably more important for a PMR than for a conventional roof system. Ponding on any roof heightens the risks of leakage. On a PMR, it promotes heat leakage from the interior and heightens the risk of water leakage into the interior.

The $\frac{1}{4}$-in. slope recommended for conventional membranes (occasionally compromised to $\frac{1}{8}$ in. or even 0) should be considered a minimum for a PMR, to ensure positive drainage of the entire roof surface. The importance of drainage for a PMR may be a handicap on reroofing projects where the existing deck is inadequately sloped.

Besides positive slope, other design features help to promote good drainage. The most obvious is proper drain elevation, which is more critical for a PMR than for a conventional built-up membrane. Drains must be depressed below the level of the membrane, well below the elevation of the insulation's top surface. Chamfering the bottom edges of insulation boards promotes good drainage.

Extruded polystyrene foam in the higher density ranges (2.3 pcf or more) exhibits substantially better moisture-absorptive resistance than other insulating materials that have been tried on PMRs. Products satisfying ASTM D578, Type VI, with minimum density of 1.8 pcf, have proven adequate on well-drained roofs. Independent laboratories and field tests on both sides of the Atlantic substantiate this superiority of extruded polystyrene. Laboratory tests conducted by the Cold Regions Research and Engineering Laboratory of the U.S. Army Corps of Engineers indicate nearly 10 times as much water absorption by beadboard polystyrene compared with extruded polystyrene (2.4 lb/ft^3) for 160 days of submergence in water.[4] In field tests conducted by Swedish investigators at the Chalmers University of Technology in Gothenburg, molded, beadboard polystyrene (MEPS) absorbed 5 to 6 percent moisture (by volume), roughly 100 times the 0.05 percent moisture found in extruded polystyrene after both test samples were exposed for 1 year as components in PMRs surfaced with loose aggregate. Polyisocyanurate foam is also far more water-absorptive than extruded polystyrene.

Resistance to freeze-thaw cycling, the second major property required by a PMR insulation, depends on the insulation's ability to accommodate the expansion of freezing water entrapped in or near its top surface. (As indicated earlier, the bottom surface should never drop as low as freezing temperature.) The property required for freeze-thaw cycling resistance is elasticity—the ability of the closed-cell foam to yield under the expansion of freezing water without brittle cracking, spalling, or flaking of the insulation material. Obviously, the less water absorbed by the insulation, the less hazard exists from the absorbed water's expansion.

Reporting on their research into freeze-thaw cycling, Dow researchers Epstein and Putnam claim that extruded polystyrene performs the best of a number of commonly used insulating materials. Under their test method, ASTM C666-73 (freeze in air, thaw in water, designed for concrete cylinder testing), most tested insulations began to disintegrate after between 97 and 458 freeze-thaw cycles, with water absorption of up to 89 percent (by volume). The Dow researchers claim that ASTM Test Method C666-73 is actually less severe than some PMR roof exposures. For example, the test does not allow one side of the insulation to retain water or remain at 70°F. And because both sides of the insulation are at the same temperature, there is no vapor-driving force tending to force vapor from the warm bottom surface toward the cold top surface, the normal, cold-weather situation for a PMR.

Two remaining properties required of a PMR insulation, compressive strength and dimensional stability, disqualify fewer materials than water-absorptive and freeze-thaw cycling resistance. Few currently marketed roof insulating materials lack the required compressive strength because this property is required to just about the same degree for a conventional roof's insulation.

Dimensional stability is less important for a PMR than for a conventional roof's insulation. A high coefficient of thermal expansion-contraction poses no particular problem. The most important aspect of dimensional stability for a PMR insulation is low long term net contraction or expansion under countless wetting-drying cycles (like the long term shrinkage of organic felts under such cycles). Despite its high coefficient of thermal expansion-contraction, extruded polystyrene satisfies this requirement.

Thermal quality of PMR insulation

The thermal resistance of wet insulation naturally decreases with increasing moisture content. According to the Swedish investigators, polystyrene board loses about 10 percent of its thermal resistance in a conventional roof system exposed to the same weather conditions. (However, if the insulation in a conventional roof gets wet—through membrane leakage or condensed water vapor—then its thermal resistance could drop even lower than a PMR's insulation because evaporation of water in the confined insulation in a conventional built-up roof system is much more difficult than in a PMR.) A designer might reduce published R values by 10 percent for PMRs in wet climates, by 5 percent in moderate climates, and by 2 percent in dry climates.

Because of the many complex variables affecting heat transfer through the roof, the so-called thermal efficiency of a PMR may exceed 100 percent.[5] This means that the actual heat gains and losses may be *less* than those calculated from theoretical heat-flow, steady-state, heat-transfer equations, based on the U value and the inside and outside air temperatures. In other words, the roof's thermal performance in retarding heat flow and cutting heating- and cooling-energy bills is actually better than the conventional steady-state heat-flow formulas indicate. (Judged by this criterion, a conventional roof's thermal performance might also exceed 100 percent, depending especially on its surface color.)

In addition to conducted heat flow, the single factor considered in conventional heat-flow calculations for roof design, actual heat flow depends on many other factors: evaporation, condensation, water flow over the membrane, solar radiation intensity and duration, snow cover, and wind.

In cold weather, the forces *increasing* a roof's thermal efficiency are solar radiation (which raises roof-surface temperature and reduces the temperature gradient through the roof) and snow cover (which in sufficient depth can add an insulating blanket to the roof's components). Among the forces *reducing* a roof's thermal efficiency are rainwater evaporation (which transfers heat energy from the roof surface into the evaporating water) and cold rainwater or melt water flowing across the roof to the drains. Although detrimental during heating season, these evaporative/convective heat loses are of course helpful during the cooling season because they reduce heat gain.

According to Aamot, thermal efficiency greater than 100 percent was observed about 85 percent of the time in a heat-metered roof in Hanover. On cold, bright days, with intense, undiffused solar radiation, thermal efficiency sometimes exceeded 300 percent. Thermal efficiencies less than 100 percent are observed during rainfall, when the flowing water carries off roof-surface energy, and for several days after, as the evaporating residual moisture dries the insulation and cools the roof surface.

Refinements in locating PMR insulation

From the tested performance of PMRs, Aamot worked out a general thermal-design theory that indicates the benefits derived from putting some insulating material under the protected membrane. (This is normally done only for such highly humidified interiors as pulp and paper mills.) The thermal performance of such a PMR is higher in a cold climate because internal cooling is reduced. Gypsum board, required as a substrate for fire protection on steel decks, serves this purpose.

Placing additional insulation on top of a conventional roof—in effect, converting it into a PMR—can benefit the roof system in two ways: (1) by the obvious increase in thermal resistance and (2) by raising the membrane's cold-weather temperature, thereby reducing the threat of moisture condensation at the underside of the membrane. Note that any such addition requires a structural review to ensure that the additional dead load of the PMR insulation and ballast is not excessive.

Surfacing Protected Membrane Roofs

Loosely laid aggregate or pavers resist flotation forces plus wind uplift and scour. For design techniques to avert these hazards, see Chapter 7, "Wind Uplift." As that chapter indicates, it may be necessary to specify pavers instead of loose gravel for vulnerable corner and other peripheral areas subjected to high wind velocities and uplift pressures. As an additional advantage, pavers ensure uniform weight distribution, whereas loose aggregate can vary.

In addition to their greater wind-uplift protection, pavers may be economically justified to serve three other purposes:

- As a foot-traffic surface for terraces or plazas, or for a roof requiring frequent maintenance trips for access to mechanical equipment and so forth.
- Easier access to repair the roof (because removing pavers is more convenient than removing loose aggregate)
- Better appearance

When specifying pavers as a PMR ballast, the designer should also consider an underside air space—corrugations or channeling—to facilitate evaporation from the top surface of the insulation. Ribs in the top surface of the insulation serve this purpose (see Fig. 15.4).

Figure 15.4 Ribbed polystyrene foam board has channels in top surface to allow air circulation for drying pavers. (*Dow Chemical.*)

Ballast weight against flotation is about 5 pcf/in. insulation thickness. Water weighs 5.2 psf/in. and polystyrene foam weighs about 0.2 psf/in., making 5 psf/in. a minimum figure to prevent flotation. Typically, 2-in. (50-mm) pavers at 22 psf (107 kg/m^2) are installed at perimeters and to serve as walkways on PMRs. Concrete compressive strength should be a minimum 3000 psi; the concrete should also be air-entrained for freeze-thaw resistance. Stone meeting ASTM D448, sizes 57, 5, 4, 3, or 2, is acceptable, applied at a rate of 10 to 20 psf, depending on wind-uplift requirements and "hiding power." (Larger stones require greater thickness to screen ultraviolet radiation.) Larger stones (No. 4, 3, or 2) are used at drains and scuppers to avoid washout.

The principle of "rafting," i.e., using a porous mat over the insulation boards or foamed polystyrene boards with a thin, integral concrete topping and tongue-and-groove edges, permits slightly reduced ballast weight because the boards tend to float in large units rather than as individual boards. The rafting principle springs from the reduced probability of a large area being subjected to full flotation pressure. Successful rafting prevents displacement of individual boards.

As an alternative to concrete pavers or loose stone ballast, designers can specify a composite panel, weighing about $4\frac{1}{2}$ psf, with a $\frac{3}{8}$-in. topping of latex-modified concrete mortar bonded to the extruded polystyrene. These panels have tongue-and-groove interlocking side joints to provide wind-uplift resistance and rafting, as mentioned above. A heavier concrete topping is available for greater resistance to rooftop traffic.

These paver tongue-and-groove joints could become unhinged if the membrane below were to billow upward from internal air pressure. Loose-laid, single-ply membranes require an air infiltration barrier underneath. This is accomplished either by sealing all deck joint openings or by installing and sealing a plastic film vapor retarder, followed by a mechanically fastened rigid board. Perimeters and other locations where the interlocking joints are interrupted require strapping or additional pavers to prevent floating or wind loss.

Life-cycle Costing

Protected membrane roofs generally, although not always, cost more than conventional roof systems, as substantiated by the previously cited Canadian survey of architects and roofers. Increased cost for a PMR may result from one or more of the following factors:

- Additional cost for extruded polystyrene compared with less expensive insulations
- Additional cost for handling ballast weight, compared with lighter, adhered systems
- Additional cost for leveling board on steel deck

Benefits, with probable cost savings, offered by a PMR, are:

- Longer projected service life
- Avoidance of wind-uplift stresses on the structure (as a result of the ballast)
- Greater protection, functional enhancement (as a roof terrace), and improved roof appearance of pavers
- Need for only one membrane, which doubles as a vapor retarder
- Reduced installation work and fewer weather constraints on application of loose-laid PMR system

Some converts to PMR roofs find the increased first cost economically justified on a life-cycle (long term) cost basis. From a survey of conventional built-up maintenance costs in Alaska, Aamot estimates annual maintenance cost at 5 percent of initial cost, vs. 1 percent for a PMR. At interest rate = 12 percent for a 20-year life cycle, a 4 percent annual maintenance saving with 10 percent cost escalation is equivalent to a 66 percent increase in initial cost [computed on a present-worth (PW) basis].* In other words, a PMR is more economical than a conventional built-up roof system if its initial cost is less than 1.66 times the conventional roof system's initial cost.

Although the dramatic savings estimated for an Alaskan location may not hold for more moderate climates, the owner concerned with long term economy should investigate the PMR concept.

---

*On a 20-year life cycle, the PW of the 4 percent annual maintenance-cost saving (expressed as a percentage of initial cost C) is computed as follows:

$$PW = 0.04C \frac{a(a^2-1)}{a-1}$$

where $a = (1 + m)/(1 + i) = (1 + 0.10)(1 + 0.12) = 0.982$
m = maintenance-cost escalation = 10%
i = interest rate = 12%
n = number of years = 20

$$PW = 0.04C \times 16.62$$
$$= 0.66C$$

Waterproofed Deck Systems

Under the economic pressure of soaring land costs and the consequent need for more efficient use of building space, waterproofed plazas and roof terraces (called waterproofed decks for our purpose) have become more popular over the last several decades. Underground storage areas, access tunnels, and parking garages, even computer rooms and office space, are located below plazas, planters, and sidewalks.

Waterproofed deck systems differ greatly from roof systems in their performance criteria. In conventional roof systems, the membrane must resist ultraviolet degradation and extreme temperature variations. Like a PMR, a waterproofed deck's membrane experiences far less temperature change than an exposed roof membrane, and it is permanently shielded from ultraviolet radiation. Its primary performance criterion is to resist continuous exposure to moisture. For earth-covered membranes, resistance to vegetation root penetration is another performance criterion. And because a waterproofing membrane may be far less accessible than a roof membrane, its anticipated service life must equal the service life of the building, not merely the normal 10 to 20 years anticipated for a roof membrane. Reroofing is an inconvenience; rewaterproofing is, by comparison, an economic holocaust, as it costs about 10 times the cost of a roof tearoff to remove up to 8 ft of earth fill and plantings or to jackhammer a concrete topping and haul it away.

A waterproofed deck system with a separate wearing course contains some or all of the following components (see Fig. 15.5):

- Deck
- Membrane
- Protection boards
- Aggregate percolation stratum or plastic drainage panel
- Insulation
- Flashing
- Wearing surface

Designers of waterproofed decks should be alert to two ASTM standard guides (C981 for built-up membranes and C898 for liquid-applied membranes), both under revision in 1995. The following discussion incorporates material from these standards.

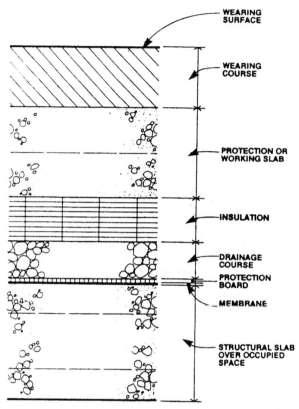

Figure 15.5 Basic components of membrane waterproofing over a framed structural slab. Insulation and drainage course are often reversed from the above order. (*American Society for Testing and Materials.*)

Structural decks

Cast-in-place, monolithic structural concrete slabs make the best substrates for waterproofed deck systems. Their continuity gives them an advantage over precast concrete. Because of joint irregularities, precast concrete decks usually require a continuous concrete topping (normally 2 in. thick) to provide a jointless top surface. The rotation of the bearing ends of precast structural members can open and close joints, and so expansion joints may be required for accommodation (see Fig. 15.6). Standard-weight concrete (150 pcf) is preferable to lightweight structural concrete (100-pcf minimum density), as it experiences less ultimate deflection attributable to long term creep. Lightweight insulating concrete is unsuitable as a deck substrate.

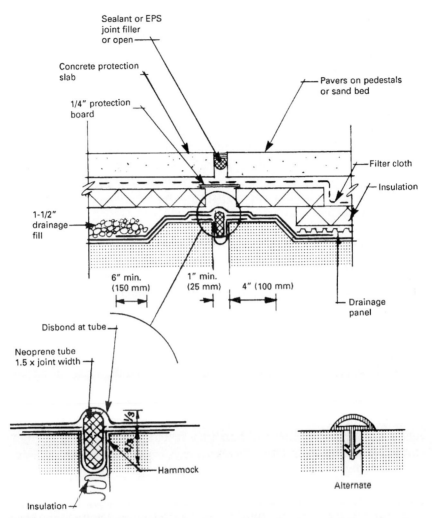

Figure 15.6 Expansion joint at structural concrete framing is carried through waterproofing system. (*American Society for Testing and Materials.*)

Membranes

Membranes for waterproofed deck systems include conventional BUR, modified-bitumen sheets, single-ply sheets, and liquid-applied elastomers. As in roof membranes, glass-fiber felts have replaced organic felts in waterproofed deck membranes because of their greater strength and moisture resistance. Glass-fiber felts can be alternated with woven glass fabric, which offers greater flexibility than glass-fiber felts and is more easily molded to corners and other substrate irregularities. Membranes range from a minimum of three to a maximum of five plies, depending on the required degree of waterproofing.

The bitumen of choice is coal tar pitch (ASTM D450, Type I or II), because of its self-healing property. Even the softest asphalt (ASTM D449, Type A) can hydrolyze in the presence of water containing the high- or low-pH chemicals in soil poisoners and fertilizers.

Felt plies can be shingled or phased (i.e., laid in continuous planes). Ply-on-ply construction results in continuous films of interply bitumen. Moisture penetration through a lap thus leads only to the next layer, not through the entire membrane cross section, as in a shingled pattern. Some five-ply membranes have three shingled plies of alternating felts and fabric, crossed with two shingled plies of felt and fabric at right angles. This arrangement produces a strong, highly waterproof BUR membrane.

As in roof construction, however, single-ply elastomers, weldable thermoplastics, and modified-bitumen sheets are rapidly replacing the more labor-intensive BUR membranes. A popular modified-bitumen sheet is a self-adhering single ply, laminated to a polyethylene backing. Called "peel-and-stick," these sheets are applied over a concrete substrate primed with either a solvent or an emulsion-type primer. They must be protected from ultraviolet degradation within a few days of application.

Among elastomeric single-ply sheets, butyl rubber has an advantage over EPDM in its much lower moisture absorptivity, a more important property for a waterproofing membrane than EPDM's much greater ultraviolet resistance. PVC, in thicker sheets than PVC roof membranes for improved puncture resistance, is another common single-ply waterproofing membrane.

These single-ply elastomeric and thermoplastic sheets are either fully adhered or loose laid. When loose laid, however, they are nonetheless adhered in grid strips, usually 10 ft each way, to form watertight compartments that aid in restricting leakwater migration and simplify leak detection.

Liquid-applied membranes include hot and cold polymer-modified asphalt, single-component asphalt or coal tar-extended urethane, and two-component urethane elastomer. For proper performance, these membranes must contain minimum 65 percent solids to reduce pinholing. Their resistance to crack propagation from the deck below depends on thickness, normally a minimum 60 mils. (The thicker the membrane, the greater the elongation over the crack.)

Liquid-applied membranes' advantages include localization of leaks (compared with compartmentalization of leaks with loose-laid sheet membranes). Weighing against this advantage, however, is the need for extremely rigorous preparation of the substrate, which must be dry and largely dust-free, with patched cracks.

Moisture is a prime enemy of liquid-applied membranes. It causes urethanes to foam and hot-applied modified bitumen to froth. Single-component liquid-applied coatings are most vulnerable to high humidity. They require exposure to air with minimum absolute humidity for sufficient time to assure full curing. Premature application of protection boards may delay, or even prevent, full curing. In contrast, two-component coatings are relatively unaffected by humidity. They also have these advantages: greater latitude for low-temperature application, long shelf life, and noncuring in the can. Their countervailing disadvantage: carefully controlled field mixing is required to assure complete blending.

Dust is also a problem for liquid-applied coatings. Pinholing can be caused by dust, or by overly rapid application, which can entrain air bubbles in the coating film.

Hot-applied rubberized asphalts, heated in a jacketed kettle, are often reinforced with polyester or woven glass. This process requires two applications, which minimize pinholing and thin spots via a thicker, more stable membrane. Since they cure as soon as they cool, these coatings are less vulnerable to rain or sudden temperature drops, which jeopardize the slower-curing cold-applied coatings.

All of these liquid-applied coatings are highly elastic, are capable of bridging small cracks, and have very low permeability. They are easily applied to contoured surfaces; they are generally self-flashing. Their chief drawback is their great dependence on rigorous field supervision and application, requisites for uniform thickness and film integrity.

Protection boards

All waterproofed membranes require protection from construction damage and ultraviolet exposure as soon as practicable after installation and flood testing. The most common material is an asphalt-core laminated panel. Prefabricated plastic drainage panels may also serve as protection, but it is safer to use protection boards as well.

Drainage design

Drainage of a waterproofed deck system should incorporate all components, from the wearing surface down to the membrane, according to ASTM C981.[6] Plaza drainage should minimize saturation of the wearing course, which might disintegrate under freeze-thaw cycling.

At the membrane level, drainage is required to avoid the following:

- Hydrostatic pressure from accumulated drain water
- Freeze-thaw cycling of trapped water
- Reduction of the insulation's thermal resistance

An absolute minimum slope of 1 percent, and preferably 2 percent, is recommended to assure positive drainage. Either a percolating gravel stratum or plastic drainage panels can provide the pervious medium facilitating water flow to drains. In very cold climates, where snow meltwater flows over the membrane, locating the drainage course *above* the insulation may reduce the probability of condensation below the membrane. But as a countervailing disadvantage, this location may impair drainage, which is promoted by a *below*-insulation location.

Designers of waterproofed decks should specify multilevel drains, designed to permit differential movement between the strainer at the wearing course and the drain body embedded in the structural slab. A detail allowing relative movement between wearing course and structural slab prevents rupture of the drain body or connected pipes (see Fig. 15.7).

At the wearing surface, drainage is accomplished via (*a*) an *open-jointed* system that rapidly filters drain water down to the membrane level, or (*b*) a *closed-joint* system sloped to surface drains. Open-jointed systems include pavers on pedestals or placed directly on ribbed insulation boards. Closed-joint systems have either a mortar-setting bed or caulked joints. An intermediate compromise is provided by bricks or stones in a sand bed. This compromise is inferior to an open-jointed system, but superior to a mortar-set system.

Aggregate percolation vs. open subsurface drainage

Open-jointed surfacing units—precast concrete, tile, or masonry units—usually sit on ribbed insulation that conducts water downward to the waterproofing membrane, where it runs to drains. Fine gravel, clean coarse sand (held on a No. 30 sieve), or no-fines (i.e., porous) concrete can serve as the percolating stratum underlying the open-jointed surfacing units. Loose aggregate can accommodate freeze-thaw cycling of entrapped water, with freezing expansion contained harmlessly within the aggregate interstices.

In a frequently specified and generally superior alternative to loose aggregate percolation, the surfacing units can sit on pedestals, on insulation, or directly on protection boards, providing faster, less obstructed subsurface drainage. Surfacing soffits can also be grooved,

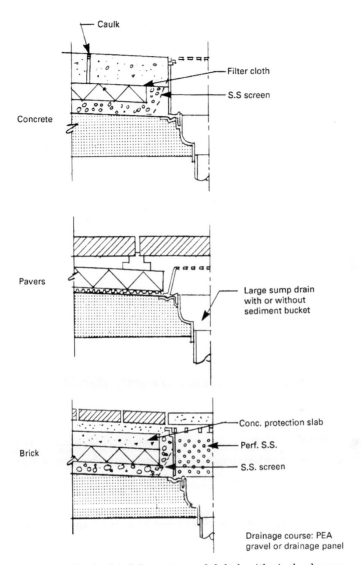

Figure 15.7 Drain detail for waterproofed deck with single-ply membrane varies with different wearing surfaces: concrete (top); pavers (middle); brick (bottom). (*American Society for Testing and Materials.*)

corrugated, or otherwise contoured to create open paths for subsurface drainage.

This alternative pedestal-supported, open-jointed system has several advantages:

- Faster, more efficient drainage, with better surface-dirt removal
- Ventilation drying of subsurface areas
- Easier access for maintenance and repair of subsurface components

In return for these benefits, this alternative poses a more complex design/construction problem and probably has a higher initial cost.

Monolithic (closed-joint) construction changes the waterproofed deck design philosophy, demoting the membrane from the primary to the secondary line of waterproofing defense. The membrane nonetheless is an essential component of the system because the surfacing of waterproofed decks is normally exposed to traffic loads, drastic temperature changes, and freeze-thaw cycling, with expansion of water trapped in cracks, holes, or other imperfections. Insulation may be interposed between the membrane and the wearing surface material, with a setting bed for the surfacing units: precast ceramic or masonry tiles or manufactured conglomerates. Surfacing should slope away from adjoining walls and expansion joints to direct water away from them, both at and below the surfacing.

Insulation

The key question on insulation concerns its location: *above* or *below* the membrane? Until the energy crisis of the early 1970s, the key question was whether to specify insulation. But the old practice of simply relying on the aggregate or earth fill to insulate the occupied space is seldom justifiable, unless the space is neither heated nor cooled. Gravel, earth, and coarse sand provide high heat capacity, thus stabilizing heat gains or losses. But their thermal resistances (R values) are low compared with those of insulating materials of equal thickness.

Generally, the PMR concept—insulation above membrane—is the better design for a waterproofed deck, for several reasons. Protection of the membrane's waterproofing integrity is even more important in a waterproofed deck than in a roof system. Moreover, because the insulation in a waterproofed deck system normally carries higher traffic loads than in a roof system, it already requires high compressive strength. By making the insulation moisture-resistant as well, you can conveniently satisfy the requirements for the above-membrane location.

The above-membrane location for insulation in waterproofed deck construction limits the material choice to essentially the same as that for a PMR. As a practical matter, that choice is currently (1995) limited to extruded polystyrene board, the only material possessing the two required properties of adequate compressive strength and moisture resistance. Foamglass is the only other material satisfying both the compressive strength and moisture resistance required for below-membrane use, but its vulnerability to freeze-thaw cycling makes it too risky as an above-membrane insulation, unless the climate is mild enough to preclude freeze-thaw cycling.

Another possible location for the insulation, under the deck, has several disadvantages:

- Subjection of membrane and deck to extreme daily and seasonal temperature cycling
- Possible condensation of upward-migrating water vapor in the insulation
- Fire resistance required

(For further discussion of the below-deck location for insulation, see Chapter 5, "Thermal Insulation.")

Flashing

Like roof systems, waterproofed deck systems require flashing at terminations, penetrations, and expansion joints. In contrast with roof systems, where base flashing installation follows membrane installation, flashing at waterproofed deck terminations may be installed prior to membrane application. In the plaza field, at rising walls and expansion joints, the horizontal leg of base flashing should be at least $1\frac{1}{2}$ in. above the structural slab to direct water flow away from the joint (see Fig. 15.8). Material for expansion-joint covers should be elastomeric supported on neoprene tubes, to prevent distortion from water or earth-fill pressure. A continuous hammock under the tube serves as a vapor retarder and bitumen dam. It can also be sloped to a drain, to provide a secondary means of protection. Some experts recommend installation of a sheet metal gutter under expansion joints.

Internal and exterior corners and similar transitions must be reinforced with at least one membrane ply or liquid-applied fillet. Raising this flashing height above the membrane is good practice, like the similar practice for expansion joints.

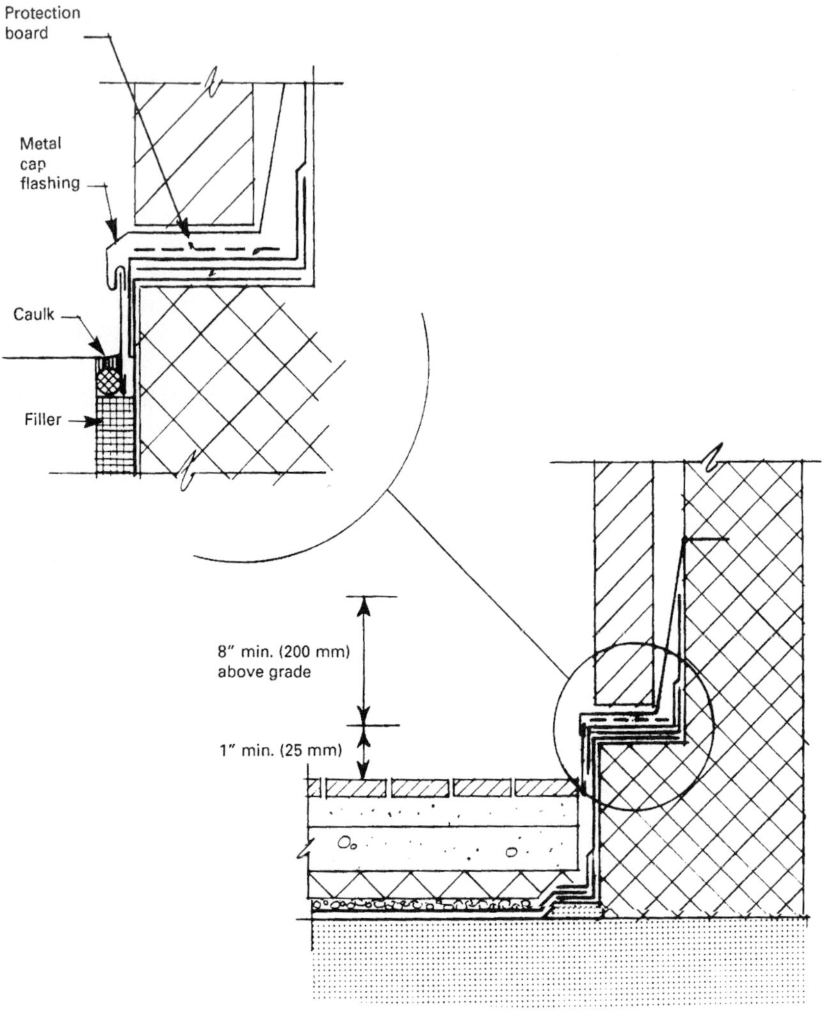

Figure 15.8 Waterproofed deck flashing detail at intersection with brick-faced wall. (*American Society for Testing and Materials.*)

Penetrations are flashed like roof penetrations. Individual pipes should be spaced a minimum of 6 in. apart. Avoid ganged pipes in pitch pockets.

Base flashing must extend above the wearing course, preferably 8 in. This is critical if the wearing course is a closed system.

Wearing surface design

In heavily trafficked locations, waterproofed deck systems have the following surfacing design requirements:

- Structural strength to bear traffic loads
- Durability under heavy wear and weathering
- Esthetic appearance (on plazas and roof terraces)
- Heat reflectivity (to avoid summer temperature buildup)

The first two criteria are mandatory; the latter two are optional.

Cast-in-place or precast concrete, ceramic, or masonry units generally satisfy these requirements. Dark colors—especially black—are normally avoided because of their high heat absorption. (See Table 15.1 for solar absorptance values of several surfacing materials.)

Joints on a roughly 10-ft maximum grid are normally required to accommodate thermal contraction and expansion in surfacing units. They can experience daily temperature extremes of 80°F and annual extremes of 180°F in the most severe climates. Use of snow-melting equipment (embedded electrical-resistance cables or hot water or steam piping) complicates the problem of thermal design of the surfacing. Rapid temperature changes accompanying intermittent operation of the snow-melting system or expansion of corroding embedded pipes can crack surfacing slabs.

Plaza "furniture"

Planters, reflecting pools, fountains, and other plaza "furniture" should be installed above the waterproofing membrane. These items are individually waterproofed. Trees should be planted in separate concrete containers to avoid damage to the membrane from root penetration or damage from landscaping shovels when trees must be replaced or replanted.

TABLE 15.1 Solar Reflectance of Surfacing Materials

Surfacing material	Solar reflectance*
White or light cream brick	0.50–0.70
Yellow, buff brick or stone	0.30–0.50
Concrete, red brick or tile	0.20–0.35
Green grass	0.25
Crushed rock	0.20
Gravel-surfaced bituminous roofing	0.15
Asphalt paving	0.05–0.10

*Most tabulated values are computed from the *ASHRAE Handbook and Product Directory, 1977 Fundamentals* table 3, p. 2.9, which tabulates values for solar absorptance.

Detailing

The location and extent of waterproofing should be indicated on drawings. Cross sections should be drawn at ¾ in. = 1 ft, details at a minimum 1½ in. = 1 ft. Waterproofing extent should be indicated with heavy dashed lines. Special conditions—cants, fillets, chamfered corners, etc.—require details. Shop drawings are mandatory.

Flood testing

Flood testing, for which a new ASTM standard is due in 1996, is important for waterproofed decks because a failed waterproofing system is generally more destructive and more expensive to correct than a failed roof. If tearoff-replacement of a failed roof system costs $6 psf, its counterpart for a failed waterproofed deck can cost $60 psf. It is thus advisable to flood-test a waterproofed deck after the membrane and flashing have been applied and before the above-membrane components are applied.

Flood-test with a minimum 1-in. water depth over the entire surface. Plug drains and use permanent or temporary curbs to retain water for 24 h. Wood dams adhered to the waterproofing membrane can contain the flood-test water.

When drains are not connected, make special provisions for pumping and disposal of water. Cover the membrane with protection boards as soon as possible after drying the membrane surface.

Alerts (for PMRs)

1. Provide slope for positive drainage, ¼ in. minimum, ½ in. if practicable.

2. Locate roof drains within the heated part of the building. (Exterior drains periodically freeze shut in cold climates.)

3. Depress drains below membrane elevation.

4. Check roof structural load, especially when reroofing existing buildings, for both strength and deflection under ballast load.

5. Flood-test the completed membrane before installing insulation and surfacing. (If it is impracticable to flood-test *before* the insulation goes down, flood-test afterward.)

6. Reduce the theoretical dry factor of insulation by 2 to 10 percent (to allow for moisture-caused reduction in thermal resistance and interior cooling from rainwater flow).

7. Specify two insulation layers, with staggered joints, for best thermal performance.

8. Design for ventilation of insulation covered with loose aggregate or concrete pavers.

9. Use glass-fiber felts for a built-up bituminous PMR membrane.

10. Pay particular attention to air-seal and securement requirements when using lightweight concrete insulating panels.

11. Do not specify an adhered membrane on top of lightweight insulating concrete.

12. Design loose aggregate size to resist wind uplift and scour. (Specify heavy pavers in high wind areas or along roof perimeter.)

13. Always incorporate a filter fabric between the stone ballast and the top surface of the insulation. But do not use filter fabric under pavers.

14. Note flashings' increased vulnerability to damage, especially at cants adjacent to pavers.

Alerts (for Waterproofed Decks)
Design
Deck
1. Provide minimum $\frac{1}{8}$-in. slope (preferably $\frac{1}{4}$-in.) in structural deck.

2. Slope deck away from walls.

3. Specify two-level drains for closed-joint waterproofed deck systems.

4. Over precast concrete decks, provide a 3000-psi concrete topping, minimum 3 in. thick, reinforced with welded wire fabric. Precast concrete units must be securely tied together (to prevent relative movement and consequent cracking of the topping).

Built-up bituminous membranes
1. Specify priming of concrete decks for built-up bituminous membranes, with primer satisfying ASTM D41 for asphalt membranes, ASTM D43 for coal-tar-pitch membranes, and provide overnight drying.

2. For coal-tar-pitch waterproofing membranes, specify (*a*) coal tar pitch, per ASTM D450, Type I, 25- to 30-lb/square mopping; (*b*) coal-tar-pitch felts, per ASTM D227.

3. For asphalt waterproofing membranes, specify (*a*) asphalt per ASTM D449, Type I, 25-lb/square interply mopping; (*b*) asphalt primer per ASTM D41, 1 gal/square; (*c*) felts, per ASTM D2178, Type IV.

4. For either coal-tar-pitch or asphalt waterproofing, specify fabrics per ASTM D1668 (treated glass fabrics).

5. Specify two plies of glass fabric reinforcing at corners (90 and 270°).

Liquid-applied membranes
1. Cold liquid-applied membranes shall conform to ASTM C898.

2. Obtain the manufacturer's signed approval confirming the suitability of the product for the project use. Also require manufacturer's certification of the applicator.

3. Specify firm wood-float or comparable light trowel, power machine finish for concrete deck surface.

4. Require the concrete contractor, not the roofing contractor, to repair defective concrete surface. Require concrete patching with epoxy mortars.

5. Require water or paper curing of concrete decks, with minimum 28 days' curing time before membrane is applied.

6. Specify two coats, with reinforcement, for hot-applied rubberized asphalt.

Field

Concrete decks
1. Check concrete-deck surface with 10-ft straightedge. Permit maximum "gradual" offset of $\frac{1}{4}$ in./ft (per American Concrete Standard 301-72).

2. Deck surface must be clean, smooth, and free of loose particles, grease, oil, and other foreign matter. Curing compound shall be compatible with waterproofing.

Sheet membranes
1. Maximize sheet size (to minimize vulnerable field-spliced lap seams).

2. Reinforce substrate joints or cracks.

3. Lap joints a minimum of 3 in., or manufacturer's recommendation.

4. Do not reposition sheet after contact adhesion is made. Cut out and patch wrinkles or fishmouths.

References

1. M. C. Baker, "Protected Membrane Roofs in Canada—Results of a Survey," *Nat. Res. Counc. Can., Div. Build. Res.,* October 1973.
2. Haldor W. C. Aamot and David Schaefer, "Protected Membrane Roofs in Cold Regions," CRREL Rept. 76-2, U.S. Army Corps of Engineers, Cold Regions Research and Engineering Laboratory, Hanover, N.H., March 1976, p. 8.
3. M. C. Baker and C. P. Hedlin, "Protected Membrane Roofs," *Can. Build. Dig.,* no. 150, June 1972, p. 150–4.
4. Chester W. Kaplar, "Moisture and Freeze-Thaw Effects on Rigid Thermal Insulations," Tech. Rept. 249, U.S. Army Corps of Engineers, Cold Regions Research and Engineering Laboratory, Hanover, N.H., April 1974, pp. 7, 9.
5. Aamot and Schaefer, op. cit., p. 11.
6. "Standard Guide for Design of Built-Up Bituminous Waterproofing Systems for Building Decks," ASTM C981.

Chapter

16

Sprayed Polyurethane

Sprayed-in-place polyurethane foam (SPF) can sometimes solve reroofing problems that are virtually insoluble by more conventional remedies. It substitutes a field-manufactured, seamless foam insulation for conventional factory-fabricated insulation boards. Like conventional roof assemblies, SPF roof systems require a protective waterproofing membrane. This fact was sometimes overlooked by early enthusiasts. Viewing SPF as a dual-purpose component comprising both insulation and membrane, these enthusiasts sometimes coated SPF with cheap asphalt cutbacks that soon deteriorated and exposed the underlying foam to rapid degradation. These early failures were a serious setback to the growth of the SPF industry.

SPF differs from most other technological advances in that they tend to make roofing application *less* sensitive to weather conditions. A loose-laid, ballasted system, for example, is the least sensitive to weather conditions during application. In contrast, SPF is more sensitive to weather conditions than even a conventional built-up roof system. Not only can entrapped moisture from the substrate furnish a future agent of destruction, but it can interfere with the chemically reactive field fabrication of an SPF system. As much as any roof system, SPF illustrates the no-free-lunch principle that rules technology as well as economics.

In summary, compared with more conventional roof systems, SPF offers these advantages:

- High thermal resistance per unit thickness
- Ultralightweight (less than 1 psf for a 3-in.-thick foam plus membrane coating)

- Easy transport of material (expanded at jobsite)
- Tenacious adhesion (demonstrated in recent hurricanes)
- Resistance to moisture intrusion, even when the membrane coating is damaged (also demonstrated in recent hurricanes, where airborne missiles damaged many roofs)
- Seamless construction, including flashing
- Adaptability for crickets and general slope for drainage

As always, there are offsetting disadvantages:

- Added cost of periodically recoating the liquid-applied coatings
- Extreme dependence on the applicator's skill and on the proper operation of spray-foam equipment
- High dependence on good substrate preparation
- High dependence on good weather, notably lack of wind and moisture

Readers should also heed this warning: Negligently designed and applied SPF roof systems can result in roof failures that are expensive to remedy. Failure of an SPF system requires tearoff-replacement, not re-covering or reroofing, and the tearoff process itself can be slow, tedious, and more expensive than the removal of any other type of failed roof system. (SPF's excellent adhesion becomes a huge liability in a tearoff.)

With the design and application guidelines readily available today, there is no excuse for the irresponsibly spectacular SPF failures of the past. Properly designed, applied, and maintained (including possibly two membrane coatings), SPF roof systems can readily provide 20-year service lives.

Historical Background

The basic technology from which SPF roof systems evolved sprang from the introduction of Freon (circa 1960) as a blowing agent to produce low-density polyurethane foam with high thermal resistance. In 1964 came a significant advance in application procedure, a new generation of plural-component, heat-controlled equipment developed for spraying SPF. Advances in both SPF and reliable dispensing equipment opened the door to many commercial applications. Sprayed polyurethane roof systems were introduced in the late 1960s.

As previously indicated, early applications led to some spectacular failures. Inexperienced contractors, ignorant (or oblivious) of the

Figure 16.1 This large blister, several square feet in area, resulted from spraying a wet substrate. On this large project, thousands of square feet of roof area were blighted by inadequate substrate preparation, which is of paramount importance for successful SPF application.

required precautions for SPF application, sometimes applied SPF to dirty, disintegrating, or wet substrates. The result: horrendous blistering (see Fig. 16.1). Rapid oxidation (rusting) of SPF surfaces in the presence of sunlight requires prompt completion of lifts and coating application on the same day. Some early contractors failed in both these respects, with resulting delamination of the foam and peeling of the tardily applied coating (see Figs. 16.2 and 16.3). As a result of these spectacular failures, which continued into the early 1980s, SPF's reputation suffered severely

Attempts to economize on coatings also posed a major problem in SPF's early days. Acrylic house paints incompatible with urethane foam, aluminum-pigmented asphalt paints, hot asphalt and gravel—all produced dismal results. Even when good liquid-applied coatings were specified, they were sometimes applied in inadequate thickness.

In the early 1970s, improved flexible coatings—notably single-packaged, moisture-cured urethane coatings—solved the membrane coating problem for designers knowledgeable enough to specify them. In the mid-1970s, elastomeric coatings were further augmented with the introduction of aliphatic urethanes and acrylics. Together with silicones, a survivor of first-generation SPF coatings, these basic types constitute the predominant coatings in use in the mid-1990s.

Figure 16.2 Erosion and peeling of this SPF roof's coating resulted from a combination of (a) an excessively rough SPF coating substrate (verge of popcorn), (b) rusting (i.e., oxidation) of the SPF's surface, caused by several days' delay in applying the coating (which should be applied on the *same* day as the SPF), and (c) inadequate coating thickness.

Technical publications—notably, NBS Technical Note 778, by Cullen and Rossiter—promulgated guidelines for specifying and applying SPF.[1] But these publications were not heeded widely enough to stem the tide of SPF's early failures.

What finally saved SPF was the formation of the Polyurethane Foam Contractors' Division (PFCD) of the Society of the Plastics Industry in the early 1980s. The PFCD developed guidelines for application, specification, and detailing for foam and coating. An indispensable link in the delivery chain was institution of an accreditation program for SPF applicators. This critical step recognized the extreme dependence of SPF roof systems on applicators' skills. As previously stated, an SPF system is field-fabricated, unlike a single-ply system, where virtually all the manufacturing and quality-control functions (with the important exception of field lap seaming) are performed in the factory. Association with the Society of the Plastics Industry also enabled the PFCD to address environmental issues, such as stratospheric ozone depletion by CFC molecules.

Figure 16.3 Delamination of this blister was caused by a delay of a day or more between SPF spraying operations. This delay can be inferred by an adhesion failure at the lift plane (between separate spraying operations). Rapid deterioration of SPF exposed to sunlight spoils its adhesion. To assure good adhesion at lift planes requires tight scheduling. Spraying and coating operations must occur on the same day.

Materials

Urethane foam

SPF is spray-applied to its substrate as a liquid mixture that, within a few seconds, expands from 20 to 30 times its original liquid volume to an ultralightweight cellular mass less than 5 percent the density of water (see Fig. 16.4). In the best-quality foam (i.e., that with the highest thermal resistance), cell size ranges down to 4 mils (0.1 mm).

The chemical reaction that produces SPF results from the nozzle-mixing of two separate liquid streams: an A (isocyanate) component and a B (hydroxyl, resin, or polyol) component. Besides these basic chemical ingredients, the sprayed field mixture also contains

- A *blowing agent* (to form foamed cells that expand the foam resin's volume)
- A *surfactant,* or *cell stabilizer* (to control cell size and cell-wall rigidity)

Figure 16.4 On a typical reroofing project, spraying of polyurethane proceeds in multiple lifts, minimum ½-in. depth, over a spudded, aggregate-surfaced BUR membrane, which must be solidly anchored and essentially dry.

- A *catalyst* (to control the reaction rate between the two chemical components)
- *Fire retardants* (to satisfy safety and code requirements)

The traditional blowing agent for SPF was a CFC (chlorofluorocarbon) until CFCs were implicated in the depletion of the earth's ozone layer. Less-depleting HCFCs were introduced in the early 1990s as successful replacements. Although carbon dioxide and other blowing agents can form cells, they lack the low thermal conductance that gives urethane foams their superior insulating quality. Initial k values of monofluorotrichloromethane (R-11)–blown foams were 0.10 to 0.15, whereas those of carbon dioxide–blown foams are 0.22 to 0.24 (roughly half the thermal resistance).

The *catalyst* controls the chemical reaction rate and also governs the polymer chain propagation, extension, and cross-linking. It also helps the resin reach final cure, a vital process that assures attainment of full strength.

The *isocyanate* component is usually a polymeric methylene diisocyanate (MDI). Because of its potential reactivity with moisture, this component is usually provided by itself as the A side of the mixture. When reacted with OH-containing molecules, a urethane linkage, through an NCO group, is formed.

The reaction can be presented symbolically as

$$\text{R-NCO} + \text{R'OH} = \text{R-NH-}\overset{\overset{\displaystyle O}{\|}}{\text{C}}\text{-OR'}$$
$$\text{Isocyanate} \quad \text{polyol} \quad \text{polyurethane}$$

The *resin* or *polyol* component may be a polyester or polyether polyol or a blend of several types. Selection of polyols determines whether the foam will be rigid or flexible.

With its closed-cell content exceeding 90 percent (in correctly sprayed foams), the cured cellular polyurethane is relatively resistant to the passage of moisture in either liquid or gaseous form. It is, however, highly vulnerable to ultraviolet degradation. It thus requires a shield against ultraviolet radiation, normally a liquid-applied elastomeric coating.

Foam properties. A major improvement in foam quality in guidelines promulgated by the PFCD has increased foam density from 1.5 to 2.5 pcf to 2.8 to 3.0 pcf (see Table 16.1). Higher foam density assures higher compressive strength (via the 2.8-pcf minimum) and higher thermal resistance.

Compressive strength (per ASTM D1621), set at 40 psi minimum, should range around 50 psi in a correctly applied 2.8-pcf foam. This is an important foam property (see Table 16.1).

Membrane coating

SPF requires a membrane coating for several reasons:

- SPF oxidizes rapidly in sunlight, the degree of degradation depending on intensity and duration of exposure. As the deterioration progresses, the foam gradually changes color: from cream to tan to burnt orange. At this last stage, the SPF is friable and thus vulnerable to erosion. Allowed to continue indefinitely, erosion can ultimately work through the entire SPF cross section.
- Despite its water-resistant, closed-cell structure, SPF is not waterproof. In roof areas subject to ponding, and even in humid areas, liquid water or vapor can penetrate the cell structure, reduce thermal resistance, and saturate and soften the SPF.

TABLE 16.1 Required Properties Promulgated by the PFCD of the Society of the Plastics Industry for Sprayed-in-Place Polyurethane Foam

Properties (sprayed in place)	ASTM Test	Value	Units
Density	D-1622	2.8–3.0	lb/ft^3
Compressive strength	D-1621	40 psi (minimum)	lb/in.^{2}
Closed-cell content	D-2856	90% (minimum)	%/value
Thermal conductivity, k*	C-177 or C-518	0.18 (maximum)	Btuh/(in. · ft^2 · °F)
		0.026	W/(m · °C)
Flammability†	E-84	<75	

*Refer to SPI Bulletin AX-113, *An Assessment of the Thermal Performance of Rigid Polyurethane and Polyisocyanurate Foam Insulation for Use in Building Construction.*
†This standard is used solely to measure and describe properties of products in response to heat and flame under controlled laboratory conditions. This numerical flame spread rating is not intended to reflect hazards presented by this or any other material under actual fire conditions.

- SPF lacks the toughness to resist roof traffic, hailstone damage, dropped tools, and other puncturing and abrasive forces. A good, tough coating resists these forces.
- SPF lacks adequate fire resistance required by building codes.

To serve its required functions, an SPF membrane coating requires the following properties:

- Dependable adhesion to the urethane foam
- Impact resistance
- Abrasion resistance
- Resistance to temperature change (i.e., flexibility at low temperature, stability at high temperature)
- Resistance to deterioration from water (liquid and vapor)
- Weathering resistance
- Maintainability (i.e., ease of repair and capability of a weathered surface to accept and retain additional coating when recoating becomes necessary)
- Durability
- Strength and elasticity (to accommodate substrate movement without rupture)
- Fire resistance

Liquid-applied elastomeric coatings, with minimum 100 percent breaking strain and recovery of original dimensions on stress release) provide the best shield for SPF. The most popular and successful over the past decade are these:

- Urethane
- Acrylics
- Silicones

Urethane elastomeric organic coatings, based upon the same basic isocyanate and polyol reaction that produces SPF, are the most popular and generally satisfactory membrane coatings, with the least number of limitations. With their cyclical molecular structure, *aromatic polyurethanes* have high solids content. This feature makes them a durable *base coat,* less vulnerable to pinholing than more volatile coating liquids. *Aliphatic* polyurethanes, with their straight-chain polymers, provide excellent ultraviolet resistance. This property qualifies aliphatic polyurethanes as an excellent *topcoat*.

Aliphatic polyurethanes illustrate a distinguishing feature of liquid-applied elastomeric coatings: i.e., they can be *single-* or *plural*-component systems. As an example of a plural-component system, an aliphatic polyurethane is normally packaged in two separate packages, because of the inherent reactivity. When the two parts are mixed, a curing process forms the polymeric coating. Single-component systems, packaged in one container, form films via solvent evaporation.

Among plural-component systems, there are two types: *standard-cure* materials, with a potlife exceeding 1 h, and *fast-cure* systems of much shorter potlife. With their longer potlives, standard-cure materials can be premixed and applied with single-component equipment. Fast-cure systems, however, require plural-component systems to prevent rapid coagulation and possible blockage of the spray lines.

Fast-cure urethane coatings attack the SPF applicator's toughest problem: rainy, windy, or cold weather. Fast-cure, low-viscosity, 100 percent–solids aromatic urethanes cure within 1 min. This rapid curing enables the contractor to apply the base coat and topcoat on the same day instead of waiting overnight to apply the topcoat. In 1986 came another improvement: weather-resistant urethane coatings capable of achieving 30- to 45-mil thickness in one application.

Acrylic coatings generally offer the lowest first cost among elastomeric coatings. Several liabilities limit their use. Classified as "breathable" (i.e., relatively vapor-permeable) coatings, acrylics are not suitable for roofs subject to ponding. (Because it creates conditions of 100% RH even when the coating resists the entry of liquid moisture, ponding promotes a vapor drive toward the underlying foam, thereby posing the ultimate threat of condensation and saturation in the SPF's closed cells.) As water-based elastomers that cure by evaporation of water, acrylics are also threatened by dew, rainfall, or other sources of moisture during application, as these may re-emulsify the coating and reduce its thickness via wash-off or impairment of film properties. Acrylics are also more sensitive than solvent-cured coatings to low temperature. Freezing temperatures can destroy the coating.

Silicone coatings, also classified as breathable, are subject to similar limitations. Ponding of silicone-coated roofs is taboo. Minimum slope should be $\frac{1}{2}$ in./ft (4 percent). Their vulnerability to bird pecking and dirt retention (resulting from surface tackiness and static electric charges) can be rectified by application of mineral-granule shielding embedded in the finished coating surface. Their high permeance also assures their virtual freedom from blistering; water vapor entrapped in substrate voids simply diffuses through the permeable silicone coating.

Silicones come in single- and plural-component packages.

Butyl, a liquid-applied coating with extremely low permeance, may be the best material when the membrane must also function as a vapor retarder—i.e., where the prevailing vapor drive is outside-in, as in humid tropical or semi tropical climates (e.g., Hawaii, Key West), or where the interior is refrigerated.

Among the losers in the trial-by-experience coating sweepstakes is the liquid neoprene/Hypalon coating popular in early SPF systems. As a low-solids coating, requiring many applications to achieve adequate DFT, neoprene/Hypalon lacks the durability of urethanes, acrylics, and silicones. Hypalon, the normal topcoat in these applications, tends to erode rapidly by "chalking," thus requiring frequent recoating.

Asphaltic coatings, emulsions and cutbacks, are another casualty of early SPF experience. They simply lacked the elasticity required to accommodate SPF's high thermal coefficient of expansion. A new generation of materials, inspired by the same chemistry impelling the development of modified-bitumen membranes, may make modified-asphalt coatings a contender.

Dry-film thickness (DFT), the coating thickness *after* curing (or loss of evaporative solvent), is critical to coating durability. Minimum recommended DFT for any coating is 25 to 30 mils.

DFT is determined by two factors: coverage and percent solids. A gallon of liquid spread over 100 ft^2 (1 roofing square) will deposit a wet film of 16 mils (0.016 in., 4.06 mm). (In the metric system, 1 liter/m^2 is 1 mm thick.) Using the *volume* proportion of solids, you can compute DFT as in the following example: Since 1 gal/square produces 16 *wet* mils, a coating of 40 percent solids (by volume) will produce $16 \times 0.40 = 6.4$ *dry* mils. To produce a 30-mil DFT would thus require nearly 5 gal/square.

A third factor, less easily quantified, also affects the required wet coverage of a liquid elastomer. The rougher the substrate foam texture, the more liquid volume is obviously required to fill the deeper, more capacious valleys in the foam surface.

As a general rule, you can consider the following estimates for additional coating liquid: 5 percent for smooth foam, 10 percent for orange peel, 25 percent for coarse orange peel, and 50 percent for verge of popcorn, the roughest (marginally) acceptable foam substrate. (See Fig. 16.5 for photos of various foam substrates.)

An irregular substrate causes other problems in addition to requiring greater quantities of liquid material to produce minimum DFT. Pinholes or voids are more likely to form. And the uneven coating

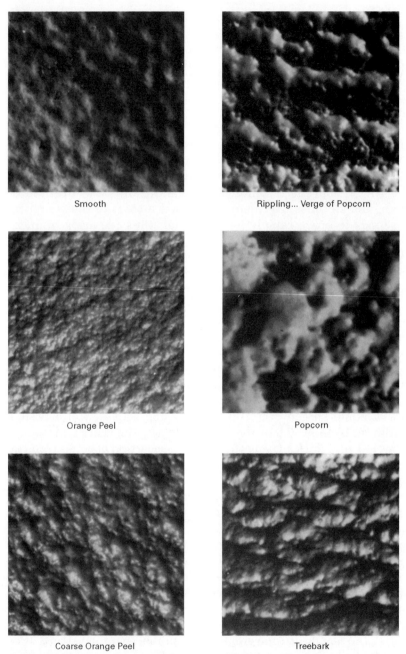

Figure 16.5 The three foam surfaces on the left (*smooth, orange peel,* and *coarse orange peel*) provide acceptable SPF substrates for liquid-applied coatings. *Rippling—verge of popcorn* (upper right) is marginally acceptable. *Popcorn* and *treebark* are unacceptable. Smooth foam requires only 5 percent additional coating liquid to fill in surface voids, whereas verge of popcorn requires an additional 50 percent. An irregular substrate makes pinholes and voids more likely. (*Polyurethane Foam Contractors' Division.*)

thickness can produce stress concentrations at thin sections, promoting cracking that would not occur on smoother substrates that permitted a more uniform coating thickness.

Coating properties. Among the major tested properties required of a liquid-applied coating material are the following:

- Tensile strength
- Elongation
- Tear strength
- Hardness
- Impact resistance
- Abrasion resistance
- Chemical resistance

Tensile strength and *elongation,* both measured by ASTM D412, are required to enable the membrane coating to resist thermal movement without splitting. Elongation, or breaking strain, minimum 100 percent, is the more important property, as the coating offers minimal resistance to substrate thermal movement.

Tear strength is important because a nick in the coating may propagate a membrane rupture in a material with low tear resistance. Via ASTM D624, tear resistance is measured by the force required to tear through a 1-in.-wide sample.

Hardness measures the membrane coating's resistance to indentation. Higher hardness means better dirt release. ASTM D2240 uses a Shore Durometer, with higher numbers indicating harder materials. The Shore A scale is the most popular of several available scales used for coatings.

Impact resistance indicates the coating's ability to resist rupture under a sudden dynamic force—e.g., a hailstone. Impact resistance is evaluated per ASTM D2794 by dropping a cylindrical weight with a cone-shaped tip through a guide tube from various heights. The height of the weight multiplied by the mass of the weight yields the impact (1 ft · lb = 1.36 joules). Impact resistance indicates a protective coating's ability to resist mechanical damage and impact from maintenance crews, hailstones, and bird pecking.

Abrasion resistance indicates the coating's ability to resist various erosion forces in service. Per ASTM D4060, the amount (by weight) of coating lost when the coating is subjected to the wearing forces of an abrasive wheel is measured.

Chemical resistance indicates a roof coating's ability to resist the corrosive effect of roof contaminants without losing its physical properties. Per ASTM D471, membranes are immersed in chemical solutions at both room and elevated temperatures to determine their performance. Both temperature and chemical concentration affect these results. (See Table 16.2 for chemical resistance and other coating properties. See Table 16.3 for the PFCD-promulgated format for specifying coating performance requirements.)

Surfacing aggregate

Loose aggregate, minimum $\frac{1}{2}$-in. diameter, applied at a rate of 500 to 600 psf (flat aggregate preferred over round), is sometimes specified as an ultraviolet and fire shield on roofs (*a*) capable of supporting the additional dead load and (*b*) subject to less rigorous exposure than normal conditions. This approach accepts wetting of the foam, relying on post-rain evaporation to dry the SPF surface and prevent water intrusion. On slopes greater than 4 percent ($\frac{1}{2}$ in./ft), liquid coatings are applied at the perimeter, where wind scour is likely to bare the foam surface. On *all* gravel-surfaced roofs, vertical flashed surfaces obviously require liquid coating. In addition to the risks of exposure and consequent ultraviolet degradation of the SPF, loose-aggregate SPF systems are subject to losses of thermal resistance from slow water intrusion into the foam cell structure.

Mineral granules embedded in a wet topcoat of liquid-applied elastomeric coating provide an alternative to loose aggregate on an uncoated SPF surface. Applied at a rate of 50 lb/square, No. 11 granules (the size used in asphalt shingles and mineral-surfaced cap sheets) improve durability, fire rating, and slip resistance (especially under wet surface conditions). They also discourage bird pecking, a mysterious but traditional problem with SPF. The extra application of coating and granules can also serve as a walkway. (Use of a contrasting color delineating walkways is a help.)

Walk pads, made of asphalt planks or reconstituted rubber, have created problems on SPF roofs. Applied to liquid coatings, these impermeable, dark-colored walk pads have tended to retain heat, to expand at a different rate from the SPF, and to trap water. Blisters within or beneath the planks have buckled or delaminated the foam and coating. Open pads, such as Yellow Spaghetti, embedded in a compatible caulk have worked better. (They permit easy escape for water and vapor.)

TABLE 16.2 Properties of Liquid-applied Elastomeric Coatings for Sprayed Polyurethane (SPF) Roof Systems

Coating material	DFT* (% solids)	Chemical resistance[†]				Tensile properties		Tear strength[‡] (lbf/in.)	Impact[§] resistance (in. · lbf)	Abrasion[¶] resistance (g lost)	Hardness (Shore A)
		Acids	Alkalis	Solvents	Salts	Strength (psi)	Elongation (%)				
Aromatic polyurethane	50–100	G	G	F/G	E	250–3500	150–600	100–500	60–160	0.015–0.2	65–95
Aliphatic polyurethane	40–85	G	G	F/G	E	1000–3500	150–350	100–200	40–160	0.1–0.25	75–95
Acrylic	50–65	F	F	P	G	200–350	170–300	30–40	20–60	0.55–0.85	40–65
Silicone	58–65	E	G	F/G	E	400–500	100–150	30–40	20–30	0.70–1.1	40–60
Butyl	40–60	E	E	P	E	200–800	150–250	30–50	15–25	0.40–0.50	45–60
Neoprene	25–30	G	G	F	G	1000–2000	400–500	100–300	20–50	0.10–0.50	60–80
Hypalon	25–35	G	G	P	G	500–800	150–350	40–80	35–60	0.40–0.60	70–80
Modified asphalt	80–92	F	F	P	E	40–200	50–150	NA	NA	NA	60–70

*DFT = dry film thickness (percent solids by *volume*).
[†]Key to rating: E = excellent; G = good; F = fair; P = poor.
[‡]Per ASTM D624.
[§]Per ASTM D2794.
[¶]Per ASTM D4060.

TABLE 16.3 Specifier Selects Required Properties, as Dictated by Service Conditions of the SPF, per Recommendations of the PFCD

Properties	ASTM Test	Value
Tensile strength	D-412	_____
Elongation	D-412	_____
Hardness Shore "A"	D-2240	_____
Tear resistance (lbf per lineal in.)	D-624	_____
Ultraviolet exposure	D-822	_____
	Atlas Carbon Arc,	
	Type F, weatherometer	_____
Moisture vapor transmission	E-96 Procedure E	_____
Fire resistance of system	E-108	_____

Detailing SPF Systems

As with all other roof systems, flashing and other construction details can make or break an SPF system. Some typical details promulgated by the PFCD are shown in Figs. 16.6 through 16.11. (For other details, see the current edition of *Spray Polyurethane Foam Systems for New and Remedial Roofing: Recommended Design Considerations and Guide Specifications,* AY-104.)

Note that the principles illustrated by these details must be adapted to the unique conditions that occur on the vast majority of projects, new or remedial.

One such principle, analogous to the principles of built-up and single-ply flashing, concerns the extension of base flashing up vertical surfaces, beyond the insulation level (see Fig. 16.7 for an example). With SPF systems, this principle simply requires the application of membrane liquid coating rather than a separate base flashing sheet, as in built-up or modified-bitumen systems.

Application Requirements

Application of a SPF roof system comprises three stages:

- Substrate preparation
- Foam spraying
- Coating application

Substrate preparation for both new and remedial (re-covering) projects requires removal of dust, dirt, and other contaminants that can spoil the vitally important adhesion of the foam to the substrate. Depending upon severity and extent, substrate contaminants can be

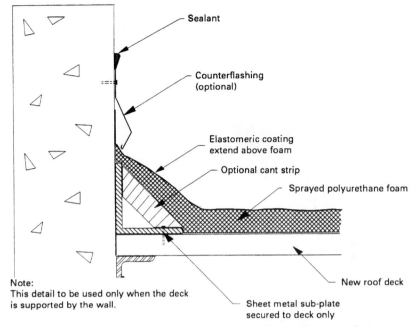

Figure 16.6 Typical flashing detail for wall-supported deck where no relative movement is anticipated. (*Polyurethane Foam Contractors' Division.*)

removed by air pressure, vacuum equipment, manual or power brooming, sandblasting, solvents, and manual scraping. For re-covering projects over gravel-surfaced BUR systems, the most difficult substrate preparation challenge, high-pressure, low-volume water wash, combined with vacuum, has proved successful.

Liquid moisture, like dirt and dust, impairs adhesion. It also does additional mischief, as it disrupts the chemical foaming reaction between the isocyanate and the polyol. Water reacting with the *A*, isocyanate, component imbalances the mix ratio in the worst way, producing an excess of the *B*, polyol, component. Such imbalance can create a soft, spongy foam with defective cell structure, vulnerable to moisture absorption, weaker, and less thermally efficient than a correctly mixed foam. And in addition to this physical impairment of foam quality, substrate moisture can promote growth of disastrous domed blisters like those depicted in Figs. 16.1 and 16.3. Loss of adhesion combined with entrapment of water vapor within the void formed between substrate and foam can create these blisters.

Coping with moisture problems on new substrates—for example, steel decks—is obviously much easier than on old BUR systems, which often retain entrapped moisture in the insulation. After

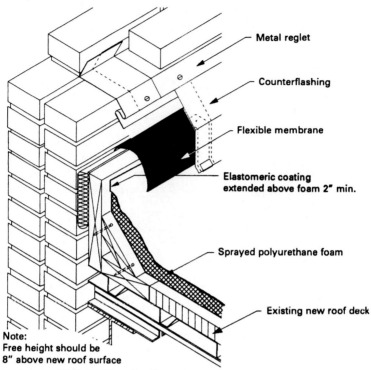

Figure 16.7 Wall flashing detail requires an expansion joint to isolate the roof system when it is supported on roof framing adjacent and parallel to the wall. (*Polyurethane Foam Contractors' Division.*)

removal of loose gravel, dirt, and residue, the PFCD recommends the following procedure:

- Investigation (with moisture-detection equipment, if necessary) to determine moisture presence within the roof assembly.
- Removal of any saturated insulation discovered by the foregoing investigation.
- Inspection of the existing system for anchorage of felts and insulation. Anchor poorly adhered components.
- Cutting out, or fastening, of defective membrane areas—blistered, buckled, wrinkled, fishmouthed.
- Removal from the BUR surface of soft mastic and other materials jeopardizing foam adhesion.
- Removal or reanchoring of loose base flashing, counterflashing, and gravel stops.
- Repair of defective expansion joints.

Sprayed Polyurethane 423

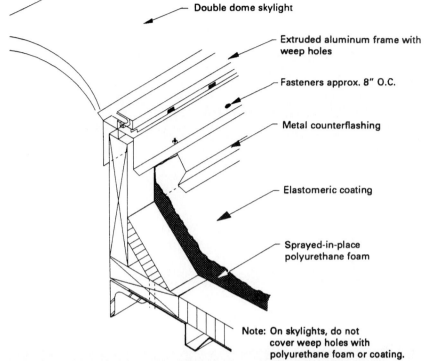

Figure 16.8 Flashing detail at skylight, hatch, or smoke vent. (*Polyurethane Foam Contractors' Division.*)

- Relocation of electrical conduit and mechanical ducts or raising them above the finished SPF surface.
- Masking of lightning rods prior to SPF application. (Do not embed lightning rod cables.)

Major substrates for new construction include the following:

- Metal deck
- Structural concrete
- Wood
- Miscellaneous (e.g., gypsum board, isocyanurate board)

Metal decks (22-gage minimum) require sandblasting of unprimed or unpainted steel surfaces. Scrape or wire-brush loose rust and unsound primer from shop-primed steel surfaces. Clean galvanized steel, aluminum, and stainless steel as recommended by the manufacturer issuing the warranty. Cleaning solutions can remove grease, oil, and other adhesion-impairing contaminants.

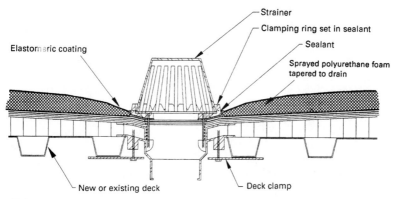

Figure 16.9 Detail for tapering SPF at roof drain (new or remedial). (*Polyurethane Foam Contractors' Division.*)

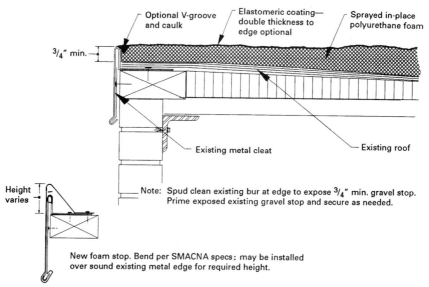

Figure 16.10 New and remedial roof edge detail. (*Polyurethane Foam Contractors' Division.*)

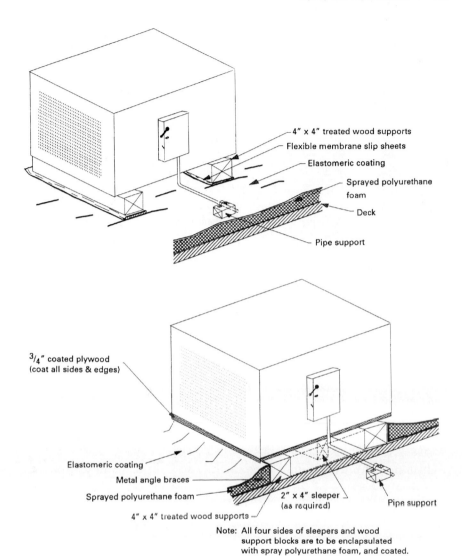

Figure 16.11 Remedial airconditioner details. (*Polyurethane Foam Contractors' Division.*)

Ribbed metal decks require a cover board or rib-filler material prior to foaming. (See Fig. 16.13 for one solution to this problem.)

Structural concrete (cast-in-place or precast) must be free of surface laitance and chemical release agents. After dirt removal, oil, grease, and other contaminants must be removed with a cleaning solution. Precast deck joints exceeding $\frac{1}{4}$-in. width require grouting or caulking. Priming is mandatory for all structural concrete surfaces, and

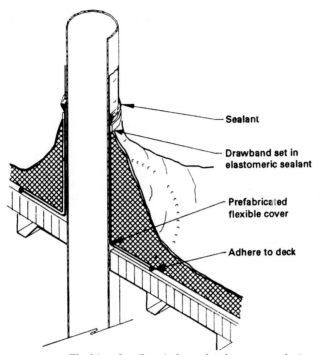

Figure 16.12 Flashing detail at independently supported pipe penetration requires a prefabricated flexible cover between pipe and SPF to permit relative movement. (*Polyurethane Foam Contractors' Division.*)

cast-in-place concrete must cure for a minimum of 28 days prior to primer application.

Lightweight insulating concrete and poured gypsum are not satisfactory substrates for SPF, because of (*a*) low cohesive strength and (*b*) excessive moisture.

Plywood (minimum $\frac{1}{2}$ in. thick) is limited to maximum 18 percent moisture content (per ASTM D4442 or D4444). All untreated and unpainted surfaces require exterior-grade primer (to minimize moisture absorption and promote good adhesion). Joints greater than $\frac{1}{4}$ in. wide require taping or filling with sealant.

Tongue-and-groove sheathing and planking require a minimum $\frac{1}{4}$-in.-thick plywood, or other suitable cover board, to serve as an SPF substrate.

Miscellaneous substrates (e.g., gypsum boards, isocyanurate boards) are generally anchored to ribbed metal decks. They must be butted at joints, without gaps. Joints over $\frac{1}{4}$ in. wide require caulking. Power brooming is banned, because it could damage the materials. These materials also require special care to prevent moisture absorption.

Foam application

To produce satisfactory SPF, the foam-spraying operation must conform to constraints imposed by the following factors:

- Weather conditions
- Lift requirements
- Cure
- Surface finish

Weather conditions are more restrictive for SPF application than for any other type of roof system. Spraying during rainy weather is obviously taboo. Temperature is generally limited to (*a*) minimum 40°F (4°C), and (*b*) a temperature within 5°F (3°C) of dew-point temperature. (Check manufacturer's recommendations for possibly more severe restrictions.)

Wind speed is limited to 12 mph (19 km/h) without wind screens. *With* windscreens, maximum windspeed is 20 to 25 mph (32 to 40 km/h).

Lift, the foam thickness sprayed per pass, is limited to a minimum of ½ in. Total foam thickness is a minimum of 1 in., with +¼ in., 0 tolerances. Adequate curing time is required before roof traffic is permitted. Full foam thickness must be completed in a single day, plus a base coat for the ultraviolet protection. On projects where the total roof area requires more than a single day's operations, break the roof area into segments; do not spray a larger area than can be foam-sprayed and base-coated in a single day. If bad weather forces a delay of more than 24 h, inspect the foam for ultraviolet degradation.

Surface finish must be smooth, orange peel, coarse orange peel, or, at worst (and only marginally acceptable), verge of popcorn for the substrate to be satisfactory for coating application. (This last surface condition requires an exceptionally large volume of additional coating material—50 percent compared with only 10 percent for orange peel—to assure adequate coating DFT. The irregular substrate represented by verge of popcorn also opens the door to coating defects—notably cracking.)

Defective foam can result from failure to maintain tight control of spraying operations. Tolerances in the ratio of the A (isocyanate) and B (polyol) components range around 2 percent (by volume). Outside these tight tolerances, SPF quality can suffer. (As indicated earlier, excessive A component is less serious than excessive B, which can produce soft, spongy foam.)

Cream time, measured in seconds required for the foaming reaction to start after the A and B components are mixed and sprayed on the substrate, depends upon two basic factors:

- The particular A and B component mixture
- Control of material temperature in the spray equipment

Within the proper 4- to 8-s cream-time range, a combination of short cream time and warm temperature produces a coarse orange peel surface, or worse. On the other hand, an excessively long cream time may produce an overly dense skin on a smooth surface, plus excessively thin lifts. Long cream time also makes the spraying of vertical surfaces difficult. The delayed foaming reaction causes the foaming SPF to run and sag. At the correct cream time, foam sprayed onto vertical surfaces rises straight out (i.e., horizontally), with no visible sag. Surface rippling from wind is another defect associated with excessive cream time.

Coating application

Liquid elastomeric coatings require a minimum of two coats, to avoid pinholes. The base coat should be darker in color, so that the applicator can visually confirm total coverage by the top coat.

According to PFCD recommendations:

- Apply the base coat on the same day as the foam. (Delay 2 h to allow the foam to cure and to minimize "off-gassing," which would create pinholes.)

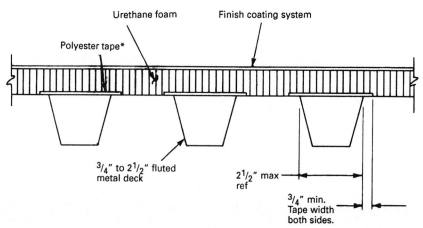

Figure 16.13 Steel deck ribs can be taped to prevent SPF from filling ribs. (*U.S. Navy.*)

- Apply the base coat in uniform thickness, with the application rate governed by the texture of the foam surface. (The more irregular the surface, the greater the quantity of liquid base coat.)
- After the coating cures, inspect for pinholes, thinly coated areas, uncured areas, and other defects.

Coating defects come in three basic categories:

- Pinholing
- Blistering
- Cracking

Pinholing normally results from defects in the foam substrate. Liquid coatings tend to flow into pinholes, blowholes, and crevices in the foam substrate. Though the coating may initially appear to cover the substrate holes, pinholes often form later, sometime after application of the liquid coating. Air that is slightly compressed and trapped in the substrate holes by the liquid coating can later force its way out through the coating as it begins to dry or cure. Surface tension, or viscosity of the liquid coating, then prevents the flow of liquid into the air-release hole. Additional coating may not fill the hole. A defective substrate can thus produce a defective coating, another lesson on the interdependence of roof-system components.

Blistering in coatings has several causes. The choice of a nonbreathing (i.e., low-permeance) coating can promote blistering. (A breathable coating would release water vapor that may be entrapped within voids at the substrate-coating interface.) A rusted (i.e., ultraviolet-degraded) foam surface or an irregular surface—e.g., popcorn or tree bark—can also cause blistering, from lack of adhesion and later heat-induced pressure within the void. Vapor migration from the interior toward the exterior in cold climates can promote coating blisters.

Cracking (crazing, "crows' feet," "mud checking") can result from several factors: an irregular foam surface, excess wet-film thickness, or "puddling," improper exterior temperatures, or exposure of the sprayed coating to excessive moisture before drying or curing. Nonuniform coating thickness, inevitable over an irregular foam substrate, creates stress concentrations in the thinnest coating cross sections. Thermal contraction resulting from cold temperatures or drying shrinkage promoted by heat can aggravate the cracking tendency. Excessive application can produce puddling, especially adjacent to vertical surfaces—parapets, vent pipes, and other vertical surfaces can shed the coating liquid and aggravate disparities in coating thickness.

Quality Assurance

As a field-fabricated component, SPF requires especially rigorous quality control. Even when robotic equipment applies the foam, quality control is essential. Quality control measures should include the following:

- Check the site for proximity of parking lots and other areas vulnerable to damage from wind-conveyed spray.
- Check air intakes and exhausts for possible shielding or shutdown during spraying operations.
- Verify adequate slope for drainage, making provision for additional foam to eliminate depressions that would produce ponding.
- In cool weather, consider a dark-colored primer, to raise substrate temperature and accelerate surface drying.
- Include in jobsite records job name, date, weather conditions, roof plan sketch, material types (primer, foam, base coats, topcoats, color), material quantities, batch numbers, placement location of respective batches, etc.
- Store materials at proper temperatures. Required storage conditions are posted on drums.
- Check equipment to assure that pumps and preheaters are functioning correctly.
- Examine first base coat application for pinholes.
- Take slit samples of foam, using a sharp utility knife, at the PFCD-recommended rate: 10 samples for the first 100 squares, plus 1 for each additional 25 squares. Measure foam thickness at the slit (since it requires patching). Examine test specimens with an optical comparator for average and minimum thickness of base coat and topcoat.
- Cut core-test specimens through the entire foam thickness, at a minimum rate of 2 per 100 squares. Note the number of foam passes (i.e., lifts), adhesion, total foam thickness, cell uniformity, cell size, presence of soft or spongy foam, presence of moisture or degradation, and color differences.

Maintenance

The speed with which ultraviolet radiation degrades foam makes regular maintenance inspection especially urgent for SPF roof systems. This urgent necessity is, however, counterbalanced by the relative ease of repairing minor coating or foam defects.

When coating erosion has progressed only to the first appearance of pinholes, it is easy to recoat an SPF system and restore it to full performance. Punctures or bird pecks should be detected and repaired as needed. Wipe the puncture clean and fill it with a generic caulk. (The caulk must be compatible with the coating system—i.e., silicone caulk for a silicone-coated roof.)

The PFCD has commented on bird pecking as follows:

> A small percentage of SPF roofing installations experience some bird damage. The result is small holes ranging from $\frac{1}{4}$ to $\frac{1}{2}$-in.-dia and occasionally larger. It is extremely unusual for these bird pecks to cause leaks in the roof system. The small holes can be easily repaired with a caulking gun and an appropriate sealant. Since this condition rarely causes the roof to leak, the damage does not represent failure of the system, but should be considered a maintenance item.[2]

Moisture surveys can detect wet or saturated foam. These findings should be verified by cuts, cores, or probes, to determine the precise location of the moisture—i.e., whether it is in the original roof system or within the SPF.

For *minor surface coating problems*—coating degradation or soiling, chalking, small eroded areas exposing foam—the surface can be power-washed, scoured, and then recoated. Remove blisters (and asphaltic repairs) prior to recoating.

Moderate problems—large areas of degraded coating, foam erosion to $\frac{1}{4}$-in. depth, extensive pinholing, wet foam, foam blisters up to 25 percent of roof area—can be repaired by removing bad foam both vertically and laterally until sound foam is reached. Clean and prime this area. Then spray a minimum $\frac{1}{2}$-in. lift of new foam, taking care to match the thickness of the adjacent foam and provide slope for drainage. Prime and coat the foam to restore the required coating DFT.

Major problems—defective areas exceeding 50 percent of the total roof area—require total removal and replacement.

Alerts

General

1. Consult PFCD *and* manufacturer's literature for all limitations—weather, substrate conditions, material storage, foam and coating properties, etc.—for design and application.

2. Survey site conditions, especially for reroofing projects, before specifying SPF.

Design

1. Limit foam thickness to 2 in. per lift for robotic application, 1½ in. per lift otherwise.

2. Require a minimum ¼-in./ft (2 percent) slope.

3. Consider a vapor retarder under conditions that would require one with an alternative system (see Chapter 6, "Vapor Control").

Application

1. Require PFCD certification of applicator, plus listing as an approved applicator by the particular manufacturer.

2. Require continuous monitoring of installation, in conformance with PFCD recommendations, and immediate repair or replacement of defective SPF system components.

Maintenance

1. Institute a regular maintenance-repair program, with periodic (6-month maximum interval) inspections of the roof. Inspect monthly if roof is subjected to vandalism or bird pecking.

2. Plan recoating every 8 to 15 years.

Reference

1. W. C. Cullen and W. J. Rossiter, *Guidelines for Selection and Use of Foam Polyurethane Roofing Systems,* NBS Technical Note 778, Center for Building Technology, Institute for Applied Technology, NIST, 1973.
2. "Bird Damage Statement," Polyurethane Foam Contractors Division, The Society of the Plastics Industry, Inc., Washington, D.C., June 1991.

References on SPF

PFCD—SPI

Guide to Selection of Elastomeric Protective Coatings over SPUF	PFCD-PC1	1994	SPI Stock #AY-102
	PFCD-SPUF1	4/89	SPI Stock #AY-106
Introduction to SPUF Systems		4/89	SPI Stock #AY-107
Blisters in SPUF—Causes/Repair	PFCD-BL1	4/89	SPI Stock #AY-105
SPUF Buyer's Checklist	PFCD-CL1	4/94	SPI Stock #AY-104
SPUF for New and Remedial Roofing	PFCD-GS1	6/90	SPI Stock #AY-110
Aggregate-Surfaced SPUF	PFCD GS-3	3/91	SPI Stock #AY-118
Moisture Vapor Transmission	PFCD-MVT	3/91	SPI Stock #AY-119
Glossary of Terms Common to SPUF	PFCD-Gloss		
Renewal of SPUF Foam and Coating Roof Systems	PFCD-GS-1	2/93	SPI Stock #AY-122

CERL

Evaluation of SPUF BUREC (U.S. Dept. of the Interior)	Report M-297	11/81	M. J. Rosenfield
Maintenance and Repair of Sprayed Polyurethane Foam Roofing	Report R-94-18	12/94	Swihart & Alumbaugh

References on SPF (Continued)

NCEL

Tech. Note N-1496	Investigation of SPUF-1	7/77	Alumbaugh & Keeton
CR 79.004	Fire Tests of SPUF on Steel	1978	Rhodes & Castino
PO No. 79-MR-461	Principles of SPUF Application	6/80	Coultrap
Tech. Data Sheet 82-17	SPUF System	1982	Alumbaugh & Conklin
Tech. Note N-1643	Thermal Conductivity of Weathered SPUF	9/82	Zarate & Alumbaugh
Tech. Note N-1656	Experimental PUF II	1/83	Alumbaugh, Keeton, Humm
Tech. Note N-1683	UL Tests—SPUF on Metal	12/83	Alumbaugh & Conklin
Tech. Note N-1691	Guidelines for Maintenance of SPUF	3/84	Alumbaugh, Conklin, Zarate
Tech. Note N-1742	SPUF III Field Inspections	1/86	Conklin, Zarate, Alumbaugh
UG-0011	Users Guide to SPUF	4/87	Coultrap, Alumbaugh, Humm
Tech. Note N-1815	Investigation of SPUF II	9/90	Alumbaugh, Humm, Keeton

SPI-Polyurethane Division

Disposal of Containers	Guidelines for Disposal of Containers and Wastes from Polyurethane Processing		
Rigid PUF and Polyisocyanate	Assessment of Insulating Properties	10/88	AX-151
MDI-Based Polyurethane Foam Systems	Guidelines for Safe Handling and Disposal	1992	SPI Stock #AX-119
Diisocyanates	Effects of Diisocyanates and Medical Personnel Hyperreactivity and Other Health		AX-150

MDI

	Hazardous Materials Transportation Regs for MDI		AX-165

Chapter 17

Metal Roof Systems

For building owners fed up with recurring roof problems, skeptical about the new materials as well as about traditional BUR, and ready to pay a high initial cost in return for a durable, low-maintenance roof system, a metal system may provide the answer. With the advent (circa 1970) of standing-seam connection and the "floating-roof" concept, a new vista opened for metal roof systems. Previously limited to positively sloped preengineered buildings or to hand-crafted and soldered-seam membranes for flat roofs, metal roof systems became practicable for reroofing all kinds of roofs. Their light weight keeps them competitive where structural limitations in the roof framing eliminate ballasted systems from consideration.

Where they are practicable, metal roofs may offer the lowest long term (i.e., life-cycle) cost of any competing roof systems, despite an initial cost ranging up to three times the initial cost of other competing roof systems. Offsetting the high initial cost are (a) probably longer service life and (b) almost certainly lower operating and maintenance (O & M) costs.

Maintenance costs are reduced by the use of coil-applied, durable, corrosion-resistant treatments and coatings. Metal roofs are thus spared the photo-oxidative chemical reactions that, to one degree or another, attack bituminous, elastomeric, and thermoplastic membranes. Metal roofs can reduce operating costs via energy savings. Unlike more conventional reroofing systems, where insulation thickness is limited by curb and parapet heights or other geometric constraints, metal reroofing systems can be designed for virtually any thickness of below-deck thermal insulation. And they can be readily designed to provide durable, heat-reflective surfaces, more readily maintained near pristine efficiency than aggregate or coated surfaces on conventional roof membranes.

Since roof designers, perhaps better than economists, know that there is no such thing as a free lunch, the distinctive limitations of metal roof systems will hold no surprises. Metal roof systems have great difficulty coping with the following problems:

- Multiple rooftop openings
- Complex building shapes
- Custom-designed flashing details

Progress has accelerated in manufacturing techniques, coatings, clip details, and sealant technology. When this manual's second edition was published in 1982, metal roof systems accounted for a trivial proportion of reroofing projects. Today, they account for two-thirds of a billion square feet, according to one trade survey.[1]

Historical Background

Metal has been used in roofs since antiquity. Early lead and copper roofs provided beauty, function, and durability, with some having service lives that exceeded a century. These early lead and copper roofs required a great deal of handwork and skilled craftsmanship, and consequently were very expensive. Medieval European cathedrals were often roofed with lead and/or copper.

In modern times, steel and later aluminum entered the markets for commercial and residential roofing.

Over the past two decades, several notable developments have extended the range of low-sloped roof problems solvable by metal roof systems. These developments include:

- Improved panel production
- Expanded corrosion-protection techniques
- New sealant technology
- The "floating-roof" concept

Each of the foregoing developments removed a design obstacle to metal roof systems. Production of strong, thin steel sheets in coiled rolls extended the range of uses and made metal roofs more economical. Corrugating the cross-sectional shape into efficient structural shapes (i.e., with high section moduli) buttressed the basic production improvements as a means of extending metal roofs' range of uses.

Expanded corrosion protection—through galvanizing (i.e., zinc plating), aluminizing, zinc-aluminum alloy coatings, and modern organic-coating technology—extended the service life as well as the architectural range of metal roof systems. Improved corrosion resistance offered greater long-term economy, and organic coatings added new aesthetic possibilities in durable, nonmetallic colors.

Mechanical seaming and advances in sealant technology reduced the minimum slope requirement to the $\frac{1}{4}$ in./ft (2 percent) recommended (or required!) for competing systems (see Figs. 17.1 and 17.2). Through new techniques for enhancing weather resistance at seams and flashings, new sealant technology contributed to long term economy. Reduced slope reduces building volume, especially in buildings with large roof areas.

The floating-roof concept, introduced by Butler Manufacturing Co. in 1969, included the first movable clip, a motorized seaming machine, as well as trim, ridge, and penetration details designed to accommodate the accumulating thermal expansion. These developments solved a major problem confronting metal roof systems: the vastly increased thermal movement of the exposed deck compared with the minimal thermal movement in a more conventional roof system's insulated steel deck. Experiencing annual temperature ranges of 200°F (110°C) and daily ranges of up to 120°F (70°C), an exposed metal roof must be capable of accommodating movements of up to 2 in. (100 mm) in 200 ft (60 m). A variety of clip designs have been developed by different manufacturers to hurdle this obstacle and extend the range of uses for metal roof systems.

Metal Roof Concepts and Terminology

Metal roofing systems must be divided into different types based upon their function.

Structural standing-seam metal roof denotes a panel that can span more than 3 ft (0.9 m) and resist gravity and wind-uplift loading.

Architectural metal roofing panels and flashings have as a primary purpose the aesthetic enhancement of a building structure. They function best on relatively steep slopes (greater than 25 percent) and are usually supported by a solid roof deck such as plywood or OSB. While significant weather protection may be furnished by the panels themselves through water shedding, water protection is also furnished by felts or other waterproof underlayments on the subdeck.

In addition to the basic division into *structural* and *architectural,* metal roof systems are also classified as *fixed* or *floating.* A *fixed* system may be through-fastened to its longitudinal laps or seams, and

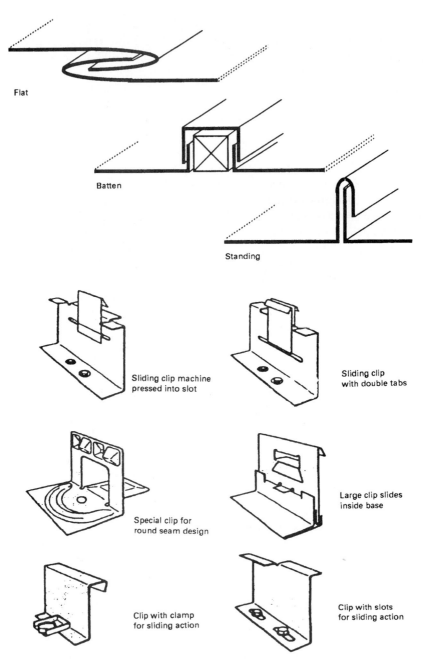

Figure 17.1 Early seams for metal roof systems (top) featured simple details compared with contemporary standing-seam clip details, which must accommodate thermal movement and transfer wind loading to structural framing. (*Rob Haddock, top; Exteriors Magazine, bottom.*)

Metal Roof Systems   439

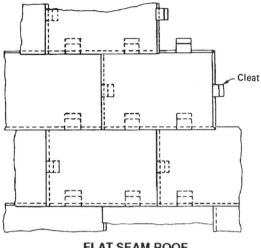

FLAT SEAM ROOF

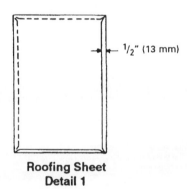

**Roofing Sheet
Detail 1**

Figure 17.2 Flat seam roofing has folded edges that engage cleats anchored to the deck. Copper, "dead-soft" stainless, and terne metals are most common in these systems. (*Sheet Metal and Air Conditioning National Association.*)

possibly by "purlin roll" in the longitudinal direction of the panel. These systems require expansion joints at close intervals [i.e., 30 ft (9 m) or less].

A *floating* system is generally fixed at a single location along the panel length (i.e., eave, center, or ridge). Thermal movement in the longitudinal direction is accommodated by sliding between the fixed structural member and the moving panel. The attachment clip may slide at its base relative to the structural member. It may be a two-piece clip that allows limited movement between sections. Alternatively, the panel may slide relative to the clip. Expansion (or contraction) in the transverse direction is accommodated through flexure of the trapezoidal or vertical legs of the seam.

A metal panel or flashing system may be designed to shed water (hydrokinetic), or it may be intended to be waterproof (hydrostatic). Low-slope seam closures require soldered joints, use of sealants, and/or hemmed seams. Flashings on low-slope structural panels may be hydrokinetic, especially at details designed to accommodate thermal movement. Positive drainage away from these flashings is essential, even though the roof system is regarded as hydrostatic.

This chapter is limited to structural, waterproof systems, primarily with low slope (see Fig. 17.3).

Design Factors

Thermal movement

In contrast with conventional roof systems, where the deck is normally subjected to a narrow temperature range, metal roof systems can experience extreme annual, and even daily, temperature variations. Thermal movement in the longitudinal direction can accumulate. On very long runs, expansion joints with a step may be required (Fig. 17.4). The upslope side of the step is through-fastened, like an eave detail, whereas the downslope side duplicates a perpendicular transition with a flexible closure.

Adequate provision should be provided for expansion within gutters.

Drainage

The permitted minimum slope of 2 percent means that there will be times in a heavy rain or when snow accumulates on the roof when the seams and laps could become submerged. Closures must provide a watertight system, not just a watershed.

Unlike membrane roofs, where water can follow the slope of the uninterrupted surface, in standing-seam roofs the flow of water is channeled in the direction of the ribs. As a consequence, roof slope must always parallel the ribs to ensure proper drainage, and to prevent water from flowing against the rib closures. At downslope obstacles, the ribs must be truncated short of the curb to allow diversion of water around the roof projection.

When exterior drainage is impracticable, interior gutters must be provided. In most instances, the gutter will be difficult to replace, as the ends of all panels lie directly on top. For this reason, the gutter

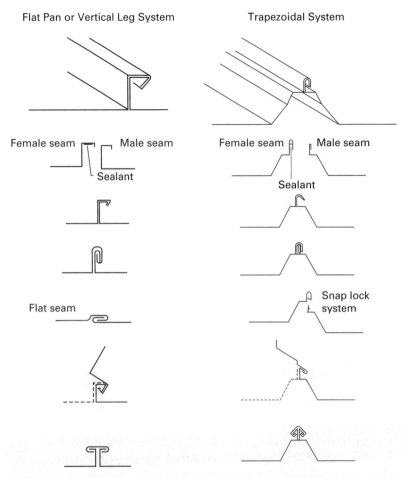

Figure 17.3 Standing-seam metal roof systems come in two basic types: vertical leg (left) and trapezoidal (right). Each panel has a male and female cross section, with a factory-applied sealant in the female leg. (*Paul D. Nimtz.*)

should be fabricated of highly corrosion-resistant material—e.g., stainless steel. For less corrosion-resistant metal, specify an elastomeric membrane liner.

Condensation

Metal panel systems may be vulnerable to condensation problems. High interior humidities may dictate use of a composite deck (structural subdeck), with a film vapor retarder installed under the thermal insulation.

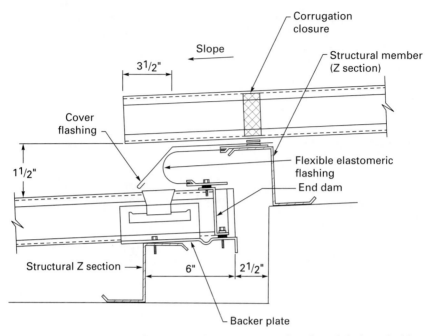

Figure 17.4 Expansion joint features an elevation step, with right-to-left slope shedding water over the step. (*BHP Steel Building Products.*)

In warm, humid climates, a plastic film directly beneath the metal panel may also be desirable to avoid sweating and possible backside corrosion of the panel.

Condensation at interior gutters must also be considered. In cold climates, the gutter may contain icewater, which will result in condensation and dripping. An alternative to heat cable is to install a drip pan under the gutter to catch the condensation.

Penetrations

Flashing of penetrations varies with size. For small penetrations, specify flexible boots that can accommodate movement (Fig. 17.5). Large curbs supporting heavy equipment must be supported by the structural framing, not the metal panels. These large curbed openings require a double curb; an interior curb supports the unit, and an outer secondary curb provides a vertical termination for the panels. A water-shedding counterflashing allows the two curbs to move differentially (Fig. 17.6).

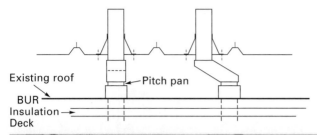

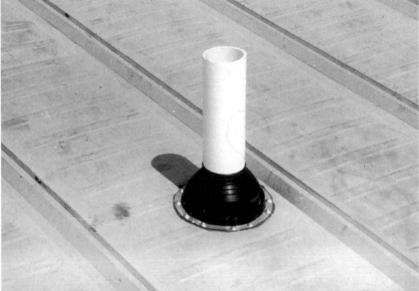

Figure 17.5 For small vent stacks in reroofing projects, hole is cut to allow minimum 1-in. clearance around stack. If existing stack is vertically aligned with standing seam above, elbow it over to the flat part of the panel. Rubber boot flashing is set in sealant and fastened to panel. (*Roofing Industry Educational Institute.*)

Wind-uplift resistance

Wind-uplift loading of the entire panel is transmitted to the holddown clips and fasteners anchoring the clip to the structural framing members. In structural panel designs, the clip must allow longitudinal movement, yet simultaneously resist concentrated uplift forces. Spacing of clips and their anchorage to the substrate are also critical. In nonstructural designs, clips are frequently little more than a narrow strip of metal bent over the panel edge. In hurricanes, these clips are often stressed beyond the steel's yield strength, resulting in severe wind damage.

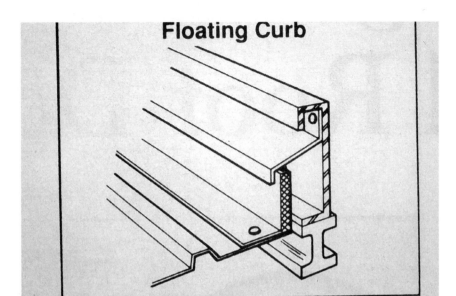

Figure 17.6 Floating curb. (*Roofing Industry Educational Institute.*)

Seam closures and sealants may perform differently under wind loading.

Corrosion Protection

The formidable job of shielding exposed metal roof panels from corrosion is entrusted to two basic techniques: *metallurgical* coatings and *organic* polymer coatings. (See Table 17.1 for tabulated corrosion-resistance properties of building panel materials.)

Metallurgical coatings for steel come in four types:

- Galvanized
- Aluminized
- Aluminum-zinc-coated
- Metallic-coated

Galvanized steel relies on a zinc coating for corrosion protection. Coating thickness (the thicker the better) comes in three grades: G30, G60, and G90, with total zinc content 0.30, 0.60, and 0.90 oz/ft^2,

TABLE 17.1 Compatibility of Metal Building, Roofing and Siding Materials

Panel, Flashing or Accessory Material	Galvalume	Prepainted Galvalume	Galvanized	Prepainted Galvanized	Aluminum-coated Type II
Galvalume	Compatible	Compatible	Compatible, although galvanized may have a shorter life and will eventually have an adverse effect on Galvalume.	Compatible, although galvanized may have a shorter life and will eventually have an adverse effect on Galvalume.	Compatible, although interior resistance of aluminum-coated steel to standing-water and cut-edge corrosion could result in rust-staining of adjacent Galvalume materials.
Prepainted Galvalume	Compatible	Compatible	Compatible, although galvanized may have a shorter life and will eventually have an adverse effect on Galvalume.	Compatible, although galvanized may have a shorter life and will eventually have an adverse effect on Galvalume.	Compatible, although inferior resistance of aluminum-coated steel to standing-water and cut-edge corrosion could result in rust-staining of adjacent prepainted Galvalume materials.
Galvanized	Compatible, although galvanized may have a shorter life and will eventually have an adverse effect on Galvalume. Small areas of unpainted galvanized subject to water runoff from Galvalume panels should be avoided.	compatible	Compatible	Compatible, although small areas of unpainted galvanized subject to water runoff from painted panels should be avoided.	Compatible, although galvanized may have a shorter life and may eventually have an adverse effect on aluminum-coated steel. Small areas of unpainted galvanized subject to water runoff from aluminum-coated steel should be avoided.
Prepainted Galvanized	Compatible, although galvanized may have a shorter life and will eventually have an adverse effect on Galvalume.	Compatible	Compatible	Compatible	Compatible, although inferior resistance of aluminum-coated steel to standing-water and cut-edge corrosion could result in rust-staining of adjacent prepainted galvanized materials.
Aluminum-coated Type II	Compatible, although inferior resistance of aluminum-coated steel to standing-water and cut-edge corrosion could result in rust-staining of adjacent bare or prepainted Galvalume panels.	Compatible, although inferior resistance of aluminum-coated steel to standing water and cut-edge corrosion could adversely affect the galvanized coating and/or result in rust-staining of adjacent bare or prepainted galvanized panels.	Compatible, although inferior resistance of aluminum-coated steel to standing water and cut-edge corrosion could adversely affect the galvanized coating and/or result in rust-staining of adjacent bare or prepainted galvanized panels.	Compatible	Compatible
Copper	Not compatible - direct contact or exposure to water runoff may seriously affect Galvalume.		Compatible		Not compatible - direct contact or exposure to water runoff may seriously affect aluminum-coated steel.
Lead	Not compatible - direct contact or exposure to water runoff may seriously affect Galvalume.		Compatible		Not compatible - direct contact or exposure to water runoff may seriously affect aluminum-coated steel.
Stainless Steel	300 series grades are compatible. 400 series grades with > 1.0 mil zinc or cadmium coatings are compatible.		300 series grades are compatible. 400 series grades with > 1.0 mil zinc or cadmium coatings are compatible.		300 series grades are compatible. 400 series grades with > 1.0 mil zinc or cadmium coatings are compatible.
Graphite	Not compatible - avoid contact.		Compatible		Not compatible - avoid contact.
Plastics	Compatible		Compatible		Compatible

Courtesy BIEC International, Inc.

respectively. Computed per ASTM A525, a thickness of 1.7 mils for a 1-oz coating yields a two-sided protective coating of 0.76 mil (0.019 mm) coating for G90.

For *aluminized steel,* ASTM Specification A-463 requires 0.65 oz/ft^2 of aluminum metal [approximately 1.4 mils (0.036 mm)] per side.

For *aluminum-zinc alloy coated steel,* Galvalume satisfies ASTM A792, which requires a mixture of 80 percent aluminum by volume. This yields approximately 0.55 oz/ft^2 or 0.9 mil (0.023 mm) per side. Galfan is about 95 percent zinc, 5 percent aluminum by weight, per ASTM A875.

Metallic-coated steel provides rust resistance through two mechanisms: *sacrificial* and *barrier* coatings. *Sacrificial* coatings—i.e., zinc—have a higher electrical potential than the underlying steel and will preferentially oxidize in the presence of an oxidizing medium and an electrolyte. When the zinc is consumed, the underlying, less active steel will then rust. The zinc oxide that forms is water soluble (sometimes called "white rust") and washes away.

Barrier coatings provide a hard, physical impediment to corrosion. The aluminum coating on aluminized steel forms a corrosion-resistant, insoluble aluminum oxide surface. Aluminized steel lacks "throw," a term describing a scratched surface's ability to heal by movement of atoms of the protective film at the site of potential corrosion. The coatings on Galvalume Sheet and Galfan contain both zinc and aluminum to achieve the combination of "throw" and sacrificial *and* barrier properties. These zinc-aluminum alloys have proven their durability, with a 20-year corrosion warranty available from manufacturers of Galvalume sheet.

Most *aluminum* panels for roofing use thicknesses of 0.032 to 0.040 in. and meet ASTM B209.

Organic polymer coatings and films are factory-applied to the steel, in a process called coil coating. They are capable of accommodating subsequent roll-forming and seaming operations without cracking or delaminating.

Common polymer coatings for coil steel have been rated by the National Coil Coaters Association (see Tables 17.2 and 17.3).

Metal Roofing Accessories

Clips have evolved from the simple hand-bent strips of sheet metal of nonstructural panel systems into a variety of engineered details that accommodate thermal movement of structural metal panel systems. At the same time, they must resist wind uplift. Many have a flat base screwed down to the substructural support system. A second compo-

TABLE 17.2 Properties of Common Polymer Coatings for Coil Steel.

GENERIC COATING TYPE	Resistance and Durability Properties													Weathering Properties				
	Impact Resistance	Mar Resistance (fingernail test)	Metal Marking Resistance	Resistance to Pressure Mottling in Coil	Solvent Resistance (Aliphatic Hydrocarbons)	Solvent Resistance (Aromatic Hydrocarbons)	Solvent Resistance (Ketone or Oxygenated)	Grease and Oil Resistance	Stain Resistance (household agents and foodstuffs)	General Chem. Resistance (acids, alkalis, etc.—spot tests)	Resistance to Water Immersion	Humidity Resistance	Abrasion Resistance	Gen. Corrosion Resistance (Industrial Pollution)	Corrosion Resistance (salt spray)	Gloss Retention (5 year Florida, 45° south)	Chalk Resistance (5 year Florida, 45° south)	Color Retention (5 year Florida, 45° south)
Epoxy-ester*	1	NA	NA	NA	4	5	3	NA	NA	NA	4-5	4-5	NA	NA	NA	NA	NA	NA
Epoxy*	1	2-3	3	3	4	5	4-5	2-3	2-3	NA	2	3	NA	2	2	1-2	1-2	1-2
Alkyd-Amine	2	3-4	3-4	4	4	3	3	4	3	3	3-4	3	3	3	3-4	2-3	2-3	2-3
Polyester	3	3-4	3-4	3	4	4	3	4	3	3	3	3	3	3	3	1-2	1-2	1-2
Solution Vinyl	4	2	2-3	3	3	3	1	2	2	3	2	3	3	3	3	1-2	1-2	1-2
Organosol-Vinyl	4-5	2	2	2-3	3	3	1	3	2	3-4	3	4	4	3-4	3	2	2	2
Plastisol-Vinyl	2-3	4	4	4	4	4	4	3	3	2	4	4	3	2-3	2-3	3	2-3	2-3
Thermoset Acrylic	2	4	3	4	4	4	4	3	3	3	4	4	3	3	3	3	3	3
Silicone Acrylic	2-3	3-4	4	4	4	4	4	3	3-4	3-4	3	4	3	3-4	3-4	3-4	3-4	3-4
Silicone Polyester	4	3	3	3-4	4-5	4-5	4	4	4	4-5	5	5	3-4	3	3	4-5	4-5	4-5
Poly-Vinylidene Fluoride	5	3	3	4	4-5	4-5	3	4	4	4-5	4	4	4	4	3	4	4-5	4
Poly-Vinyl Fluoride Film Laminate	4	3	2	2	4	3	1	4	4	3	4	4	4	4	3	2	3	3
Poly-Vinyl Chloride Film Laminate	4	3	3	4	4-5	4	3	4	4	3	4	4	4	3	2	3	3	2
Acrylic Film Laminate	3-4	3	4	4	4-5	4-5	3-5	3	3	4	2	2	2-3	3	3	3-4	3-4	3-4
Acrylic Latex-Waterborne																		

*Primer

Rating key
5. Excellent
4. Very good
3. Good
2. Fair
1. Poor or not possible

TABLE 17.3 Properties of Coatings for Steel.

GENERIC COATING TYPE	Thermoplastic	Thermoset	Ease of Application	Cure Temperature PMT	Color and Gloss Retention (doubletime in oven)	Unit Finishing Cost Range	Film Hardness—Pencil	Film Flexibility—T-Bend	Film Adhesion	Ability to Fabricate after Aging	Ability to Achieve High Gloss (above 85 units 60°)	Adaptability to Embossing (with metal)
Epoxy-ester*	✓		4	420	NA	L	NA	2T	3	NA	3-4	3-4
Epoxy*	✓		4	435	NA	L	NA	2-3T	4	NA	3	4
Alkyd-Amine		✓	4	435	2	L	F-H	2T	2	4	1-2	2
Polyester		✓	4	435	3	M	F-H	1-3T	3-4	5	3-4	3-4
Solution Vinyl	✓		3-4	400	2	M	F-H	1T	3	3-4	4	4
Organosol-Vinyl	✓		4	400	2	M	F	1T	3	2	4	4
Plastisol-Vinyl	✓		2-3	440	2-3	MH	HB	0T	2	2	4-5	5
Thermoset Acrylic		✓	4	440	4	LM	F	3T	2	3	3	3
Silicone Acrylic		✓	4	450	4	MH	F-H	3T	3	3-4	3	3
Silicone Polyester		✓	4	450	4	MH	F-H	2T	3	4	4	3
Poly-Vinylidene Fluoride	✓		4	465	4	H	H-2H	1T	2	1	4	4-5
Poly-Vinyl Fluoride Film Laminate	✓		4	400	NA	VH	F-H	1T	3	1	4	4
Poly-Vinyl Chloride Film Laminate	✓		4	400	NA	H	H	1T	1	3	4	4
Acrylic Film Laminate	✓		4	400	NA	H	H	2T	2	3	4	4
Acrylic Latex-Waterborne		✓	2-3	430	4	MH	HB-F	2T	3	2-3	4	3

*Primer

SOURCE: National Coil Coaters Association

nent, initially centered on the clip base, engages the upturned edge of the structural panel and "rides" with the metal panel as it moves longitudinally with thermal change. Other systems have a fixed rib; slippage occurs between the clip and the panel.

Sealants are required for a variety of rooftop conditions to assure watertight metal roof systems. They have been divided into three types:

- End lap sealing
- Pumpable caulk
- Exterior caulk[2]

End lap sealing features isobutylene-isoprene copolymer tape and/or combination tape and nonskinning, nonhardening butyl caulk. The tape must adhere to Galvalume, polyurethane, polyvinyl fluoride, and siliconized polyester metal in hot or cold climates, as well as to surfaces that are lightly coated with oil, wax, and mill lubricants.

Pumpable caulk is normally a liquid sealant, used either to seal pipe flashings, closure strips, and other panel connections by itself or to augment tape sealants. A nonskinning, nonhardening butyl caulk with at least 15 percent polymer is recommended.

Exterior caulk, applied to exterior joints of flashing, gutter, and roof curbs, should be a polyurethane sealant meeting Federal Specification TT-S-00230c, Type II, as well as ASTM C920, Type S, Grade NS, Class 25. Acid-curing silicones should be avoided with Galvalume.

A wide variety of *fasteners* for metal roof systems are available to serve different needs. *Tapping* or *self-tapping* screws make their own threads as they are installed. They are not designed to drill their own holes. (Charts are provided to show the correct diameter hole for various sizes of self-tapping screws.)

A typical numerical designation, such as $\frac{1}{4}"-14\times\frac{3}{4}"$, Type AB hex, gives a physical description of the screw. The first number refers to the fastener diameter, and the second, to the number of threads per inch. The lower the number, the coarser the thread. The final number is the length of the fastener from the underside of the head to the end of the point. Most of these fasteners have a hex head.

Type A, B, or AB refers to the thread form on the fastener. A *Type A* thread is used in joining two relatively thin pieces of steel, where it is meant to increase the stripping torque (i.e., the force required to strip out the threads in the hole in the metal sheets). The point on the Type A fastener is used to extrude the hole on installation, thus giving a larger metal surface to form threads in and a higher stripping torque. The *Type B* tapper is used when attaching relatively thick pieces of steel, such as 16-gage steel purlins through ½-in. steel beams. It has a finer thread, and thus a lower *tapping torque* (the force it takes to tap threads into the steel). The tapping torque must not be so high as to cause the screw to break or the installing tool to stall out. The *Type AB* screw is used in steel of intermediate thickness (26 gage through 12 gage).

Tapping screws are generally made of a heat-treated carbon steel. They are also available in 300 and 400 series stainless steel. The carbon steel fasteners are zinc- or cadmium-plated.

Self-drilling or Teks screws differ from self-tapping screws in that drill points are provided to enable the screws to drill their own holes. They have hex washer heads to provide a better bearing surface for the screw gun socket by eliminating wobble. Typically they might be described as 12–14×1" self-drilling No. 2. The number after "self-drilling" refers to the size of the point. The numbers run from 1 to 5; The higher the number, the thicker the steel that the fastener can handle. A No. 1 point is used for stitching two pieces of 26-gage material together. It has a coarse thread and drills a small hole so that it can have 100 percent thread engagement in the steel. A No. 2 point is used for 16-gage to 12-gage purlin material. It has a longer flute length and does not give 100 percent thread engagement so that the tapping torque doesn't go too high. The pilot point is important, as the drill point must penetrate the full thickness of the steel before the threads begin to tap.

Fasteners are used for the following connections:

- Subpurlins to structural members (reroofing)
- Subpurlins to steel deck
- Panel clips to purlins (or joists)
- End laps (exposed to weather)

When subpurlins are connected to structural framing in reroofing, a self-tapping length greater than the 4-in. maximum length of self-drilling screws may be required. (J bolts constitute another alternative.)

Subpurlin-to-steel-deck connections are made with insulation fasteners in predrilled holes, to prevent the screw from engaging the thicker purlin.

Panel-clip-to-purlin connections are typically made with self-drillers with a No. 2 drill point and/or AB tapping screws. The screw is applied through the purlin, generally 16 gage. When this connection is made to bar joists, a self-driller with a No. 4 or 5 drill point or a Type B tapping screw is required for the thicker chord metal of the bar joists.

Weather-exposed end laps are fastened with Type AB tapping or self-drilling stitch screws with a No. 1 drill point. When two thin sheets of steel are clamped together, a sealer washer is required. It should be driven snugly, but not damaged by overdriving.

Thermal Insulation

Structural metal roof systems are generally insulated with flexible glass-fiber blankets because of their easy handling and economy. The simplest application consists of suspending the blankets from purlin to purlin just before the metal panels are installed. A laminated plastic film faces the interior of the building, and may serve as a light-reflectance ceiling surface as well as a vapor retarder. The glass batts are compressed when the panel is laid on the purlin, resulting in a significant loss in thermal efficiency. This can be improved considerably by the use of a thermal spacer block, usually of relatively rigid plastic foam, placed on top of the purlin at the clip locations (Fig. 17.7). When an even greater R value is needed, it may be possible to use a clip with greater height to provide the needed offset. NAIMA 202, "Standard for Flexible Fiber Glass Insulation Used in Metal Buildings," provides detailed information on applicable documents and U values (North American Insulation Manufacturers Association, 44 Canal Center Plaza, Alexandria, VA 22134).

A second option is to suspend a ceiling from the bottom flanges of zee purlins. The suspended ceiling may be a combination of boards overlaid with batt insulation or a more rigid glass board suspended by a grid or by wires strung between joists.

When reroofing with a structural metal roof system over a conventional roof membrane, a plenum is usually created, as one virtue of structural metal is its ability to provide additional slope. Unfaced glass-fiber batts may be laid directly on top of the old roof, as the attic space can allow moisture to evaporate provided it is not trapped by a plastic film on the batt. Caution must be exercised here, as the additional insulation over the old BUR might degrade a Factory Mutual Class 1 system to Class 2. Building codes may require that the attic created have an access hatch if the clearance height is 30 in. (760 mm) or greater. (Refer to Factory Mutual *Loss Prevention Data Sheet 1-31* for further information.)

Underlayments

Many architectural metal roof systems are water-shedding rather than waterproof, and water may penetrate beneath the panel on occasion. This is handled by the proper use of underlay materials. One of

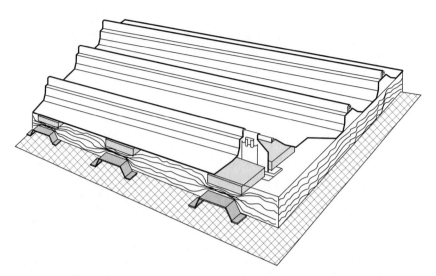

Figure 17.7 *Thermal spacer blocks* isolate metal panels from framing members below. This reduces heating and cooling energy loss through thermal "bridges." It also reduces the threat of condensation on interior metal surfaces cooled by conductive heat loss through poorly insulated spots. (*American Iron and Steel Institute.*)

the oldest is No. 30 asphalt-saturated felt (ASTM D226, Type II). The felt can be shingled from eave to ridge, or on steeper slopes and with batten seam systems aligned parallel to the slope. More recent innovations are the use of peel-and-stick bituminous membranes such as Grace Ice and Watershield, introduced in 1978. These latter products are usually SBS-polymer-modified bitumens. When a release film is removed during application, they will adhere to a plywood or OSB deck. They will also self-seal when penetrated by fasteners during the application of the metal panel and accessories.

Vapor retarders

Moisture from within a building may be driven into the thermal insulation by its vapor pressure. When flexible blankets are used, a laminated vapor retarder installed on the interior surface provides moisture protection. Tables 17.4 and 17.5 give the materials currently used as well as a summary of ASTM specification C-1136, "Standard Specification for Flexible, Low Permeance Vapor Retarders for Thermal Insulation."

Low permeance in a vapor-retarder system is important, but if the edges of the retarder are not sealed airtight, it will not be acceptable. Flexible retarders may be sealed by taping of laps and penetrations, or

TABLE 17.4 Material Properties of Vapor Retarders Currently in Use in Flexible Insulation Blankets with Metal Roof Systems

	Vinyl	FSK	PSK	PSK-VR	PSF	VSF	VSP
WVTR (U.S. PERM) ASTM E-96 Procedure A	1.0	0.02	0.02	0.09	0.02	0.02	0.02
Foil thickness (in.)	N/A	0.00035 0.0005	N/A	N/A	0.0003	0.00035 0.0005 0.0007	N/A
Film type	Vinyl	N/A	Polypropylene	Polypropylene	Polypropylene	Vinyl	Vinyl/polyester
Film thickness (mils)	3.0 & up	N/A	1.5	1.5	1.5	0.8–3.0	3.0/0.5
Kraft weight (lb/3000 ft^2)	N/A	30	11–30	11	N/A	N/A	N/A
Scrim reinforcement	None	Light to heavy	Heavy	Heavy	Heavy	Light to heavy	Heavy
Color	Bright white	Silver	Bright white	Bright white	Bright white	Bright white	Bright white
Cold weather handling	Poor	Good	Good	Good	Good	Varies	Varies

SOURCE: Frank Bitting, Lamtec Corp.

TABLE 17.5 Vapor-Retarder Physical Property Requirements, per ASTM C-1136

Physical property	Type			
	I	II	III	IV
Permeance, max, permissible ($ng \cdot Pa^{-1}s^{-1}m^{-2}$)	0.02 (1.15)	0.02 (1.15)	0.10 (5.75)	0.10 (5.75)
Puncture resistance, min. beach units (metric units)	50 (58)	25 (29)	50 (58)	25 (29)
Machine direction tensile, min., lb/in. width (N/mm width)	45 (79)	30 (5.3)	45 (7.9)	30 (5.3)
Cross direction tensile, min. lb/in. width (N/mm width)	30 (5.3)	20 (3.5)	30 (5.3)	20 (3.5)

SOURCE: Frank Bitting, Lamtec Corp.

by rolling and stapling the selvage edges of the film. Another approach is to install the vapor retarder on top of a subdeck, where it is supported and easier to seal from the top side. The use of rigid boards with factory-installed vapor-retarder facings may also be successful.[3]

Detailing Metal Roofing

As with any other roofing system, flashings are the most critical component. Since metal has the unique properties of channeling water and magnified thermal movement, details must be designed with these factors in mind (see Fig. 17.8). Unless it is corroded through its cross section, the metal membrane is no source of leaks. Leakage in metal roof systems occurs predominantly at panel joints and terminations.[4]

Exterior gutters. The design should ensure that water cannot back up under the panels (see Fig. 17.9). When a trapezoidal closure of foam or plastic is used, it must be adequately sealed to resist wind-driven water or dislodgement by birds or insects.

Gables or rakes. Since structural panels are designed as free-floating membranes, the gable trims must be designed to move with the panel. Cleats attached to the wall panels allow the trim to expand with the panel.

Figure 17.8 End laps, minimum 6 in., have a backup strap and sealant to maintain a watertight connection that can accommodate thermal movement. (*Roofing Industry Educational Institute.*)

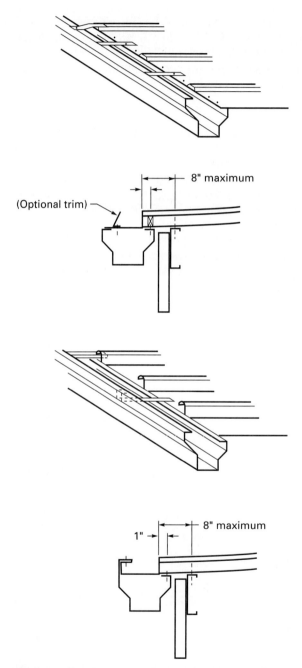

Figure 17.9 Exterior gutters. (*Roofing Industry Educational Institute.*)

Ridge assemblies. In many structural metal roof systems, the panels are through-fastened at the eave. All thermal movement consequently accumulates at the ridge. Panel closures attached to the metal panels guard against water intrusion. The metal ridge cover flexes as the panels converge and diverge.

End-wall transitions. When the panel spans parallel to a wall, a two-piece flashing is used for this *parallel* transition. The base assembly is anchored to the roof panel and becomes part of it. This upturned base piece is then counterflashed by a second piece of metal attached to the wall (see Fig. 17.10). *Perpendicular* or high-side transitions may use nonmetallic membranes, affixed to the top of a panel closure and bridging to the wall. This acts as an air seal, as well. The flexible membrane is usually protected with a metallic apron flashing against foot traffic or moisture accumulation.

Penetrations. Pitch pans should never be used with metal roof systems. Small round penetrations can be flashed with rubber boot jacks. Care should be taken to make sure that most of the penetration stays within the flat pan area, rather than in the rib area.[5]

Roof curbs. These are of two types. *Nonstructural* units have the same configuration as the roof panel and float with the roof system.

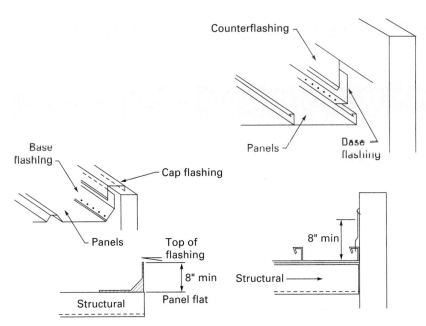

Figure 17.10 End-wall transitions. (*Roofing Industry Educational Institute.*)

These are suitable for lightweight units such as skylights and smoke hatches. *Structural units,* supporting heavier equipment, must transfer the load directly to the structural framing. A double curb is used. The structural curb is fixed, and a second, floating curb moves with the metal panel. Closure is accomplished in much the same way as in a high-side slope transition. Because of the channeling effect of the standing seams, water draining toward the curb must be diverted around it. This is accomplished by a custom-made end cap and water diverter.

Metal-Roof-System Defects

Aesthetic problems

Oilcanning is a rippling effect in the panel surface caused by buckling stress. Production antidotes to oilcanning include:

- Reducing panel width
- Using well-tuned roll-forming equipment
- Using a panel profile with stiffening flutes in the plane area
- Using tension-leveled coil stock with close camber tolerances
- Designing for thermal movement

A *dogleg* indicates a shift in the upper-tier alignment of a metal panel roof. It results from a lack of modularity of panels during installation, the upper edge of a panel being either slightly wider or slightly narrower than the lower edge. Though visually objectionable, doglegs are normally not a functional problem.

Functional problems

Fasteners are a major source of leaks in metal panel roofs because of relative movement between the panel and the fasteners, which are restrained by structural members below. The resultant force elongates the fastener holes.

To prevent leakage through these elongated holes, the fastener can be covered with a metal or plastic dome-shaped cap, over which a patch of elastomeric material is installed. Elastomeric coatings or mastics reinforced with nonwoven synthetic mat are also used to cover suspect fasteners, metal laps, and penetrations.

Leakage may also occur as a result of embrittlement or over-torquing of the elastomeric washer at or in the fastener head. Washers suspected of leaking may be tested with a seal tester. Faulty seals can be mended by the previously described cap cover patch or by removing the fastener and replacing it with a new one.

If fasteners are backing out because of strip-out of the threads, replacement with a new fastener one diameter larger will normally solve the problem.

Omitted or misplaced sealants can also cause leakage at end laps, flashings, or penetrations. A flat feeler gage inserted in the joint can determine the location and presence or absence of sealant. While washers stop water penetration where a fastener penetrates the upper sheet of a metal lap, the hole in the lower sheet is unprotected. Sealant must be installed on the wet side of the lap. If the feeler gage finds that the sealant is on the dry side of the hole, the lap needs to be opened up and new sealant installed. Application of caulk to the exposed lap is rarely successful. Precut sealant tape is available, with beads properly spaced to bracket the fastener. The second bead prevents "dishing" of the metal pan during fastener installation. Penetrations require flexible connections, and flashing must be detailed to permit movement.

Corrosion. The aluminized and zinc-aluminum alloys are very durable, but they may be affected by acid rain, by condensate that had run through copper coils, or by ferrous scraps, chips, or filings left on the roof. When the metallic coatings begin to fail, the service life of the panel has expired. Field-applied coatings can, however, be applied to salvage and rejuvenate the system.

References

1. *Metal Construction News Contractor Survey—1994,* Modern Trade Communications Inc., Skokie, IL.
2. *Jim Palmer Metal Building News,* September 1993.
3. Refer to "Condensation Fact Sheet," Metal Building Manufacturers Association, 1300 Sumner Ave., Cleveland, OH 44115-2851.
4. Paul D. Nimtz, PDN Associates, "Standing Seam Metal Roofing," *Metal Architecture,* September 1989.
5. Paul D. Nimtz, *Metal Architecture,* September 1989.

Chapter 18

Field Inspections

Inspection is one of the weaker links in the roof-construction chain. Many owners are unaware of the need for rigorous field inspection during construction and of the benefits of periodic visual inspection and moisture-detection surveys during the roof's service life.

Field inspections fall into three classifications:

- Quality assurance
- Maintenance
- Moisture surveys

Quality Assurance

The terms *quality assurance* and *quality control* have different meanings in the industry. *Quality assurance* refers to measures taken by the *owner* to ensure that the system is installed as specified. *Quality control,* on the other hand, refers to measures taken by the contractor or manufacturer, notably in training the applicator.

Quality assurance (QA) procedures can include meetings prior to start-up to review planned installation practices, preparation of check sheets to be used during the QA audit inspections, taking of roof test cuts and other field sampling techniques to verify compliance, and visual observations conducted during roof visits.

Though the basic purpose of QA is to assure construction of a good roof, there is also a secondary purpose: to eliminate poor work as a factor in the analysis of any subsequent roof problems. Weather and jobsite conditions should be noted, as they may help explain problems that become apparent months or even years later, such as blister, wrinkle, or ridge formation; curled or shrinking insulation; or poor attachment of components.

QA inspections

There are three basic arrangements for roof inspection:

1. Manufacturer's representative (on roofs covered by manufacturer's warranty)
2. Architect's or owner's representative
3. Roof consultant

Manufacturer's representative. Under the terms of the manufacturer's warranty, the manufacturer's representative inspects the installation of insulation, membrane, and flashing. If the inspector doubles as a sales representative, he or she may find it awkward to criticize the work of a customer whose continued good will is necessary for future material sales. Some manufacturers circumvent this obvious conflict of interest by providing a separate technical service inspector with no direct sales responsibility. Manufacturer's inspections tend to be infrequent; they usually occur more during start-up, occasionally during installation, and again at final inspection before issuance of the warranty.

Architect's or owner's representative. Inspection by either the architect's or the owner's representative or by a roof consultant eliminates the conflict of interest. On large projects, a full-time architect's or owner's inspector can inspect the roofing application. On smaller projects, a roving inspector, retained by the architect, may make periodic site visits. As jack-of-all-inspection-trades, however, the architect's inspector is unlikely to be master of roofing (See Fig. 18.1).

Roof consultant. The roof consultant usually offers inspection as part of a service that includes advising on roof design and specifications. However, consultants range from professional engineers or architects who perform this complete service to those who function more like a conventional testing laboratory, inspecting and reporting on the roofer's work. Under any arrangement for inspector's services, a

Figure 18.1 Quality assurance (QA) inspector verifies application's conformance with specifications. (*Roofing Industry Educational Institute.*)

major responsibility concerns the recording of the inspector's observations of defective roof areas (see Fig. 18.2).

Qualifications of a QA inspector. The QA inspector should be familiar with the specifications and details of the roof system *before* setting foot on the roof. Nothing will destroy a roofer's respect for an inspector more quickly than an early display of naiveté or ignorance.

A successful inspection program generally entails the following:

- A preapplication conference, including the inspector (owner's or architect's representative), general contractor, roofer, sheet-metal contractor, and materials manufacturer or agent, held at least two days before application begins.
- Notification to the inspection agency at least two days before application of vapor retarder, insulation, or membrane.
- Intense inspection as each component is installed, especially at the start.
- Formal issuance of inspection reports, promptly distributed to all parties concerned, on the progress of the work and its concordance with or deviation from the specifications. Results of roof cut analysis and/or moisture surveys should also be distributed.

Figure 18.2 Inspector uses measuring wheel to determine area dimensions of defective roof area (and later notes them on roof plan). (*Roofing Industry Educational Institute.*)

- An irregular schedule of inspector appearances.
- Issuance of formal notice by the contractor when ready for final inspection.

Documents and equipment. At the jobsite, the inspector should have access to the following:

- Complete contract documents—specifications and drawings
- Foreman's installation manual, if available
- List of approved subcontractors and material suppliers

- Copies of approved shop drawings
- Roof plans
- Specimens of approved submitted materials for visual check

Jobsite equipment should include the following:

- Thermometer (to check bitumen temperature)
- Camera (to record field conditions and practices)
- Level, straightedge, ruler, or measuring tape
- Whisk broom, knife, marker, pouch to collect samples
- Seam probe
- Clipboard, checklists
- Moisture meter

The inspector should follow a formal routine when making the inspection, using a detailed checklist. On projects of ample size and complexity, preparation of a special project checklist can facilitate the inspector's work. (Even on small projects, a checklist is a great help.)

Material storage checklist

1. Labels:
   a. Bitumen: EVT data provided? Correct material on jobsite? Samples collected?
   b. Felts: Brand name correct? UL, FM, code, or approval agencies identified? Batch numbers available? Samples collected?
   c. Liquid materials: Labels correct? Bearing asbestos? Flash point? Storage and shelf-life information? Date or batch code?
   d. Sheet, single-ply membranes: Panel number? Fire-retarded sheet? Correct thickness? Surface meets visual standard provided? (Especially with EPDM because of the frequency and severity of pockmarks, ASTM recommends that a visual standard be established and agreed upon in advance by purchaser and seller.)
   e. Modified bitumens: Proper designation and reinforcement? Fire-retarded? Approval agencies? Surfacing and burn-off backing?
   f. Aggregate or ballast: Meets submittal standard for color, texture, roundness, fracture, dust, sieve size?
   g. Thermal insulation: Proper designation, thickness, R factor, approval agencies, facers? Proper size? [Measure to $1/64$ in. (0.4 mm) as reference against possible future shrinkage.] Flatness? (Look for warpage, concavity at edges.)

h. Mechanical fasteners: As specified for type, length, thread, polymer coating, stress plate? (Pull-out tests needed in re-cover applications.)
   i. Accessories: Expansion joint covers, cant strips, termination bars, sealants, etc., as specified?
2. Storage conditions:
   a. Are all materials stored above ground?
   b. Is ventilation provided by slitting plastic wraps and covering with tarpaulins (see Fig. 18.3)?
   c. Are materials in original, unbroken containers with identification intact? Are the materials intact (no crushed ends, broken corners)?
   d. Is the roofing material dry? Dry is defined as a product at or below the equilibrium moisture content (see Table 18.1).
   e. Are roll goods stored properly? (BUR goods on end, on closed-faced pallets; others per manufacturer's instructions. See Fig. 18.4.)

Note: Reject defective material *before* application starts, if possible (to give roofer more time to replace rejected material).

Figure 18.3 Roofing materials should be stored on elevated pallets and tarped for weather protection. (*Roofing Industry Educational Institute.*)

TABLE 18.1 Equilibrium Moisture Contents of Roof-System Materials

Material	Equilibrium moisture content (% by weight @ 75°F, 90% RH)
Fiberboard	15
Perlite board	5.0
Faced glass fiberboard	0.5–0.8
Isocyanurate foam	2.9
Polystyrene beadboard	2.0
Polystyrene extruded	0.8
Phenolic foam	23.4
Cellular glass	0.1–0.5
Gypsum	1–2.5
Steel	0
Wood	7.5–18.0
Concrete	1–2
Vermiculite concrete	6.6
Organic felt, asphalt-saturated	2.4–4.0
Glass felt, asphalt-treated	0.1–1.0

SOURCE: Carl G. Cash, "Moisture and Built-up Roofing," Proceedings of the 1985 International Symposium on Roofing Technology, NRCA, 1985, p. 423.

Figure 18.4 Felt rolls laid on their sides in snow (instead of on elevated, tarp-covered pallets) guarantee future trouble—e.g., fishmouths and possible blistering from retained moisture. Inspector should reject these rolls. (*Roofing Industry Educational Institute.*)

Deck preparation

1. Drains: Correct location when deck is fully loaded or deflected? Flow to drains blocked by curbs, expansion joints? Overflow drains or scuppers provided? Overflow-drain downspouts independent of primary drainage?
2. Nailers: Correct elevation? Correct thickness and width ($\frac{1}{2}$ in. wider than metal flanges)? Adequately anchored to resist anticipated forces?
3. Cant strips: Correct dimensions, fire-retarded, installed per specifications?
4. *Before* roofer starts application, check deck for slope, smoothness, and cleanliness.
5. Check *panelized decks* for joint tolerances of $\frac{1}{8}$-in. (3.2-mm) vertical gap, $\frac{1}{4}$-in. (6.35-mm) horizontal gap plus the following:

   For steel decks:
   a. Dishing = $\frac{1}{16}$ in. (1.6 mm) maximum measured over three adjacent flange surfaces (i.e., two deck flutes).
   b. Unequal deflection at side laps? [Minimum of one side lap fastener for spans 6 ft (1.8 m) or less, two side lap fasteners at one-third points for deck spans over 6 ft.]
   c. End laps properly crimped to ensure maximum $\frac{1}{16}$-in. break at overlap? End laps located at supports (i.e., not cantilevered over supports, which allows two units to deflect differentially)?
   d. If more than one rib has been cut for an opening, has an angle-brace been installed to stiffen the deck?

   For wood decks:
   a. Plywood or OSB joints stripped with felt? H clips at unsupported joints (to prevent differential deflection)?
   b. Board sheathing or plank covered with rosin-sized paper, no. 15 saturated felt, kraft-paper laminate? Base sheet nailed correctly? (Do not allow solid or sprinkle mopping to wood or plywood decking.)

   For precast concrete decks: If $\frac{1}{8}$-in. vertical misalignment tolerance is exceeded at joints between units, require leveling of the deck surface with flexible grout or apply minimum 2-in. (51-mm) concrete topping.

   For lightweight insulating concrete fill: Coated alkali-resistant base sheet in place with proper laps, fasteners, and spacing? Is a provision for downward drying in place (i.e., slotted steel subdeck for perlite and vermiculite concrete)?

   For poured gypsum deck: Verify that fastener is appropriate and withdrawal strength is adequate.
6. Deck structurally supported at openings (HVAC equipment, skylights, scuttle access openings, drains, large vent openings)?

Application checklists

Before application starts, check the local weather forecast. If precipitation is expected, make sure materials are protected.

Insulation application

1. Check for specified mechanical fasteners: proper length, pattern, stress plate size, and spacing (see Figs. 18.5 and 18.6). Check spacing requirements for
   a. Building corners
   b. Perimeter strip
   c. Interior roof area
2. Check for insulation-board joint pattern: staggered or aligned (see Fig. 18.7).
3. Check insulation boards' bearing ($1\frac{1}{2}$ in. minimum) on steel deck flanges. Prohibit cantilevering of boards over steel deck flutes.
4. Check that paper-wrapped edges of glass-fiber insulation lie on top flanges of steel deck, not perpendicular to flanges.
5. Before roofing over wet decks, test for residual water. If specifications call for adhesion of components to a concrete deck with hot bitumen, pour some hot bitumen onto the surface and look for frothing and bubbling (evidence of excessive moisture). If the

Figure 18.5 QA inspector verifies type and spacing of mechanical fasteners. (*Roofing Industry Educational Institute.*)

Figure 18.6 Fastener patterns are critical for wind-uplift resistance. This single-ply system requires preliminary anchorage of roof insulation prior to membrane application. The membrane is then anchored with fasteners and stress plates in the membrane lap seams. (*Roofing Industry Educational Institute.*)

Figure 18.7 Use of two layers of insulation boards, with staggered joints, helps prevent membrane stress concentration (and possible splitting) and eliminates thermal bridges, which can leak energy at single-layer insulation joints. (*Roofing Industry Educational Institute.*)

bitumen can be easily peeled off the deck, it is too wet. In the absence of hot bitumen, place a glass plate or piece of single-ply membrane on the deck, and hold it down with dunnage. If condensation is found under the glass or membrane, additional drying is needed.
6. Concrete decks that will be hot-mopped require primer.
7. Spot-check insulation-board anchorage by manually prying a few random boards with exposed edges. Reject boards that come loose easily. Examine EPS for evidence of cell collapse.
8. Verify that wrapped edges of glass-fiber insulation run parallel to a steel deck's flute direction (see Fig. 5.10).

Membrane application

1. Bituminous:
    a. Check membrane materials: bitumen (asphalt, coal tar pitch); felt or modified roll goods; number of felt plies and pattern (base sheet installation, MB or cap sheet overlaps, all shingled plies, head lap).
    b. Check that EVT information is available and being followed.
    c. Require fire extinguishers near kettles and roof torches.
    d. Felt-laying order: On lower slopes [½ to 1 in./ft (4 to 8 percent), depending upon products used], apply felts shingle fashion, perpendicular to slope, starting at lowest point.
    e. On slopes 8 percent and greater, require backnailing and install nailers of the same thickness as the roof insulation. Run felts parallel to slope and nail through the back edge of the felts into a wood nailer.
    f. When an asphalt tanker is on the site, maintain asphalt temperature below the finished blowing temperature [FBT; approximately 490 to 500°F (286 to 292°C)] and feed asphalt into a pumping kettle for further heating before pumping it to the roof.
    g. When ambient conditions are 40°F or colder, store coated roll goods (including modified bitumens) in heated storage for at least 12 h prior to application.
    h. Check for continuous, uniform interply hot moppings, with no dry voids between reinforcing layers. Ensure that all hot-applied layers are installed in a single day, without phasing.
    i. Verify that burn-off film of polyethylene on the back side of heat-fused modified bitumens is completely melted away, not just shriveled.
    j. For mopped systems, mopping bitumen should be visible at the edge of all laps.
    k. Require immediate repair of application defects: fishmouths, blisters, ridges, splits, omitted envelopes, and so on (see Fig. 18.8).

Figure 18.8 Fishmouth (top) was evidently created by debris on insulation substrate. Fishmouth (bottom) should be split open and firmly embedded in bitumen. The slit should be patched. (*Roofing Industry Educational Institute.*)

l. Terminate the day's work with a complete edge seal (see Fig. 18.9).
  m. Systems designed for aggregate surfacing may be glaze-coated with bitumen, with aggregate and the rest of the flood coat delayed until a defined roof area is completed. (This practice is less critical with asphalt–glass-fiber systems than with organic-coal-tar systems.)
  n. Laps in modified-bitumen cap sheets must be offset from laps in the underlying base sheet (see Fig. 18.10).
2. Nonbituminous systems:
  a. Elastomeric sheets should be unrolled and allowed to relax (retract) for 30 to 60 min before perimeter attachment.
  b. Talced surfaces should be cleaned, using appropriate solvents and absorbent wipers. Primers may also be required. Check for drying of primer.
  c. Adhesives for bonding rubber to rubber (splicing adhesives) have solvents that must be allowed to evaporate. Open time is checked with tack and "push" tests. Adhesive is ready when it does not transfer to a dry finger and does not flow when the skinned surface is pushed.

Figure 18.9 Temporary water cutoff seals edge of just completed roof area against weather until work resumes. (*Roofing Industry Educational Institute.*)

Figure 18.10 Snapped chalk lines enable BUR applicator to get correct exposure and alignment of felts. Chalk lines may also be required when 36-in.-wide base sheets are the underlayment for meter-wide modified-bitumen cap sheets, since manufacturers require offsets between laps. (*Roofing Industry Educational Institute.*)

    d. Lap seams for elastomeric membranes should be field-checked by inspector (see Fig. 18.11).
    e. Heat-weldable sheets are laboratory tested by cutting strips perpendicular to the side laps. (Weld strength should exceed the strength of the fabric-polymer bond.)
    f. Some plasticized membranes must avoid contact with asphalt-treated substrates and unsurfaced EPS, as the plasticizer may be extracted. "Slip" or "separator" sheets of treated paper, fleece, or plastic film are used to separate the membrane from the substrate.
    g. Mechanically anchored systems require proper placement, either in the laps, in nailing tabs located under the factory-formed seams, or in the field of the sheet. These field fasteners are weatherproofed with patches over the fasteners or continuous membrane batten strips.
    h. Exposed reinforcing fabric found at the nonselvage edges (i.e., end laps, cut holes in the membrane, etc.) should be sealed with the specified lap sealant (see Fig. 18.11).
    i. Sealant should be allowed to firmly skim over before ballast is applied.
    j. Wrinkles in flexible membranes are common and are normally ignored, unless the wrinkle crosses a field seam, where it might

Figure 18.11 QA inspector probes single-ply field lap seam with thin, blunt tool, pressed against exposed ply lap edge. (*Roofing Industry Educational Institute.*)

create a water-leaking defect. If wrinkles are large enough to fold over, it may be preferable to pull the membrane into several smaller wrinkles.

3. Flashing and accessories:
    a. Check application of primer to metal, masonry, and other porous or absorptive surfaces for bituminous flashing.
    b Avoid combustible cant strips when torch application is planned. All wood and other combustible surfaces should be covered with an appropriate fire-barrier sheet. Air intakes should be blocked or shut off so that smoke and flames are not drawn into the interior.
    c. Top-nail flashings to avoid gravity slippage.
    d. Metal termination bars may be used to restrain polymeric flashings. Elongated slots will be less likely to result in fastener backout than round holes. Check for caulking of the top edge of the "term-bar" to prevent water penetration behind the bar and the flashing membrane.
    e. Check for specified treatment of wood nailers and cants, as well as gravel-stop metal.
    f. Check flashing heights at vertical intersections for proper dimensions: minimum 8 in., maximum 12 in., 4-in. minimum vertical counterflashing lap.

g. Thermally scan all torch-applied flashings at the end of the workday to detect hot spots (potential fire sites).

Visual Roof Surveys

Roofs should be inspected at routine, scheduled intervals. Replacing a neglected roof is an expensive undertaking, a drain on the owner's nerves as well as his or her finances. There are additional factors that can increase the replacement cost: extra care necessary because the building is occupied; additional labor involved to get the old material off the roof and moved to a disposal area; elevating pipes and other equipment so that roofing work can proceed; possible shutdown of rooftop equipment due to fume intake or interference with roofing operations.

Another reason for routine inspection is water leakage, which can disrupt production or cause equipment malfunction. Long-term leakage may damage structural components of the building as well as roof-system components. A well-planned and well-executed program of inspection and maintenance can anticipate problems and correct them before the entry of water into the building creates a serious difficulty.

Figure 18.12 Maintenance inspector may require reinforced fabric and mastic to make minor repairs, as well as tools for cleaning, measuring, marking, and photographing defects. (*Roofing Industry Educational Institute.*)

A roof should be treated as a piece of capital equipment that must be kept functioning. Just as a motor requires occasional lubrication and brush replacement to be trouble-free over its life span, a roof requires maintenance to perform as intended.

Roof management begins at building concept, and continues through the installation and all the way to ultimate replacement. Roof inspection and preventive maintenance are key components of this management program. They are anticipated by the quality assurance functions conducted during installation. These include

- Records of materials installed
- Plans and specifications followed
- Certifications
- Submittals and deviations
- Minutes of prejob conferences
- Guarantees and warranties

The as-built roofing file becomes the historical file, a cornerstone of a roof maintenance program.

Visual roof inspections should be performed at least twice a year: in the early spring to note and correct damage that occurred during the winter, and in the early autumn to prepare the roof to withstand the onslaught of another winter. Additional inspections should take place after severe weather—e.g., a hailstorm or hurricane. They are also needed after workers have repaired rooftop equipment.

Basic inspection tools are

- Tool box or carrying bag
- Work gloves
- Measuring tape
- Clipboard
- Roof plan
- Camera
- Marking crayon or paint

An inspector who may make test cuts or perform minor repairs as a part of the survey needs the following materials (see Fig. 18.12):

For BUR:

- Sharp knife
- Roof mastic

Figure 18.13 Repair of single-ply membrane starts with thorough cleaning of weathered surface. (*Roofing Industry Educational Institute.*)

- Trowel
- Broom to remove dirt and debris
- Reinforcing fabric, i.e., treated jute, glass fiber, polyester
- Claw hammer, spud bar, or chisel to remove aggregate

For single-ply:

- *EPDM*—Scrub pad, absorptive rags, primer, contact adhesive or tape, fresh membrane material, lap sealant, metal roller, knife, spatula
- *PVC and other weldable copolymers*—Hot-air welding gun, power source, clean rags, detergent and water, polymer sheeting, silicone roller (see Fig. 18.13).
- *Hypalon*—Scrub pad, solvent and primer, fresh Hypalon sheeting; alternatively, use adhesive and fresh membrane
- *Modified bitumens* (torching grades)—Wire brush, knife, L.P. torch, spatula or round-tip trowel, sheets of torching-grade MB (for mopping grade, see BUR repairs)

Check lists, detailed for each specific project, are helpful, if not essential.

Nondestructive Moisture Detection

Nondestructive moisture-detection surveys—via infrared (IR) thermography, nuclear backscatter, or capacitance (impedance) meters—have made accurate location of wet insulation economically feasible (see Fig. 18.14). In 1995, these surveys cost between $0.05 and $0.10 psf. Before the advent of these nondestructive survey techniques, the identification of roof areas plagued with wet insulation was necessarily based on incomplete information. Confronted with this problem, an owner has three basic choices: tearoff-replacement, re-covering of the old system (which remains in place), and living with the problem, with ad hoc repairs. Nondestructive moisture surveys provide data for a better informed decision.

Periodic nondestructive moisture surveys are like periodic medical checkups for a human being, with early detection of moisture problems analogous to early cancer detection, according to Wayne Tobiasson, research engineer with the U.S. Army Cold Regions Research and Engineering Laboratory (CRREL), Hanover, New Hampshire, and a pioneer in thermographic detection of roof mois-

Figure 18.14 Modern moisture detection utilizes portable, easily operated equipment. Capacitance meter (left) is wheeled across roof in a serpentine pattern until readings indicate wet areas. Extent of wet area is determined by rolling the detector until readings fall back to "dry" background level. Nuclear device (right) is placed at prelocated grid intersection points (generally at 5- or 10-ft spacing). Operator records readings for later analysis and plotting of wet areas on roof plans. (*Roofing Industry Educational Institute.*)

ture.[1] A cancer caught in its early stages, like a roof system gaining moisture, may require only minor surgery for its removal and restoration of the patient's health. Allowed to gain moisture undetected, however, a water-saturated roof can result in premature death of a roof system, just as a neglected cancer kills its human victim.

As an example of the benefits of nondestructive moisture surveys a nuclear moisture-meter survey was credited by a major building owner with permitting a $62,000 repair program instead of a $210,000 tearoff-replacement project.

Because of its many leaks, plus a widely blistered membrane and other defects, the 14-year-old, 70,000-ft^2 roof system was apparently headed for total replacement at a cost of $3 psf before the owner bought the idea of periodic moisture surveys. Readings were plotted on a grid and then computer-analyzed for moisture content. The survey revealed 13 percent of the roof area wet. Spot repairs with removal of identified wet areas, plus replaced flashing and overall general repairs, were able to prolong the roof's life for many years, postponing tearoff-replacement expense to some remote future date at a $148,000 saving.[2]

Benefits of nondestructive moisture detection

Accurate detection and location of roof areas containing wet insulation can alert the owner to the need for remedial action to avert the following problems:

- Leakage fed by water-soaked insulation
- Cooling- and heating-energy waste
- Decay of roof-system materials
- Ever-increasing expansion of the area of wet insulation

Water trapped in insulation can be a voluminous source of leakwater. In a porous insulation like fiberglass over a poured concrete deck, entrapped water can feed slow leaks through small cracks in the concrete for months, even years. If continually replenished by ponded rainwater over an imperfectly watertight membrane, this entrapped water can feed slow leaks perpetually.

Wasted heating and cooling energy is a direct result of wet insulation. Especially on large roofs covering low, sprawling buildings, these energy losses can be extremely costly. Heat loss through wet roof areas has been measured by heat-flux sensors at three times the heat-loss rate through dry roof areas.[3] Decay of roof-system materials, freeze-thaw breakdown, and loss of adhesive bond are additional consequences of water's presence.

Expansion of areas with wet insulation is a result of neglecting this problem. Regardless of whether the moisture source is liquid leakwater from above or condensing water vapor from below, the process extending the areas of wet insulation will continue. Extension of the wet insulation areas by condensing water vapor is promoted by the following process: Areas with wet insulation are cooler and consequently at lower pressure than dry areas when the sun bakes the roof surface. This situation creates a horizontal dry-to-wet vapor-pressure gradient, which impels water-vapor flow toward cooler, lower-pressure wet areas. Upon reaching the cooler wet insulation, the migrating water vapor condenses, thus extending the wetted areas. This process can be especially serious with porous insulation, which facilitates water-vapor flow.

Nondestructive testing thus enhances the benefits of more commonplace, routine maintenance and repair programs. It can enable early detection of entrapped moisture or a roof system that is just starting to experience serious moisture problems, possibly averting the need to replace the roof system. These nondestructive testing techniques can delineate existing roof areas with entrapped moisture or even signal impending moisture problems 9 or 10 months later.

Limitations

The new moisture-detection techniques have several limitations:

- Core or probe samples should be taken to validate the survey results, regardless of which nondestructive technique is used.
- Nondestructive surveys are limited to certain economies of scale, with higher unit costs for small roofs. Survey cost can nonetheless be below $1000.

Nondestructive moisture-detection survey data can translate into benefit-cost ratios of 10 or more in energy-cost savings and averted replacement costs.

Principles of nondestructive moisture detection

Infrared (IR) thermography, or "imaging," depends on the emanation of IR radiation from all bodies at an intensity varying, in accordance with the Stefan-Boltzmann equation, with the fourth power of the absolute temperature. As temperature rises, IR radiation gains intensity, with higher frequency and shorter wavelength. A "red hot" body radiates energy in the red portion of the visible spectrum—about 0.7 μm (= 0.0007 mm). In the normal range of nocturnal terrestrial temperatures—say, between 32 and 86°F (0 and 34°C)—the emitted IR wavelengths are much longer, ranging around 10 μm (= 0.01 mm). Because the emitted IR energy occurs in the invisible part of the electromagnetic spectrum, it can be detected only by sensitive media: semiconductor crystals, such as mercury-doped germanium or indium antimonide, as the thermograph sensor. Moreover, to attain the required sensitivity, these detectors must be cryogenically cooled to temperatures below −300°F (−207°C).[4]

The thermograph indicates areas of wet insulation by showing contrasting tones for areas with different temperatures—a darker tone for dry areas at lower temperature, and thus emitting less IR radiation at longer wavelength, and the lighter tone for wet areas at higher temperatures (see Fig. 18.15). Because of the radiative subcooling of roof surfaces on clear nights, with little or no cloud cover to absorb and reradiate escaping IR radiation beamed out into space, clear, cool nights provide the biggest contrast between wet- and dry-roof-area temperatures.

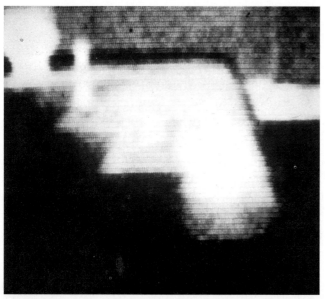

Figure 18.15 Infrared thermogram indicates wet insulation boards in roof corner. (*Roofing Industry Educational Institute.*)

An IR camera depicting these temperature differences is basically a TV camera, with the IR radiation sensed by the camera and converted into a video signal that displays the warm, wet areas as lighter shades and cooler areas as darker shades.

Because thermography depends on locating roof areas with wet insulation by their temperature differences from roof areas with dry insulation, the sharpest depiction occurs when the temperature difference between wet and dry areas is greatest. Based on extensive experience, from several hours after sunset to several hours before sunrise on a clear, cool night after a clear, sunny day is the best time to run a thermographic moisture-detection survey (see Fig. 18.16). But thermographic surveys are practicable year-round in all parts of the globe, as long as the roof surface is not wet. In summer, transient solar heating raises the roof-surface temperature and consequent roof-temperature differences. But steady-state heat flow during the winter-heating season is equally satisfactory because the interior heat energy creates temperature differentials between wet and dry areas. (During the heating season, wet-insulation areas gain more heat than dry areas because of the wet insulation's loss of thermal resistance.) Obviously, conducting nighttime rooftop surveys during the winter is physically more challenging than conducting them during mild weather.

Figure 18.16 Nighttime is best for thermographic survey because of complicating thermal effects produced by moving shadows cast on the roof by parapets, chimneys, and so on, variable solar radiation, and other factors not present at night. (*Tremco.*)

The *nuclear moisture meter* works on a totally different principle. A neutron generator with a radioactive source emits high-energy ("fast") neutrons aimed at the target area. Some neutrons collide with atoms in this material and are reflected back to the vicinity of the neutron emitter. Neutrons that hit hydrogen atoms are slowed by sharing some of their energy with the hydrogen atoms and can be counted by the instrument. The number of returning "slow" neutrons indicates the number of hydrogen atoms in the tested material. And the number of hydrogen atoms, which constitute two-thirds of the atoms in water, becomes an index of the quantity of water in the tested area.

Because hydrogen atoms abound in a membrane roofing system (bitumen, EPDM, PVC, cellulose-fiberboard, etc.), the nuclear meter will always detect hydrogen and give a base reading. This datum level of hydrogen is established for dry areas, and the excess count is then assumed to indicate the presence of water.[4]

With the nuclear moisture meter, readings are taken on a premeasured grid. The intervals typically are 5 ft, but 4- to 10-ft spacings may be selected depending upon the precision and speed desired. A pattern of wet insulation can be inferred from the readings and plotted as shown in Fig. 18.17. The nuclear devices are easily portable, and use solid-state electronics. Count time is adjustable, from a few seconds to one-half minute. Like the choice of wider grid spacing, increasing count-time speed sacrifices precision.

Other moisture-detection techniques

Another nondestructive moisture-detection technique, the *capacitance or impedance meter*, was borrowed from the paper industry. As with nuclear detection, a grid layout can be used for measurements, but a wheeled version allows a continuous reading as the meter traverses the roof (Fig. 18.14). When abnormally high readings are found, these areas can be intensely scanned by taking readings in a pattern radiating from the suspected spot like the spokes of a wheel. The capacitance meter depends upon the tremendous difference between the dielectric constant of water (80) and the constants of most dry roofing materials (which range from 1 to 4). This technique resembles nuclear detection, with sensors utilizing capacitance circuits roughly paralleling the nuclear backscatter gages. Variations in dielectric constant from a norm obtained in the dry roof areas indicate varying quantities of entrapped water. Capacitance meters are rendered ineffective when electrically conductive materials are present between the wet insulation and the meter. These would include foil surfacings on insulation boards and black EPDM rubber. A different capacitance

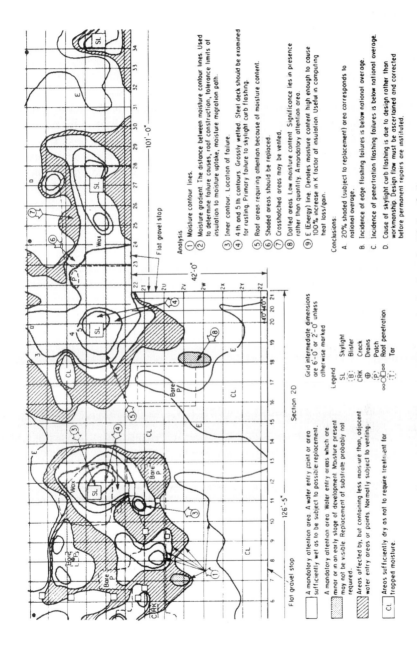

Figure 18.17 Roof-surface grid, at 6-ft spacing in this example, is plotted with insulation moisture contents inferred from nuclear meter readings. Shaded areas indicate areas of wet insulation recommended for removal and replacement. (*Gammie Nuclear Service Co.*)

device, grounded to the steel roof deck, is available for surveying black EPDM.

Electrical-resistance moisture meters exploit the principle that changes in electrical resistance indicate water content. With its inevitable impurities, water is a good conductor, and electrical resistance in nonconductive insulating materials varies inversely with their moisture content. This method is more accurate for determining the moisture content in wood than in roofing materials, but it can nonetheless indicate the presence or absence of water. Obviously it is less accurate than the other three moisture-detection techniques discussed.

Electrical-resistance moisture meters are small, battery-activated, hand-held instruments weighing only 18 oz or so. Since the probes must puncture the membrane to get a reading, these meters are not nondestructive. The probe holes should be promptly patched. These meters are invaluable, however, in verifying that anomalies discovered by the NDE methods are actually wet insulation.

Each of the moisture-detection methods discussed can be subject to errors, but cross-checking by using a distinctly different physical principle eliminates most of these false readings. When a nondestructive survey gives poor results, it is generally because the verification step was omitted.

References

1. Wayne Tobiasson, Charles Korhonen, and Alan Van den Berg, "Hand-Held Infrared Systems for Detecting Roof Moisture," *Proc. Symp. Roofing Technol.*, NBS-NRCA, September 1977, p. 265.
2. Paul Geffert, "Diagnosing Roof Problems with a Nuclear Moisture Meter," *Plant Eng.*, Apr. 28, 1977, pp. 213ff.
3. Charles Korhonen and Wayne Tobiasson, "Detecting Wet Roof Insulation with a Handheld Infrared Camera," U.S. Army Corps of Engineers, Cold Regions Research and Engineering Laboratory, Hanover, N.H., 1978, p. A14. See also Wayne Tobiasson and John Ricard, "Moisture Gain and Its Thermal Consequence for Common Roof Insulations," *Proc. Symp. Roofing Technol.*, NBS-NRCA, April 1979.
4. H. W. Busching, R. G. Mathey, W. J. Rossiter, and W. C. Cullen, "Effects of Moisture in Builtup Roofing—A State-of-the-Art Survey," NBS Technical Note 965, July 1978, p. 24.

Chapter

19

Reroofing and Repair

A building owner's decision to tear off and replace a leaking roof system often follows this scenario: A prolonged series of futile attempts at repair, sometimes stretching out for years, harasses the owner with continuing irritation, inconvenience, functional handicaps, and sometimes serious economic loss (e.g., lost rentals of motel rooms). Finally, these problems overwhelm any resistance to the cost of reroofing. The owner decides to reroof—often in the dead of a northern winter, when roofing application problems oscillate between the difficult and the impossible.

So, as the first advice on reroofing, building owners should attempt to anticipate the need for reroofing. This does not mean that they must become soothsayers prophesying the fate of their roofs. It merely implies the exercise of greater prudence and control in recognizing a perennially troublesome roof that is leaking moderately in summer and autumn as an almost inevitably more severe problem in late winter, when it it much more difficult to repair, replace, or re-cover.

A re-covered, or even totally replaced, roof has less chance of lasting out a 20-year service life than a new roof built with comparable care and skill. This is a second important point for roof-troubled building owners. Despite its high cost, re-covering or tearoff-replacement generally requires major compromise with roof-design principles. Working on the clean slate of a new building, the roof designer faces far fewer geometric constraints (discussed in detail later in this chapter). Tearoff is a slow, dismal job, requiring much greater care than new construction (see Fig. 19.1).

The difficulty of tearoff-replacement needs emphasis because building owners seldom comprehend the complexities of reroofing, and they are often shocked by the high cost. A reroofing project is almost

Figure 19.1 This tearoff-replacement project started with gasoline-powered cutter slicing old aggregate-surfaced membrane into roughly 10-ft^2 segments (top), which were pried off roof (middle) and carted away (bottom). The project *should* have started (but did not) with removal of aggregate, including power vacuuming, to reduce risk of tracking aggregate into new, replacement roof areas.

always more difficult to carry out successfully than a new project of similar scope. Aggregate-surfaced roofs in particular are an extremely difficult problem. Keeping the old aggregate out of the new membrane, where it can form blister-forming voids and puncture felts, is always a difficult, and sometimes a herculean, task. It takes good planning and careful execution by the roofing contractor to solve this problem because the aggregate has a remarkable affinity for the workers' bitumen-covered shoes and equipment wheels.

Re-covering or Tearoff-Replacement?

After investigation and analysis, the roof designer faces the first major decision: whether to re-cover the existing membrane or to go all the way with total tearoff-replacement. Tearoff-replacement offers two major opportunities:

- Inspection and repair of the deck
- Possible application of tapered insulation (depending on flashing heights) to slope an existing dead-level membrane that ponds water

Accompanying these two advantages are two disadvantages:

- Greater disruption of building operations
- Substantially higher cost

For some sensitive building occupancies—e.g., top-floor computer rooms, laboratories, or telephone equipment—tearoff-replacement may be an unacceptable risk. Removal of the entire existing roof system may expose the roof to intolerable leakage if the roofer is caught by a sudden, unanticipated rainstorm. Prefabricated decks—e.g., steel, plywood, tongue-and-groove boards, precast concrete—are obviously a greater hazard than poured-in-place concrete. Joints between prefabricated deck units provide easy, open access for water entry into the building. But even poured-in-place concrete and gypsum decks will leak through cracks caused by shrinkage or thermal contraction.

Thus the decision to accept a tearoff may depend on building use. For most uses, the owner can accept some slight risk of leakage. (The owner has normally been doing that for some time prior to the decision to reroof.) But for some operations—e.g., sensitive production or laboratory operations—a tearoff replacement may require protection of the contents from dust, dirt, and water. Replacement of substantial areas of deck will definitely require a shutdown for several days (or weeks) for reroofing operations. The owner must weigh the immediate costs of a shutdown against the future benefits of owning a thoroughly rebuilt roof. So, as one early question, the owner must decide whether to accept a tearoff.

Against the difficulties and greater expense of a tearoff-replacement, the roof designer must consider the feasibility of re-covering. Re-covering requirements differ for different kinds of roof systems: conventional built-up bituminous membranes, protected membrane roofs, loose or adhered single-ply sheet synthetics, or sprayed polyurethane foam plus fluid-applied membrane coating. But re-covering does carry several general requirements:

1. The structural deck must be sound.
2. The existing roof system must be adequately anchored.
3. Existing insulation should be strong enough to resist traffic and normal impact loads, and it should be essentially dry.
4. The existing membrane must form a reasonably smooth surface or be economically repairable into a smooth surface.

Whether all the foregoing requirements are mandatory for all re-covering systems is debatable. Compromise is an inherent aspect of most reroofing projects. A loose-ballasted roof system requires a stronger deck than an adhered system to carry an additional 10 to 25 psf of loose aggregate or pavers. But a loose system with a vapor-permeable membrane can tolerate more moisture in the existing insulation than an adhered system susceptible to blistering. If wet insulation is accepted, the designer must decide how much insulating value to sacrifice from wet insulation. And the cost of each alternative method is a ubiquitous factor in any reroofing design decision.

In any event, the foregoing rules should be considered highly desirable for any type of re-covering of a roof system. Violation of any of these rules for re-covering should, at the least, prompt consideration of total tearoff-replacement. Planning a reroofing project is an exercise in running through the entire spectrum of possible solutions and balancing liabilities and advantages of systems A, B, C, and possibly D against each other. The solution to a reroofing problem will seldom, if ever, appear as definitely clear as the decision on a new project, where the designer has greater control of all the conditions.

Investigation and Analysis

In a paper titled "Technical Aspects of Retrofitting," R. L. Fricklas, Director, the Roofing Industry Educational Institute (RIEI), recommends that investigative analysis proceed through the following five stages:

- Nature and condition of existing roof-system components
- Roof-system history
- Roof-system environment (interior or relative humidity, exterior surface contaminants, structural loads, vibrations, and other unusual conditions)
- New performance criteria (especially for thermal insulation or heat-reflective coatings)

- Improved drainage (many existing roofs pond water, and, if at all practicable, reroofing must correct this condition to justify the owner's capital investment)

Preliminary investigation

Start by examining the original plans and specifications, if available, to identify roof-system components: deck, vapor retarder (if any), insulation, and membrane (number of plies, type of felts, bitumen, polymer type, and so on). Unfortunately, many building owners have lost the specifications; you are lucky to find a set of drawings. Check whether the architectural drawings call for slope. The structural roof framing plan can help determine the roof's load-carrying capacity. You should find the design live load on the roof framing plan.

Be forewarned that drawings and specifications often fail to show the final roof system. They are not always emended to reflect job-change orders.

Besides checking drawings and specifications, check for subsequent modifications to the roof. Even an oral record, from the maintenance person or building superintendent, is better than nothing. As part of this inquiry into modifications, also ask about possible changes in building use or additions of rooftop equipment. For example, a change in building use might raise interior relative humidity, thereby increasing water-vapor migration into the roof system.

Next, interview the building superintendent about the roof system's history. Successful reroofing requires proper identification of leak sources. Wall leaks are sometimes mistaken for roof leaks, as is dripping condensation.

Leak-detection technique

A leak-detection interview proceeds as an elimination process, designed to eliminate suspects, first in big batches, then in smaller ones as you converge onto the solution. For example, if the building leaks *only* when it rains, that fact indicates a true exterior leak. But if it leaks at other times—e.g., after high-humidity occupancy, followed by cold weather—the "leaks" could consist of condensation dripping down from the insulation. Leaks when the wind is blowing from certain directions can be identified as roof flashing leaks, or sometimes wall leaks, where wall flashing is defective or even totally omitted. Persistent leakage irrespective of rainfall may indicate water-saturated insulation, especially prevalent under ponded membranes.

When it is necessary to link a leak to a definite roof defect, local flood testing may be required. A hose carried up to the rooftop, or simply a bucket or two of water poured on a suspected flashing leak source, can simulate rain in order to check the interior for leakage. Note horizontal offsets from the leak source in prefabricated deck units. Steel decks will generally leak at end laps, not side laps, where the flute forms a trough capable of retaining substantial amounts of water.

Visual inspection and analysis

Interior inspection of the deck soffit through removal of a ceiling panel (or other access mode) should seek the following information:

Steel.   Rusting? Differential deflection at side or end laps? Excessive deformation? Sound welds? Do rooftop components—HVAC, other equipment, access hatches, and so forth—have their own structural angle supports?

Wood.   Rotting? Warped? Shrunk? Excessive joint gaps? Unanchored?

Structural concrete.   Cracks over $\frac{1}{8}$ in.? Excessive deflection in evidence?

Precast concrete.   Excessive joint gaps? Differential deflection at adjacent units?

Poured gypsum.   Excessive deflection of subpurlin bulb tees? Cracking? Evidence of excess moisture?

Corrugated steel supporting lightweight insulating concrete. Underside venting slots or side laps? Efflorescence on steel? (Make note to check deck surface carefully during top-side inspection.)

Structural wood fiber.   Excessive deflection? Differential deflection between adjacent units? Excessive joint gaps?

During this interior inspection, generally applicable to all deck types, note also the following:

- Evidence of foundation settlement (bearing walls' cracks and so on).
- Changes in deck type or span direction. (These lines should have an expansion-contraction joint.)
- Drain locations and drain leader accessibility.
- Location of rooftop HVAC units and their supply ducts or chiller pipes.

- Areas of most severe leakage (or condensation drippage).

Roof-surface inspection should start with a general observation of ponding. Obviously, the best time to observe the roof's ponding is shortly after rainfall. But the sedimentary deposits of darkened areas (algae in warm climates) indicate ponding. And the building superintendent or maintenance person can also report on ponding.

In your visual assessment of ponding, note particularly the following:

- Evidence of ponding along parapet walls, gravel-stop edges, expansion joints, penthouses, HVAC supports, and other flashed components.
- Location of drains. (Are they, in accordance with the generally diabolical law governing drain elevations, at the roof's high points? Are they blocked or constricted?)
- General flashing-height elevations.

These observations can give you a quick, general impression of the scope of the problem: whether you can economically correct the roof's drainage problem, raising flashed components out of low spots and so forth.

Your attention should then turn to these specific items:

- Damaged, unadhered, wrinkled, or deteriorated flashings?
- Splits in membrane at gravel-stop strips?
- Damaged, inadequately sealed counterflashings?
- Membrane blisters, splits, ridges, wrinkles, shrinkage, chalking, crazing?
- Flashing conditions at HVAC units, vents, skylights, parapets, hatches, and so forth?
- Pitch-pan condition?
- Drains—intact lead seal: clamping bolts tight?

Field-test cuts

Have sample roof test cuts taken down to the deck for (1) visual field inspection and (2) laboratory analysis, if required.

A visual inspection should reveal the following information:

- How many membranes?

- Condition of deck—e.g., has lightweight insulating concrete suffered freeze-thaw disintegration? Is it wet?
- Is there a vapor retarder?
- Is insulation well-adhered to deck (or to vapor retarder)?
- Type, condition (wet or deteriorated), thickness of insulation?

Laboratory analysis of sample test cuts in bituminous systems can refine and confirm your visual observations. From a 40×6 in. sample cut across the felts, you can get a laboratory determination of the number of plies, laps, and so on for better assessment of the existing membrane's (or membranes') condition. From core samples, the laboratory can give you moisture contents of lightweight insulating concrete fill or other insulating material, or even the membrane. Photographs can provide a permanent visual record of conditions.

Large-scale moisture surveys

On projects of sufficient size to justify the expense, recent advances in nondestructive techniques for locating wet insulation in membrane roof systems have made accurate surveys of large roof areas economically practicable (see Chapter 18). Regardless of the nondestructive testing technique—nuclear, capacitance, or infrared—core samples are required to verify the results. By coordinating data from nondestructive roof surveys over large areas with core samples and construction data, you can get an accurate picture of the moisture content over the entire roof area.

Assessment for Remedial Action

With the results of your investigation and test-cut sample reports, you should now have a general idea of what is economically practicable and technically feasible.

Tearoff-replacement is indicated if your investigation has turned up any of the following:

- Extensive ponding
- Deteriorated deck
- Wet and/or deteriorated insulation
- Poor anchorage between deck and insulation (or other roof-system component) and no practicable way to mechanically anchor them; i.e., the deck is structural concrete (cast-in-place or precast)
- An essentially irreparable membrane surface—blistered, wrinkled,

and deteriorated beyond economically practicable repair into a smooth surface for re-covering

Re-covering the existing membrane may be practicable if the scope of the foregoing problem (or problems) is restricted to minor areas. If there are only a few rotted plywood panels, they can be replaced and reinsulated, and the remaining roof can be re-covered. If a full-scale moisture survey indicates wet insulation in 10 percent of the roof area, that segment can be removed and replaced and the remaining roof covered. Determining the break-even point for economical partial replacement vs. total tearoff is a complex question to be answered for each specific project. But whenever you start approaching one-quarter removal of random areas of wet insulation or membrane, you are probably moving close to an economic decision for total tearoff-replacement.

New performance criteria

Most reroofing projects require a decision about the practicability of three major improvements:

- Additional thermal-insulating quality
- Providing slope for positive drainage
- Meeting code-mandated snow-load provisions

If the rooftop geometry permits the raising of flashing heights, you can usually provide slope for positive drainage through the addition of tapered insulation or lightweight fill, or by adding a new structural framing and metal roof system.

Most roofs that leak water leak heating and cooling energy as well, and the vast majority of existing roofs, like walls, are drastically underinsulated. Consider the benefits of double-layered insulation.

Other improvements you may consider, especially in association with improved drainage, are the practicality of

- Relocating or remounting rooftop equipment
- Reconstructing (or adding) expansion joints, where advisable
- Changing the roof surface
- Adding new drains (or scuppers)

This last-cited change may not be a voluntary option. Building codes now require double drains as a safeguard against blockage of primary drains. Overflow drains should connect to drain lines independent of the primary drain lines. (The Uniform Building Code requires installation of overflow drains with the inlet flow located 2 in. above the roof's low point, or overflow scuppers three times the cross-sectional area of the roof drains in adjacent parapet walls, with the inlet flow line 2 in. above the low point of the adjacent roof.)

If the existing membrane has aggregate surfacing, it may be wise to change to a smooth surface, or a mineral surface for modified-bitumen membranes, for industrial buildings. If cement, food, or other material is regularly discharged on the roof, a smooth-surfaced roof permits easier removal than an aggregate-surfaced or loose-ballasted roof. However, remember that a smooth-surfaced membrane is a *water-shedding* rather than a *water-resistant* membrane, requiring a minimum $\frac{1}{4}$-in. slope. Consider adding wood walkways or special mineral-surfaced treads on paths that get the heaviest roof traffic (around HVAC units and so on).

When the existing slope is inadequate, it is seldom practicable to achieve $\frac{1}{2}$-in. slope, or even the recommended $\frac{1}{4}$-in. minimum for aggregate-surfaced roofs. (With a drain just 24 ft away, it takes at least 7-in.-thick insulation to provide for $\frac{1}{4}$-in. slope: 6-in. allowance for slope, plus 1-in. minimum thickness at or near the drain.) You are lucky to get $\frac{1}{8}$-in. slope with tapered insulation, and often you would settle for $\frac{1}{16}$ in. And because most rooftop components are located without the slightest thought of placing them clear of low spots (much less at high spots, where they should be), flashed components often sit within 2 ft or less of a drain, thus ensuring that they will always be exposed to the hazards of ponded or flowing rainwater.

As a possible solution when it is impracticable to use tapered insulation to provide slope, consider the following alternatives:

- Drainage via (1) an electronically controlled siphon system, (2) solar-powered siphons, or (3) additional drains
- Use of a coal-tar-pitch membrane

Reroofing Specifications

The architect should prepare detailed specifications and drawings for a reroofing project. The specifications should contain the following:

- Time limit for job completion
- Acceptable membranes (and surfacing)
- Acceptable insulation (and required thermal resistance)

- Provision for unit costs for insulation and/or deck replacement
- Provision for contractor proposals on alternative details
- Responsibility for disposal of debris and waste material

Isolating the new membrane

On re-covering projects, never adhere the new membrane to the existing membrane; always isolate it with either a venting base sheet or porous insulation (e.g., fiberglass). Venting base sheet is preferably nailed, but if nailing is impracticable, specify spot mopping. For a deck that will take mechanical fasteners, you can slash venting holes through the existing membrane, mechanically anchor a felt slip sheet and, on top of the felt, a venting base sheet or porous insulation, and then apply new insulation in hot-mopped Type III asphalt. As an alternative, you could specify a ballasted or mechanically fastened single-ply system.

For re-covering projects in areas with high vapor-pressure gradient, slash holes in the old membrane, so that it vents upward-migrating water vapor to the venting base sheet or porous insulation (see Chapter 6, "Vapor Control" and Fig. 19.2).

Make sure that the venting base sheet is above the dew-point temperature. (See Chapter 5, "Thermal Insulation," for the computation technique for locating the dew point in the roof cross section.)

Flashings

For all reroofing projects—re-covering as well as tearoff-replacement—replace flashings and gravel stops, for several reasons:

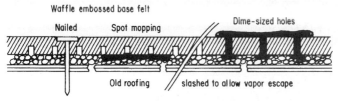

Figure 19.2 On re-covering projects, with a high vapor-pressure gradient from a humid interior, holes are sometimes slashed in the old membrane remaining in place, to let it vent. Diagram shows three different techniques for anchoring the base sheet, with its waffle-grooved vent ducts: nailing, the most dependable technique (left); spot mopping (center); dime-sized, grid-patterned holes (right), through which hot asphalt flows to adhere the base felt to the existing membrane below. (*R. L. Fricklas, Roofing Industry Educational Institute.*)

- Flashings generally deteriorate faster than the membrane.
- Edge details and flashings are a major source of water entry.
- Roof-assembly-thickness change requires new wood nailer edge strips (thicker for overall tapered insulation or merely for tapered fiberboard strips designed to raise edges and flashed components out of water).

For re-covering projects, pressure-relief vents are sometimes recommended. If they have any utility, it would be where insulation is sandwiched between two impermeable membranes—i.e., the new and original membranes (see Fig. 19.3).

Surface preparation of existing membrane

For re-covering projects, require a smoothly prepared, clean surface on the existing membrane for application of the new system, via the following steps:

- Power brooming and vacuuming to remove all loose aggregate from the surface
- Cutting out of ridges, blisters, loose felts, and other surface projections
- Airblasting or vacuuming to remove dirt, dust, and other loose materials
- Application of asphalt cutback primer for surfaces designed to receive hot-mopped asphalt (e.g., for application of a layer of porous insulation)

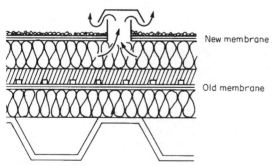

Figure 19.3 Re-covering projects sometimes incorporate pressure-relief vents for new insulation sandwiched between old and new membranes. (*Roofing Industry Educational Institute.*)

Existing aggregate-surfaced membranes require special attention in the specifications. Highlight the roofer's responsibility to keep this aggregate out of the new membrane. Aggregate sticks to the bitumen coating on workers' shoes and equipment wheels. Without special precautions by the roofing contractor, the old aggregate will be tracked into areas of new roof application and get trapped in the new roof system: at the deck-insulation interface, at the insulation-membrane interface, and between felt plies. Entrapped aggregate punctures membranes and creates voids, thus posing a tremendous threat to membrane integrity. Warn the contractor in the specifications that the bid price must include the cost of necessary precautions to prevent aggregate entrapment. These precautions may include:

- Laying plywood walkways over old work areas to prevent workers and equipment from picking up aggregate
- Establishing "aggregate-removal" stations to keep workers and equipment "uncontaminated" when they enter new roof-application areas
- Arranging work patterns to avoid aggregate contamination
- Requiring special gravel-vacuuming equipment to ensure proper removal of aggregate from areas to be reroofed (see Fig. 19.4).

Alerts

For replacement bituminous roof membranes

1. Favor tearoff replacement over re-covering if the existing roof system fails to satisfy (and cannot be repaired to satisfy) the following requirements:

    - Sound structural deck
    - Secure insulation-to-deck anchorage
    - Reasonably dry, firm insulation
    - Smooth membrane surface, suitable as substrate for new membrane

2. Consider upgrading of remedial roof system for (*a*) improved thermal performance and (*b*) improved drainage. (If it is impracticable to provide slope through tapered insulation or fill, consider other methods of drainage: electronically controlled siphon or solar-powered siphon systems.)

500   Chapter Nineteen

Figure 19.4 Field vacuuming of an old gravel-surfaced BUR membrane removes much dirt and debris as well as loose aggregate and pulverized bitumen. Such a surface may be suitable for sprayed-in-place polyurethane foam or re-cover board. (*Roofing Industry Educational Institute.*)

3. Check for conformance of the new system with Factory Mutual and Underwriters' Laboratories requirements (which may be more stringent than when the original roof system was applied).

4. Favor mechanical fastening for steel decks in tearoff-replacement projects because of (*a*) more dependable anchorage for wind-uplift and membrane-splitting resistance and (*b*) risk of excessive hot-mopped bitumen added to bitumen possibly on deck from original application.

5. On all reroofing projects (re-covering and tearoff-replacement), require removal and replacement of all flashing.

6. On tearoff-replacement projects, require removal of old systems down to the deck. (Leaving an old vapor retarder in place on a steel deck can (*a*) obscure water in flutes and (*b*) obscure defective side lap or end lap fastening. See Fig. 19.5.)

7. On re-covering projects:

    - Always isolate new membrane from existing membrane with either a slip sheet or a porous insulation layer.

Figure 19.5 Tearoff-replacement of a roof system on a steel deck requires cleaning, metal patching, and wire brushing before installation of new roof system. (*Roofing Industry Educational Institute.*)

- Always provide for ventilation of the old system (to accommodate suspected and probable moisture migrating as vapor upward toward the new roof system).
- Always reanchor existing insulation suspected of being inadequately adhered to the deck. (If you cannot fulfill this condition, tearoff-replacement becomes your rational choice.)

8. Take special precautions to prevent entrapment of old aggregate from existing membrane in the new membrane.

For loose-laid synthetic single-ply membranes

1. Check the roof's structural framing and deck to ensure that it can carry the additional gravel or paver ballast load for loose-laid, ballasted roof systems.

2. Before specifying a synthetic single-ply system as a remedial membrane over an existing bituminous membrane substrate, check for the synthetic material's compatibility with asphalt or coal tar pitch.

3. Require the following preparation of an existing built-up roof system as a substrate for a new synthetic membrane:

   a. Remove *all* aggregate if the new membrane is applied directly on top of the existing membrane. (If a protective insulation board is applied to the existing membrane as substrate for the new membrane, remove only *loose* aggregate.)
   b. Repair defects in the existing membrane (blisters, ridges, splits, fishmouths, and so on).
   c. Apply new materials (protective insulation or membrane) only after the existing built-up substrate is clean, dry, smooth, and free of loose particles. (Consider application of a primer or a single-ply base sheet to improve the existing substrate.)
   d. For loose-laid synthetic membranes, require a smooth, firm substrate. (If protective insulation board is applied over existing roofs, it can be laid loose.)

For sprayed polyurethane foam (plus fluid-applied coating)

Substrate preparation (replacement roof after tearoff down to existing deck)
1. Remove moisture, grease, oil, loose particles, dust, and rust. Use special treatment—e.g., wire brush, commercial sandblasting, or chemical treatment—where required to prepare substrate for good adhesion.

2. Prime or seal all substrate surfaces to receive sprayed foam. Check primer manufacturer for the specific primer to apply to the substrate. Let the primer dry and cure, per manufacturer's directions. The primer must be dry to the touch before foam spraying starts. Avoid asphalt cutback primers because they contain high-boiling-point solvents that delay drying. Chlorinated rubber, which generally dries to a hard film within 30 min, is a generally suitable primer for structural decks (poured-in-place or precast, new or existing) and plywood. (Steel decks, with good factory coating, do not normally require a primer.)

Substrate preparation (reroofing over existing built-up membranes). To qualify as a satisfactory substrate for application of sprayed foam insulation, an existing built-up membrane must be essentially dry and well anchored. If investigation of the roof indicates only small, isolated areas with soft, wet insulation, remove these areas and replace with dry material. But do not risk placing a sprayed-foam system over wet insulation.

After investigating the roof system for entrapped moisture and adequate anchorage (taking test cuts, if necessary), proceed to prepare the existing membrane surface as follows:

1. Cut and patch all existing blisters, buckles, wrinkles, fishmouths, and soft-spot punctures. (Remove wet insulation from soft areas.)

2. *Anchor loose sections* to be sprayed.

3. Renail or otherwise secure all loose base and counterflashing, flanges, gravel stops, vent pipes, pitch pockets, and scuppers.

4. Repair membrane splits, removing gravel and cleaning an area 12 in. on each side of the split. Mop- or torch-apply modified bitumen membrane over split area.

5. Vacuum and power broom all loose aggregate from the surface.

6. Airblast the surface to remove dirt, dust, and other loose material. Repeat once or twice if required to ensure a clean substrate for foam application.

7. Apply a chlorinated rubber, if required by presence of residual dust remaining after airblasting.

Weather. Before starting foam-spraying operation, check the weather report to ensure high probability of the following weather conditions:

- Low probability of rain (limit area planned for spraying to applicator's ability to protect substrate from wetting in event of sudden rain)
- Air temperature 40°F or above
- Dewpoint 5°F or more below ambient temperature
- Maximum 12-mph wind (or shielding precautions)

Take precautions to prevent overspray and coating of surrounding buildings, automobiles, shrubs, and so forth.

Spray limits. Spray in minimum ½-in.-thick lifts. Allow a maximum 4-h period between lifts. Discoloration—progressing from yellow to orange rust, sometimes accompanied by powder formation—may occur if the foam surface remains exposed for a day or more. When such discoloration occurs, skin the foam surface and apply a new surface coating before applying either (1) another foam lift or (2) a membrane coating.

Chapter 20

Roof-System Specifications

Roof-system specifiers frequently take refuge in brief, vague roofing specifications that in effect simply require a 20-year guarantee by manufacturer X, or equal, in an attempt to shunt responsibility from themselves onto the roofing contractor or manufacturer. These ineffectual, incomplete specifications often lie at the root of a roof-system failure. Indiscriminate substitutions are allowed—an inferior, cheaper system for a superior, more expensive system—and from then on everything slides downhill, with the contending parties ultimately battling to avert legal responsibility for a leaking roof doomed before its application to premature failure.

"Roofing may be coal, tar, pitch, or specially processed asphalt," says one architect's roofing specification. Judging from this slovenly edited specification, you could expect a slovenly designed roof, and without disappointment. The school-building roof constructed from this design was not made of coal, but it lacked slope for positive drainage and had improperly flashed, flat skylights, only 2 in. above the roof and located near the maximum deflection areas of long-span metal roof deck, where ponded water soon leaked into the school classrooms.

A common specification states: "Roofs and flashing shall be a complete system for 20-year warranty as specified. However, manufacturers considered equal will be considered by the Architect for approval."

Skimpy membrane specifications have launched many roofing failures. The specifier who writes a specification so devoid of technical content is effectively abdicating responsibility. The brief, offhanded specification expresses the designer's disdainful attitude toward the roof. It is the kind of specification you usually find on prematurely failed roofs.

Writing specifications for contemporary roof systems is a daunting challenge of exponentially increasing complexity. The roof designer is confronted with a bewildering array of materials, systems, and details—with choices between hot-mopped asphalt or torching of modified-bitumen membranes; between welding, taping, or adhering of single-ply lap seams; between adhered, mechanically fastened, or ballasted wind-uplift resistance; among SPF, metal, PMR, or more conventional systems; among complex flashing details, levels of fire resistance, and a host of other aspects. There are some 500 systems listed in the *NRCA Roofing Materials Guide*. The specifier can seldom, if ever, use unedited standard specifications (or manufacturer's specifications). In architect Justin Henshell's words, "Every project has unique conditions impossible to cover in a standard specification."[1]

Specifications and Drawings

Both words and pictures are indispensable for describing a roof system. It is impracticable to convey an accurate description of roof construction details in words alone.

Ideally, the drawings show the *form* of the construction, whereas the specifications establish construction *quality*. Drawings show what is to be done; specifications show how it is to be accomplished. Specifications should amplify but not repeat information shown on drawings. The architect's drawings graphically portray design details, location, and dimensions. The specifications establish standards for material quality and work. To serve their complementary functions, drawings and specifications should dovetail like a snugly fitted, two-part jigsaw puzzle, with no overlaps and no gaps. When information is erroneously duplicated on drawings and specifications, possible conflicts can cause incorrect construction, higher costs—at the least, confusion and delay.

Drawing requirements

The drawings (minimum scale = $\frac{1}{4}$ in. for plans and $1\frac{1}{2}$ in. = 1 ft for details) should include the following:

1. Scope—location and types of roofing, change of roof levels on plans, elevations and details
2. Slopes and drains
3. Walkways

4. All penetrations—e.g., vent stacks, curbs, ventilators, skylights, mechanical equipment, scuttles
5. Details of drains, cants, openings, crickets, eaves, inside and outside corners
6. Details of base flashing, cap flashing, dunnage support flashing, expansion joints and transitions, vent stack flashing, pitch pockets
7. Wood nailers to receive flashing

Specification requirements

The specification should provide the following information:

1. Scope of the work: roof membrane, insulation, and vapor retarder should be combined in one section. Sheet metal and accessories are usually in separate sections.
2. Selection of a roof system compatible with a reputable manufacturer's published specification.
3. Coordination of the roofing specification with other relevant specifications.
4. Responsibilities of the different members of the construction team—owner, architect, manufacturer, general contractor, roofing and sheet metal contractor, mechanical contractor, plumber.
5. Description of vapor retarder, insulation, felt, bitumen, single-ply membrane, and flashing materials, with tests and standards for acceptance.
6. Maximum and minimum roof slopes (coordinated with deck specifications and structural drawings).
7. Methods of applying all components, with limiting conditions, e.g., heating of bitumen, tack time of adhesives.
8. Provisions for inspection; requirements for bitumen; anchorage requirements for vapor retarder, insulation, and membrane; T joints on single-ply systems.
9. References to (but not repetition of) mandatory provisions already stated in the general conditions.

Writing Style

Roofing specifications must be written in clear, concise prose. Here are several rules:

1. Favor short, simple sentence structure over long, complex structure.
2. Give directions, not suggestions, in the imperative mood whenever practicable.
3. Do not parrot legal jargon ("said" and "same" as identifying pronouns) or long-winded, all-purpose safety clauses copied from old specifications. (Cluttering a specification with needless words obscures the essential information and reduces chances that the specifications will be read.)
4. Avoid indefinite expressions—e.g., "reasonable," "best quality," "or equal."

Specifying Methods

There are two basic classes of specifications: *prescriptive* and *performance*. A prescriptive specification describes the means for achieving desired but normally unstated ends. On the other hand, a performance specification explicitly states the results to be achieved: e.g., a weathertight, fire-resistant, wind-resistant roof system with specified thermal resistance, with appropriate numbers and referenced test standards.

There are three varieties of prescriptive specifications: proprietary, reference standards, and descriptive. Normally, the specifier chooses between the first two, leaving descriptive specifications for exotic roofing projects.

Proprietary specifications are easily written, short, and precise. Their major disadvantage concerns the difficulty of defining "equal" products for substitutions. The notorious "or-equal" provision has spawned a host of roofing failures, entangled in a corresponding host of disputes or lawsuits, resulting from substitution of inferior roofing membrane materials for superior, originally specified products. Obviously, the specifier relying on a proprietary specification should know the product. If the contractor requests a substitution, the specifier should demand evidence that this substituted product will meet the same performance standards met by the specified product. Architects and owners have suffered in the past from substitution of inferior coated-felt membranes with two or three plies for four-ply, saturated-felt membranes.

Specifying by performance criteria is the ideal goal toward which specifications for roof systems and other buildings are evolving. But, as the discussion of performance standards in previous chapters has indicated, complete performance standards for roof systems remain a long way in the future. Performance standards for particular requirements—e.g., fire and wind resistance—are well established, and performance requirements for other attributes can be incorporated into specifications, in conjunction with reference standards. The specifier can thus call for a membrane that meets the performance requirements of a UL Class A roof covering, a minimum TSF of 100°F, maximum linear coefficient of thermal expansion-contraction of 40×10^{-6}°F, and other attributes listed in the "Preliminary Performance Criteria for Bituminous Membrane Roofing," *Build. Sci. Ser.* 55, NBS. Along with these performance criteria, the specifier can specify ASTM (or other) standards—e.g., Type I asphalt, per ASTM D312, and so on. And for the roof system as a whole, the specifier can incorporate the requirement of UL Class 90 wind-uplift resistance and a 1- or 2-h fire time-temperature rating, per ASTM standard furnace test E119.

Master Specification

MASTERSPEC, a national computerized master-specification system, places special emphasis on roofing because of the high incidence of roofing failures. MASTERSPEC follows the standard CSI format, a standardized procedure that not only reduces the probability of a specifier's omitting a required item but offers greater convenience to the contractor preparing a bid. CSI has "spectext," similarly computerized.

Of the two basic specifying methods in the MASTERSPEC system, Justin Henshell favors *broadscope* over *narrowscope,* because broadscope reduces the probability of conflict. The only justification for using narrowscope is when multiple low-slope roof systems are specified on the same project, according to Henshell. Few projects have more than two types of membrane roof systems. (Limiting these types is desirable to simplify future maintenance operations.)

Others, however, disagree with this advice, generally favoring narrowscope.

Division of Responsibility

When assigning responsibility, the specification writer should follow these general principles:

1. The *architect* bears ultimate responsibility for the design of all components in the roof system and its final inspection and acceptance.
2. The *manufacturer* (or manufacturers) of the various roofing components is responsible for furnishing materials that conform with

the specifications and for furnishing accurate technical data about the physical properties, chemical composition, and other information pertaining to the material and its compatibility with adjoining materials.

3. The *general contractor* is responsible for:
   a. Coordinating the work of the roofing contractor with that of other subcontractors penetrating, adjoining, or working on the roof
   b. Ensuring that the roofing subcontractor follows the specifications
   c. Ensuring the protection of stored roofing materials and components from moisture and other hazards before and after installation
   d. Providing walkways, where required, to protect the roof from traffic damage
   e. Installing a satisfactory deck
   f. Ensuring that all roof-penetrating elements and perimeter walls are in place and that perimeter nailing strips, cants, curbs, and similar accessories are installed *before roofing work starts*

4. The *roofing contractor* is responsible for:
   a. Following the specifications for all roof-system components: vapor retarder (if specified), insulation, membrane, and flashing. (If the roofing contractor finds any specified material or procedure impracticable or contrary to good roofing practice, the architect should be so informed. No material or procedural substitutions are acceptable without a formal change order.)
   b. Inspecting the roof deck surface and either (1) accepting it as suitable for application of the roofing system or (2) notifying the general contractor about deficiencies requiring correction before the roof-system application can proceed.
   c. Upon completion of work, giving to the general contractor an "alert" report recommending steps (e.g., walkway protection) that the general contractor should take to protect the roof system from abuse during the remainder of the project.

Temporary Roofing

Omitting specification provision for temporary roofing can create colossal roofing problems, particularly on projects requiring roof protection during cold winter weather. As discussed in Chapter 10, "Elements of Built-up Membranes," a temporary roof (two-ply built-up membrane or single-ply coated felt) is advisable when the job is hampered by any of the following:

- Mandatory in-the-dry work within the building before the weather permits safe application of a permanent roof system
- Prolonged rainy, snowy, or cold weather
- Necessity of storing building materials on the roof deck
- Large volume of work on the roof deck by trades other than the roofing contractor

The specification should require removal of temporary roofing (including insulation boards required to provide a substrate over steel deck). Some manufacturers permit repair of temporary roofs left in place as vapor retarders. But the safest rule on temporary roofs is to remove them, including their price in roofing contractors' bids. A vapor retarder subjected to the rigors of construction traffic and weathering before it is covered with insulation may lose its vapor-retarding ability.

Here is the Associated General Contractors (AGC)–National Roofing Contractors Association (NRCA) advice to remove a temporary roof:[2]

> In general no attempt should be made to retain a temporary roof as the base for the permanent roofing system. The fact that permanent roofing could not be properly installed at the time is evidence that the temporary roof is probably imperfect. Its imperfections may contribute to failure of the permanent roofing at a later date. Temporary roofing should be removed before installing permanent roofing. In general "phased construction" of roofing application cannot be recommended.

Special Requirements

Some roofing projects—notably tearoff-replacement—create special problems. When these special problems are anticipated, the specifier should make special note of them in the specification. For example, when an old aggregate-surfaced membrane is to be removed, a prudent specifier concerned about the potential damage to the new membrane (mopping voids and felt punctures) can insert into the specification the following clause:

> Take precautions required to prevent blowing or tracking of aggregate from existing membrane to be removed into new work areas where aggregate can be trapped within the new membrane. Assure that aggregate is not tracked into new work areas on workers' shoes or equipment wheels. Aggregate within the new membrane shall be cut out and the membrane patched as specified for test cuts.

Such clauses can answer possible claims by the contractor that it is impracticable to prevent the tracking of aggregate, that the cost of preventing it was not included in the bid, and that it does not make any difference anyhow. If the specification contains an explicit clause warning the contractor to beware of tracking aggregate, the discovery of entrapped aggregate in the new membrane will constitute an incontrovertible specification violation.

When other serious problems can be anticipated, the specifier should insert similar clauses highlighting these problems and forcing the contractor to pay attention to them. These clauses can also help the owner if the project winds up in litigation. Judges and juries, in their vast ignorance about roofing technology, need all the help the specifier can give in identifying good roofing practice.

Specifying New Roofing Products

In today's hazardous legal environment, architects specifying new products of any kind run heavy malpractice risks, and in the roofing area, these risks rise exponentially. Although it may be legally safer to stick with the old methods, the poor performance of the traditional practices spurs architects/engineers to try new materials and methods. Ultraconservatism in specifying materials is no answer for a designer who wants to maintain professional utility.

The ultimate answer to the design professional's dilemma in dealing with new products is the systems-building approach, with its more rational apportionment of responsibilities among designer, contractor, and manufacturer and its reliance on performance standards with practical testing methods and certification procedures attesting to the building subsystem's—e.g., roof system's—satisfactory performance. The advent of performance standards for roof systems is, however, probably a decade or two in the future. Meanwhile, the contemporary roof specifier must cope with today's roofing industry in its glorious fragmentation and divided responsibilities.

When specifying a new roof-system product, the architect should judge a new product guilty until proven innocent. It is the manufacturer's responsibility to establish the fitness of this product for its intended use, and it is the architect's responsibility to document research of this product every step of the way. From a longer list drawn up by attorney John S. Martel comes the following list of investigatory steps:

1. Obtain from the manufacturer a list of projects on which the product has been previously used, including the owner, architect/engi-

neer, and roofing contractor. By interviewing these principals, you can ascertain whether the conditions on their projects are similar to your project's.
2. Ask the manufacturer's representative about conditions in which the product is not recommended, about its failures as well as its successes. (This advice is not so naive as it might appear. Perhaps the manufacturer will not candidly inform you about the product's shortcomings or failures. But later, in possible litigation involving the product's failure, the architect/engineer may be questioned in cross-examination as to whether there was at least an attempt to get this information. An admission that there was not cannot help but weaken the defense.)
3. Notify the manufacturer in writing how you intend to use the product.
4. Request relevant technical data from the manufacturer. (If it cannot be supplied, you have, ipso facto, an excellent reason for not specifying the product.)
5. Insofar as practicable, investigate the manufacturer's past performance, seeking answers to such questions as these:
    a. Does the manufacturer warrant the product's performance?
    b. What is the company's past record for manufacturing and marketing reliable roofing?
    c. Has the manufacturer produced previous roofing products?

Before specifying a new product, the architect should inform the client (preferably in writing) about possible risks, as well as the advantages, of using the new product. If the client is unwilling to accept these risks, specify a more conventional material.

Quality Assurance

Though negligent design is responsible for a substantial minority of failures, defective materials and application together account for the majority of roof-system failures. Quality assurance thus poses a big problem for the specifier.

Henshell proposes the following general rules:

- Require a preroofing conference to establish mutual understanding of the specifications.
- Require ASTM and/or UL labels on materials, where practicable.
- Select single-ply manufacturers with proven track records for the specified system and material.
- Require certification, licensing, or approval of applicators by sin-

gle-ply manufacturers. (Refer to *NRCA Roofing Materials Guide* for applicator training offered by various manufacturers.)
- Refer to application tolerances for BUR and modified bitumen recommended by ARMA and NRCA.
- Require *continuous* inspection of BUR projects, plus prequalifying of contractors per AIA form A-305.

This last-cited requirement is necessary because BUR manufacturers are less likely than single-ply manufacturers to require training for licensing, according to Henshell. The benefit-cost ratio of continuous inspection on BUR projects is higher than the corresponding ratio for a manufacturer's warranty.[3]

The following excerpt from Henshell's specification for a building at Baltimore-Washington Airport shows how quality assurance can be handled for a modified-bitumen roof system.

1.03 QUALITY ASSURANCE

A. All work under this Section shall be performed by a single firm with a minimum of three years experience in the installation of not less than five SBS modified bitumen roofs. The foreman or crew chief must be able to read and communicate in English and be able to read construction drawings and specifications.

B. The manufacturer shall have been manufacturing and actively marketing the specified roof system in the United States for a minimum of ten years.

C. All materials shall be in manufacturer's unopened packages, wrappings or containers, clearly labeled with all pertinent information. Labels on uncured materials shall include date of manufacture, shelf life and open time.

D. Materials improperly stored or become wet, warped or damaged shall be identified, conspicuously marked as rejected and removed from the job site.

E. The membrane manufacturer shall provide a field advisor for a minimum of 16 working hours. He shall be certified in writing by the manufacturer to be technically qualified in design, installation, and servicing of the required products. Personnel involved solely in sales do not qualify. The field advisor shall be present at the beginning of the actual membrane installation to render technical assistance to the Contractor regarding installation procedures of the system and answer questions that may arise.

On most built-up membrane projects there is no attempt to check conformance of the materials with the specifications, despite this membrane's unusually high dependence on material quality and proper application. Checking of materials should start with laboratory analysis of the bitumen, to make sure that the specified material is used. Material for test sampling should be drawn from the kettle and sent to a laboratory for testing. If the tested asphalt violates the specification, the owner should order the roof torn off and replaced.

Such draconian measures might be avoided by requiring that cartons be labeled, and that tankers provide certified bills of lading. If doubt remains, test cuts can be ordered, per ASTM D3617, and analyzed.

The need for such rigorous treatment of BUR membranes stems from the widespread practice of substituting Type III for Type I asphalt in the flood coat of low-sloped roofs. Use of Type III asphalt for the entire roof, flood coat as well as interply moppings, is convenient for the roofer; it is common practice in some regions, notably Florida. But Type III asphalt lacks Type I asphalt's cold-flow property and durability. Thus its substitution can shorten a membrane's service life. (See Chapter 10, "Elements of Built-up Membranes," for further discussion of the deleterious effects of Type III asphalt on built-up roof construction.) On large projects the architect may want to check other properties—e.g., penetration, ductility—to ensure conformance of specified asphalt with ASTM D312 provisions.

Coordination

Lack of coordination among the three design disciplines—architect (or roof consultant), structural engineer, and mechanical engineer—is at the root of some roof failures. Conflicts often occur when the sections on metal decks or precast concrete tee framing (the structural engineer's province) or the sections on fans, vent stacks, and roof drains (the mechanical engineer's province) are written by outside consultants and not coordinated by the roof specifier (architect or roof consultant).

As an example of the kind of errors that can occur, the structural engineer often specifies a sump pan for a heavily insulated metal deck. A sump is required only for insulation thickness of less than 1 in., or where gypsum board substrates are provided under PMRs. Since it is difficult to create a smooth transition between the roof system and the sump pan, the sump pan is often filled with insulation.

The roof specifier must also notify the structural engineer about wind-uplift requirements. Deck welds must be spaced at 6 in. (rather than the usual 12 in.) at the perimeter to meet FM Zone 2 wind criteria.

Problems with roof drains can result from the roof specifier's failure to work with the mechanical engineer. Cast-iron drains can promote ponding on projects with below-deck insulation or PMRs. Cast-iron drain flanges project more than $3/8$ in. above the deck when the drain is dropped into a sleeve in the deck. The resultant ponding around the drain occurs at the worst possible place, at a vulnerable flashed joint. (Ponding is less serious in the membrane field.)

Test Cuts

Some roofing consultants periodically take test cuts from BUR membranes to determine approximate quantities of roof-system components and to check for membrane quality—presence of entrapped moisture, interply mopping voids, bare felt spots, and so on. Test samples taken during roofing-application operations are examined and weighed in the field and replaced if practicable, and the roof is repaired with at least an equal number of felts. These samples are normally taken *before* flood coating and aggregate surfacing are applied, per ASTM Standard D3617, "Standard Recommended Practice for Sampling and Analysis of New Built-up Roof Membranes." This standard recommended at least one specimen per roof, plus one per 400 m^2 (4300 ft^2).

Field-test sampling shows whether the roofer's application is meeting specified requirements: e.g., *average* interply mopping weight of ± 15 percent of specified weight. If mopping weights are found to be out of tolerance or voids are discovered in the interply moppings, these findings signal a warning to both the roofer and the owner that the system has built-in potential trouble. The field report is a helpful document to the owner in the event of future litigation, a seldom improbable destiny for a roofing project.

Roof test cuts are opposed by some roofing contractors. Test cuts weaken the membrane, and their results are often misinterpreted by ignorant owners or architects who demand harmfully overweight bitumen moppings and force contractors to apply excessive bitumen. So go the arguments against test cuts during application.

These two criticisms are easily answered. Small test cuts can be replaced and patched. And the solution to erroneous interpretation of test cuts is not abolition of test cuts, but correct interpretation. Critics of test cuts generally advocate a "trust-us" policy that emphasizes experience in observing application operations as the best guide to a good membrane. Experience *plus* test-cut findings are better than experience alone.

Opposition to test cuts extends to the post-construction phase, when test samples are often needed to determine the quality of work or materials. Obscurantist lawyers typically ask whether it is not possible that the roof is better than the test samples indicate. That is, of course, a possibility. It is also possible that the roof is *worse* than the samples indicate.

The basic argument against the validity of random test sampling focuses on the statistical validity of extrapolating from, say, four 1-ft^2 test cuts on a 100,000-ft^2 roof. How, ask these critics, can you infer the quality based on a test sampling of only 0.004 percent of the roof's surface area?

To answer this question, I present the following statistical example. Assume that you find four test samples with no adhesion of insulation to deck, and that a satisfactory roof system must have a least 90 percent satisfactory samples (thus indicating that 90 percent of the roof is well adhered). Under these conditions, the probability that you would get four unsatisfactory (and no satisfactory) samples = $(1-0.9)^4 = 0.0001$. In other words, the odds are about 10,000 to 1 against such a combination. They are, moreover, about 270 to 1 against finding three or more bad samples and 18 to 1 against finding two or more bad samples out of four in a roof where, as postulated, 90 percent of the samples are satisfactory.*

Moral: You do not have to test a major portion of a roof to make virtually certain inferences about its general condition.

For test cuts through existing roof systems, you can have a laboratory analysis of several properties of both insulation and membrane. A laboratory-analyzed test cut can yield the following information:

- Number of felt plies
- Weight and defects in interply bituminous moppings (e.g., mopping voids that grow into blisters)
- Evidence of entrapped foreign materials—e.g., pieces of aggregate from a previously removed membrane on a tearoff-replacement project)
- Flood-coat weight
- Quantity of embedded aggregate

Test-cut size presents another decision. A 12×12 in. test cut, a convenient size for easy packaging in a plastic bag, is normally sufficient. With a 6-in.-wide×40-in.-long sample, cut across the 36-in.-wide felt widths, you can check for correct 2-in. felt laps and lap exposures. But poor felt laps constitute a relatively uncommon violation of good roofing practice. The greater convenience of the smaller samples usually

---

*The probability of two or more bad samples is calculated by the binomial law as the sum of the coefficients for the terms containing p^2, p^3, and p^4, representing, respectively, 2, 3, and 4 bad samples in which $p = 0.1$ is the probability of a bad sample and $(1-p) = 0.9$ represents the probability of a good sample. Then

$$P = (6 \times 0.9^2 \times 0.1^2) + (4 \times 0.9 \times 0.1^3) + 0.1^4$$
$$= 0.0486 + 0.0036 + 0.0001$$
$$= 0.0523$$

justifies the 12×12 in. size. Moreover, a 12×12 in. sample also can be analyzed for number of felt plies, material, and weight—everything, in fact, except the felt laps.

Alerts

General

1. Specify roofing materials and components preferably to conform with ASTM standards or federal specifications as a minimum, with any additional requirements appended to the selected standards. Using standards from a single source simplifies the task of checking specifications and promotes consistency. Make sure that the specifications are understood and available for reference.

2. Avoid duplication of material properties covered by specifications, e.g., "aggregate shall be $\frac{1}{4}$ to $\frac{5}{8}$ in. in size, clean, and free from dust and foreign matter." Instead, require conformance with ASTM Specification D1863.

3. Check other specification sections for inclusion of information necessary for other roof components—e.g., Section 0610, "Rough Carpentry." Include wood nailers in the roofing section for single-ply membranes and for anchorage of metal flanges and devices requiring connection with bituminous flashing.
   Specify the *furnishing* of sheet metal elements in Section 07600. Specify their *installation* in the roofing section.

4. Require the manufacturer of flashing materials to approve compatibility with roofing materials in coefficient of thermal expansion, adhesion, elasticity, resistance to sunlight, and other relevant properties.

New products

1. Commit the manufacturer to a specification—either its own or *written* approval of a specification written by the architect/engineer. (This procedure helps establish an implied warranty of the product's suitability, undermining the manufacturer's possible argument that the product was improperly used.)

2. Include in the general or supplemental conditions a clause requiring the roofing contractor to become familiar with all specified products and, if the contractor disapproves of any, to submit objections in writing.

General field

1. Require a joint job inspection and conference by the roofing contractor, general contractor, architect's representative, manufacturer's representative, and building inspector before application starts.

2. Require the general contractor to have perimeter walls and roof-penetrating building components in place *before roofing work starts*. (Installation of these elements after roofing work starts requires needless patching and repairs that multiply the chances for leaks and other modes of premature failure.)

3. Require the roofing contractor to approve the deck substrate as satisfactory *before* work is started. Do not, however, make the roof contractor responsible for items beyond his or her control—e.g., for checking the adequacy of steel deck welds.

4. Require the general contractor to give the architect ample notice of intent to start roofing operations (usually 2 days or more).

5. Require the general contractor to coordinate the work of the roofing subcontractor, mechanical subcontractor, or other subcontractors, with each assuming full responsibility for any damage that may be inflicted on another's work.

6. Require subcontractors installing rooftop equipment to notify the architect and general contractor of any damage they cause to the membrane. Require offending subcontractors to finance the necessary repairs made by the roofing subcontractor.

7. Set tolerances for application in accordance with relevant association documents: e.g., ARMA, *BUR Systems Design Guide*; NRCA/ARMA, *Quality Control Recommendations for Polymer-Modified Bitumen Roofing*; and NRCA, *Quality Control in the Application of Built-up Roofing*.

Technical

For technical recommendations for the various roofing components, consult the Alerts in the various technical chapters: Chapter 4, "Structural Deck," Chapter 5, "Thermal Insulation," Chapter 6, "Vapor Control," Chapter 10, "Elements of Built-up Membranes," and so forth.

As a guide to writing the specification, use the specification work sheets provided by the AIA or CSI.

References

1. Justin Henshell, "Specifying Membrane Roofing: A Systematic Approach," *The Construction Specifier*, November 1987, p. 102.
2. AGC-NRCA, *Roofing Highlights*, p. 8.
3. Justin Henshell, "Specifying Membrane Roofing: A Systematic Approach," *The Construction Specifier*, November 1987.

Chapter

21

Roofing Guarantees and Warranties

The comparatively high probability of premature roof failure makes financial responsibility for repairing or replacing defective roof systems a major concern for building owners. This financial responsibility can be covered via one or more of the following:

- Manufacturer's system (or membrane) warranty: for 5 to 20 years
- Manufacturer's material guarantee: usually for 5 to 10 years
- Roofer's guarantee: for 1, 2, or, less frequently, 5 years

Roofing bonds (backed by a surety) have disappeared, but they are discussed first because of their historical importance as the earliest method of contracting for financial responsibility. By the early 1980s, roofer's and manufacturer's guarantees or warranties had virtually eliminated roofing bonds. By the 1990s, roofing bonds were extinct.

Manufacturer's Bond

Historically, the manufacturer's bond appeared in response to a challenge created within the roofing industry. In the early days of built-up roofing, the manufacturer doubled as roof applicator and exercised total control over the whole roof construction process—from factory production in the factory to field installation. As the volume of built-up roofing increased during the latter part of the nineteenth century and the early part of the twentieth, manufacturers found their dual role as fabricator-applicator economically impractical. Thus was born the independent roofing contractor, and what was once an integrated responsibility was split in two.

Under the new arrangement, roofing quality frequently declined. Lured by the newly opened opportunities, inexperienced (and often unethical) roofers entered the business and lowered the previous standards set by the manufacturer-applicators. These new roofers sometimes moved from place to place, leaving behind a trail of leaking roofs.

In response to growing roofing troubles, materials manufacturers established standards for manufacturing quality, membrane design, and field practices. In 1905, the old Barrett Company originated the specification roof, with its prescription for number of plies, quantity of bitumen, and best application procedures. In 1916, Barrett established a network of approved roofers and, under prescribed conditions, began guaranteeing roofs applied by these roofers. This original roofing bond guaranteed the built-up membrane for 10 years against leaks attributable to material failure or faulty application.

Roofing bonds were sometimes mistakenly considered insurance policies for roofs in case of total failure. This gross error could have been corrected simply by reading the bond's conditions. Moreover, under a manufacturer's *bond* (as opposed to a *guarantee*), the top coverage available in 1980 was about $60/square, a minor fraction of the cost of tearoff-replacement, which could have exceeded $350/square.

System warranties with *no* monetary limits were available in 1995.

The mechanics of the manufacturer's bond was as follows: The roofing materials manufacturer issued the bond, which was backed by a surety company pledged to assume the manufacturer's potential liability for a failed roof. The bond was designed to assure the owner that:

1. The manufacturer's materials were used.
2. The membrane and base flashing (if included in a supplementary flashing bond) were installed by a manufacturer-approved roofer.
3. The manufacturer's representative had inspected the installation, during application and after completion.

Especially during periods of high inflation, like the early 1980s, a roofing bond was an extremely poor investment. Consider the following example: a 20-year manufacturer's bond with $10/square liability limit, $6/square premium. Invested at 12 percent compound interest, this $6 premium would take less than 5 years to exceed the $10 liability limit.* Thus a 12 percent interest rate made this roofing bond a poor investment unless you collected the full amount before the roof reached its fifth year. If you owned 100 roofs, you would have to recover payment on the 100 bonds' total liability limit at an average rate of 4.5 years for the bond premium investment to prove more profitable than investing the total premium at 12 percent interest. This example is an extreme one, but it nonetheless shows the need for rational economic analysis. A 5-year bond with unlimited liability at a slightly larger premium made far more economic sense than the type of bond just analyzed.

The higher liability limits—unlimited in some instances—of manufacturer's guarantees generally offer better protection than the older bonds. Most roofs that perform satisfactorily throughout the first 5 years of service life continue to do so for many more years. Thus the guarantees' shorter period, coupled with higher liability, focuses the owner's protection on the period when it is most needed. The typical 5-year guarantee with a renewable option for another 5, for a total of 10, should normally offer adequate protection. Some guarantees run as long as 20 years.

Note too that the second half of a 20-year guarantee is worth far less than the first half (which is equivalent to a 10-year guarantee). On a present-worth basis, the only rational way to calculate such value, a 10-year guarantee for a $20 liability limit carries a $6.44 present-worth value at the end of its 10-year period at 12 percent interest. At the end of 20 years, this value drops to $2.07. Fixed long term liability limits grow increasingly unattractive when money depreciates at a rapid rate.

Like roofing bonds, warranties have their financial absurdities, too (generally with a heads-I-win-tails-you-lose bias favoring the warrantor). One EPDM manufacturer's warranty agrees to fix leaks (subject, of course, to a lot of exclusions) for a fee of 3 cents psf for a 10-year warranty period. For 6 cents psf, you get a 15-year warranty. You pay *twice* the fee to extend the warranty by a mere 50 percent. In view of industry experience—i.e., that trouble usually occurs early in a roof project—this extended warranty represents an absurd cost to the

---

* $6(1 + 0.12)^n = 10$

$1.12^n = 10/6 = 1.667$

$\log 1.12^n = \log 1.667$

$n = \dfrac{\log 1.667}{\log 1.12} = 4.5 \text{ years}$

owner. Anything promised in the future, especially the *remote* future, should be discounted. Moreover, if problems occur on a roof older than 10 years, you are probably looking at replacement rather than leak repair, which is all the warranty covers.

Manufacturer's Warranty

A manufacturer's warranty, backed by the company itself instead of a surety company, has replaced the old roofing bond. These warranties may be satisfactory when dealing with manufacturers that have been in business for decades or those with vast financial resources. But to rely on the permanence of newer, smaller companies in this volatile industry, which has seen tumultuous changes in the past decade, is something else. The value of a guarantee or warranty should be assessed in hard-headed *financial* terms, not in *psychological* terms. (The old roofing bond lingered like a dinosaur who refused to become extinct because of the psychological comfort it afforded diffident specifiers. They sometimes inserted the phrase "20-year bondable roof" in their specifications even when no bond was purchased, in a vague attempt to cover themselves as having specified a high-quality roof membrane. Some specifiers even urged owners to take out bonds despite their recognized poor coverage.)

Like a manufacturer's bond, a warranty establishes a direct contractual relationship between owner and manufacturer, irrespective of the involvement of the architect, general contractor, and roofer in its procurement. The warranty itself is the contract document defining the parties' rights and responsibilities. The dictionary definition of *warranty*—"a written guarantee of the integrity of a product and of the maker's responsibility for the repair or replacement of defective parts"—is more ideal than real, according to attorney Peter Goetz, of the New York City law firm Goetz, Fitzpatrick & Flynn, specialists in construction law. A warranty is neither an insurance policy nor a maintenance contract, warns Goetz.

Licensed applicator agreements represent the manufacturers' attempts to assure quality in the installation of their products. Approved roofing contractors are licensed, via the applicator agreement with the manufacturer, after completing training programs designed to assure their competence in installing the manufacturer's roof system. Licensed applicator programs are especially important for exotic roof systems like sprayed polyurethane foam (SPF).

The licensed applicator agreement between manufacturer and roofing contractor is a critical legal document. In almost all cases, even where the roofing company does not issue its own guarantee to the owner, it is responsible for maintaining the roof for 2 years pursuant to the licensed applicator agreement, according to attorney Stephen M. Phillips of the Atlanta law firm Hendrick, Phillips, Schemm & Salzmann.

Analyzing the warranty

The NRCA's annually published *Roofing Materials Guide* provides an indispensable and voluminous guide to roughly 100 manufacturer's guarantees or warranties. It also discusses 27 aspects of these warranties. Though the publication's primary audience is roofing contractors, these items are of vital importance to owners. They *may* become of vital importance to the roof designer, roofing contractor, and general contractor as well if roofing problems occur.

Since many of these 27 items have a narrow technical focus (though they are by no means insignificant on that account), this discussion will focus on the more important items:

- Scope of coverage (material *and* application or material *only*)
- Monetary limits on remedy
- Exclusions (legal remedies and conditions)
- Nullification provisions
- Determination of warranty applicability

Scope of coverage is obviously of paramount importance. It can include materials only, exclusive of workmanship (i.e., application), or it can include both materials *and* workmanship. A warranty that excludes workmanship is probably not worth any premium charged for it. (Some warranties are "free.") On the other hand, a warranty including both materials and workmanship indicates a much greater faith in its system on the part of the manufacturer. In the event of a roof-system failure, pinpointing defective materials as the exclusive cause is much more difficult than attributing the failure to defective material *or* application.

Monetary limits on the remedy (usually limited to repair of leaks) are usually narrower when the scope of coverage is limited to materials only. In such instances, the monetary limit is the cost of replacing the defective material, *exclusive* of the labor cost of removing this defective material and installing new material. (This is another reason why material warranties are close to worthless.)

Monetary caps in material-and-workmanship warranties often limit the manufacturer's liability to the original cost of the installed roof system. Since it is set in *nominal* (not *real,* or inflation-corrected) terms, a fixed liability limit is of constantly depreciating value. After

5 years at an average 3 percent inflation rate, a $100,000 original cost is worth only $86,260 (= $100,000/1.03^5).

Exclusions come under two classifications: *legal* and *technical*. NRCA item 9 warns about legal exclusions. Warranty documents frequently serve as a liability-limiting device, restricting the warrantor's liability to the narrow scope of warranty provisions. If, for example, the warranty is a so-called "exclusive" warranty, issued in lieu of all other warranties, the claimant may be barred from seeking recovery based on breach of any other warranties, express or implied. Adding insult to injury, an exclusive warranty may also seek to bar other legal remedies that would otherwise be available. It could, for example, preclude a claim based on other theories of liability—e.g., negligence or breach of contract. "Sole and exclusive remedy" is a phrase hazardous to the owner's interest.

Technical exclusions, listed under NRCA items 12 and 13, may include the following:

- Natural disasters and "acts of God" (lightning, tornadoes, etc.)
- Hailstone damage
- Abuse, misuse, vandalism
- Damage by structural failure (foundation settlement, excessive deflection)
- Failure of material not supplied by warrantor (e.g., metal accessories)
- Repairs or alterations made without warrantor's approval
- Change in building use without warrantor's prior approval
- Roof traffic
- Rooftop storage of materials
- Ponding

The wind-coverage exclusion, listed as separate NRCA item 13, may make the warranty totally worthless for wind-inflicted damage. NRCA references the Beaufort scale (see Fig. 21.1) in its questions to warrantors. Some manufacturers' warranties specifically exclude gales, strong gales, windstorms, hurricanes, and/or tornadoes as natural disasters not covered by the warranty. A warranty excluding gales (39- to 46-mph wind velocity) is obviously close to worthless in regard to wind damage. (Try proving that wind velocity never exceeded 39 mph!)

Determination of warranty applicability, NRCA item 11, is closely related to the exclusionary provisions. The phrase "manufacturer's

Beaufort Scale

Beaufort Number	International Description	Miles Per Hour	Specifications
0	calm	less than 1	calm; smoke rises vertically
1	light air	1-3	direction of wind shown by smoke but not by wind vanes
2	light breeze	4-7	wind felt on face; leaves rustle; ordinary vane moved by wind
3	gentle breeze	8-12	leaves and small twigs in constant motion; wind extends light flag
4	moderate breeze	13-18	raises dust and loose paper; small branches are moved
5	fresh breeze	19-24	small trees in leaf begin sway; crested wavelets form on inlet islands
6	strong breeze	25-31	large branches in motion; whistling heard in telegraph wires; umbrellas used with difficulty
7	moderate (or near) gale	32-38	whole trees in motion; inconvenience in walking
8	gale (or fresh gale)	39-46	breaks twigs off trees; generally impedes progress
9	strong gale	47-54	slight structural damage occurs
10	storm (or whole gale)	55-63	trees uprooted; considerable damage occurs
11	violent storm	64-72	accompanied by widespread damage
12	hurricane	73*-136	devastation occurs

*The U.S. uses 74 statute mph as the speed criterion for hurricane.

Figure 21.1 The Beaufort scale of wind velocities sometimes serves as a reference for wind-uplift provisions in roof-system warranties.

determination" indicates that the manufacturer reserves to itself the right to determine whether a leak is covered or not. Even if flatly erroneous, and not merely debatable, the manufacturer's determination may be binding, provided it was made in "good faith." If the questionnaire entry for this item is "neutral," then a neutral party would decide whether the warranty is applicable. Obviously, a prudent owner would count the phrase "manufacturer's determination" a potent negative for that manufacturer's warranty.

Nullification provisions, discussed under NRCA item 14, are the thermonuclear weapons in the warrantor's arsenal, whereas the previously discussed exclusions are merely conventional armaments. Unlike exclusions, which merely eliminate certain specific leaks from coverage (e.g., those from hailstone damage), nullification provisions can blast away the entire warranty protection.

Here are some clauses that may nullify all protection from a manufacturer's warranty:

- Repairs, alterations, or additions without manufacturer's prior approval
- Owner's failure to pay all bills for installation and materials
- Lack of inspection at time of application
- Failure to notify manufacturer of building ownership transfer within a required time period

- Assignment (i.e., transfer of ownership) of warranty without manufacturer's written approval
- Failure to properly maintain roof (or to follow manufacturer's instructions)
- Change in building use
- Owner's failure to repair leaks excluded by warranty
- Excessive roof traffic
- Use of roof as work deck
- Flooding of roof
- Defective application of manufacturer's materials (in materials-only warranty)
- Owner's failure to repair damaged roof within specified time limits, via approved applicator

Warranty checklist

Owners, architects, and others concerned with roof warranties can profitably follow attorney Peter Goetz's checklist as a guide to investigating a roof-system warranty. Here are its highlights:

1. Nature of warrantor

    - Is warrantor a manufacturer or a middleman marketer? (It is obviously better to deal directly with the manufacturer.)
    - Does warrantor have a fully staffed warranty department?
    - What is warrantor's record of warranty service?
    - Does warrantor have a fully funded warranty fund?
    - What is warrantor's company worth?

2. Type of warranty

    - Does it cover material *and* workmanship, or materials only? (This is the most important question, other factors being equal.)
    - Cost ceilings on repairs?
    - Leak remedies? repair? maintain in "watertight condition"?

3. Coverage

    - Leak remedies? repair? maintain in "watertight condition"?

- Covered components (membrane, insulation, fasteners, vapor retarder, flashing, counterflashing)?
- Consequential damages?

4. Exclusions

    - Beware of standard exclusions—e.g., "acts of God."
    - Definition of "ponded water," "substrate movement," "environmental fallout," and "gale-force wind" (require specific mph figure for wind damage).

5. Implied warranties

    - Is Uniform Commercial Code (UCC) excluded by warranty? (UCC exclusion is a significant negative factor for owners.)

6. Warranty costs? (Premium charged or "free"?)

7. Legal details

    - Deductible for warranty-covered repairs?
    - Can warrantor recover legal costs from owner?
    - Time limits on filing lawsuit?
    - Roofer required to be a warranty signatory?
    - Clearly defined dispute resolution procedure?

8. Roofer's duties to assure issuance of warranty

    - Install roof per specifications?
    - Changes approved in writing?
    - Complete manufacturer's punch list?

9. Owner's responsibility to maintain warranty in force

    - Know required maintenance procedures?
    - Notify warrantor of change in building use or ownership?

10. Procedure when roof leaks

    - Roofer to get financial commitment from warrantor?

A survey conducted by attorney Goetz of major building owners (each owning an average of 144 buildings) demonstrates the laxity of even presumably sophisticated owners in their investigation of warrantors. The majority of these surveyed owners (54 percent) failed to run a Dun & Bradstreet credit check on their warrantors. And a high percentage (44 percent) reported difficulty in forcing warrantors to honor their warranties.

This same survey nonetheless demonstrates a collectively high regard for warranties as a factor in the selection of roof systems. On an ascending scale of 1 to 5, the warranty scored an average of 4 as its importance factor. Among the reasons cited for this high rating, some owners see a warranty as a means of avoiding fly-by-night roofing contractors. Others see a warranty as a test of manufacturers' faith in their products.

Note, also, that the apparent paradox created by the large number of complaints vs. the great importance assigned to the warranty really betrays no inconsistency. What the survey does reveal is the building owners' failure to live up to their convictions—i.e., to follow through with a thorough investigation of a quality-assurance program that they implicitly admit requires such investigation. It's something like the electorate's demand for a balanced budget, reduced taxes, and higher government spending.

Roofer's Guarantee

The normal 1- or 2-year guarantee may supplement or, in some cases, replace a long term manufacturer's guarantee. Manufacturers often require a 2-year guarantee from the roofing subcontractor. After an 18-month inspection by the manufacturer, defects uncovered by this inspection must be corrected by the roofing subcontractor before the manufacturer's guarantee takes effect. Failure of the roofing subcontractor's guarantee to the manufacturer does not relieve the manufacturer of liability under the bond or guarantee.

The typical roofer's guarantee requires the roofer to repair leaks resulting "solely from faults or defects in workmanship applied by or through the roofer." It excludes the following: all damage attributable to lightning, windstorm, hailstorm, or other unusual phenomena of the elements; foundation settlement; failure or cracking of the roof deck; defects or failure of substrate; vapor condensation under the membrane; faulty construction of parapets, copings, chimneys, skylights, and so on; fire; or clogging of drains. Like the manufacturer's warranty or guarantee, this roofer's guarantee excludes liability for damage to building contents or other parts of the structure.

Occasionally, owners negotiate broader coverage and longer terms in contractors' guarantees—5 years on insulation, vapor retarders, and roof sumps as well as the roofing membrane.

Roof Maintenance Program

As a part of the specification requirements, complete and deliver to Owner the following proposal. The Proposal is conditioned on acceptance by the Owner within 30 calendar days of receipt of Proposal.

Form of Proposal

Project: Date of Roofing Completion:

Roof Section(s):   Roofing Contractor:

Owner: Name:

Address: Address:

Date of Proposal: Phone:

1. The undersigned Contractor agrees, beginning at the termination of the 2-year guarantee on the roofing installation covered in the construction contract, to provide the following services at the listed rates:

    a. Inspect the entire roof area twice each year; once during the months of April or May, and once during the months of September or October. Following each inspection visit, the contractor shall prepare and deliver to the Owner a Report of Condition, including recommendations for needed repair or maintenance work. Include an estimate of a not-to-exceed cost of recommended work.

    Cost of Inspection & Reports: $ Each.

    b. Upon authorization by the Owner, to complete the repair and maintenance work on a Time & Material basis as follows, such material and labor costs to include overhead, equipment, employee benefits, transportation, etc.

    Material: Cost plus   %

    Labor: Cost plus   %

2. Terms of the Agreement shall be 10 (or other) years, and shall be cancellable by either party upon providing written notice to the other party.

3. Payment for services included — Inspection & Reports and authorized repair/maintenance work — shall be upon completion and acceptance of the work. Terms shall be net 30 days from receipt of invoice.

Contractor: _____   Accepted:

        By: _____   Owner: _____

                                                                   By: _____

                                                                 Date: _____

Figure 21.2 Roof-maintenance program can be negotiated via a form like the above. (*The late A. L. "Pete" Simmons, Roofing Consultants, Inc.*)

Roof-Maintenance Program

Complementing the roofer's guarantee and starting with its expiration date, a roof-maintenance program can be negotiated between owner and roofing contractor (see Fig. 21.2). As part of the proposal for roof application, the contractor includes a proposal for semiannual (spring and fall) inspection of the roof, with a report to the owner on the roof's condition, plus recommendations for repair or maintenance work.

Glossary of Roofing-related Terms

Absorption Ability of a porous solid material to hold relatively large quantities of gases or liquid.

Addition polymerization Polymerization in which monomers are linked together without the splitting off of water or other simple molecules.

Adhesion The state in which two surfaces are held together by interfacial forces, which may consist of molecular forces or interlocking action or both. Measured in shear and peel modes.

Adhesive failure A separation of two bonded surfaces that occurs at the interface between the adhesive and the material being bonded.

Adsorption The adhesion of an extremely thin layer of molecules (of gases or liquids) to the surface of solids or liquids with which they are in contact.

Aggregate (1) Crushed stone, crushed slag, or water-worn gravel used for surfacing a built-up roof. (2) Any granular mineral material.

Aging The effect on materials of exposure to an environment for an interval of time. The process of exposing materials to an environment for an interval of time.

Air lance A device used to test, in the field, the integrity of field seams in plastic sheeting. It consists of a wand or tube through which compressed air is blown.

Aliphatic polyurethane coating Normally used as a topcoat, a material that contains a specific class of isocyanates based on a long, straight-chain molecular structure noted for its high tensile strength, high gloss, color-stable properties, and ultraviolet resistance.

Alkalinity The capacity of water to neutralize acids, a property imparted by the water's content of carbonates, bicarbonates, hydroxides, and occasionally borates, silicates, and phosphates. Expressed in milligrams of calcium carbonate equivalent per liter.

Alligatoring Shrinkage cracking of the surfacing bitumen on a built-up roof, producing a pattern similar to an alligator's hide. The cracks may or may not extend through the entire surfacing bitumen thickness.

Alloy, metallic A material that has metallic properties, composed of two or more chemical elements, at least one always being metal.

Alloys, polymeric A blend of two or more polymers, e.g., a rubber and a plastic, to improve a given property, e.g., impact strength.

Angstrom (Å) A unit of measurement equal to 10^{-10} meter (0.0000000001 meter). Generally used to measure the wavelength of certain types of electromagnetic radiations; e.g., the red line of cadmium is defined as 6438 Å.

Annealing Heating to and holding at a suitable temperature, then cooling at a suitable rate, for such purposes as reducing hardness, improving machinability, facilitating cold working, producing a desired microstructure, or obtaining other desired mechanical or physical properties.

Annual value A uniform annual amount equivalent to the project costs or benefits taking into account the time value of money throughout the study period. (Syn.: *annual worth, equivalent uniform annual value.*)

Annually recurring costs Costs incurred in a regular pattern each year throughout the study period, normally for operation and maintenance (O & M).

Anodic metallic coating A coating that prevents corrosion of exposed areas of the base metal by galvanic protection. The coating becomes the anode and undergoes increased corrosion so that the base metal cathode is protected from further corrosion until all the nearby coating metal is gone.

ANSI American National Standards Institute.

Antidegradant A compounding material used to retard deterioration caused by oxidation, ozone, light, and combinations of these.

Note: *Antidegradant* is a generic term for such additives as antioxidants, antiozonants, and waxes.

Antioxidant A substance that prevents or retards oxidation of material exposed to air.

Approval drawings Approval drawings may include framing drawings, elevations, and sections through the building as furnished by the manufacturer for the approval of the buyer. Approval by the buyer affirms that the manufacturer has correctly interpreted the overall contract requirements for the system and its accessories, and the exact location of accessories in the building.

Architectural drawing A drawing that shows the plan view and/or elevations of the finished building for the purpose of showing the general appearance of the building, indicating all accessory locations.

Aromatic polyurethane coating A single- or plural-component polyurethane coating that is a polymer based on cyclical molecular structure. It is normally used as a base coat for aliphatic topcoat systems and as an ultraviolet-resistant topcoat.

Asbestos A group of natural fibrous impure silicate materials.

Ash   The incombustible material that remains after a substance has been burned.

Asphalt   A dark brown to black cementitious material whose predominating constituents are bitumens that occur in nature or are obtained in petroleum processing.

Asphalt, air blown   An asphalt produced by blowing air through molten asphalt at an elevated temperature to raise its softening point and modify other properties.

Asphalt, steam blown   An asphalt produced by blowing steam through molten asphalt to modify its properties.

Asphalt felt   An asphalt-saturated felt.

Asphaltene   A high-molecular-weight hydrocarbon fraction precipitated from asphalt by a designated paraffinic naphtha solvent at a specified temperature and solvent-asphalt ratio.
   Note: The asphaltene fraction should be identified by the temperature and solvent-asphalt ratio used.

ASTM   American Society for Testing and Materials.

Atactic   A chain of molecules in which the position of the side methyl groups is more or less random (*amorphic; low crystallinity*).

Atomic number   The number of protons in the nucleus of a chemical element.

Atomic weight   The mass of an atom of an element, compared to the mass of carbon 12.

Average   In statistics, the mean average is the sum of the values from a series of tests divided by the number of tests performed. The median would be the middle value of a series of numbers. The mode would be the most common number of a set of numbers.

AWS   American Welding Society.

Backup plate   A rigid plate to support an end lap to provide uniform compression.

Background count   The reading of a test device before the material to be tested or identified is introduced.

Backnailing   "Blind" (i.e., concealed by overlapping felt) nailing of roofing felts to a substrate in addition to hot mopping to prevent slippage.

Backscatter   The number of neutrons reflected back as contrasted to passing through a substance.

Ballast   Loose aggregate, concrete pavers, or other material designed to prevent wind uplift or flotation of a loose-laid roof system.

Banbury mixer   A heavy-duty batch mixer with two counterrotating rotors. Used mainly in the rubber industry.

Barrier protection   Protection from the environment by a physical, inert barrier. If the barrier is broken, the underlying base metal is unprotected. Contrasts with **anodic coatings,** which, if breached, continue to protect the underlying base metal.

Base time The date to which all future and past benefits and costs are converted when a present-value method is used (usually the beginning of the study period).

Base sheet A saturated or coated felt placed as the first ply in a multi-ply bituminous roofing membrane.

Batten A raised rib in a metal roof, or a separate part or formed portion in a metal roofing panel.

Beaufort scale A scale in which the force of the wind is indicated by numbers from 0 to 12. No. 7 is "near gale" at 32–38 mph. No. 9 is "strong gale" at 47–54 mph.

Bill of materials A list of items or components used for fabrication, shipping, receiving, and accounting purposes.

Bitumen (1) A class of amorphous, black or dark-colored, (solid, semisolid, or viscous) cementitious substances, natural or manufactured, composed principally of high-molecular-weight hydrocarbons, soluble in carbon disulfide, and found in asphalts, tars, pitches, and asphaltites. (2) A generic term used to denote any material composed principally of bitumen. (3) In the roofing industry there are two basic bitumens: asphalt and coal tar pitch. Before application they are either (a) heated to a liquid state, (b) dissolved in a solvent, or (c) emulsified.

Bituminous (adj.) Containing or treated with bitumen. Examples: bituminous concrete, bituminous felts and fabrics, bituminous pavement.

Bituminous emulsion A suspension of minute globules of bituminous material in water or in an aqueous solution.

Blanket insulation Fiberglass insulation in roll form, often installed between metal roof panels and the supporting purlins.

Blind rivet A small headed pin with an expandable shank for joining light-gage metal. Typically used to attach flashing, gutter, etc. Applied from one side, with a stem that pulls against material on the blind side.

Blister An enclosed pocket of air–water vapor, trapped between membrane plies or between membrane and substrate.

Blister (polyurethane foam) An undesirable rounded delamination of the surface of a polyurethane foam whose boundaries may be either more or less sharply defined.

Block copolymer An essentially linear copolymer in which there are repeated sequences of polymeric segments of different chemical structure.

Block or board thermal insulation Rigid or semirigid thermal insulation preformed into rectangular units.

Blocking (1) Wood built into a roofing system above the deck and below the membrane and flashing to (a) stiffen the deck around an opening, (b) act as a stop for insulation, (*c*) serve as a nailer for attachment of the membrane or flashing. (2) Wood cross-members installed between rafters or joists to provide support at cross-joints between deck panels. (3) Cohesion or adhesion between similar or dissimilar materials in roll or sheet form that may interfere with the satisfactory and efficient use of the material.

Blocking, wood   Treated wood members designed to help prevent movement of insulation.

Bloom   A visible exudation of efflorescence on the surface of a material.

Blowing agent   A compounding ingredient used to produce gas by chemical or thermal action, or both, in the manufacture of hollow or cellular articles.

Blueberry   A small bubble or blister in the flood coating of a gravel-surfaced membrane.

BOCA   Building Officials and Code Administrators, International, Inc. Author of the National Building Code.

Bodied solvent adhesive   An adhesive consisting of a solution of the membrane compound in solvent, used in the seaming of membranes.

Bond   The adhesive and cohesive forces holding two roofing components in intimate contact.

Boot   A bellows-type covering to exclude dust, dirt, moisture, etc., forming a flexible closure.

Breaking strain   Percent elongation at which a sheet or other tested component ruptures under tensile force.

Breaking stress   Stress (in force per linear or area unit) at which a sheet or other tested component ruptures under tensile force.

British thermal unit (Btu)   The heat energy required to raise the temperature of one pound of water by one degree Fahrenheit.

Brooming   Embedding a ply by using a broom to smooth it out and ensure contact with the adhesive under the ply.

Btuh   Btu per hour.

Builder/contractor   A general contractor or subcontractor responsible for providing and erecting metal building systems.

Building code   Published regulations and ordinances established by a recognized agency describing design loads, procedures, and construction details for structures. Usually applies to a designated political jurisdiction (city, county, state, etc.). Building codes control design, construction, quality of materials, use and occupancy, location, and maintenance of buildings and structures within the area for which the code was adopted. (*See* **Model codes.**)

Built-up roofing (BUR)   A continuous, semiflexible membrane consisting of plies of saturated felts, coated felts, fabrics, or mats assembled in place with alternate layers of bitumen, and surfaced with mineral aggregate, bituminous material, or a granule-surfaced sheet.

Bull   Roofer's term for flashing or plastic cement.

Butyl rubber A synthetic rubber based on isobutylene and a minor amount of isoprene. It is vulcanizable and features low permeability to gases and water vapor and good resistance to aging, chemicals, and weathering.

By-product nuclear material Secondary radioactive material derived from nuclear refining processes in the manufacture of nuclear fuels. This type of material is used in nuclear moisture meters.

Calender A machine with two or more rolls, operating at selected surface speeds and controlled temperatures, for sheeting, laminating, skim coating (topping), and friction coating to a controlled thickness or surface characteristic, or both.

Calorie The amount of heat required to raise one gram of water one degree Celsius.

Camber A predetermined curvature designed into a structural member to offset the anticipated deflection under design load.

Canopy Any overhanging or projecting roof structure with the extreme end usually unsupported.

Cant strip A beveled strip used under flashings to modify the angle at the point where the roofing or waterproofing membrane meets any vertical element.

Cap flashing *See* **Flashing**.

Cap sheet A granule-surfaced coated felt used as the top ply of a built-up roofing membrane.

Capacitance The ratio of the charge to the potential difference between two conducting elements separated by a nonconductor.

Capillary action That action that causes movement of liquids by surface tension when they are in contact with two adjacent surfaces, such as panel side laps.

Capital cost The cost of acquiring, substantially improving, expanding, changing the functional use of, or replacing a building or building system.

Catalyst A substance that causes or changes the rate of a chemical reaction.

Cationic emulsion An emulsion in which the emulsifying system establishes a predominance of positive charges on the discontinuous phase.

Caulk To seal joints, seams, or voids by filling with a waterproofing compound or material.

Caulking A composition of vehicle and pigment, used at ambient temperatures for filling joints, that remains plastic for an extended time after application.

Cavity wall A wall built of hollow masonry units arranged to provide a continuous internal air space.

Centipoise Unit measurement of viscosity=centistokes×specific gravity.

Centistoke Unit measurement of viscosity—i.e., resistance to flow.

Chain scission Breaking of chemical bonds between carbon atoms by ultraviolet photo-oxidation resulting in embrittlement and cracking; a reversal of the asphalt-blowing polymerization process that produces long hydrocarbon chains.

Chalk resistance A measurement of performance for paint systems; the ability to resist a dusty/chalky appearance over time.

Chalking A powdery residue on the surface of a material resulting from degradation or migration of an ingredient, or both.

Channel mopping *See* Strip mopping, under **Mopping.**

Charging current The transient current charging a capacitor.

Chlorinated polyethylene (CPE) A family of polymers produced by chemical reaction of chlorine on the linear backbone chain of polyethylene. The resultant rubbery thermoplastic elastomers presently contain 25 to 45 percent chlorine by weight and 0 to 25 percent crystallinity. CPE can be vulcanized but is usually used in a nonvulcanized form.

Chlorosulfonated polyethylene (CSPE) A family of polymers that are produced by polyethylene reacting with chlorine and sulfur dioxide. Present polymers contain 25 to 43 percent chlorine and 1.0 to 1.4 percent sulfur. They are used in both vulcanized and nonvulcanized forms. Most membranes based on CSPE are nonvulcanized. The ASTM designation for this polymer is CSM. It is best known by the DuPont trademark Hypalon.

Closure strip A resilient strip of a material such as neoprene, flat on one side and formed to the contour of ribbed sheets on the other, used to close openings created by joining metal sheets and flashings.

Coefficient of thermal expansion The change in length per unit of length for a unit change in temperature. (Thus the coefficient per °F must be multiplied by 1.8 for the coefficient per °C.)

Coal tar A dark brown to black cementitious material produced by the destructive distillation of coal.

Coal tar felt A felt saturated with refined coal tar.

Coal tar pitch A dark brown to black, solid cementitious material obtained as residue in the partial evaporation or distillation of coal tar.

Coated fabric A fabric impregnated and/or coated with a plastic material in the form of a solution, dispersion hot melt, or powder. (The term also applies to materials resulting from the application of a preformed film to a fabric by means of calendering.)

Coated sheet (or felt) (1) An asphalt felt that has been coated on both sides with harder, more viscous asphalt. (2) A glass-fiber felt that has been simultaneously impregnated and coated with asphalt on both sides.

Coating weight The weight of a coating on a surface (both sides), usually expressed in ounces per square foot or grams per square meter.

Coil coating The application of an organic finish to a coil of metal using a continuous process.

Cold flow Slow deformation, under gravitational force, at or below room temperature. (*See* **Creep**.)

Cold-process roofing A continuous, semiflexible membrane consisting of plies of felts, mats, or fabrics laminated on a roof with alternate layers of roof cement and surfaced with a cold-applied coating.

Cold working Deforming metal plastically at a temperature lower than the recrystallization temperature.

Collector box Transition piece between a gutter and downspout to facilitate water flow.

Color retention The ability to resist fading; the measurement of performance for paint systems.

Compound An intimate admixture of polymer(s) with all the materials necessary for the finished product.

Condensation The conversion of water vapor or other gas to liquid as the temperature drops or atmospheric pressure rises. (*See also* **Dew point**.)

Condensation polymerization Polymerization in which monomers are linked together with the splitting off of water or other simple molecules.

Conductance, thermal The thermal transmission in unit time through a unit area of a particular body or assembly having defined surfaces, when unit average temperature difference is established between the surfaces. $C=[W/(m^2 \cdot K)]$; $C=[Btu/(h \cdot ft^2 \cdot °F)]$.

Conductivity (electrical) The reciprocal of electrical resistance.

Conductivity, thermal The thermal transmission, by conduction only, in a unit time through a unit area between two isothermal surfaces of an infinite slab of a homogeneous material of unit thickness, in a direction perpendicular to the surface, when unit temperature difference is established between the surfaces. $k=[W/(m^2 \cdot K)]$; $k=[Btuh/(in. \cdot ft^3 \cdot °F)]$.

Contractor *See* **Builder/contractor**.

Coping A covering on top of a wall exposed to the weather, usually sloped to carry off water.

Copolymer A mixed polymer, the product of polymerization of two or more substances at the same time.

Counterflashing Formed metal or elastomeric sheeting secured on or into a wall, curb, pipe, rooftop unit, or other surface, to shield the upper edge of a base flashing and its associated fasteners.

Counting interval The time period during which meter detector tubes are measuring nuclear backscatter. Since isotopic discharges (disintegrations) occur randomly, a mean average can be obtained after a suitable interval. Precision improves by 30 percent when the time segment is doubled. Normal counting intervals range to 30 s.

Coverage The surface area to be continuously covered by a specific quantity of a particular material.

Covering The exterior roof and wall covering for a metal building system.

CPM Counts per minute.

Cream time The time, measured in seconds at a given temperature, when the A and B components of a polyurethane foam compound will begin to expand after being mixed through the spray gun.

Creep The dimensional change with time of a material under load, following the initial instantaneous elastic deformation. Creep at room temperature is sometimes called cold flow.

Creep modulus The ratio of initial applied stress to creep strain.

Creep strain The total strain, at any given time, produced by the applied stress during a creep test.

Note: The term *creep,* as used in this method, reflects current plastics engineering usage. Plastics have a wide spectrum of retardation times, and the elastic portions of strain cannot be separated from the nonelastic in practice.

Cricket A relatively small elevated area of a roof constructed to divert water from a horizontal intersection of the roof with a chimney, wall, expansion joint, or other projection.

Cross-linking A general term referring to the formation of chemical bonds between polymeric chains to yield an insoluble, three-dimensional polymeric structure. Cross-linking of rubbers is **Vulcanization.**

CRT Cathode-ray tube, used as TV picture tubes, display on IR cameras, etc.

Curb A raised member used to support roof penetrations such as skylights, hatches, etc.

Cure To change the properties of a polymeric system into a more stable, usable condition by the use of heat, radiation, or reaction with chemical additives.

Note: Cure may be accomplished, for example, by removal of solvent or cross-linking.

Curie The official unit of radioactivity, defined as 3.70×10^{-10} disintegrations per second. (*See* **Roentgen.**)

Curing *See* **Vulcanization.**

Curled felt BUR membrane defect characterized by a continuous, open longitudinal seal with the top felt rolled back from the underlying felt.

Cutback Solvent-thinned bitumen used in cold-process roofing adhesives, flashing cements, and roof coatings.

Cutoff A detail designed to prevent lateral water movement into the insulation where the membrane terminates at the end of a day's work, or used to isolate sections of the roofing system; usually removed before the continuation of the work.

Daily standard count The counts per minute with a nuclear device resting on a calibrated reference standard. Daily refers to the historical need to calibrate older instruments each day.

Dampproofing Treatment of a surface or structure to resist the passage of water in the absence of hydrostatic pressure.

Dead-level Absolutely horizontal, or zero slope. (*See* **Slope**.)

Dead-level asphalt A roofing asphalt conforming to the requirements of Specification D312, Type I.

Deck The structural surface to which the roofing or waterproofing system (including insulation) is applied.

Degree-days The difference between a reference temperature (usually 65°) and the mean temperature for the day times 24 h times the number of days in the period. Degree-days are used to compare the severity of cold or heat during the heating or cooling season.

Delamination Separation of the plies in a membrane or separation of insulation layers after lamination.

Denier A unit used in the textile industry to indicate the fineness of continuous filaments. Fineness in deniers equals the mass in grams of a 9000-m length of the filament.

Depth of measurement The maximum thickness of a roof system upon which a given moisture survey method is effective.

Design loads The "live load" (i.e., superimposed load) that a structure is designed to resist (with appropriate safety factor) plus the "dead load" (i.e., weight of permanent loads).

Dew point The temperature at which water vapor starts to condense in cooling air at the existing atmospheric pressure and vapor content.

Dielectric constant A number defining the relative efficiency of a dielectric material for passing lines of electric force compared to that of vacuum. The number will be greater than 1.

Dielectric seaming *See* **Heat seaming**.

Differential price escalation rate The expected percent difference between the rate of increase assumed for a given item of cost (such as energy) and the general rule of inflation.

Diffusion The material permeation of two or more substances due to the kinetic energy of their molecules, so that a uniform mixture or solution results. Diffusion occurs with all forms of matter: more rapidly for gases, more slowly for liquids and for solids in solution.

Discount factor A multiplicative number (calculated from a discount formula for a given discount rate and interest period) that is used to convert costs and benefits occurring at different times to a common time, usually the present.

Discount rate The rate of interest reflecting the investor's time value of money, used to determine discount factors for converting benefits and costs occurring at different times to a base time. The discount rate may be expressed in nominal or real (inflation-corrected) terms.

Discounting A technique for converting cash flows that occur over time to equivalent amounts at a common time.

Double pour Doubling of the flood-coat, graveling-in operation to provide additional waterproofing integrity for a BUR membrane.

Downspout A conduit used to carry water from the gutter of a building to the ground or storm drain.

Dry (n.) A material that contains no more water than one would find at its equilibrium moisture content.

Duck-board A boardwalk or slatted flooring laid on a wet, muddy, or cold surface.

Eave The line along the sidewall formed by the intersection of the planes of the roof and the wall.

Eave height The vertical dimension from the finished floor to the eave.

Edge stripping Application of felt strips cut to narrower widths than the normal felt roll width to cover a joint between flashing and built-up roofing.

Edge venting The practice of providing regularly spaced protected openings at a roof perimeter to relieve water-vapor pressure in the insulation. (It is of doubtful efficacy.)

Efflorescence A deposit or encrustation of soluble salts, generally white and most commonly consisting of calcium sulfate, that may form on the surface of stone, brick, concrete, or mortar when moisture moves through and evaporates on the masonry. It is often caused by free alkalies leached from mortar, grout, or adjacent concrete.

Elasticity The property of matter by virtue of which it regains its original size and shape after removal of stress.

Elastomer A macromolecular material that returns rapidly to its approximate initial dimensions and shape after subsequent release of stress.

Electric charge A physical phenomenon caused by an isolated imbalance between the number of protons and electrons in a substance.

Electric force field The invisible forces created by electric charges of opposite polarity.

Electrode potential The potential of a half cell as measured against a standard reference half cell.

Electrolyte A current-conducting liquid, usually a solution.

Electron-volt (eV) An extremely small unit used in measuring the energy of nuclear constituents. It is the energy developed by an electron falling through a potential difference of 1 volt.

Embedment (1) The process of pressing a felt, aggregate, fabric, mat, or panel uniformly and completely into hot bitumen or adhesive to ensure intimate contact at all points. (2) The process of pressing granules into coating in the manufacture of factory-prepared roofing, such as shingles.

Emissivity A quantity characterizing the radiant emittance of a substance equal to the ratio of the power of its radiation to the power of the radiation of a blackbody (the perfect emitter and absorber) at the same temperature, area, and solid angle of emission. The ratio of radiant energy emitted from a surface under measurement to that emitted from a blackbody at the same temperature.

Emulsion A dispersion of fine particles or globules of a liquid in a liquid. Asphalt emulsions consist of asphalt globules, an emulsifying agent such as bentonite clay, and water.

End lap The overlap where one panel or felt nests on top of the end of the underlying panel or felt.

Envelope A continuous edge seal formed by extending one ply of felt beyond the edge of the assembly. After other plies or insulation are in place, the extended ply is turned back and adhered.

E/P (elastoplastic) Pertaining to polymeric materials, including the thermoplastic and elastomeric categories.

EPDM A synthetic elastomer based on ethylene, propylene, and a small amount of a nonconjugated diene to provide sites for vulcanization. EPDM features excellent heat, ozone, and weathering resistance and low-temperature flexibility.

Epichlorohydrin rubber A synthetic rubber that includes two epichlorohydrin-based elastomers of saturated high-molecular-weight, aliphatic polyethers with chloromethyl side chains. The two types include a homopolymer (CO) and a copolymer of epichlorohydrin and ethylene oxide (ECO). These rubbers are vulcanized with a variety of reagents that react difunctionally with the chloromethyl group, including diamines, urea, thioureas, 2 mercaptoimidazoline, and ammonium salts. This rubber offers excellent oil resistance.

Equilibrium moisture content (1) Moisture content of a material stabilized at a given temperature and relative humidity, expressed as percent moisture by weight. (2) The typical moisture content of a material in any given geographical area.

EVA A family of copolymers of ethylene and vinyl acetate used for adhesives and thermoplastic modifiers. They possess a wide range of melt indexes.

EVT (equiviscous temperature) The temperature at which the viscosity of an asphalt is appropriate for application. Viscosity units are generally expressed in centipoise or centistokes. Tolerance on EVT is usually ±25°F (±14°C).

Exotherm Heat generated in a chemical reaction.

Expansion joint A structural separation between two building elements that allows free movement (expansion or contraction) between elements without damage to the roofing or waterproofing system.

Expected total error Where different portions of a testing procedure have measurement errors, the accumulated effects of these individual errors.

Exposure (1) The transverse dimension of a roofing element not overlapped by an adjacent element in any roofing system. The portion not overlapped by an adjacent element in any roofing system. The exposure of any ply in a membrane may be computed by dividing the felt width minus 51 mm (2 in.) by the number of shingled plies; thus, the exposure of 914-mm (36-in.)-wide felt in a shingled, four-ply membrane should be 216 mm (8½ in.). (2) The time during which a portion of a roofing element is exposed to the weather.

Extra steep asphalt *See* **Super-steep asphalt.**

Extractables Components or substances removable from a solid or liquid mixture by means of an appropriate solvent.

Extruder A machine with a driven screw that forces ductile or semisoft solids through a die opening of appropriate shape to produce continuous film, strip, or tubing.

Fabric A woven cloth of organic or inorganic filaments, threads, or yarns.

Fabric reinforcement A fabric, scrim, etc., used to add structural strength to a two-or-more-ply polymeric sheet. Such sheeting is referred to as "supported."

Fabrication (1) The manufacturing process performed in a plant to convert raw material into finished metal building components. The main operations are cold forming, cutting, punching, welding, cleaning, and painting. (2) The creation of large panels of rubber from smaller calender-width sheets as in EPDM.

Fallback Reduction in bitumen softening point, sometimes caused by refluxing or overheating in a relatively closed container.

Fascia A decorative trim or panel projecting from the face of a wall, serving as a weather closure.

Felt A flexible sheet manufactured by the interlocking of fibers through a combination of mechanical work, moisture, and heat, without spinning, weaving, or knitting. Roofing felts are manufactured from vegetable fibers (organic felts), glass fibers (glass-fiber felts), or polyester fibers (synthetic-fiber mats).

Felt mill ream The mass in pounds of 480 ft^2 of dry, unsaturated felt, also termed *point weight*.

Fiberglass insulation Blanket insulation, composed of glass fibers bound together with a thermoset binder, faced or unfaced, used over or under purlins to insulate roofs and walls; semirigid boards, usually with a facer.

Field The job site, building site, or general market area.

Fill As used in textile technology, the threads or yarns in a fabric running at right angles to the warp. Also called *filler threads*.

Filler strip *See* **Closure strip.**

Film Sheeting having a nominal thickness not greater than 10 mils (0.25 mm).

Film badge A device to measure the total radiation exposure of users of nuclear measuring instruments.

Fin A sharp, raised edge capable of damaging a roof membrane.

Fine mineral surfacing Water-insoluble inorganic material, more than 50 percent of which passes the 500-micrometer (No. 35) sieve, used on the surface of roofing.

Fishmouth (1) A half-cylindrical or half-conical opening formed by an edge wrinkle or failure to embed a roofing felt. (2) In shingles, a half-conical opening formed at a cut edge.

Flash point The temperature at which a test flame ignites vapor above a liquid surface.

Flashing The system used to seal membrane edges at walls, expansion joints, drains, gravel stops, and other places where the membrane is interrupted or terminated. Base flashing covers the edges of the membrane. Cap flashing or counterflashing shields the upper edges of the base flashing.

Flashing cement A trowelable mixture of cutback bitumen, mineral stabilizers, and fibers.

Flat asphalt A roofing asphalt conforming to the requirements of Specification D312, Type II.

Fleece Mats or felts of usually nonwoven fibers.

Flood coat The top layer of bitumen used to hold the aggregate on an aggregate-surfaced, built-up roofing membrane.

Fluid-applied elastomer An elastomeric material, fluid at ambient temperature, that dries or cures after application to form a continuous membrane. Such systems normally do not incorporate reinforcement.

Fluorocarbon films Substituted ethylene polymers, featuring outstanding formability, heat resistance, color retention, and resistance to solvents and chalking.

Framed opening Frame work (headers and jambs) and flashing which surround an opening in the wall or roof of a building, usually for field-installed accessories such as overhead doors or powered roof exhausters.

"Free carbon" in tars The hydrocarbon fraction precipitated from a tar by dilution with carbon disulfide.

Frequency The number of vibratory cycles per unit of time; equals the speed of light divided by the wavelength for electromagnetic radiation.

Friability The tendency of a material or product to crumble or break into small pieces easily.

Gable roof A ridge roof that terminates in gables.

Galvalume Trademark for steel coated with aluminum-zinc alloy for corrosion protection.

Galvanic cell A cell in which chemical change is the source of electrical energy. It usually consists of two dissimilar conductors in contact with each other and an electrolyte.

Galvanized steel Steel coated with zinc for corrosion resistance.

Glass felt Glass fibers bonded into a sheet with resin and suitable for impregnation in the manufacture of bituminous waterproofing, roofing membranes, and shingles.

Glass mat A thin mat of glass fibers with or without a binder.

Glass transition The reversible change in an amorphous polymer or in amorphous regions of a partially crystalline polymer from (or to) a viscous or rubbery condition to (or from) a hard and relatively brittle one.

Glaze coat (1) The top layer of asphalt in a smooth-surfaced built-up roof assembly. (2) A thin protective coating of bitumen applied to the lower plies or top ply of a built-up membrane when application of additional felts or of the flood coat and aggregate surfacing is delayed.

Gloss A subjective term describing the relative amount and nature of mirrorlike reflection from a surface.

Grain A weight unit equal to 1/7000 lb, used in measuring atmospheric water-vapor content.

Gravel Coarse, granular aggregate, with pieces larger than sand grains, resulting from the natural erosion of rock.

Gravel stop A flanged device, usually metallic, designed to prevent loose aggregate from washing off the roof and to provide a continuous finished edge for the roofing.

Grout A mixture of cement, sand, and water used to fill cracks and cavities. Often used under base plates or leveling plates to obtain uniform bearing surfaces.

Gutter A channel member installed at the eave of the roof for the purpose of carrying water from the roof to the drains or downspouts.

Haunch The deepened portion of a column or rafter, designed to accommodate the higher bending moments at such points. (Usually occurs at connection of column and rafter.)

Header A horizontal framing structural member of a door, window, or other framed opening.

Headlap The minimum distances, measured at 90° to the eave along the face of a shingle or felt as applied to a roof, from the upper edge of the shingle or felt to the nearest exposed surface.

Heat capacity The amount of energy required to raise the temperature of a unit substance 1°F (or 1°C).

Heat seaming The process of joining two or more thermoplastic films or sheets by heating areas in contact with each other to the temperature at which fusion occurs. The process is usually aided by a controlled pressure. In dielectric seaming, the heat is induced within films by means of radio-frequency waves.

Heat transfer The transmission of thermal energy from a location of higher temperature to a location of lower temperature. This can occur by conduction, convection, or radiation.

High-speed neutron High-velocity neutrons such as would emanate directly from atomic nuclei such as the radioactive isotopes used in roof moisture meters. (*See* **Thermalization**.)

Hip roof A roof which rises by inclined planes from all four sides of the building. The line where two adjacent sloping sides of a roof meet is called the hip.

Homopolymer A natural or synthetic high polymer derived from a single monomer.

Holiday An area where a liquid applied material is missing, a void.

Hood A cover, usually of lightgage metal, over piping or other rooftop equipment.

Hot-dip metallic coating An adherent protective coating applied by immersing steel in a molten bath of coating material.

"Hot stuff" or "hot" A roofer's term for hot bitumen.

Humidity The amount of moisture contained in the atmosphere. Generally expressed as percent relative humidity (the ratio of the vapor pressure to the saturation pressure for given conditions times 100).

Humidity test A test involving exposure of specimens at controlled levels of humidity and temperature.

Hydrocarbon An organic chemical compound containing mainly the elements carbon and hydrogen. Aliphatic hydrocarbons are straight-chain compounds of carbon and hydrogen. Aromatic hydrocarbons are carbon-hydrogen compounds based on the cyclic or benzene ring. They may be gaseous (CH_4, ethylene, butadiene), liquid (hexene, benzene), or solid (natural rubber, naphthalene, cispolybutadiene).

Hygroscopic Attracting, absorbing, and retaining atmospheric moisture.

ICBO International Conference of Building Officials, author of the Uniform Building Code.

Incline The slope of a roof expressed in percent or in the number of vertical units of rise per horizontal unit of run.

Inelastic scattering Scattering of particles as a result of collisions in which part of the kinetic energy is lost as heat or radiation. (*See* **Thermalization**.)

Infrared spectrum Those wavelengths of the electromagnetic spectrum which are by convention called infrared, generally considered to be wavelengths from just beyond the visible (0.77 micron) to about 3000 microns (longer wavelengths).

Inorganic (adj.) Comprising matter other than hydrocarbons and their derivatives, or matter not of plant or animal origin.

Insulation *See* **Thermal insulation**.

Interlace The intermixing of more than one complete field of scan lines. Instead of scanning each line sequentially from the top to the bottom, some lines are left as spaces and are filled in on the display by a subsequent field.

Internal pressure Pressure inside a building, a function of wind velocity, building height, and number and location of openings.

Internal rate of return (IRR) The compound rate of interest that, when used to discount study-period costs and benefits of a project, will make the two equal.

Inverse square law A process where the intensity of radiation decreases as the square of the distance—that is at twice the distance, the intensity would be 1/4.

Inverted display The transposition of the grey scale (i.e., black for white) in a CRT display.

Isocyanate A highly reactive chemical grouping composed of a nitrogen atom bonded to a carbon atom bonded to an oxygen atom, $=N=C=O$; a chemical compound, usually organic, containing one or more isocyanate groups.

Isotherm A contour or line on the display screen depicting equal apparent temperature. On imaging scanners, these areas are electrically highlighted on the display as bright white.

Isotherm thermogram A picture of a thermal image showing areas of equal apparent temperature.

Isothermal unit A unit of thermal measurement common to a particular IR system. It must be converted to temperature by correcting for instrument settings, detector output, emissivities, and ambient conditions.

Joist Any of the small timbers or metal beams arranged parallel from wall to wall to support a floor, ceiling, or roof of a building.

Kesternich test A test that simulates acid rain conditions by subjecting samples to a sulfur dioxide atmosphere as well as condensing moisture.

Kick-out (elbow, turn-out) A lower downspout section used to direct water away from a wall.

Laitance An accumulation of finer particles on the surface of fresh concrete due to an upward movement of water (as when excessive mixing water is used).

Lap The dimension by which a felt covers an underlying felt in BUR membrane. Edge lap indicates the transverse cover; end lap indicates the cover at the end of the roll. These terms also apply to single-ply membranes.

Lapped joint A joint made by placing one surface to be joined partly over another surface and bonding the overlapping portions.

Layer (plywood) A single veneer ply or two or more plies laminated with parallel grain direction. Two or more plies laminated with parallel grain direction form a "parallel laminated layer."

Leno fabric An open fabric in which two warp yarns wrap around each fill yarn in order to prevent the warp or fill yarns from sliding over each other.

Life-cycle cost (LCC) method A technique of economic evaluation that sums over a given study period the costs of initial investment (less resale value), replacements, operation (including energy use), and maintenance and repair for an investment decision (expressed in present or annual value terms).

Life-cycle costing An analytical technique that systematically compares economic alternatives over the useful life of the asset.

Light reflectance The percentage of light incidence that is not absorbed by the surface.

Live load All loads, including snow, exerted on a roof except dead, wind, and lateral loads.

LN_2 Liquid nitrogen, the reference temperature material used by many IR imaging systems. Its boiling point is $-196°C$ or $-324°F$.

Loose-laid membrane An unadhered roofing membrane anchored to the substrate only at the edges and penetrations through the roof and ballasted against wind uplift by loose aggregate or pavers.

Low-speed neutron A high-speed neutron that has been slowed down after repeated collisions with hydrogen atoms. (*See* **Thermalization**.)

Macromolecule A large molecule in which there is a large number of one or several relatively simple chemical units, each consisting of several atoms bonded together.

Maintenance and repair cost The total of labor, material, and other related costs incurred in conducting corrective and preventive maintenance and repair on a building, on its systems and components, or on both.

Masonry Anything constructed of materials such as bricks, concrete blocks, ceramic blocks, and concrete.

Mastic Caulking or sealant normally used in sealing roof-panel laps.

Membrane A flexible or semiflexible roof covering or waterproofing whose primary function is the exclusion of water.

Memory Tendency of a material to regain a previous configuration—notably, the tendency of glass-fiber felts not to lie flat on their substrate after unrolling; the retraction of single-ply roll goods which were stretched during production or winding.

Mer The repeating structural unit of any high polymer.

Mesh The square opening of a sieve.

Metal flashing Frequently used as through-wall cap flashing or counterflashing. *See* **Flashing**.

Micron (also called micrometer) A unit of measurement equal to 10^{-6} meter (0.000001 meter) or 10,000 Angstroms.

Mineral fiber Inorganic fibers of glass, asbestos, or mineral wool (slag).

Mineral granules Natural or synthetic aggregate, ranging in size from 500 μm (1 μm=10^{-6}m) to ¼-in. diameter, used to surface BUR or modified bitumen capsheets, asphalt shingles, and some cold-process membranes.

Minimum detectable temperature difference A quantification of the smallest temperature difference between one point and another on an object that can be discerned using an infrared sensor.

Model codes Codes established to provide uniformity in regulations pertaining to building construction. Examples are the Uniform Building Code, published by the ICBO; the National Building Code, by BOCA; and the Standard Building Code, by SBCCI.

Modulus of elasticity The ratio of stress (nominal) to corresponding strain below the proportional limit of a material, expressed in force per unit area based on the minimum initial cross-sectional area.

Moisture conduction Migration by wicking as contrasted to vapor movement.

Moisture contour map A map with lines connecting continuous levels of moisture. When drawn by computer, the wettest areas are often indicated by the darkest symbols and the driest areas left blank.

Mole run A meandering ridge in a membrane not associated with insulation or deck joints.

Monochrome image (grey image) A thermal image with a blending of grey tones from dark to light but without the presence of isotherms.

Monomer A simple molecule which is capable of combining with a number of like or unlike molecules to form a polymer.

Mop-and-flop A procedure in which roof components (insulation boards, felt plies, cap sheets, etc.) are initially placed upside down adjacent to their ultimate locations, are coated with adhesive, and are then turned over and adhered to the substrate.

Mopping Application of hot bitumen with a mop or mechanical applicator to the substrate or to the plies of a built-up or modified-bitumen roof. There are four types of mopping: (1) solid, a continuous coating; (2) spot, in which bitumen is applied in roughly circular areas, generally about 460 mm (18 in.) in diameter, leaving a grid of unmopped perpendicular area; (3) strip, in which bitumen is applied in parallel bands, generally 200 mm (8 in.) wide and 300 mm (12 in.) apart; and (4) sprinkle, in which bitumen is shaken on the substrate from a broom or mop in a random pattern.

Mud cracking Surface cracking resembling a dried mud flat.

Nail-type concrete anchor A hammer-driven fastener with spiral or annular rings that provide pullout strength.

Nailer A wood member bolted or otherwise anchored to a nonnailable deck or wall to provide nailing anchorage of membrane or flashing.

Nailing (1) Exposed nailing of roofing wherein nail heads are bare to the weather. (2) Concealed nailing of roofing wherein nail heads are concealed from the weather.

National Coil Coaters Association The association composed of North American coil coaters charged with the promotion of the use of coated coils.

Needle punched A mechanical entanglement of dry-laid (usually cross-lapped, carded staple fiber) webs in which barbed needles achieve, through multiple punches, mechanical bonding.

Neoprene Synthetic rubber (polychloroprene) used in liquid or sheet-applied elastomeric roofing membranes or flashing.

Neutral sealants Acid-free and amine-free sealants.

Neutron A fundamental particle of matter having a mass of 1.009 but no electric charge, a constituent of the nucleus of all elements except hydrogen.

Neutron absorption The process by which a neutron is "captured" by an atom of the target material, thereby transforming its nucleus to the next higher isotope of the target.

Neutron sources Neutrons may be produced by reactors, accelerators, or certain radioactive isotopes. In most portable gaging applications, neutrons are produced by the reaction between alpha particles and beryllium, with radium 226 or americium 241 being the source of alpha particles.

Newton (N) The SI unit of measure for force.

Nitrile rubber A family of copolymers of butadiene and acrylonitrile that can be vulcanized into tough, oil-resistant compounds. Blends with PVC are used where ozone and weathering resistance are important requirements in addition to its inherent oil and fuel resistance.

Nondestructive testing (NDT) Methods for evaluating the strength or composition of materials without damaging the object under test.

Nonwoven fabric A structure produced by bonding or interlocking of fibers (or both) by mechanical, thermal, or solvent means (or combinations thereof).

Norm The meter CPM reading of hydrogen found in a roof system in its driest area (lowest reading). (*See* **Background count**.)

Nylon The generic name for a family of polyamide polymers characterized by the presence of the amide group—CONH. Used as a scrim in fabric-reinforced sheeting.

Off-ratio mix A system in which the mixture of isocyanate and resin does not conform to the manufacturer's recommended mixing ratio. The acceptable ratio for most systems has the two components combined in equal volumes.

Olefin An unsaturated open-chain hydrocarbon containing at least one double bond: ethylene or propylene.

Olefin plastics Plastics based on polymers made by the polymerization of olefins or copolymerization of olefins with other monomers, the olefins being at least 50 mass percent.

One-on-one *See* **Phased application.**

Operational amplifier A device used in electronic instrumentation to amplify electrical signals.

Organic (adj.) Composed of hydrocarbons or their derivatives, or matter of plant or animal origin.

Organic coatings Coatings that are generally inert or inhibited. They may be temporary (e.g., slushing oils) or permanent (paints, varnishes, enamels, etc.).

Organic content Usually synonymous with volatile solids in an ashing test. A discrepancy between volatile solids and organic content can be caused by small traces of some inorganic materials, such as calcium carbonate, that lose weight at temperatures used in determining volatile solids.

Osmosis Diffusion of fluids through a semipermeable membrane or porous partition.

Panel clip An independent clip used to attach roof panels to the substructure.

Panel creep Tendency of the transverse dimension of a roof panel to gain in modularity due to spring-out or storage distortion.

Parapet The portion of a wall above the roof line.

Pascal The SI unit of measure for force per unit area (N/m^2).

Peak The uppermost point of a gable.

Penetration The consistency of a bituminous material expressed as the distance in tenths of a millimeter (0.1 mm) that a standard needle or cone vertically penetrates a sample of material under specified conditions of loading, time, and temperature.

Percent elongation In tensile testing, the increase in the gage length, measured after fracture of the specimen within the gage length.

Percent water by volume

$$\frac{\text{Volume of water in sample}}{\text{Volume of sample}} \times 100$$

Percent water by weight

$$\frac{\text{Sample weight wet} - \text{Sample weight dry}}{\text{Sample weight dry}} \times 100$$

Perlite An aggregate used in lightweight insulating concrete and in preformed perlite insulating board, formed by heating and expanding siliceous volcanic glass.

Perm (vapor transmission) A unit to measure water-vapor transmission—one grain of water vapor per square foot per hour per inch of mercury pressure difference [1 Perm=1 grain/(h · ft^2 · in.Hg)].

Permeability (1) The capacity of a porous medium to conduct or transmit fluids. (2) The amount of liquid moving through a barrier in a unit time, unit area, and unit pressure gradient not normalized for but directly related to thickness. (3) The product of vapor permeance and thickness (for thin films,

ASTM E96; for those over ⅛ in., ASTM C355). Usually reported in perm inches or grain/(h · ft² · in.Hg) per inch of thickness.

Permeance The rate of water vapor transmission per unit area at a steady state through a membrane or assembly, expressed in ng/(Pa · s · m²)[grain/(ft² · h · in.Hg)].

Petroleum pitch A dark brown to black, predominantly aromatic, solid cementitious material obtained by the processing of petroleum, petroleum fractions, or petroleum residuals.

pH (1) The negative log of the hydrogen ion concentration, a measure of acidity and alkalinity. (2) A measure of the relative acidity or alkalinity for water. A pH of 7.0 indicates a neutral condition. A greater pH indicates alkalinity, and a lower pH, acidity. A one-unit change in pH indicates a tenfold change in acidity and alkalinity.

Phased application The installation of a roofing or waterproofing system during two or more separate time intervals; a roofing system not installed in a continuous operation.

Phenolic plastics Plastics based on resins made by the condensation of phenols, such as phenol and cresol, with aldehydes.

Picture framing A rectangular pattern of ridges in a membrane over insulation or deck joints.

Pig spout A sheet metal flashing designed to direct the flow of water through the face of the gutter rather than through a downspout.

Pinhole A tiny hole in a film, foil, or laminate comparable in size to one made by a pin.

Pitch *See* **Incline, Coal tar pitch,** or **Petroleum pitch.**

Pitch pocket A flanged, open-bottomed container placed around a column or other roof penetration and filled with hot bitumen, flashing cement, or pourable sealer.

Plastic A material that contains as an essential ingredient one or more organic polymeric substances of large molecular weight. It is solid in its finished state and at some stage in its manufacture or processing into finished articles can be shaped by flow.

Plastic cement *See* **Flashing cement.**

Plasticizer Material, frequently solventlike, incorporated in a plastic or a rubber to increase its workability, flexibility, or extensibility. Adding the plasticizer may lower the melt viscosity, the temperature of the second-order transition, or the elastic modulus of the polymer. Plasticizers may be monomeric liquids (phthalate esters), low-molecular-weight liquid polymers (polyesters), or rubbery high polymers (E/VA). The most important use of plasticizers is with PVC, where the choice of plasticizer dictates under what conditions the membrane may be used.

Plastisols Mixtures of resins and plasticizers that can be cast or converted to continuous films by the application of heat.

Ply A layer of felt in a roofing membrane; a four-ply membrane should have a least four plies of felt at any vertical cross section cut through the membrane.

Ply (plywood) A single veneer lamina in a glued plywood panel.

Plywood A flat panel built up of sheets of wood veneer called plies, united under pressure by a bonding agent to create a panel with an adhesive bond between plies as strong as or stronger than the wood. Plywood is constructed of an odd number of layers with the grain of adjacent layers perpendicular. Layers may consist of a single ply or of two or more plies laminated with parallel grain direction. Outer layers and all odd-numbered layers generally have the grain direction oriented parallel to the long dimension of the panel.

Pointing (1) Troweling mortar into a joint after masonry units are laid. (2) Final treatment of joints in cut stonework. Mortar or a puttylike filler is forced into the joint after the stone is set.

Polyester fiber The generic name for a manufactured fiber in which the fiber-forming substance is any long-chain synthetic polymer composed of an ester of a dihydric alcohol and terephthalic acid. Scrims made of polyester fiber are used for fabric reinforcement.

Polyisobutylene The polymerization product of isobutylene, varying in consistency from a viscous liquid to a rubberlike solid, with corresponding variation in molecular weight from 1000 to 400,000.

Polymer A macromolecular material formed by the chemical combination of monomers having either the same or different chemical composition. Plastics, rubbers, and textile fibers are all high-molecular-weight polymers.

Polyol A polyhydric alcohol, i.e., one containing three or more hydroxyl groups.

Polypropylene $(C_3H_5)_n$; a synthetic thermoplastic polymer with a molecular weight of 40,000 or more.

Polyvinyl chloride (PVC) A synthetic thermoplastic polymer prepared from vinyl chloride. PVC can be compounded into flexible and rigid forms through the use of plasticizers, stabilizers, filler, and other modifiers; rigid forms used in pipes; flexible forms used in manufacture of sheeting.

Pond A roof area that retains water instead of draining after rainfall.

Ponding Water in low or irregular roof areas that remains longer than 48 h after the cessation of rainfall.

Pot life The working time once a product has been reacted (catalyzed).

Prepainted coil Coil steel which receives a paint coating prior to the forming operation.

Present worth (or value) In life-cycle cost analysis, the procedure of discounting expenses at some future time to the present time to permit comparison of alternatives in equivalent terms.

Press brake A machine used in cold-forming metal sheet or strip into desired cross section.

Prestressed concrete Concrete in which the reinforcing cables, wires, or rods in the concrete are tensioned before there is load on the member, holding the concrete in compression for greater strength.

Preventive maintenance The regular, scheduled inspection for and the repair of normal, expected breakdown of materials and equipment.

Prime coat The first liquid coat applied in a multiple-coat system.

Primer (bituminous) A thin liquid bitumen applied to a surface to improve the adhesion of heavier applications of bitumen and to absorb dust.

Protected membrane roof (PMR) A roof assembly with insulation on top of the membrane instead of vice versa, as in conventional roof assembly (also known as inverted or upside-down roof assembly).

Puncture resistance An index of a material's ability to withstand the action of a sharp object without perforation.

Quantized grey scale A display scale for an infrared scanner which assigns discrete grey tones to particular temperature increments on the screen display.

Racking To stretch or strain by force, such as by thermal or wind action.

RAD Unit of absorbed dose of ionizing radiation equal to an energy of 100 ergs per gram of irradiated material. (*See* **REM**.)

Radioactivity Natural or artificial nuclear transformation. The energy of the process is emitted in the form of alpha, beta, and gamma rays.

Raggle *See* **Reglet**.

Rake The sloped edge of a roof at the first or last rafter.

Rake angle An angle fastened to purlins at the rake for attachment of endwall panels.

Rake trim A flashing designed to close the opening between the roof and the endwall panels.

Raspberry *See* **Blueberry**.

Real discount rate The rate of interest reflecting that portion of the time value of money related to the real (inflation-corrected) value of money over time.

Re-covering The process of covering an existing roof system with a new roof.

Reentrant corner An inside corner of a surface, where stress concentrations may occur.

Reference level A reading or image associated with normal or dry conditions.

Reference temperature A temperature of known value used as a basis to determine other temperatures.

Reglet A groove in a wall or other surface adjoining a roof surface for the attachment of counterflashing.

Reinforced membrane A roofing or waterproofing membrane reinforced with felts, mats, fabrics, or chopped fibers.

Relative humidity The ratio of the mass per unit volume (or partial pressure) of water vapor in an air-vapor mixture to the saturated mass per unit volume (or partial pressure) of the water vapor at the same temperature, expressed as a percentage.

Relative saturation

$$\frac{\text{Volume of water in sample}}{\text{Maximum volume of water sample could hold}} \times 100$$

REM (Roentgen Equivalent Man) A figure used to adjust radiation measured in rad to account for the differing effects on man.

Remedial roofing The repair of selected, isolated portions of the roof system to return the roof to uniform condition. This normally involves the removal of wet materials along with correction of the original cause of the problem.

Reroofing The replacement of a defective existing roof system with a new system.

Resistance, thermal *See* **Thermal resistance.**

Retrofit The modification of an existing building or facility to include new systems or components.

Ridge The highest point on the roof of the building, a horizontal line running the length of the building.

Ridge cap A transition of the roofing materials along the ridge of a roof. Sometimes called *ridge roll* or *ridge flashing*.

Ridging An upward, tenting displacement of a membrane, over an insulation joint.

RIEI The Roofing Industry Educational Institute.

Rockwell hardness A test for determining the hardness of a material based upon the depth of penetration of a specified penetrator into the specimen under certain arbitrarily fixed conditions of test.

Roentgen The unit of quantity or dose for x-rays, gamma rays, etc., that will produce as a result of ionization one electrostatic unit of electricity in one cubic centimeter of dry air.

Roll goods A general term applied to rubber and plastic sheeting, usually furnished in rolls.

Roll roofing Coated felts, either smooth or mineral-surfaced.

Roof cement *See* **Flashing cement.**

Roof covering The exposed exterior roof skin.

Roof curb An accessory used to mount and level units (such as air conditioning and exhaust fans) on the sloped portion of the building roof.

Roof jack An accessory used to cover pipes (such as vents or flues) that penetrate the roof panel.

Roof overhang A roof extension beyond the endwall/sidewall of a building.

Roof seamer A machine that crimps panels together or that welds laps of E/P systems using heat, solvent, or dielectric energy.

Roof slope The angle a roof surface makes with the horizontal, measured as the number of inches of vertical rise in a horizontal length of 12 in. (or as a ratio such as 1:48, or as a percent).

Roof system An assembly of interacting components designed to weatherproof, and normally to insulate, a building's top surface.

Rubber A material capable of quickly recovering from large deformations, normally insoluble in boiling solvent such as benzene, methyl ethyl ketone, and ethanol toluene azeotrope. A rubber in its modified state retracts within 1 min to less than 1.5 times its original length after being stretched to twice its length.

Sacrificial protection Reducing the extent of corrosion of a metal in an electrolyte by coupling it to another metal that is electrochemically more active in the environment, i.e., galvanic protection.

Saddle A small structure that helps to channel surface water to drains. Frequently located in a valley, a saddle is often constructed like a small hip roof or like a pyramid with a diamond-shaped base. (*See also* **Cricket**.)

Sandwich panel A panel assembly used as covering; it consists of an insulating core material with inner and outer skins.

SBA Systems Builders Association (formerly MBDA).

SBC Standard Building Code. (*See* **SBCCI**.)

SBCCI Southern Building Code Congress International, Inc. (*See* **Model codes**.)

Scanning-line frequency The number of lines scanned by an infrared scanner per second. The number of lines in a TV picture divided by the number of times per second that the picture is repeated.

Scarf To scrape or abrade a surface to remove degraded or wet polyurethane foam.

Scrim A woven, open-mesh reinforcing fabric made from continuous-filament yarn. Used in the reinforcement of polymeric sheeting.

Scupper A channel through a parapet, designed for peripheral drainage of the roof, usually a safety overflow to limit accumulation of ponded rainwater caused by clogged drains.

Sealant Any material used to close up cracks or joints to protect against leaks. Lap sealant is applied to exposed lap edges in E/P systems.

Sealing washer A metal-backed rubber washer assembled on a screw to prevent water from migrating through the screw hole.

Seam strength The strength of a seam of material measured in either shear or peel mode, reported either in absolute units, e.g., pounds per inch of width, or as a percent of the sheeting strength.

Self-drilling screw A fastener that drills and taps its own hole; used as a fastener for attaching panels to purlins and girts.

Self-tapping screw A fastener that forms receiving threads when turned into a previously drilled hole. It is used for attaching panels to purlins and girts and for connecting trim and flashing.

Selvage An edge or edging which differs from the main part of (1) a fabric, or (2) granule-surfaced roll roofing.

Service life Anticipated useful life of a building, building component, or building subsystem (e.g., roof system).

Shark fin Curled felt projecting upward through the flood coat and aggregate of a BUR membrane.

Shear The force tending to make two contacting parts slide upon each other in opposite directions parallel to their plane of contact.

Sheeting A form of plastic or rubber in which the thickness is very small in proportion to the length and width and in which the polymer compound is present as a continuous phase throughout, with or without fabric.

Shelf life The maximum safe time to store a fluid or uncured construction material before use.

Shingle (1) A small unit of prepared roofing designed for installation with similar units in overlapping rows on inclines normally exceeding 25 percent. (2) To cover with shingles. (3) To apply any sheet material in overlapping rows like shingles.

Shingling (1) The procedure of laying parallel felts so that one longitudinal edge of each felt overlaps and the other longitudinal edge underlaps an adjacent felt. (*See also* **Ply.**) Normally, felts are shingled on a slope so that the water flows over rather than against each lap. (2) The application of shingles to a sloped roof.

SI The international system of metric units (Système International d'Unités).

Side lap The continuous overlap of closures along the side of a panel.

Sieve An apparatus with square apertures for separating sizes of material.

Sill The bottom horizontal framing member of an opening such as a window or door.

Single slope A sloping roof with one surface. The slope is from one wall to the opposite wall of a rectangular building.

Siphon break A small groove to arrest the capillary action of two adjacent surfaces.

Skylight A roof accessory to admit light, normally mounted on a curbed, framed opening.

Slab A semifinished steel product, intermediate between ingot and plate, with the width at least twice the thickness.

Slippage Relative lateral movement of adjacent felts (or sheets) in a roof membrane. It occurs mainly in roofing membranes on a slope, sometimes exposing the lower plies or even the base sheet to the weather.

Slope The tangent of the angle between the roof surface and the horizontal plane, expressed as a percentage or in inches of rise per foot of horizontal distance. (*See also* **Incline**.)

Smooth-surfaced roof A roof membrane without mineral aggregate surfacing.

Soffit The underside covering of any exterior overhanging section of a roof, gable, or sidewall.

Softening point The temperature at which a bitumen becomes soft enough to flow as determined by an arbitrary, closely defined method.

Softening-point drift Change in the softening point during storage or application. (*See also* **Fallback**.)

Solid mopping *See* **Mopping**.

Spectral absorptance The quantity characterizing the ability of a substance to retain incident radiation. It is equal to the ratio of the amount of radiation absorbed by a substance to that absorbed by a blackbody for a given wavelength.

Spectral reflectance The quantity characterizing the surface of a substance equal to the ratio of its reflected incident radiation to that reflected by a blackbody for a given wavelength.

Sprinkle mopping *See* **Mopping**.

Spud To remove the roofing aggregate and most of the bituminous top coating by scraping and chipping.

Spudder A heavy steel implement with a dull, bevel-edged blade for removing embedded aggregate from a BUR membrane surface.

Spunbonded A generic name for nonwoven fabrics formed directly from polymer chips, spun into continuous filaments which are laid down and bonded continuously, without an intermediate step.

Spunlaced A hydroentangled nonwoven fabric; a dry-laid staple fabric is mechanically bonded by a water jet which entangles the individual fibers.

Square A roof area of 9.29 m^2 (100 ft^2), or enough material to cover 9.29 m^2 of deck.

Stack vent A vertical outlet designed to relieve pressure exerted by water vapor between a membrane and the vapor retarder or deck.

Stainless steel An alloy of steel which contains a high percentage of chromium. Also may contain nickel or copper. Has excellent resistance to corrosion.

Standing seam A type of watertight seam featuring an upturned rib, which may also be structural. It is made by turning up the edges of two adjacent metal panels and then folding them over in one of a variety of ways.

Standing water test An evaluation in which test panels are alternately submerged in aqueous solutions and dried in air.

Starting platform A movable platform used to support a seaming machine as it begins to roll-seal a metal seam.

Steep asphalt A relatively viscous roofing asphalt conforming to the requirements of Specification D312, Type III.

Strain Deformation under stress.

Strawberry *See* **Blueberry.**

Stress (1) A measure of the load on a structural member in terms of force per unit area (kips per square inch, MPa). (2) The force acting across a unit area in solid material in resisting the separation, compacting, or sliding that tends to be induced by external forces. Also the ratio of applied load to the initial cross-sectional area, or the maximum stress in the outer fibers due to an applied flexural load.

Stress concentration A condition in which stress is highly localized, usually induced by an abrupt change in the shape of a member or at a substrate joint (e.g., between insulation joints).

Stress relaxation The time-dependent change in the stress resulting from application of a constant total strain to a specimen at a constant temperature. The stress relaxation at a given elapsed time is equal to the maximum stress resulting when the strain is applied minus the stress at the given time.

Strikethrough A term used in the manufacture of fabric-reinforced polymeric sheeting to indicate that two layers of polymer have made bonding contact through the reinforcing scrim.

Strippable films Added protection sometimes applied to continuous strip in the coil-coating process. It is applied after prime and top coats to resist damage prior to and during erection.

Stripping (strip flashing) (1) The technique of sealing a joint between metal and bituminous membrane with one or two plies of felt or fabric and hot- or cold-applied bitumen. (2) The technique of taping joints between insulation boards or deck panels.

Substantial completion The stage in the progress of the work when it is sufficiently complete for the owner to occupy or utilize the work for its intended use.

Substrate The surface upon which a roof component is placed (structural deck or insulation).

Super-steep asphalt A high-viscosity roofing asphalt conforming to the requirements of Specification D312, Type IV.

Supported sheeting *See* **Fabric reinforcement.**

Surface cure Curing or vulcanization which occurs in a thin layer on the surface of a manufactured polymeric sheet or other items.

Surfactants Surface active agents that reduce surface tension when dissolved in water or water solutions, or that reduce interfacial tension between two liquids or between a liquid and a solid.

Susceptibility When not otherwise qualified, the degree of change in viscosity with temperature.

Tack-free A film is considered tack-free when the finger, with a slight pressure, will not leave a mark. The surface will not be sticky.

Tapered edge strip A tapered insulation strip used to elevate the roofing at the perimeter and at penetrations of the roof.

Tar boils Bubbles of moisture vapor encased in a thin film of bitumen, also known as blueberries, blackberries, etc.

Tear strength The maximum force required to tear a specified specimen, the force acting substantially parallel to the major axis of the test specimen. Measured in both initiated and uninitiated modes. The obtained value is dependent on specimen geometry, rate of extension, and type of fabric reinforcement. Values are reported as stress, e.g., pounds, or stress per unit of thickness, e.g., pounds per inch.

Tearoff Removal of a failed roof system down to the structural deck surface.

Tensile strength (1) The maximum tensile stress per unit of original cross-sectional area applied during stretching of a specimen to break (units: SI metric, megapascal or kilopascal; customary, pounds per square inch). (2) The longitudinal pulling stress a material can bear without tearing apart. (3) The ratio of maximum load to original cross-sectional area. Also called *ultimate strength*.

Tensile test A test in which a specimen is subjected to increasing longitudinal pulling stress until fracture occurs.

Therm A unit of heat commonly used by utilities, equivalent to 100,000 Btu.

Thermal block A spacer of low-thermal-conductance material, designed to prevent formation of a thermal bridge.

Thermal bridge Interruption of a layer of thermal insulation by a material of high thermal conductivity (e.g., metal).

Thermal conductance (C) The rate of heat flow, in Btus per hour, through a square foot of a material or combination of materials whose surfaces have a temperature differential of 1°F [W/(m$^2 \cdot$ °C)].

Thermal conductivity (k) The rate of heat flow, in Btus per hour, through a square foot of material exactly one inch thick whose surfaces have a temperature differential of 1°F [W/(m$^2 \cdot$ °C)].

Thermal insulation A material designed to reduce the conductive heat flow.

Thermal resistance (R) Resistance to heat flow; the reciprocal of conductance (C).

Thermal shock A stress-producing phenomenon resulting from sudden temperature drops in a roof membrane—when, for example, a rain shower follows brilliant sunshine.

Thermalization of fast neutrons The process of reducing the energy of a neutron to a level where it is in equilibrium with its environment. Generally, thermal neutrons have energy levels in the 0.01 to 0.3 electronvolt range. Thermalization occurs when the energy of fast neutrons is partially absorbed by moderators or hydrogen atom collisions.

Thermogram A visible-light record of the display of an infrared camera system via a Polaroid print, 35-mm film, or videotape.

Thermography A technique for producing heat pictures from the invisible radiant energy emitted from stationary or moving objects at any distance and without in any way influencing the temperature of the objects under view. The electronic generation and display of a visible image of an infrared spectrum.

Thermoplastic Capable of being repeatedly softened by increase in temperature and hardened by decrease in temperature. The thermoplastic form allows for easier seaming, both in the factory and in the field.

Thermoplastic elastomers Polymers capable of remelt, but exhibiting elastomeric properties; related to elasticized polyolefins. They have a limited upper temperature service range.

Thermoplastic resin A material with a linear macromolecular structure that will repeatedly soften when heated and harden when cooled.

Thermoset A material that will undergo (or has undergone) a chemical reaction by the action of heat, catalysts, ultraviolet light, etc., leading to a relatively infusible state.

Thread count The number of threads per inch in each direction, with the warp mentioned first and the fill second; e.g., a thread count of 20×10 means 20 threads per inch in the warp direction and 10 threads per inch in the fill direction.

Through-wall flashing A water-resistant membrane or material assembly extending totally through a wall and its cavities, positioned to direct any water within the wall to the exterior.

Time value of money The time-dependent value of money stemming both from changes in the purchasing power of money (that is, inflation or deflation) and from the real earning potential of alternative investments over time.

Toggle bolt A two-piece assembly consisting of a threaded bolt and an expanding clip that can fit through a drilled bolt hole, then spring outward to provide anchorage from the blind side.

Trim The lightgage metal used in the finish of a building, especially around openings and at intersections of surfaces. Often referred to as *flashing*.

Tuck pointing The filling in with fresh mortar of cut out or defective mortar joints in masonry.

TV-compatible An infrared camera system whose frame rate and line frequency are compatible with closed-circuit TV systems.

UBC Uniform Building Code. (*See* **ICBO** and **Model codes**.)

Ultimate elongation The elongation of a stretched specimen at the time of break. Usually reported as a percent of the original length. Also called *breaking strain*.

Unsupported sheeting A polymer sheeting one or more plies thick without a reinforcing fabric layer or scrim.

Valley gutter A channel used to carry off water at the intersection of two sloping roof planes.

Vapor barrier *See* **Vapor retarder.**

Vapor migration The flow of water vapor from a region of high vapor pressure to a region of lower vapor pressure.

Vapor pressure The pressure exerted by a vapor that is in equilibrium with its solid or liquid form.

Vapor retarder A material that resists the flow of water vapor.

Vent An opening designed to convey water vapor or other gas from inside a building or a building component to the atmosphere.

Ventilator An accessory usually used on the roof that allows air to pass through.

Vermiculite An aggregate used in lightweight insulating concrete, formed by heating and expanding a micaceous mineral.

Viscoelastic Characterized by changing mechanical behavior, from nearly elastic at low temperature to plastic, like a viscous fluid, at high temperature.

Viscosity Index of a fluid's internal resistance to flow, measured in centistokes (cSt) for bitumens. (Water has a viscosity of roughly 1 cSt, light cooking oil 100 cSt.)

Vulcanization An irreversible process during which a rubber compound, through a change in its chemical structure, e.g., cross-linking, becomes less plastic and more resistant to swelling by organic liquids, and elastic properties are conferred, improved, or extended over a greater range of temperature.

Warp In textiles, the lengthwise yarns in a woven fabric.

Water-vapor transmission (WVT) Water-vapor flow normal to two parallel surfaces of a material, through a unit area, under the conditions of a specified test such as ASTM E96. Customary units are grains/(h · ft^2).

Waterproofing Treatment of a surface or structure to prevent the passage of water under hydrostatic pressure.

Weatherometer An instrument used to subject specimens to accelerated weathering conditions, e.g., rich ultraviolet source and water spray.

Wicking The process of moisture movement by capillary action, as contrasted to movement of water vapor.

Wind uplift Upward-acting pressure resulting from air motion across a roof.

Yield strength (1) The longitudinal stress a material can bear before plastic deformation (i.e., elongation under constant stress). (2) The stress at which a material exhibits a specified reduction in the constant stress/strain ratio in the elastic range or the stress at which a material exhibits a specified deviation from proportionality of stress and strain.

Appendix A

TABLE A.1 SI Conversion Factors

To convert from	To	Multiply by
°F	°C	0.556 (°F−32)
°C	°F	1.8(°C)+32
mm	in.	0.03937
in.	mm	25.400
g/m²	lb/square (100 ft²)	0.0205
lb/square (100 ft²)	g/m²	48.83
pcf (lb/ft³)	kg/m³	16.02
kg/m³	pcf	0.0624
psf (lb/ft²)	kg/m²	4.882
kg/m²	psf	0.205
psi (lbf/in.²)	kPa	6.895
kPa	psi	0.1450
lbf/in.	kN/m	0.1751
kN/m	lbf/in.	5.710
Perm (vapor permeance) [grain/(ft² · h · in.Hg)]	ng/(m² · s · Pa)	57.45
R [°F/(Btuh · ft²)]	R [K/(W · m²)]	0.176
R [K/(W · m²)]	R [°F/(Btuh · ft²)]	5.678
U [Btuh/(ft² · °F)]	U [W/(m² · K)]	5.678
U [W/(m² · K)]	U [Btuh/(ft² · °F)]	0.176
grain	g	0.0648
Btu	J	1055
gal/square (100 ft²)	L/m²	0.408

Key to abbreviations:
°F = degrees Fahrenheit
°C = degrees Celsius
K = kelvin (= °C + 273.15)
Btuh = Btu per h
lbf = pound · force
N = newton (= 0.2248 lbf)
kN = kilonewton = 10^3 N
m = meter
mm = millimeter
g = gram
kg = kilogram (= 10^3 g)
ng = nanogram (= 10^{-6} kg = 10^{-9} g)
Pa = pascal (= N/m²)
kPa = 10^3 Pa
W = watt (= 3.414 Btuh)

TABLE A.2 Important ASTM Standard Specifications for Roofing and Waterproofing

ASTM designation	Standard specification for
	Bitumens
D312	Asphalt Used in Roofing
D449	Asphalt Used in Dampproofing and Waterproofing
D450	Coal Tar Pitch for Roofing, Dampproofing, and Waterproofing
	Felts
D224	Smooth-Surfaced Asphalt Roll Roofing (Organic Felt)
D226	Asphalt-Saturated Organic Felt Base Sheet Used in Roofing and Waterproofing
D227	Coal-Tar Saturated Organic Felt Used in Roofing and Waterproofing
D249	Asphalt Roll Roofing (Organic Felt) Surfaced with Mineral Granules
D371	Asphalt Roll Roofing (Organic Felt) Surfaced with Mineral Granules; Wide Selvage
D2178	Asphalt Glass Felt Used in Roofing and Waterproofing
D2626	Asphalt-Saturated and Coated Organic Felt Base Sheet Used in Roofing
D3909	Asphalt Roll Roofing (Glass Felt) Surfaced with Mineral Granules
D4601	Asphalt-Coated Glass Fiber Base Sheet Used in Roofing
D4897	Asphalt-Coated Glass Fiber Venting Base Sheet Used in Roofing
D4990	Coal Tar Glass Felt Used in Roofing and Waterproofing
	Fabrics
D173	Bitumen-Saturated Cotton Fabrics Used in Roofing and Waterproofing
D1327	Bitumen-Saturated Woven Burlap Fabrics Used in Roofing and Waterproofing
D1668	Glass Fabrics (Woven and Treated) for Roofing and Waterproofing
	Primers and Cements
D41	Asphalt Primer Used in Roofing, Dampproofing and Waterproofing
D43	Creosote Primer Used in Roofing, Dampproofing and Waterproofing
D2822	Asphalt Roof Cement
D3019	Lap Cement Used with Asphalt Roll Roofing
D3747	Emulsified Asphalt Adhesive for Adhering Roof Insulation
D4022	Coal Tar Roof Cement
D4586	Asphalt Roof Cement—Asbestos Free
D5643	Coal Tar Roof Cement—Asbestos Free
E102	Saybolt-Furol Viscosity of Bituminous Materials at High Temperature
	Surfacing
D1227	Emulsified Asphalt Used as a Protective Coating for Roofing
D1863	Mineral Aggregate Used on Built-up Roofs

TABLE A.2 **Important ASTM Standard Specifications for Roofing and Waterproofing** (*Continued*)

ASTM designation	Standard specification for
D2823	Asphalt Roof Coatings
D2824	Aluminum-Pigmented Asphalt Roof Coatings
D3805	Application of Aluminum-Pigmented Asphalt Roof Coatings
D4479	Asphalt Roof Coatings—Asbestos Free
	Single-ply Membrane
D3468	Liquid-Applied Neoprene and Chlorosulfonated Polyethylene Used in Roofing and Waterproofing
D4434	Poly (vinyl Chloride) Sheet Roofing
D4637	Vulcanized Rubber Sheet Used in Single-Ply Roof Membrane
D4811	Nonvulcanized Rubber Sheet Used as Roof Flashing
D5019	Reinforced Non-Vulcanized Polymeric Sheet Used in Roofing Membrane
	Metal Roofing
E1514	Structural Standing Seam Steel Roof Panel Systems
E1592	Performance of Sheet Metal Roof Systems by Static Air Pressure
	Polyurethane Foam Roofing
D5469	New Spray Applied Polyurethane Foam Roofing Systems
	General Application
D36	Softening Point of Bitumen (Ring and Ball)
D61	Softening Point of Pitches (Cube-in-Water Method)
D2523	Testing Load-Strain Properties of Roofing Membranes
D2829	Sampling and Analysis of Built-up Roofs
D3105	Elastomeric and Plastomeric Roofing and Waterproofing Materials
D3617	Sampling and Analysis of New Built-up Membranes
D3746	Comparative Impact Resistance of Bituminous Roofing Systems
D3805	Uplift Resistance of Steel Deck, Insulation and BUR, Field Testing
D4402	Viscosity Determinations of Unfilled Asphalts Using the Brookfield Thermoset Apparatus
D5036	Application of Aluminum-Pigmented Roof Coatings
D5076	Measuring Voids in Roofing Membranes
D5081	Hiding Power of Roofing Aggregate
D5082	Application of Intermittently Attached PVC Roofing
D5100	Adhesion of Aggregate in Hot Bitumen
D5147	Testing Modified Bituminous Sheet Material
D5405	Time-to-Failure of Single-Ply Joints
D5602	Static Puncture Resistance
D5635	Dynamic Puncture Resistance
D5636	Low-Temperature Unrolling of Felt or Sheet Roofing
E84	Surface Burning Characteristics of Building Materials
E96	Water Vapor Transmission of Materials in Sheet Form
E108	Fire Tests of Roof Coverings
E119	Fire Tests of Building Construction and Materials

TABLE A.3 Important ASTM Standard Specifications for Thermal Insulation

ASTM designation	Standard specification for
C208	Insulating Board (Cellulosic Fiber)
C552	Cellular Glass Block and Pipe Thermal Insulation
C578	Preformed, Block-Type Cellular Polystyrene Thermal Insulation
C591	Preformed Cellular Urethane Thermal Insulation
C726	Mineral Fiber Roof Insulation Board
C728	Perlite Thermal Insulation Board
C984	Perlite-Urethane Composite
C1013	Membrane-Faced Rigid Cellular Polyurethane Roof Insulation
C1050	Cellular Polystyrene–Cellulosic Fiber Composite Board Insulation
C1289	Faced Rigid Cellular Polyisocyanurate Thermal Insulation

ASTM designation	Standard test method or recommended practice for measuring
C165	Compressive Properties of Thermal Insulation
C177	Steady-State Thermal Transmission Properties by Means of the Guarded Hot Plate
C203	Breaking Load and Calculated Flexural Strength of Preformed Block-Type Thermal Insulation
C236	Thermal Performance by Guarded Hot Box
C518	Steady-State Thermal Transmission Properties by Means of the Heat-Flow Meter
C855	Thermal Resistance Factors for Preformed, Above-Deck Roof Insulation

Appendix B

Index and Abstract of ASTM Test Methods for Roofing Materials*

C-203   Breaking Load and Flexural Properties of Thermal Insulation

A bar of test material is tested as a simple beam. Breaking load and flexural strength can be determined.

C-303   Density

By measuring the volume and dry weight of a specimen, density is calculated.

C-836   Liquid Applied Elastomeric Waterproofing

This is a materials specification, but it includes a crack-bridging test method using an Aymar automatic compression and extension machine.

D-36   Softening Point of Bitumen (Ring-and-Ball Apparatus)

Specimens are heated in water or glycerin. The end point is when the steel ball sags 1 in. (25 mm) (see Fig. B.1).

---

*Complete Test Methods available from ASTM, 100 Bar Harbor Drive, West Conshohocken, PA 19428-2959

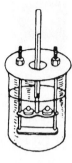

Figure B.1 ASTM D36 ring-and-ball softening-point test apparatus.

D-92 Flash and Fire Point by Cleveland Open Cup

The test cup filled with sample, the temperature is raised, and a small test flame is passed across the cup. The lowest temperature at which the vapors above the surface of the liquid ignite is the flash point. The fire point is when the liquid ignites and burns for at least 5 s (see Fig. B.2).

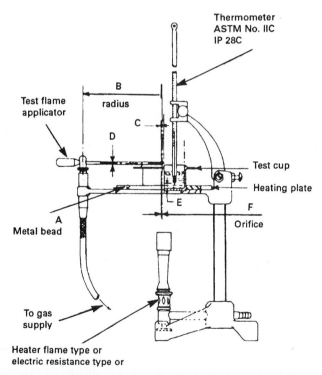

Figure B.2 ASTM D92 flash-point test apparatus.

D-95 Water in Petroleum Products by Distillation

Bituminous material is refluxed with water-immersible solvent. Water is collected in a trap (see Fig. B.3).

D-146 Testing Bitumen Saturated Felts and Woven Fabrics

Describes testing strength (1×6 in. strips at room temperature), pliability, loss on heating, etc.

D-228 Test Methods for Asphalt Roll Roofing, Cap Sheets and Shingles

Includes moisture, pliability, behavior upon heating, and analysis of composition.

D-297 Chemical Analysis of Rubber

This method covers the analysis of many types of rubber products to determine the amount and type of nonrubber constituents as well as to identify IR, CR, NR, IIR, NBR, and SBR types of polymers.

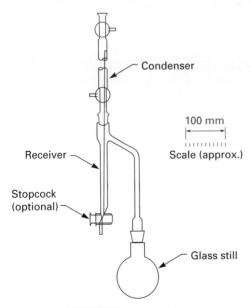

Figure B.3 ASTM D95 test apparatus.

D-412 Rubber Properties in Tension

Measurements of tensile stress at given elongations, tensile strength, ultimate elongation, and tensile set are included. Rate of jaw separation is usually 8.5 mm/s (20 in./min), and specimens may be dumbbell, straight, or cut ring specimens (see Fig. B.4).

D-413 Rubber Property—Adhesion to Flexible Substrate

An adhesion test consisting of applying a force sufficient to strip a test specimen from an adhered surface. The results are reported as (1) the average force required to cause separation at a definite rate or (2) the average rate of separation caused by a known force.

D-471 Rubber Property—Effect of Liquids

Procedures for immersing test specimens in liquids under definite conditions of temperature and time. Results are reported as change in weight, volume, tensile, elongation, or hardness.

D-518 Rubber Deterioration—Surfacing Cracking

Three procedures are used to estimate the comparative ability of soft rubber compounds to withstand the effects of normal weathering or exposure to ozone following Method D-1149.

Procedure A—straight specimens

Procedure B—looped specimens

Procedure C—tapered specimens

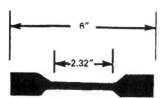

Figure B.4 Dumbbell specimen for tensile testing of elastomeric sheets, per ASTM D412.

D-543 Resistance of Plastics to Chemical Reagents

This method covers testing of plastic materials for resistance to chemical reagents. Included are methods for changes in weight, dimensions, appearance, and strength.

D-568 Rate of Burning and/or Extent and Time of Burning of Flexible Plastic in Vertical Position

This is a small-scale laboratory screening procedure for comparing the relative rate of burning of plastics. Specimens are suspended vertically and exposed to a gas flame at the lower end.

D-570 Water Absorption of Plastics

Specimens are immersed in room-temperature water or boiling water, blotted dry, and weighed. Increase in weight is reported as increase in weight percent.

D-573 Test for Rubber Deterioration in an Air Oven

Tensile properties and hardness measured before and after air oven exposure.

D-618 Conditioning Plastics for Testing

Establishes standard laboratory conditions of temperature (73.4°F) and humidity (50%), as well as procedures for conditioning.

D-624 Tear Resistance of Rubber

Three different die shapes are offered: Die A, a razor-nicked crescent specimen; Die B, a razor-nicked crescent specimen with tap ends; and Die C, an unnicked 90° angle specimen (see Fig. B.5). The rate of jaw separation is 20 in./min and may be run at elevated temperatures. The resistance to tear is reported in newtons (or pounds-force) required to tear a specimen 1 m (1 in.) in width.

D-638 Tensile Properties of Plastics

Used for materials thicker than 1 mm (0.04 in.). (See D-882 for thin flexible sheeting.) Determines tensile properties of dumbbell-shaped test specimens.

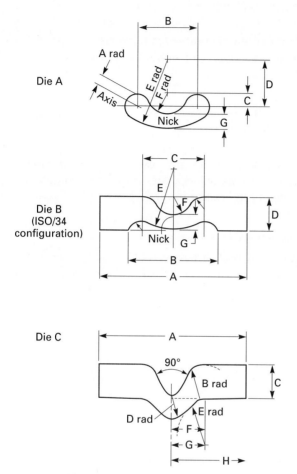

Figure B.5 Three die shapes are used for rubber tear resistance.

D-696 Coefficient of Linear Thermal Expansion of Plastics

Uses a vitreous silica Dilatometer to measure changes in length, usually between −22 and 86°F.

D-746 Cold Brittle Point

The specimen is clamped so that it acts as a cantilever beam. After conditioning, the specimen is struck by a striking arm at 6.5 ft/s.

This method has been found useful for specification purposes, but does not necessarily measure the lowest temperature at which the material may be used.

Definition: Brittleness temperature is the temperature at which 50 percent of the specimens would fail in the specified test.

D-750 Rubber Deterioration in Carbon Arc of Weathering Apparatus

Specimens of rubber are held in a jig under a specified strain. The primary criterion is to establish resistance to light aging by decrease in elongation at break or change in tensile strength.

D-751 Testing Coated Fabrics

Describes methods to measure length, width, etc. Breaking strength is tested by the grab method (Procedure A) at 12 in./min or by the cut strip method (Procedure B, 1-in.-wide strips). Elongation is measured by the same method. Tests are also included for bursting strength, tearing strength, hydrostatic resistance, coating adhesion, tack-tear resistance, seam strength, etc.

D-756 Determination of Weight and Shape Changes under Accelerated Conditions

Useful for interior (unexposed) conditions of heat and humidity on plastic parts. Uses air oven and saturated salt conditions.

D-816 Methods of Testing Rubber Cements

This standard covers two groups of test methods. The first includes procedures in which the quality of a bond is evaluated, such as adhesion strength, bonding range, softening point, and cold flow. The second includes procedures applicable to the adhesive itself, such as viscosity, stability, cold brittleness, density, or plastic deformation.

D-822 Tests on Coatings Using Carbon Arc

Panels are exposed to ultraviolet and moisture. Typical cycles are at 145°F and include water spray.

D-882 Tensile Properties of Thin Plastic Sheeting

Tests sheeting less than 1 mm (40 mils) in thickness. Thin strips are pulled by the pendulum or load cell method.

D-1002 Strength Properties of Adhesives in Shear by Tension Loading (Metal to Metal)

Two strips of metal are bonded together and pulled in vertical shear.

D-1004 Initial Tear Resistance of Plastic Film and Sheeting

Covers determination of the tear resistance of flexible film and sheeting at very low rates of loading (2 in./min). The test is designed to measure the force to initiate tearing.

The resistance to tear is only partially dependent upon thickness. Therefore, the tearing resistance is expressed in maximum newtons (or pounds-force) of force to tear the specimen.

D-1042 Dimensional Changes in Plastics under Accelerated Service Conditions

This method provides a means for measuring dimensional changes in plastic specimens as a result of exposure to service conditions. Marks are scribed on the specimen and measured with a measuring microscope.

D-1149 Surface Ozone Cracking in a Chamber (Flat Rubber Specimens)

This method covers the estimation of the resistance of vulcanized rubber to cracking when exposed to an atmosphere containing ozone. The rubber specimens are kept under a surface tensile strain, and the ozone content in the test chamber is kept at a fixed value. Specimens described in method D-518 are used. (Procedure A, 20 percent elongation; Procedure B, bent loop, 18 percent elongation; Procedure C, tapered specimen, 10, 15, and 20 percent elongation.)

D-1171 Rubber Deterioration—Surface Ozone Cracking Outdoors or Chamber (Triangular Specimens)

The method covers estimation of the relative ability of rubber or synthetic elastomer compounds used for automotive parts to withstand weathering conditions. Procedures are given for preparing triangular-cross-section specimens, for mounting them in a strained condition around specified mandrels, and for rating the effect of exposure as evidenced by the appearance of minute surface cracks.

D-1203 Loss of Plasticizer from Plastics (Activated Carbon Methods)

Specimens are heated to 70°C (Method A) in contact with activated carbon, or to 100°C (Method B) using a wire cage so that direct contact is avoided. Weight loss is expressed as a percent, and all lost weight is assumed to be plasticizer.

D-1204 Linear Dimensional Changes of Non-Rigid Thermoplastic Sheeting or Film at Elevated Temperature

This method is particularly applicable to specimens made by the calender or extrusion process. It covers measurement of changes resulting from exposure of the material to specified conditions of elevated temperature and time. It uses a convection oven and dusted paper to ensure freedom of movement. Measurements are made after specimens are allowed to cool to laboratory standard conditions.

D-1310 Flash Point of Liquid by Tag Open-Cup Apparatus

This procedure determines the flash point of liquid between 0 and 325°F (flash in the presence of a flame).

D-1499 Operating Light and Water-Exposure Apparatus (Carbon-Arc Type) for Exposure of Plastics

This covers variables for practice G-23, such as black panel temperature of 145°F and water spray.

D-1621 Compressive Properties of Rigid Cellular Plastics

The method provides information on the behavior of cellular materials under compressive loads. Deformation data can be obtained, and from a completed curve it is possible to compute the compressive stress at any load and to compute the effective modulus of elasticity. Compressive strength reported under Procedure A is the stress at the yield point if a yield point occurs before 10 percent deformation. In the absence of such a yield point, it is the stress at 10 percent deformation. Under Procedure B, substitute 2 percent strain for 10 percent deformation.

D-1622 Density of Rigid Cellular Plastics

Specimens are measured and weighed and density is calculated.

D-1653 Permeability of Organic Coatings

A free or supported film is sealed to a test cup, and measured by either the dry or wet cup method. Water vapor transmission, permeance, and permeability are reported.

D-1709 Impact Resistance of P.E. Film by Free-Falling Dart Method

Two procedures are described:

A—Uses $1\frac{1}{2}$-in.-diameter hemisphere dart dropped from 26 in.

B—Uses 2-in.-diameter hemisphere dart dropped from 60 in.

D-1790 Brittleness of Plastic Film by Impact

This test was devised by SPI specifically for plasticized vinyl films. It determines the temperature at which plastic film less than 10 mils (0.25 mm) exhibits brittle failure under specified impact conditions.

D-1876 Peel Resistance of Adhesives (T-Peel Test)

See Fig. B.6.

D-1922 Propagation Tear Resistance of Plastic Film and Thin Sheeting by Pendulum Testing

A 20-mm (0.8-in.) slit is cut into the specimen prior to testing. The force in grams required to propagate tearing across a film or sheeting specimen is measured by a calibrated pendulum device.

D-2136 Low Temperature Bend Test—Coated Fabrics

This method covers the effect of bending coated fabrics at low temperature. Flat specimens are placed on a bending jig at desired temperature and rapidly bent. Fractures or coating cracks constitute failure.

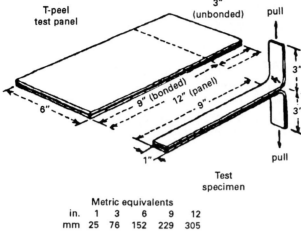

Figure B.6 T-peel resistance test.

D-2137 Rubber Property—Brittleness Point of Flexible Polymers & Coated Fabrics

This method covers the evaluation of long time effects such as crystallization, low temperature, incompatibility of plasticizer, etc. Data obtained by this test method may be useful in predicting behavior of flexible polymeric materials at low temperature only where the conditions of deformation are similar to the test method. Samples are clamped as a cantilever beam and impacted with a striker.

D-2240 Rubber Property—Durometer Hardness

This procedure covers Type A Durometers for measuring soft materials and Type D for harder materials. This method permits measurements of either initial indentations or indentations after specified periods of time. The specimen is placed on a hard, horizontal surface and the Durometer pressure foot is applied with sufficient pressure to obtain firm contact. The more resistance to indentation, the higher the Durometer reading.

D-2523 Load Strain Properties of Roofing Membranes

Dog-bone specimens 10×1 in. are tested for tensile strength and elongation at low temperatures (usually 30, 0, −30°F) and low rates of jaw separation (0.05 in./min) (see Fig. B.7).

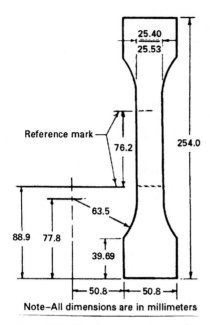

Figure B.7 Dog-bone specimens for low-temperature tensile and elongation tests.

D-2565 Operating Xenon Arc-Type (Water-cooled) Light Exposure Apparatus for Exposure of Plastics

Tests are conducted at a black panel temperature of 63°C (145°F) and controlled RH.

D-2794 Impact Test on Organic Coatings

A steel punch with a hemispherical head rests on a coated steel panel. A weight inside a guide tube is dropped from varying heights until failure is noted.

D-2824 Specification for Aluminum-Pigmented Asphalt Roof Coatings

This includes tests for nonvolatile matter and consistency (see Fig. B.8).

D-2829 Sampling BUR

This practice is a guide for removing test specimens from built-up roofing systems in the field and determining the approximate quanti-

Appendix B   583

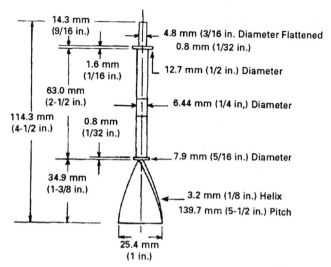

Figure B.8 ASTM D2824 propeller rotor used with Stormer viscometer.

ties of the components, including insulation, plies of felt interply bitumen, top coating, and surfacing (see Fig. B.9).

D-2856   Open Cell Content of Rigid Cellular Plastics

An air pychnometer is used following Boyle's law, which states that the decrease in volume of a confined gas results in a proportionate increase in pressure.

D-2939   Testing Emulsified Bitumens Used as Protective Coating

Tests include density, brush or spray application, solids, ash, flammability, wet flow, firm set, heat test, flexibility, and resistance to water and direct flame.

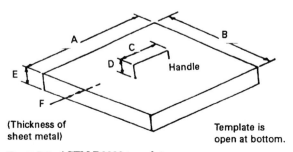

Figure B.9 ASTM D2829 template.

D-2990 Test Methods for Tensile, Compressive and Flexural Creep and Creep-Rupture of Plastics (under Specified Environmental Conditions)

Specimens are measured under constant tensile, compressive, or flexural load.

D-3105 Index of Methods for Testing Elastomeric and Plastomeric Roofing and Waterproofing Materials

This index provides a reference to aid in the selection of procedures and test methods for single-plies and liquid systems. Groups of methods include those for liquid-applied systems as received; sheet-applied as received; and film characteristics of both types.

D-3045 Heat Aging of Plastics without Load

This practice defines the exposure conditions for testing the resistance of plastics to oxidation or other degradation when exposed solely to hot air for extended periods of time.

D-3389 Abrasion Resistance

This method uses a double-headed abrader on a rotary platform.

D-3746 Comparative Impact Resistance of Bituminous Roofing Systems

A missile is dropped with an impact energy of 30 J (22 lbf · ft). Damage is assessed visually after solvent extraction of the bitumen.

D-4073 Tensile-Tear Strength of Bituminous Roofing Membranes

This method determines the maximum load in pounds-force (newtons) to tear specimens when tested at 0.1 in./min (see Fig. B.10).

D-4442 Direct Moisture Content of Wood and Wood-base Materials

Oven drying and distillation procedures are described.

D-4444 Use and Calibration of Hand-Held Moisture Meters

Both conductance and dielectric meters are described.

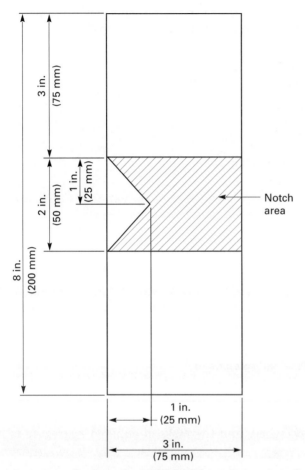

Figure B.10 Specimen for ASTM D4073 tensile-tear test of bituminous roof membrane.

D-4798 Xenon Arc Method for Accelerated Weathering of Bituminous Materials

Thin films of bitumen are applied to aluminum panels and exposed to cycles of temperature, light, and water.

D-4932 Fastener Rupture and Tear Resistance of Roofing and Waterproofing Sheets, Roll Roofing and Shingles

The force perpendicular to the fastener shank to tear the sheet is measured (tear).

The force parallel to the fastener shank to rupture the sheet is measured (rupture or head pull-through).

D-4977 Granule Adhesion to Mineral Surfaced Roofing by Abrasion

A brush with bristles is used to abrade the granule surface. The loss of weight is reported.

D-5076 Measuring Voids in Roofing and Waterproofing Membranes

Specimens are frozen and separated. A transparent sheet and flow pen are used to mark all voids, via counting an overlay grid.

D-5081 Measuring Aggregate Layer Hiding Power

The method measures the quantity of aggregate needed to provide an opaque layer. Light-sensitive paper is placed in an exposure box, covered with aggregate, and exposed to a light source.

D-5100 Adhesion of Mineral Aggregate to Hot Bitumen

Specimens are prepared using measured amounts of bituminous pour coat and mineral aggregate surfacing. After cooling, loose aggregate is removed and weighed, and the mass of adhered aggregate is reported.

D-5147 Sampling and Testing Modified Bituminous Sheet Material

Included are methods for thickness, load strain properties (area under load-elongation curve), tear strength, low-temperature flexibility (1-in.-diameter mandrel), and compound stability (suspended in oven).

D-5405 Test Method for Conducting Time-to-Failure

Tests of time-to-failure (creep-rupture) of joints fabricated from nonbituminous organic roof membrane material are included. Both T-peel and lap-shear joints are subjected to constant tensile load.

E-96 Water Vapor Transmission of Materials in Sheet Form

This method covers several procedures that are available with high humidity on one side, lower on the other. Units of permeance, permeability, and water vapor transmission are given.

Standard test conditions that have been useful are:

Procedure A—Desiccant method at 73.4°F (23°C)

Procedure B—Water method at 73.4°F (23°C)

Procedure BW—Inverted water method at 73.4°F (23°C)

Procedure C—Desiccant method at 90°F (32.2°C)

Procedure D—Water method at 90°F (32.2°C)

Procedure E—Desiccant method at 100°F (37.8°C)

E-96 Metric Conversions*,†

Multiply	By	To obtain (for the same test condition)
WVT		
g/(h · m²)	1.43	grains/(h · ft²)
grains/(h · ft²)	0.697	g/(h · m²)
Permeance		
g/(Pa · s · m²)	1.75×10^7	1 Perm (inch-pound)
1 Perm (inch-pound)	5.72×10^{-8}	g/(Pa · s · m²)
Permeability		
g/(Pa · s · m)	6.88×10^{10}	1 Perm-inch
1 Perm-inch	1.45×10^{-9}	g/(Pa · s · m)

*These units are used in the construction trade. Other units may be used in other standards.

†All conversions of mmHg to Pa are made at a temperature of 0°C.

E-108 Fire Tests of Roof Coverings

Class A, B, and C classifications resulting from intermittent flame, spread of flame, burning brand, and flying brand tests (see Fig. B.11).

G-21 Determining Resistance of Synthetic Polymeric Materials to Fungi

Specimens are inoculated with suitable organisms and exposed under conditions favorable to growth. Tests for changes in properties are conducted after exposure is concluded.

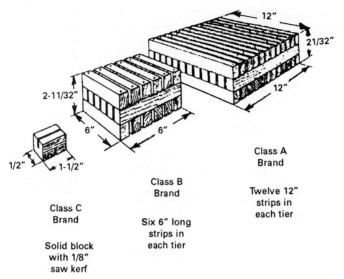

Figure B.11 Wood test blocks for fire tests per ASTM E108.

G-23 Practice for Operating Carbon Arc for Exposure of Non-Metallic Materials

The report should include type of light source, wattage, type of filters, time of exposure, etc.

G-53 Light- and Water-Exposure Apparatus (Fluorescent UV—Condensation Type)

This simulates water as rain or dew, as well as ultraviolet exposure. A typical cycle may be 4 h ultraviolet at 60°C, 4 h condensation at 50°C.

Newly Issued ASTM Specifications

D-5635 Dynamic Puncture Resistance

A chisel-headed impact is struck on the roof membrane specimen. Energy to puncture is reported in joules (see Fig. B.12).

D-5602 Static Puncture Resistance

A roof-membrane test specimen is set on a thermal insulation substrate and subjected to a predetermined static puncture force for 24 h (see Fig. B.13). Puncture is reported in newtons (N) to puncture.

Appendix B 589

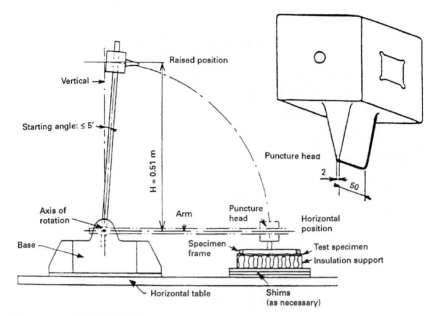

Figure B.12 ASTM D5635 dynamic puncture setup.

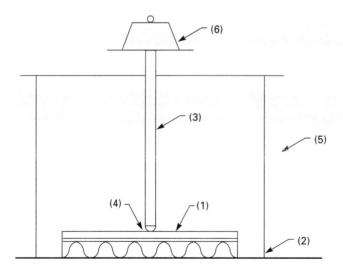

(1) Membrane test specimen
(2) Insulation substrate
(3) Movable rod
(4) Ball bearing
(5) Framework supporting the movable rod and load
(6) Load

Figure B.13 ASTM D5602 static-puncture resistance test setup.

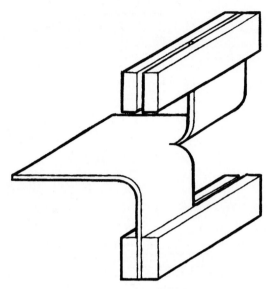

Figure B.14 Schematic of ASTM D5601 tearing resistance with sample mounted in testing machine.

D-5601 Tearing Resistance of Roofing and Waterproofing Materials

The tearing resistance of materials is measured by the tongue (single rip) method using a constant rate of extension (see Fig. B.14). Peak and average tearing force are reported.

Name Index

Aamot, H.W., 385, 388

Baker, M.C., 222
Baxter, R.P., 296
Benedict, Dan, xv
Butler, O., xv

Carothers, Bob, xv
Carlson, Jim, xv
Cash, C.G., 80, 82, 207, 221–224, 232, 247
Chaize, Alan, 23, 24, 281
Clapperton, Tony, 73
Cogneau, P., xv
Cullen, W.C., 5, 21, 226, 236, 238, 261, 263, 284, 408
Curtis, G.B., 212

Davis, Duane, 77
Dees, J.G., 76
DuPuis, R.M., 76, 112

Epstein, K.A., 383

Fabier, Bruno, 23, 24
Fricklas, R.L., 220, 490
Funk, Smith, xv

Goetz, Peter, 524, 528, 529
Gorgati, Romolo, 271
Gumpertz, Werner, 80

Haddock, Rob, 438
Henshell, J.A., xv, 234, 243, 506, 514

Kind, R.J., 148, 152

LaCosse, R., 76
Lund, C.E., 73

Martel, John, 512
Martin, J. Roy, 235
Mathey, R.G., 21, 80, 236, 238, 284
Mirra, Ed, 80
Moore, R.J., 234
Murphy, Colin, 176

Phillips, Stephen M., 525
Powell, Frank, 108
Putnam, L.E., 383

Richards, D.E., 234
Rissmiller, E.H., 225
Robinson, Jack, xv
Rossiter, W.J., 80, 408

Schuman, E.C., 250
Schwartz, T.A., 232, 247
Smith, D.J., 27
Smith, G.A. Jr., xv, 125
Smith, Phil, xv
Smith, T.J., xv, 145
Stewart, Ed, xv

Tibbets, D.C., 222
Tobiasson, W., xv, 13, 74, 86, 104, 108–111, 112, 479

Wardlaw, R.L., 148, 152
Wood Shields, Patricia, xv

Subject Index

Acrylic elastomers, fluid-applied, 407, 413–414
  illus., 88
Adhered systems, 200
Adhesives:
  cold-applied (*see* Cold adhesives)
  contact, 304, 473
  hot-mopped, 54, 77–78, 84, 128, 187, 282
    illus., 129
  pressure-sensitive tapes, 304
  single-ply joints, 335, 473, 475
  solvent-welded joints, 317
  torching, 282
Aggregate, 223
Aggregate surfacing, 15, 202, 219, 465, 473, 498
  illus., 60, 224, 229
  ballast, 70, 147, 152, 386–387, 490, 501
  benefits, 15, 33, 182, 194, 206, 219, 220
  defective, 222
  disadvantages, 220, 496, 511
  double-poured flood coat, 28, 223
  fire resistance, 182
  flood coat, 33, 206, 223
    illus., 229
  function, 15, 182, 205, 207, 219–220, 418
  grading, 221
  gravel, 15, 222, 418
  heat reflective, 227
  maximum slope, 153, 221
  properties, 15, 221, 302, 418
  PMR, 418
  removal, 220, 488, 499–511
    illus., 500
  requirements, 221
  shielding function, 15, 182, 205, 219, 221, 418
  sizing, 221
  slag, 15, 222
  undersizing hazards, 152, 221, 418
  (*See also* Ballast)
Air:
  chimney effect, 99, 116
    illus., 116
  leakage, 99, 116, 387
    effect of wind suction on, 117
    paths, 116
  thermal conductivity, 86

Air-conditioning units, rooftop, 325, 332–334, 341, 425
  recommended details, 365, 425
Alaska:
  Fairbanks, winter temperature, 377
  PMR, 374, 388
Alligatoring, causes, 3, 64, 233
  illus., 3
Aluminum coating, 227
  benefits, 226
American Institute of Architects (AIA), 228
American National Standards Institute (ANSI)
  fire test, 188
  wind-uplift design, 153–154
  (*See also* ASCE)
American Society of Civil Engineers (ASCE), 124–125, 133, 138–139, 143, 154
  wind map, 126
American Society of Heating, Refrigeration and Air-Conditioning Engineers (ASHRAE), 90, 107
American Society for Testing and Materials (ASTM), 7, 518
  bitumen requirements, 509, 515
  felt requirements, 217–218
  fire tests, 182, 184, 188, 509
  index of Test Methods, 571
  labels, 513
  roofing and waterproofing standards, table, 568
  standard specifications, 568–570
  test methods, 571–590
  thermal insulation standards, table, 570
  weldable single-ply specifications, 323
Anchorage, 5, 11, 16
  adhesives (*see* Adhesives)
  mechanical (*see* Mechanical fasteners)
Architect's representative, field inspection, 462
Architect's responsibility, 10, 17, 157, 325, 496, 505, 509, 512
Asbestos felts:
  failed, 7
  materials, 329
  moisture vulnerability, 195

593

594 Index

Asphalt, 195, 208–213, 329
  accelerated degradation, 64, 80, 215, 375
  application temperature, 54, 211, 283, 471
  cutbacks, 183, 225, 415
  vs. coal tar pitch, 115, 208, 215
  cold-weather application, 54, 128, 471
  cooling rate, 54, 128, 130, 211, 267
    illus., 55, 129–130
  degradation, 80
  equiviscous temperature (EVT), 212, 267, 465, 471
  emulsions, 183, 225
  flat (Type II, as multipurpose asphalt), 212
  manufacture, 208–211, 267, 272
    illus., 210
  overheating hazard, 208, 262, 471
  photooxidative degradation, 222, 262, 375
  properties, 208, 515
    table, 209
  slope limits, 261
    table, 265
  softening-point fallback, 262, 266
  steep asphalt drawbacks, 54, 128, 211, 261, 515
  temperature control, 212, 267, 471
  testing, 5
  types, 209, 392, 402, 515
    table, 265
  viscosity, 211, 264
  waterproofing, 15, 225, 392, 402
Asphalt cutbacks, 225
  fire classification, 183
  on SPF, 415
Asphalt emulsions, 197, 225
  advantages, 183
  application, 225
  fire classification, 183
Asphalt Roofing Manufacturers Association (ARMA):
  application tolerances, 514
  slope recommendation, 28
Asphaltenes, 210
Atactic polypropylene (APP), 203, 271
  production, 272

Backnailing, 264, 471
Ballast:
  aggregate, 201, 302, 501
    illus., 201, 303
  concrete pavers, 151, 156
  EPDM, 302
  flotation weight, 152, 162
  hazards from loss, 152
  PMR, 152, 162
  purposes, 70, 200
  requirements, 155–157
  scour, 151, 314
  shielding function, 155, 200
  specifications, 153, 302
  weight requirements, 153, 156, 165, 201, 303
    illus., 169
  wind-uplift mechanics, 147, 148
Ballasted system (see Loose, ballasted system)
Ballooning, 172
  illus., 169
Base flashing (see Flashing)
Base sheet (see Felts)

Batten bars, 167
  illus., 308
Beadboard polystyrene foam (see Polystyrene insulation)
Below-deck insulation, 64–65
Bernoulli principle, 117, 131
Bitumen kettle, equipment, 212
Bitumen-modified polyurethane coating, 329
Bitumens:
  accelerated degradation, 64, 80, 241, 290
  asphalt (see Asphalt)
  coal tar pitch (see Coal tar pitch)
  degradation, 64, 214
  effect of ultraviolet radiation, 215
  function, 14, 101, 205, 223, 283
  incompatibility between, 214
  properties, 207, 213
    table, 209
  waterproofing, 207, 214, 225, 392, 402
  (See also Asphalt; Coal tar pitch)
Blisters, 240–252, 291–294
  causes, 229, 242, 250, 293
  clues to origins, 212, 243, 249, 251, 291
  effect on service life, 241, 292
  experiment, 247
  growth, 64, 246–249
    illus., 242, 244, 245, 248, 251, 292
  incidence, 240
  interfacial, 292
  internal pressure, 246, 250, 293
  interply, 293
  limiting size, 246
  liquid coatings, 429
  mechanics, 243, 244, 245
  modified bitumen, 281, 291–294
  moisture role, 243
  origin, 212, 220, 243
  prevalence, 240
  prevention, 293
  resealing, 252, 498
  size, 241
  SPF blistering, 407
  test cut, 249
    illus., 251, 407
  types, illus., 242, 292
  urethane-promoted, 188
Blowoffs, 125
  causes, 125, 127, 145–147
  failure mode, 125
  failure plane, 127
    illus., 129
  loose gravel, 148–152, 225
  origin, 127
Blowstill, 210
Bond (see Manufacturer's bond)
Boundary layer, 135
Breaking strain:
  built-up membranes, 236, 240, 252, 296, 299
  elastomeric membranes, 252
  glass-fiber, 277
  liquid coatings, 40
  modified bitumen, 252, 281, 285–286
  polyester, 277
Brooming of felts, 267
Btu, 89
Buckling, EPDM, 311

Index   595

Building Codes, 9, 20, 190, 495
  BOCA, 28–29, 190
  fire requirements, 188, 190
  slope requirements, 5, 28, 528
  South Florida, 20
  UBC, 28, 496
  wind requirements, 20, 127
Built-up roof membrane (BUR), 13, 194–195,
    205, 268, 463
  advantages, 194, 208
  aggregate-surfaced (see Aggregate surfacing)
  breaking strain, 236, 252, 299, 339
  elements, 195, 205
  failures, 240–265
    (See also Membrane failures)
  fatigue strength, 239
  field inspection, 267
  fire classification, 183
  flood coat, 223–225
    (See also Flood coat)
  history, 194–195
  joints, 235
  market share, 193
  mineral-surfaced roofing, 207, 226
  performance criteria, 22, 237–238
  ply-on-ply pattern, 228
    illus., 230
  recommended practices, 227–228, 265–267,
    472
  service life, 6, 207
  smooth-surfaced (see Smooth-surfaced membranes)
  specifications, 227–228
  stress relaxation, 237
  surfacing, 15, 219, 250
  tensile strength, 207, 238, 240
  thermal coefficient, 238, 339
  thermal expansion/contraction, 235, 339,
    351
  thermal shock factor, 239
  thermal stress, 13, 236, 253, 339, 351
  two-ply failures, xiii, 7, 227, 251
  viscoelasticity, 237, 253
Built-up roof system:
  components, 10–11, 13, 194
    illus., 14, 206, 229
  design factors, 22
  function, 14, 206
  interdependence, 15, 205
Burning-brand test (UL), 183
Butyl rubber, 196, 304, 392

Calorimeter, 173, 187
Cants, 297, 327, 340, 343, 353
  function, 336
    illus., 327
  material, 297, 336, 475
Cap flashing, 326, 353, 355
  illus., 364
  (See also Counterflashing)
Cap Sheet, 207, 226
Capacitance moisture meters, 484
Carbon black, 299
  in EPDM, 299
Celotex (see Jim Walter)
Cement (see Flashing cement)
Centistoke (cSt), 221, 264

Charles's law, 114
Checking, 290
Chimney effect, 99, 116
Chlorinated rubber primer, 502
Chlorosulfonated polyethylene (CSPE), 318
  chalking, 320
  early use, 196
  liquid-applied, 318
  sheets, 320
Class 1 deck assembly:
  difference from Class 2, 187
  fire requirements, 173, 187
  wind-uplift requirements, 140, 166
Cleats, 341
Closed-joint system, 396
Coal tar pitch, 207–215
  vs. asphalt, 195, 213, 215
  cold flow, 214
  EVT, 213
  flood coat, 28
  production, 213
  properties, 213, 496
  slope limits, 208, 214, 264, 496
  types, 214, 402
  waterproofing, 214, 402
  (See also Bitumen)
Coated base sheet, 216, 230
Coated felts, 230–232
  blistering tendency, 231
  disadvantages, 231
  manufacture, 216
  phased construction, 231–234
  production, 219
  ridging, 259
  (See also Felts)
Cold adhesives, 54, 78, 128, 283, 328, 332, 339
  time to develop strength, 128, 304
  unreliability, 54, 78, 128, 304, 335
  vulnerability to deflection, 128, 335
Cold-flow property, 128, 214, 317
Cold Regions Research & Engineering
    Laboratories (CRREL), 13, 383
Cold-process membranes, 197
  advantages, 197
  repair, 197
  surfacing, 107
Cold-weather application, asphalt:
  hazards, 130, 211, 267
  recommended practices, 266
Color (see Surface color)
Concrete decks, 46, 373
  deflection, 51, 390, 492
  heat capacity, 70–71
  plastic flow, 51
  precast, 46, 390, 492
  primer, 471
  PMR surface preparation, 425
  shrinkage cracking, 492
  surface requirements, 52, 402, 425, 471, 502
  types, 46
Concrete insulating fill, lightweight, 55, 66–73
  advantages, 70
  aggregates, 66
  anchorage, 91
    illus., 49
  application, illus., 69
  field controls, 53, 91, 94, 236
  function, 78

596   Index

Concrete insulating fill, lightweight (*Cont.*):
  hazards, 50, 74, 91, 390
  insulating loss, 74, 86
  mechanical anchors, 75
  minimum depth, 91
  moisture absorption, 72–74, 86
  perlite aggregate, 66
  recommendations, 91–92, 94
  shrinkage cracking, 53, 256
  stress concentration, 53
  venting, 73–75, 91
  vermiculite aggregate, 66
Concrete pavers, 151, 156, 381
  illus., 380
  lightweight, 162
  primary purpose, 151, 386
  PMR, 386
  requirements, 157, 162
  secondary purposes, 386
Condensation, 84, 88, 95, 107, 441
  causes, 114
  hazards, 8, 441, 481
  prevention, 441
    (*See also* Vapor retarders)
Conduction, thermal (*see* Thermal conduction)
Conductivity, thermal (*see* Thermal conductivity)
Contractor's responsibility, 17, 510
Control joints, 237
Convection, 84–85
Cooling-load temperature, 10
Counterflashing (cap flashing), 15, 326, 332, 340, 442
  illus., 327, 333, 342, 355
Creep, 51
  concrete, 51
  thermoplastics, 317
  wood, 51
Crickets, 37, 44
  illus., 38
Curled felts, 232
Cutback (*see* Asphalt cutback)
Cutoff, 93
  detail, 93

Dalton's law of pressures, 114–115, 251
"Dead level," 30, 34
  (*See also* Slope)
Deck (*see* Structural deck)
Deflection:
  creep, 51
  ponding hazards, 51
  steel deck, 59
  structural wood fiber, 51
Delamination, 70, 294
  modified bitumen, 294
Design factors, 9, 13, 28, 70
Dew point, 84, 96, 102, 114, 497
  reroofing project, 497
  tabulated, 96
Diagonal wrinkling, 351–352
  causes, 351
  illus., 352
Diffusion, 98
Double-layered insulation, 82
  advantages, 82–83
  anchorage, 83

Double-layered insulation (*Cont.*):
  illus., 470
  pattern, 83
  ridging alleviation, 82
Double pour, 28
Drainage, 27–44
  benefits, 27
  building codes, 28–29
  design rules, 36–40, 43, 440, 496
  existing roofs, 40, 495, 499
  interior, 35, 37, 44, 515
  peripheral, 35, 44
  pipe sizing, 13, 39
    table, 43
  plans, 5
  plumbing codes, 35
  ponding risks, 30–33
  reasons for, 2, 5, 27, 34
  scuppers, 35–36, 39
  siphon system, 40, 496
  slope, 28, 43, 496
  smooth-surfaced roof, 33
  solar-powered, 43, 496
  waterproofed decks, 393–396
    illus., 395
Drains:
  clogged, 10
  field check list, 468
  location, 38, 515
  minimum number, 37
  overflow, 35, 496
  pipe sizing, 39
    table, 43
  PMR, 378
    elevation, 515
    freeze-protection, 378
    illus., 378
    location, 38, 515
  metal roofs, 440
  siphon, 40
  sump pans, 44
    illus., 38
  waterproofed deck, illus., 390
Drippage, 56, 60
Dry deck test, 52, 469–470
Dry film thickness, 415
Drying shrinkage, 55

Edge venting, 75
Elastomer, 197, 299
Elastomeric membranes (*see* Synthetic rubber membranes)
Electrical resistance moisture meters, 486
Emulsions (*see* Asphalt emulsions)
Envelope, 103, 121
  benefits, 103
Equilibrium moisture content, 73, 87, 467
  deck materials, 467
  fabrics, 467
  felts, 467
  insulation, 73, 86, 467
  wood, 467
Equiviscous temperature (EVT), 261
  coal tar pitch, 214
  necessity of, 211
  tolerances, 267
  types I and III asphalt, 211–213

Index 597

Ethylene propylene diene monomer (EPDM):
  advantages, 198
  application, illus., 302, 305
  ballast, 302
  breaking strain, 299
  chemical nature, 198, 299, 310
  failure modes, 6, 309–314
  field seaming, 302, 304
    illus., 305
  flashings, 304–307, 309–310
    illus., 311–314
  jointing technique, 302, 305, 310
  manufacture, 299–300
  mechanically fastened, 309
  "memory," 304
  permeance, 112
  properties, 301
  punctures, 309–310
  reinforced, 300
  relaxation, 473
  repair, 311–314
  sheet size, 300
  sheet thickness, 300
  shrinkage, 304, 309–310
  tearing, 309–310
  uncured, 306
  unreinforced, 300
  vulnerability to oil, 300
  (See also Single ply membrane; Synthetic rubber membranes)
Expansion/contraction joints, 81, 235, 391
  at deck-span change, 53, 235–236, 257
  defective, illus., 347–349
  details, 366, 368, 391
  function, 235–237
  locations, 53, 81, 235
  metal roof:
    illus., 442
  spacing, 5, 81, 235–236, 257
External fire resistance, 181–184
Extruded polystyrene foam (see Polystyrene insulation)

Fabric, 152, 197, 317, 391, 402
  (See also Reinforcement)
Factory Mutual (FM) Research Corporation, 23, 124–125, 181
  Approval Guide, 140, 166, 187
  calorimeter test, 173, 187
  fastener patterns, illus., 142
  fasteners, 142, 173
  fatigue test, 167
  fire classifications, 187, 380, 500
  steel deck recommendations, 57, 187
  wind design, 139–145
  wind uplift failure, 125, 338
  wind uplift tests, 154, 166–172
    fatigue, 166–168
    field, 169–172
    laboratory, 166
Failures:
  blowoffs, illus., 129, 176
  BUR, 6, 240–265
  causes, 2–5, 15, 240
  EPDM, 6
  factors, 2, 5, 10, 32, 173
  fatigue, 23, 177, 259
  flashing, 7, 31, 309

Failures (Cont.):
  hail, 205, 239, 375
  incidence, xiii, 15, 173, 240, 309
  lap-seam, 22, 309
  litigation, 1, 5, 76, 250, 512
  mechanical fasteners, 6, 168, 176
  thermal contraction, 173, 375
  polystyrene foam, 2, 16
  polyvinyl chloride, 7, 322–323
  ponding effect, 2, 32
  system, 2, 7, 15, 240
  two-ply membrane, 7
Fallback (see Softening point fallback)
Fastest-mile wind speed, 125, 127
  hurricane, Iniki, 127
Fasteners, 173–179
  backout, 177
  benefits, 54
  corrosion, 176–177
  drill points, 178
  field check list, 469
    illus., 49
  loading, 175
  patterns, 142
  pullout resistance, 176–177
  requirements, 143
  stress plates, 178–179
  thread design, 178
Felts, 215
  application, illus., 206, 229, 463
  asbestos, 6, 195, 240, 465
    (See also Asbestos felts)
  backnailing, 264, 471
  base sheet, 219, 280
  brooming, illus., 206
  coated, 75, 91, 101, 216, 230–231, 497
  combination sheets, 56
  evolution, 195
  fiberglass (see Fiberlgass felts)
  function, 14, 101, 205, 207, 219
  mineral-surfaced, 182, 219
  moisture content, 247
  organic, 6, 101, 195, 215, 219, 452
    (See also Organic Felts)
  polyester, 216
  properties, table, 217–218
  rosin paper, 56, 91, 298
  saturated, 101, 216, 452
  shingling (see Shingling of felts)
  storage, 466
    illus., 466–467
  water absorption, 195, 219, 232, 247, 255, 259
    (See also Fiberglass, Organic felts)
Fiberboard insulation, 13, 63, 76, 87, 91
  joint taping, 81–82
  moisture damage, 76, 91
  properties, table, 67–68, 70, 72
Fiberglass felts, 1, 6, 56, 195, 216, 329
  advantages, 195
  application, illus., 206
  blistering, 291
  dominance, 213
  history, 195
  moisture resistance, 195, 234
  in PMR, 379
  porosity, 214
  properties, 213, 216, 240, 291

Fiberglass felts (*Cont.*):
  table, 217
  in waterproofing membranes, 379, 391
Fiberglass formboard, 111
Fiberglass insulation, 66, 85, 93, 451
  formboard, 111
  joint taping, 81–82
  metal roofing, 451
  moisture effects, 87
  properties, 67
Field inspection, 461–486
  architect's representative, 462, 518
  checklists, 465, 469
    deck, 430, 468
    flashings, 475–476
    insulation application, 469, 471
      illus., 469–470
    membrane application, 430, 471
      illus., 463, 466, 474
    storage, 466
      underlayment, 474
  consultant, 462, 518
  documents, 464–465
  equipment, 464, 475–477
  infrared thermometers, 476
  manufacturer's representative, 462, 518
  owner's representative, 462, 518
  principles, 463, 476
  purposes, 461, 476
  records, 462, 464, 477
  requisites, 430
  vapor retarders, 510–511
  weather forecast, 430, 469
Filter fabric:
  evolution, 195
  fiberglass (*see* Fiberglass felts)
  filter function, 202, 380, 387, 401
  function, 14, 101, 205, 207, 219
  mineral-surfaced, 182, 219
  moisture content, 247
  organic, 6, 101, 195, 215, 219, 452
    (*See also* Organic felts)
  PMR, 380
  polyester, 216
  properties, table, 217–218
  rafting function, 387
  rosin paper, 56, 91, 298
  saturated, 101, 216, 452
  shingling (*see* Shingling of felts)
  storage, 466
    illus., 466–467
  water absorption, 195, 219, 232, 247, 255, 259
  weather forecast, 430, 469
Fire resistance, 16, 181–192
  ASTM tests, 16, 182–183,188
  building codes, 181, 189, 451
  burning-brand test (UL), 184
  Class 1 rating, 102, 173, 187
  conflict with conservation, 189
  flame-exposure test (UL), 183
  flame-spread test (UL), 184–186
  FM tests, 185–187
  FM classifications, 23, 181, 187–188, 451
  hazards, 181, 451
    external, 15, 181
    internal, 16, 181, 184–189, 451
  Livonia (Michigan) fire, 127–128, 185

Fire resistance (*Cont.*):
  recommended practices, 128, 190–191, 451
  time-temperature ratings (endurance), 16, 182, 188–189
  UL classification, 184
  UL tests, 23, 181, 186
Fishmouths, 228, 471–472
  illus., 472
  repair, 472
FIT System, 23, 281
Flame-exposure test (UL), 188
Flame-spread test (UL), 183–184
  parameters, 183
  qualifying conditions, 184
  slope, 183
Flashing, 15, 325
  air-conditioning units, 325, 332–334, 341, 425
  anchorage, 329, 333, 336–338
  base, 15, 326, 329, 331, 335, 421
    illus., 327
    materials, 329–330
  boots, 304
    illus., 307, 443
  cants, 327, 336, 340, 353
  caulking, 332
  cap, 15, 341
    illus., 364
  cement, 103, 328–329, 335
  clearances, illus., 371
  column, illus., 366
  composition, 328
  contours, 333, 336
  corner, 332, 344, 391
  counterflashing (*see* Counterflashing)
  curb, illus., 444
  defective details, 345, 347–349, 355, 357–358
  design principles, 325, 327, 329, 333, 440
  diagonal wrinkling, 336, 351
    illus., 352
  differential movement, 331, 333, 336, 340, 440
  drain, 332, 346, 378, 495
    illus., 370, 378, 424
  elevation, 31, 37, 334
  EPDM, 304–307, 329, 332, 344
    illus., 330
  equipment supports, illus., 365–366
  expansion-contraction joints, 335, 346
    defective, illus., 347–349
    recommended detail, 335, 366–367, 442
  failures, 15, 145, 325, 339, 346, 349, 351
  fascia gages and dimensions, 338, 353
    illus., 149–150
  field inspection, 354
  functions, 10, 15, 326–327
  gravel stops (*see* Gravel stops)
  gutter, built-in, 348, 440
  heat-reflective, 329
  height, 333–335, 353, 475
  location, 337
  materials, 321, 327, 329, 331
  metal base flashing, 321, 328, 331
    illus., 332, 366
  metal roofs, 442
  modified bitumen, 283
  NRCA details, 325, 331, 341, 348, 358–371
  penetrations, 304, 330, 333
    illus., 307, 343, 426

Flashing (Cont.):
  pipes, 329, 354, 426
    illus., 307, 331, 366–367, 443
    support, illus., 357, 371
  pitch pans, 304, 344
    illus., 345
  PMR, 381
  ponding hazard, 350
  post-construction damage, 352
  post support, 366
  recommended practices, 328–329, 334, 343–344, 353–354
  repair of EPDM, illus., 311–314
  requirements, 284,329
  reroofing, 6, 500
  sagging, 284, 329, 335, 349–350
  scuttles, 332
  sealer strips, 340, 346, 354
  single-ply, 330, 335
  skylights, 332, 341
    illus., 423
  stack detail, 330
  standard details, 359–371
  termination, 298, 304, 333
  through-wall, 341, 398
  vapor retarders, 346
  vent details, 330–331
  wall, 329, 340
    illus., 327, 333, 362–363, 422, 457
  wind failures, 145–147
  wind-vulnerable details, 146
Flashing cement, 328
Flat (Type II) asphalt:
  as multipurpose asphalt, 212
  properties, table, 209
Fleece backing, EPDM, 302
Floating roof, 437, 439–440
Flood coat:
  double-poured, 27–28, 233
  function, 206, 223, 225
  pouring illus., 14
  thickness, 206
    illus., 224
  weight, 206, 223
Flood-testing, 373, 400, 492
Flotation-ballast weight, 386
Fluid-applied coatings, 391
  acrylic elastomers, 413–414
  advantages, 392
  bird-pecking, 231
  butyl, 415
  curing, 428
  defects, 429
  disadvantages, 196, 392
  field problems, 196, 393, 429, 431
  maintenance, 430–431
  neoprene/Hypalon, 196, 318, 415
  performance criteria, 417–418
  polyurethane, 392, 407, 413
  preparation, 393, 402
  primer, 430
  properties, illus., 419–420
  quality assurance, 430
  recommended practices, 430
  requirements, 392, 402, 407
  service life, 430
  silicone rubber, 413–415

Fluid-applied coatings (Cont.):
  SPF, 413–418
  substrate, 196, 402
    limitations, 393, 416
  surfacing granules, 418
Foamed glass (foamglass) insulation, 63, 72, 86
  freeze-thaw cycling in, 72
  in PMR, 397
  as vapor retarder, 72
  water resistance, 72
Foamed plastic insulation:
  cells, 65
  foaming agents, 65
  (See also Isocyanurate insulation; Polystyrene insulation; Polyurethane insulation)
Foaming craters, 249–251
Freeze-thaw cycles:
  destructive effects, 32, 35, 70
  flashing damage, 352
  ice damming, 35
  membrane delamination, 107
  membrane weakening, 32
  metal distortion, 35
  parapet damage, 352

General contractor's responsibility, 17, 510
Glass-fiber felts (see Fiberglass felts)
Glaze coating, 233, 267
  purpose, 225, 233
  thickness, 233
  weight, 225, 233
Glossary of roofing terms, 533–565
Gradient wind, 135
Grain, 100, 113
Gravel aggregate (see Aggregate surfacing)
Gravel scour, 148
Gravel stops, 150, 328, 337
  anchorage, 146, 149, 338–339
  defective, 147, 338–339, 493
    illus., 146, 148, 339
  elevation, 338
  flange, 338
  functions, 337–338
  minimum gages, 353
  nailing, 338
  recommended details, 145
    illus., 149, 359–361, 424
  splice joints, 338
Graveling in, 234, 460
Guarantees, 521–531
  manufacturer's 17, 524
    bond, 521–523
    exclusions, 33, 526
    liability limits, 522, 525
    nullification, 527
    Present Worth value, 523
    terms, 19, 527
  roofer's, 530
    exclusions, 530
    terms, 530
  (See also Warranties)
Gusting, 134–135
  response factor, 139
Gutter, built-in, 44
Gutter, exterior, illus., 456

Gypsum deck, 46, 56
  fire classification, 48
  moisture absorption, 54, 467, 492
  shrinkage cracking, 46, 492

HCFC, 65
Heat-capacity, 70–71, 263, 295
  concrete decks, 70–71
    illus., 71
  and slippage, 263
Heat-reflectance (see Solar reflectance)
Heat-reflective surfacing, 226–227
  benefits, 226–227
  materials, 227
Heat transfer:
  calculation, 89–91
  conduction, 84, 86
  conductivity, 85, 89
  convection, 84–85
  latent, 85, 88
  overall coefficient (U), 89–90
  principles, 84–88
  radiation, 84, 482
Hood, illus., 331, 366
Hot-mopped adhesives, 83
Humidity (see Relative humidity)
Hurricanes:
  Andrew, 7, 123, 125, 127
  damage cost, 123
  Hugo, 7, 123, 125
  Iniki, 7, 20, 123, 127
  metal roofs, 125, 443
  SPF performance, 406
  wind speed, 124–125
Hypalon [see Chlorosulfonated polyethylene (CSPE)]

Ice, destructive effects, 32
Impact resistance, liquid coating, 417
Incline (see Slope)
IR (infrared) thermography, 481–483
  illus., 482–483
IR (infrared) thermometer, 212
Inorganic, 6, 14
Inspection (see Field inspection)
Insulating concrete fill (see Concrete insulating fill, lightweight)
Insulation (see Thermal insulation)
Isocyanurate insulation (see Polyurethane and Polyisocyanurate insulation)

Jim Walter Research (Celotex), 223, 225
Johns Manville (see Schuller Corp.)
Joint taping, 56, 60, 81
  benefits, 81
  disadvantages, 82
  illus., 82
"Junk science," 250–252, 256–257

Laboratory wind uplift tests, 134, 156, 166
Lap seams:
  EPDM, 302–303, 309–310
    illus., 305, 308
  failures, 309–310

Lap seams (Cont.):
  hot-mopped, 288
  modified bitumens, 288–289
  neoprene adhesive, 304
  pressure-sensitive tapes, 304
    illus., 305
  torched, 288
  welded, 315
Latent heat transfer, 86, 88
  R factor loss, 88
Leak detection, 4, 491–492
Life-cycle costing, 9, 29, 33, 435
  on Future-Worth basis, 34, 388
  metal roofs, 435
  PMR, 374, 388
  on Present-Worth basis, 383
Lightweight insulating concrete (see Concrete insulating fill, lightweight)
Liquid-applied membranes (see Fluid-applied coatings)
Litigation:
  incidence, 1, 16, 29, 250
  "junk science," 250–252, 256–257
  lawyers' obfuscation, 29–30, 516
  new products, 513
  safeguards against, 4, 512
  test cuts, 516
Livonia (Michigan) fire, 127–128, 185
Loose, ballasted roof systems:
  advantages, 70, 200
  economy, 200–201
    illus., 303
  perimeter anchorage, 154, 201
  PMR, 152, 202, 380, 382, 386
  reroofing, 200, 501–502
  structural requirements, 201–202, 490
  wind design, 147–162
  wind scour, 314
Low-sloped roofs:
  economy, 1–2
  illus., 29

Maintenance:
  program, 10, 354, 476, 531
  sample contract, 531
  scheduled inspections, 476
Manufacturer's bond, 521
  appeal, 524
  decline, 522
  duration, 522
  exclusions, 33, 526
  limitations, 522, 525
  origin, 522
  (See also Guarantee)
Manufacturer's guarantee (see Guarantees, manufacturer's)
Manufacturer's responsibility, 17, 509, 512
Manville Corporation research (see Schuller)
MASTERSPEC, 509
Mechanical fasteners:
  benefits, 54, 187, 254, 307, 500
  corrosion resistance, 177
  failure modes, 6, 127, 170, 177
  FM-approved, 173, 187
  illus., 12, 49, 83, 308–309, 319, 469
  nail holding power, 127, 177
  nail pullout resistance, 175, 177

Mechanical fasteners (*Cont.*):
  pattern, 140
    illus., 83
  problems, 6, 170, 173, 314
  spacing, 171, 474
  varieties, illus., 174
  (*See also* Fasteners)
Membrane (*see* Built-up roof membrane; Single-ply membrane)
Membrane blisters (*see* Blisters)
Membrane failures:
  alligatoring, 3, 64, 233
  blisters (*see* Blisters)
  delamination, 70, 294
  ridging (*see* Ridging)
  seams, 23
  slippage (*see* Slippage)
  splitting (*see* Splitting)
Membrane ridging (*see* Ridging)
Membrane slippage (*see* Slippage)
Membrane splitting (*see* Splitting)
"Memory":
  EPDM, 304
  fiberglass reinforcement, 278
Metal roofing systems, 435
  accessories, 443, 446, 449, 451, 457
  advantages, 435
  coatings, 435, 444–445
    illus., 445, 447–448
  corrosion protection, 444–446
  disadvantages, 436
  fasteners, 449–450
  flashings, 440, 442
    illus., 442, 444–456
  history, 436–437
  problems, 436, 441, 443, 458
  seams, illus., 438–439, 441
  sealants, 449
  thermal insulation, 435, 451
  underlayments, 451
  vapor retarders, 452, 455
Midwest Roofing Contractors Association (MRCA):
  on shrinking of EPDM, 310
    illus., 311–314
  on slope, 28
Mineral granules, 278, 418
Mineral-surfaced cap sheet, 207, 226, 496
Mineral surfaced felts, 216–217
  function, 219
  manufacture, 219
Mineralized wood-fiber deck (*see* Structural wood-fiber deck)
Modified-bitumen membranes, 14, 202–204, 269, 465, 478, 514
  advantages, 203, 269, 281
  application, 203, 282, 288–289
    illus., 204
  base sheet, 277, 281
  chemical nature, 203
  flashing, 269, 283
    illus., 283
  failure modes, 288–297
  history, 202, 270–272
  lap seams, 23, 203, 282, 288, 465, 473
    illus., 289
  low-temperature flexibility, 273
  market share, 193

Modified-bitumen membranes (*Cont.*):
  materials, 272–275
  performance criteria, 22–23, 284
    table, 286
  plasticizing techniques, 203, 271
    illus., 273–274
  problems, 282, 288
  properties, 203, 285
    table, 276
  recommendations, 297–298
  reinforcement, 272, 275, 277
  rubberizing polymers, 272, 285
  slippage, 294–296
  specifications, 280
  strain energy, 285
    illus., 287
  surfacing, 15, 278, 290
  thickness, 22, 269
  torched, 269, 282–283
    illus., 277
  versatility, 270
Moisture absorption:
  lightweight concrete insulating fill, 72–74
  structural deck, 52–53
  (*See also* Equilibrium moisture content)
Moisture detection, 479–486
  benefits, 479–481
  cost, 479
Moisture expansion, insulating materials, table, 67,68
Moisture meters, 479
  capacitance, 484
    illus., 479
  electrical-resistance, 465, 484
  nuclear meter, 484
    illus., 479
  thermographic, 483
Moisture surveys, 74, 108, 479, 485
  benefit-cost ratio, 479
  benefits, 480
  capacitance meter, 484
    illus., 479
  cost, 479, 481
  electrical-resistance meter, 465, 484, 486
  limitations, 481
  necessity, 494
  nuclear meter, 484
    illus., 479
  principles, 481, 483
  purpose, 108, 479, 480
  thermographic meter, 482
    illus., 483
Mop-and-flop, 284
Mopping:
  equiviscous temperature (EVT), 214, 261, 267
  hazards, 78, 128
  modified bitumens, 283, 288
  spot, 55
  strip, 55
  voids, 244–245, 247

Nailer, 145, 468
Nails, pullout resistance, 177
National Bureau of Standards (NBS) (*see* National Institute of Standards and Technology—NIST)

National Institute of Standards and Technology (NIST), 33, 73, 80, 112, 289
  performance criteria, 7, 21, 389
  self-drying system, 73
  slippage research, 211
  survey, 6
National Research Council of Canada, survey, 374
National Roofing Contractors Association (NRCA):
  application tolerances, 514
  flashing details, 331, 358–371
  on PVC shattering, 322
  Project Pinpoint, 5–6, 310, 315
  Roofing materials guide, 506, 514, 525
  on slope, 28
  survey, 1, 6, 194, 240
  technical bulletins, 211, 322
  temporary roofs, 511
  on viscosity, 211
Negative-pressure test, 169–171
  illus., 171
Neoprene, 196, 304
  early use, 7, 196–197
  flashing, 7, 304, 329
    illus., 306
  properties, 301
New products specifications, 22, 512–513, 518
Nuclear moisture meter, 484
  illus., 479

Open-jointed system, 394
Organic felts:
  defects, 219, 232
  history, 195
  manufacture, 216
  properties (tabulated), 215
  water-absorption, 219, 232
  water-weakening, 255
Owens-Corning, 214
Ozone depletion, 65, 408, 410

Parapets, 138
  flashing, 364
  freeze-thaw cycles, 352
  height effect (tabulated), 158–161
  wind effects, 138, 140, 153–155, 157
Partially adhered synthetic membrane systems, 200
Parting agent, 219
Pavers, concrete (*see* Concrete pavers)
Performance criteria, 7, 19, 21, 237
  built-up membranes, 19, 21, 237
  documents, 23–25
  European vs. U.S., 7
  FIT classification, 24, 281
  modified bitumen, 22–23, 281–288
  PMR, 379, 381–382
  requirements (tabulated), 25
  self-drying roof system, 109
  single-ply membranes, 19
  specifications, 508–509
  SPF, 413, 417–418
  thermal insulation, 70–74
  waterproofing decks, 389
Perlite aggregate, 66

Perlite insulation board, 66, 68, 102
  insulating loss, 87
  properties, 67
Perm, 100
Permeance, 100
  index of, 100
  roofing materials, table, 101
Phased application, 231–234, 262, 280
  defined, 231–232
  risks, 231–232
  waterproofing membrane, 228
Phenolic insulation, 65
Photooxidation (*see* Ultraviolet radiation)
Picture framing, 230
Pipe supports, illus., 371
Pitch pans, 344–346
  design rules, 344–346
  illus., 345
  problems, 344
Plastic cement, 328, 344
Plastic foam (*see* Isocyanurate insulation; Polystyrene insulation; Polyurethane insulation)
Plasticizer (*see* Polyvinyl chloride)
Plumbing Code, 35
  drainage, 35
Plywood decks, 48, 56
  H-shaped clips, 468
  moisture limit, 426
  plastic deformation, 57
Polychloroprene (*see* Neoprene)
Polyisobutylene (PIB):
  chemical nature, 198
  defects, 196
  early use, 196
Polyisocyanurate insulation, 77
  foaming agent, 77
  long-term growth, 77
Polymer, 197, 273, 299
Polymerization, 197
  cross-linking, 198
Polystyrene insulation, 16, 65
  beadboard (MEPS), 76, 86
  dimensional stability, 76–77
  extruded (XEPS), 4, 16, 76, 86, 379, 382, 397
  hazards, 4, 16, 77, 471
  moisture absorption, 86, 382–383
  in PMR, 376, 379, 382–383, 397
  shrinkage, 76
  thermal resistance loss, 86
Polyurethane coating, 413
  bitumen-modified, 392
  chemical formulation, 413
  single-component, 392, 414
  two-component, 392, 414
  weathering resistance, 407,411
Polyurethane and polyisocyanurate insulation, 65, 77, 383
  dimensional stability, 77
  facers, 300
  fire resistance, 188
  hazards, 77
  long-term growth, 77
  properties, 67, 409
  sprayed in place (*see* Sprayed polyurethane foam)
  warping, 77

Index 603

Polyurethane Foam Contractors' Association (UFCA), 408
Polyvinyl chloride (PVC) and PVC copolymer membranes, 7, 315, 318, 329, 333, 478
  advantages, 197–198
  ballasted system, illus., 317
  chemical nature, 318
  coated metal flashing, 321
    illus., 322
  failures, 7, 322–323
  fatigue testing, 166–167
  field seaming, 318
  installation technique:
    illus., 316, 319, 321
  plasticizer, 32, 318, 320, 322
  plasticizer loss, 32
  ponding hazards, 32
  reinforcement, 7, 320
  shattering, 322–323
  sheet thickness, 321, 323
  shrinkage, 32, 315
  specifications, 323
  in waterproofed decks, 392
  welded seams:
    illus., 316
Ponding:
  and blisters, illus., 242
  collapses, 5, 10, 31
  and deflection, 5, 51–52
  deleterious effects, 2, 31
    illus., 2
  effect of depth, 30
  effect on failure incidence, 27
  hazards, 5, 10, 27, 30–33
  prevalence, 27, 29
Premature failures (see Failures)
Preroofing conference, 513
Prescriptive criteria, 19–20
Pressure-sensitive tapes, 304
  illus., 305
Primer:
  asphalt cutback, 56, 401, 471
  chlorinated rubber, 502–503
  concrete decks, 56, 401, 502
  purpose, 498
Protected membrane roofs (PMR), 64, 70, 373–388
  advantages, 375–377
  application, 380
  ballast, 152, 162, 382, 386–387
    illus., 164, 386
  components, 374, 379
  concept, 373, 380, 388
  concrete pavers, 386–387
  construction sequence, 380
  design refinements, 152, 379–380, 385
  disadvantages, 378–379
  drainage, 381–383
  drains: elevation, 383
    freeze-protection, illus., 378
    location, 381
  filter fabric, 152, 380, 387
  flashings, 381
  gravel scour, 152
  heat losses, 384
  history, 373
    illus., 374
  insulation materials, 376, 379, 382–383
  insulation requirements, 380–381

Protected membrane roofs (PMR) (Cont.):
  life-cycle costing, 374, 387–388
  membrane: location, 379
    materials, 379
  number, 373
  pavers, 386
  polystyrene insulation, 376, 379, 383
  rafting, 387
  recommended practices, 162, 400–401
  reroofing, 377
  self-drying, 382
  slope, 382
  stress temperature, 64, 375
  thermal efficiency, 376, 384–385
  walk pads, 418
  waterproofed decks, 379, 389
  wind design, 162–165
Protection boards, 61, 393
Psychrometric graph, 97
Pull test, 170, 172
Puncture resistance, 23

Quality assurance, 461
  general rules, 513–515
  inspection program, 462–465
  inspector qualifications, 463
Quality control, 461, 514

Radiation (see Thermal radiation; Ultraviolet radiation)
Rafting:
  benefits, 387
  principle, 387
Rainfall, U.S. cities:
  tabulated, 41–42
Re-covering:
  aggregate-surfaced membranes, 499
  design decision, 489
  flashing, 497
    illus., 497–498
  performance criteria, 497
  requirements, 489–490, 500–501
  risks, 5–6, 487
  service life, 487
  substrate preparation, 498
  walkways, 499
Refrigerated interiors, 91, 101
  hazards, 79
  vapor pressure, 79
  vented plenum, 80
Reglet, 341
Reinforcement:
  breaking strain, 277
  composites, 277
  EPDM, 300
  fiberglass, 277
  modified bitumens, 275–278
  nonwovens, 277
  polyester, 277
  PVC 7, 320, 323
  scrim, 277
Relaxation, EPDM sheets, 310
Relative humidity (RH):
  defined, 96–98, 113

604   Index

Relative humidity (RH) (*Cont.*):
  design data, 95–96, 104–107
Reroofing, 76, 487–503
  compromises, 5, 489–490
  cost, 2, 435, 489
  design decision, 489
  difficulty, 487–488
  flashings, 497, 500
  isolating new membrane, 497, 500
  leak detection, 481, 491
  moisture surveys, 493
  performance criteria, 495
  PMR, 377
  preliminary investigation, 491–494
  re-covering (*see* Re-covering)
  recommended practices, 498, 501–502
    built-up membranes, 502
    loose, ballasted membranes, 501
    SPUF systems, 502–503
  service life, 487
  specifications, 496, 500
  system history, 490, 492
  tear off-replacement (*see* Tear off-replacement)
Ridging, 257–259, 498
  causes, 77, 82, 240, 259, 474
  history, 258
  illus., 258–259
  mechanics, 230, 258
  nature, 258
  persistence, 259
  picture framing, 258–259
    illus., 258
  prevention, 81, 230, 259
Roof-area divider, 237
  illus., 237
Roof Coatings Manufacturer's Association (RCMA), 279
Roof slope (*see* Slope)
Roof surface temperature, 32, 70, 80–81, 84, 89, 220
  color effect, 81, 89, 227, 290
  factors, 295
  graph, 14, 71, 89
  illus., 14, 89
  metal roofs, 437
  ponding effect, 32
Roof systems, 9–10, 13, 193
  conventional built-up (*see* Built-up roof system)
    illus., 11
  loose ballasted (*see* Loose, ballasted roof systems)
  PMR (*see* Protected membrane roofs)
Roofer:
  as independent contractor, 522
  responsibility, 17, 510
Roofer's cement, 328
Roofer's guarantee (*see* Guarantees, roofer's)
Roofing Industry Committee on Wind Issues (RICOWI), 7, 128
  goals, 8, 128, 130

Saddles, 37–38
Schuller (Manville) research, 212
Scuppers, 35, 36, 496
  defective, 39

Scuppers (*Cont.*):
  flow calculation, 36
  function, 35
Sealants, metal roof, 446
Self-drying roof system, 99, 108–112, 383
  design rules, 109, 382
  field problems, 111
  illus., 110–111
  NBS test program, 111
  performance criteria, 109
Shark fin, 232
  illus., 233
Shingling of felts, 228–230, 392
  benefits, 228, 230
  BUR, 237
  disadvantages, 228, 392
  EPDM, 304, 309–310
    incidence, 309
  illus., 229–230
  modified bitumen, 291
  overlaps, 228
Shrinking membrane phenomenon, 291, 310
Single-ply membranes:
  adhered system, 200, 300
  advantages, 194, 200, 299, 304, 315
  application, 473
    illus., 302, 305, 316
  chemical nature, 198, 300, 317, 501
  chlorosulfonated polyethylene (Hypalon), 198, 318, 478
  disadvantages, 300, 320
  elastomers, 198
  EPDM, 112, 197, 299, 478
    table, 301
  failure modes, 309, 322
  fatigue testing, 167–168
  flashing, 304, 321
    illus., 306–307
  fluid-applied (*see* Fluid-applied coatings)
  history, 191, 195–200, 202, 309, 318
  improvements, 304
  joint sealing, 23, 198, 300, 302, 309, 473
    illus., 305
  loose, ballasted system (*see* Loose, ballasted roof systems)
  market share, 193
  mechanically attached, 168, 200, 307
    illus., 308, 319
  modified-asphalt (*see* Modified-asphalt membranes)
  neoprene (*see* Neoprene)
  polymerization, 198, 315
  PVC, 197, 315, 322, 478
  properties, 199
  recommended practices, 473, 502
  reroofing with, 501–502
  repair, 311, 327
    illus., 311–314, 478
  second-generation, 197
  specifier's guidelines, 301
  synthetic rubber (*see* Synthetic rubber membranes)
  table, 199
  thermoplastics, 198
  thermosetting, 198
  waterproofed deck systems, 379, 392, 402
  weldable, 197, 315
Single Ply Roofing Institute (SPRI), 153, 322

Index

Skylights, 423
Slag aggregate, 15, 222
Slippage, 259–265, 291–296
  backnailing, 264
    illus., 264
  BUR, 259–265
  causes, 261–263, 295
  defined, 259
  formula, 263
  illus., 230, 260, 295
  mechanics, 230
  in modified asphalt systems, 294
  modified bitumen, 294–296
  phased application, 262
  prevention, 221, 228, 262, 264
  remedies, 264–265
  slope, 221, 260–261, 294
  timing, 260, 265
  vector diagram, 230, 261, 295
  wheeling effect, 262
Slope, 10, 27, 50
  ARMA recommendation, 28
  building codes, 28–29
  cost, 2, 30
  deck, 50
  economy, 1, 2, 29
  effect on flame spread, 183
  and fire classification, 183
  via insulating fill, 50
  limit, coal tar pitch, 265
  maximum, aggregate surfacing, 221
  metal roofs, 440
  MRCA recommendation, 28
  minimum, table, 207
  NRCA recommendation, 28
  of PMR, 382
  reasons for, 5, 27, 30, 50
  recommended, 28–29, 207
  single-ply membrane, 32
  as slippage factor, 221, 261
  smooth-surfaced roofs, 225
  structural deck, 50
  via tapered insulation, 50, 496
  waterproofed decks, 394, 400–401
Smooth-surfaced membranes, 15, 33, 225
  advantages, 15, 225, 496
  alligatored,, illus., 3
  disadvantages, 15, 226
  ponding hazard, 15, 29, 226
  slope requirement, 29, 207, 226, 496
Softening point, 209–210
Softening-point fallback, 261
Solar-powered drains, 43
Solar radiation (*see* Ultraviolet radiation)
Solar reflectance, 227, 290, 435
  deck surfacings, illus., 377, 399
Solvent-welded joints, 315
Specifications, 9, 505–519
  ASTM standard: roofing and waterproofing, 227
  coordination, 515
    thermal insulation, 569–570
  defective, 505, 513
  division of responsibility, 509, 512, 515, 518
  drawings, 506–507
  master, 509
  methods, 508–509, 512
  new products, 512, 518

Specifications (*Cont.*):
  performance criteria, 508
  quality assurance, 513
  recommended practices, 227, 496, 515, 518
  requirements, 507–509, 518
  reroofing, 496
  special requirements, 511
  statistical sampling, 517
  temporary roofing, 510
  test cuts, 516
  writing style, 507
Splitting, 240, 252–259, 296–297
  BUR, 252–259
  causes, 5, 76, 81, 240, 253–256, 296
    deck deflection, 253
    drying shrinkage, 54, 253
    illus., 251, 256, 297
    incidence, 240
    insulation movement, 5, 81, 255
    modified bitumen, 296–297
    shrinkage cracking, 54, 76, 255
    stress concentration, 5, 54, 76, 79, 81, 253, 376, 493
    stress concentration, illus., 254, 296–297
    substrate cracking, 252, 256
    thermal contraction, 5, 64, 81, 252
    thickened insulation, 81
  diagrams, 254–255
  effect of overweight moppings, 255
  effect of thickened insulation, 81, 287
  prevention, 254
Spot mopping, 55
  illus., 497
Sprayed polyurethane foam (SPF), 64, 405–434, 502
  advantages, 405–406
  application, 420, 422, 427
    illus., 421
  applicator licensing, 524
  blistering, 429
    illus., 407, 409
  coatings, 407, 411, 413, 416, 428
    illus., 419
  deck materials, 420
  density, 409, 411
  disadvantages, 405–406, 427
  field manufacture, 409, 428
  foam surface, 415–417, 427
    illus., 416
  history, 406
    illus., 4
  problems, 408, 421, 429, 431
  properties, 412, 417
    tabulated, 420
  references, 433–434
  reroofing, illus., 424–425, 502–503
  rusting, 407
  substrate preparation, 420
  surface texture, illus., 416
Steel deck roof assembly, Class 1 requirements, 173
Steel decks, 46, 423
  classification, 46, 187
  deflection problems, 57, 59
  deformation, 57, 492
  dishing, 57
  FM recommendations, 58, 187, 515
  installation, 47

Steel decks (*Cont.*):
  leak patterns, 492
  maximum span, illus., 58
  mechanical anchorage, 500
  popularity, 46, 57
  recommended thickness, 58
  rolling, 57
  side-lap fastening, 57–58
  spans, 58
  thickness, 58–60
  underside venting, 65, 75, 91, 98
  with vapor retarder, 102
Storage of materials, field inspection, 61, 465
Strain energy:
  BUR membrane, 296
  modified bitumen, 287, 296
  usable, 287
Stress concentration:
  BUR, 253–254
  causes, 253
  deck-span change, illus., 53
  illus., 53, 254, 256, 297
  modified bitumen, 296–297
Strip mopping, 55
Stripping, 56
Structural deck, 45–61
  anchorage, 11, 54
  classification, 12
  component anchorage, 54
  concrete (*see* Concrete decks)
  deflection, 11, 45, 50–52, 57, 237
  design factors, 45, 48, 468
  design rules, 60–61, 235
  dimensional stability, 12, 45, 51–52, 57
  dryness test, 52, 468
  dynamic loading, 59
  field inspection, 52
  function, 11, 45
  gypsum (*see* Gypsum deck)
  joints, 53, 56, 61, 235, 468
  leveling, 56, 468
  materials, 12, 45
  moisture absorption, 45, 52, 54, 467
  nailable and non-nailable, 12, 47
  noncombustible, 12
  oriented strand board (OSB), 46
  plywood, 12, 48, 426, 437
  precast concrete, 46, 425
  recommended practices, 55, 60, 468
  steel (*see* Steel deck roof assembly; Steel decks)
  surface, 52, 60, 468
  waterproofing (*see* Waterproofed deck systems)
  wood (*see* Wood decks)
Structural wood-fiber decks, 13, 46–47
  joints, 56, 492
  sag, 51, 54, 72, 492
Styrene butadiene styrene (SBS), 203
  advantages, 203–204
  chemical nature, 203, 271, 273
  cross-links, illus., 273
  low-temperature flexibility, 273
  rubberizing process, 273
  slippage, 295
Substrate, 45, 52, 54, 57, 63, 66, 79
Sump, 44

Surface color:
  effect on membrane temperature, 81
    illus., 93
  effect on slippage, 261–263
  solar reflectance values, 399
Surface erosion, 220
Surface temperature (*see* Roof surface temperature)
Surfacing:
  aggregate (*see* Aggregate surfacing)
  ballast (*see* Ballast)
  BUR, 219
  function, 15, 219–220
  granules, 197
  heat-reflective, 220, 295
  metallic foil, 278–279
  mineral granules, 278
  for modified asphalts, 279
  PMR, 373, 386
  smooth-surface coatings, 219
  waterproofed deck systems, 398
Synthetic rubber membranes, 15, 197, 299–314
  breaking strain, 299
  butyl, 196–197, 392, 415
  EPDM (*see* Ethylene propylene diene monomer)
  neoprene (*see* Neoprene)
  polyurethane, 414
  relaxation, 473
  repair, 311–314
  silicone, 415
System design, 9–10
System failures, xiii, 2, 15

Tapered insulation, 78, 499
  disadvantages, 50
  slope, 78, 499
Tear off, 499
Tear off-replacement, 499, 511
  advantages, 489
  conditions justifying, 490, 494, 499
  cost, 33, 400, 495
  difficulty, 487, 489
  disadvantages, 235
  recommended practices, 499–500
Temporary roof, 234, 510
  function, 234–235
  hazards of retaining, 235, 510–511
  NRCA recommendation, 511
Tenting, 177
Test cuts, 461, 477, 493
  arguments against, 516
  of existing system, 493, 502
  insulation board, 173
  random sampling, 516
  size, 173, 494, 517
  statistical sampling calculation, 517
Thermal bridge, 56
  prevention, 57, 82, 376
  steel bulb tees, 56
Thermal conductance (C), 89
Thermal conduction, 84, 86
  effect of temperature, 85
  moisture effects, 480
Thermal conductivity (k), 85, 89

Thermal conductivity (*k*) (*Cont.*):
    moisture effect, 85–86
    temperature effect, 85
Thermal efficiency, PMR, 384–385
Thermal expansion/contraction, 52, 238, 338
    bituminous membranes, 5, 21, 236, 252
        coefficient of, 236
    insulating materials, 77
    metals, 252, 328, 437, 440
    plywood, 52
    wood, 52
Thermal insulation:
    anchorage, 12, 16, 63, 78, 83, 253, 308, 465, 469
        illus., 142–143, 469–470
    ASTM standards, table, 568
    bearing of boards, 79, 254
    below-deck, 64–65, 451
    benefits, 4, 13, 63
    composite polyurethane boards, 66, 188
    compressive strength, 13, 64, 70–71, 81
    condensation, 4, 63, 72, 84
    design factors, 13, 70, 83
    design rules, 385, 451
    dimensional stability, 13, 64, 76–77
    double-layered, 82, 84, 91, 376, 495
        illus., 470
    effect on fire rating, 16, 189
    effect on membrane temperature, 63, 84
        illus., 14, 71
    effects of thickening, 64, 80, 189
    efficiency in PMR, 383, 385
    expanding wet areas, 481
    fiberglass (*see* Fiberglass insulation)
    field inspection, 465
    heat-flow calculation, 89–91, 385
    heat-reflective, 84
    insulating loss, 75, 495
    joint taping, 64, 81, 91, 298
        illus., 82
    liabilities, 4, 63–64
    lightweight concrete (*see* Concrete, insulating fill, lightweight)
    materials, 13, 64
        composite boards, 66
        fiberboard (*see* Fiberboard insulation), 13, 63, 76
        fiberglass (*see* Fiberglass insulation), 63, 72, 451
        foamglass [*see* Foamed glass (Foamglass) insulation], 63
        insulating fills, 13
        perlite board, 63
        SPUF (*see* Sprayed polyurethane foam)
    moisture absorption, 32, 64, 70, 72, 86, 383
        illus., 86
    need for, 4, 13, 63
    performance criteria, 70–74
    plastic foam (*see* Isocyanurate insulation; Polystyrene insulation; Polyurethane insulation; Phenolic insulation)
    PMR requirements, 64, 382–383
    porous, 72
    principles, 83, 85, 451
    properties, table, 67–68, 70–72
    over-refrigerated interiors, 79
    shear strength, 11, 13, 64, 71–72

Thermal insulation (*Cont.*):
    and splitting, 253
    recommendations, 91–94
    tapered, 50, 78, 495
    venting, underside, 64
    waterproofed deck, 396–397
Thermal radiation, 84
    nighttime cooling, 84
Thermal resistance, 89, 396
    cooling season reduction, 88
    defined, 89
    effect of temperature, 85
    insulating materials, table, 67–68
    loss from wetting, 86, 384
        illus., 86–87
    PMR, 387
    ratio (TRR), 87
        illus., 87
    water, 86
Thermal shock, 3, 70, 256, 376
Thermal shock factor (TSF), 239, 256
    aging effect, 239
    built-up membranes, 239
    formula derivation, 239
    meaning, 256
    significance, 256
Thermal stress, built-up roof membrane, 70
Thermographic moisture meters, 481–483
Thermoplastics, 198, 315–323
    chemical nature, 315–317
    failure modes, 322, 323
    flashings, 321
    historical, 318–321
    market share, 193
    materials, 315–318
    (*See also* PVC, CPE, and CSPE)
Thermosetting, 66, 198, 315
Thermosetting insulation fill, 66, 69, 78
    application, illus., 69
Torch-and-flop, 283
Torching, 283
    and blistering, 292
    hazards, 288
    temperatures, 292
Tunnel test (UL), 186
    illus., 186
Two-ply membranes, 7, 227, 251
    failed, 7

Ultimate strain (*see* Breaking strain)
Ultraviolet radiation, 295, 332
    accelerating factors, 32
    effects, 241
        on bitumens, 3, 15, 32, 222
        on EPDM, 15, 200, 332
        on modified bitumen, 15
        on PVC sheet, 32
    hazards, 32, 219
    protection against, 15, 200, 318
Underlayments, 46
    metal roofs, 451–452
Underwriters Laboratories, Inc. (UL), 181, 509
    Roofing Materials and Systems Directory, 166, 184
    classification of membranes, 23, 182, 500
    fire tests, 181–184, 188–189

608 Index

Underwriters Laboratories, Inc. (UL) (*Cont.*):
  wind-uplift requirements, 166, 500
  wind-uplift test, 166
Upside-down roof (*see* Protected membrane roofs)
Urethane (*see* Polyurethane coating; Polyurethane insulation)
U.S. Air Force Survey, 29

Vacuum field wind-uplift tests, 169–172
  apparatus, illus., 171
Vapor control, 95–122
  air leakage, 98–99, 115, 340
  design factors, 121
  diffusion, 88, 99, 113, 230
  field recommendations, 122
  moisture hazards, 95, 230
  recommended practices, 102, 340
  self-drying system, 99, 108
    (*See also* Self-drying roof system)
  techniques, 99–112
  vapor flow (*see* Vapor flow)
  vapor retarders (*see* Vapor retarders)
  ventilation, 99–100
Vapor flow, fundamentals, 95–99, 114–117
Vapor migration, 95, 113, 119
  seasonal reversal, 88, 110–111
Vapor permeance (*see* Permeance)
Vapor pressure, 74, 95–97, 114
  determinants, 74, 95
  graph, 97, 244
  refrigerated interiors, 74, 79–80
Vapor retarders, 4, 10, 12, 60, 100–108, 340, 452
  and blistering, 248
  calculation, 103, 118–121
  controversy, 103–108
  design criteria, 4, 91, 96, 100, 104–108, 121
  field inspection, 510–511
  field problems, 12, 98–99, 103, 108
  flashing, 103
  materials, 12, 101, 121, 452
    illus., 453–454
    perm ratings, 101
  mechanics, illus., 100, 118
  metal roofs, 441
  plastics sheets, 12, 102, 122, 453–454
  pros and cons, 4, 11, 103, 107, 121
  PMR, 377
  ratings, 121
  steel deck assembly, 102
    illus., 102
  substrate, 140
  as temporary roof, 108, 122, 235, 510–5ll
Vent, 74, 91, 112, 334, 498
Ventilation, 65, 99, 112, 498, 501
Venting, 65
  ceiling plenum, 140
  edge, 75, 91
  failure, 74, 112, 334
  illus., 497–498
  recovered membrane, detail, 497
  stack, 74, 112, 498
  of thermal insulation, 65
  topside, 112
  underside, 65, 75, 91, 98, 112

Venting base sheet, 497
Vermiculite aggregate, 66, 73
Viscoelastic, 237
Viscoelasticity, 237, 253, 270
  BUR, 237, 253
Viscosity, asphalt, 211
Vulcanized rubber membranes (*see* Synthetic rubber membranes)

Walking in, 466
Walkways, 61, 499
  illus., 266, 321
  SPF, 418
Wall flashing, details, 342
Warranties (*see* Guarantees)
Water:
  heat of vaporization, 88
  thermal resistance, 72, 88
Warranties, 17, 521–531
  checklist, 528, 530
  coverage, 525
  defects, 523
  exclusions, 526
  manufacturer's, 524
  new products, 513
  nullification, 527–528
  wind provisions, 124
Water vapor (*see* Vapor control; Vapor pressure)
Waterproofed deck systems, 389
  bitumens, 392, 402
  components, 389, 402
    illus., 390
  drainage design, 393–394, 400–401
  drains, illus., 395
  expansion joint, illus., 391
  flashings, 397, 401
    illus., 398
  flood-testing, 400
  insulation location, 396–397
  membrane design, 391
  monolithic (closed joint), 396
  open-jointed, 394
  percolating layer, 394
  protection boards, 393
  recommendations, 401–403
  single-ply membranes, 392, 402–403
  slope, 394, 400–401
  specifications, 389
  surfacing, 398–399, 401
Wearing surface, 398–399
Weathering tests, 3
Weldable thermoplastics:
  advantages, 315
  application, illus., 316, 319
  flashings, 321
  history, 318
  materials, 315, 318
  problems, 320, 322
  (*See also* PVC and copolymers)
Wet insulation, definition, 87
Wind-induced oscillation, 152, 178
  insulation boards, 172
Wind-pressure formula, 131–132
Wind speed, 20, 131, 134, 158–161
  Beaufort scale illus., 527
  critical, 151

Wind speed (*Cont.*):
  fastest-mile, 125, 133
  gusts, 134, 139
  isotach map, 126
  probabilities, 134
Wind suction:
  effect on air leakage, 117, 154
  pressure differential computation, 143–144
Wind uplift, 78, 123, 180
  anchorage techniques, 73, 140, 170, 338, 515
  ballasted roof mechanics, 147–182
  ballooning, 167, 176
  Bernoulli principle, 117, 131
  blowoffs: cause, 128, 145
    illus., 129, 147–148, 176
  building code requirements, 123, 127
  Class 1 rating, 143, 173
  computations, 132, 140, 153, 157
  critical wind speed, 152
    illus., 158–162
  defective flashing, illus., 146, 148
  design examples, 139–145, 165
  EPDM, 308
    illus., 308, 310
  failure modes, 16, 127–130, 145–147, 170, 177, 338
  field test equipment, illus., 171
  gravel scour, 138, 148, 151, 223, 314
  gusting, 134–135
  hazards, 148, 338
  hurricanes, 123, 125, 223
  loose, ballasted systems, 153–154
  mechanics, 13, 125, 130–132, 170
  metal roofing, 125
  oscillations, 168
  parameters, 132
  parapet effect, 138, 151, 153, 155

Wind uplift (*Cont.*):
  perimeter failures, 137–138, 145–147
  pressures, 123–124, 132, 144, 154
    formula, 131–132, 139
  PMR design, 162–165
  recommended practices, 140, 149, 153
  repair procedures, 165, 172
  tests, 165–172
    field, 168–172
    illus., 171
    laboratory, 134, 151, 156, 166–168
  velocity pressure, 130
    graph, 132, 136–137
    table, 141
Wall openings, 154, 157
Wind speeds, 124, 126, 131, 151
  critical, 151
  determinants, 131–139
  probabilities, 134
Wind-uplift resistance:
  concrete pavers, 151, 161–162
  hot-mopped asphalt, 78, 102, 220
  mechanical fasteners, 78, 102, 140
  metal roofing, 443
Wood decks, 16, 46, 51, 56, 468
  deflection, 51, 468
  moisture limits, 54, 61, 426
  plastic deformation, 51
  rotting, 52, 492
  (*See also* Plywood decks)
Wood-fiber decks (*see* Structural wood-fiber decks)
Wood walkway:
  illus., 266
Wrinkling (*see* Ridging)

ABOUT THE AUTHORS

C. W. GRIFFIN is a member of ASTM and the American Arbitration Association. He is a consulting engineer specializing in roofing and energy conservation and a former senior editor for McGraw-Hill's *Engineering News Record*. His books include *Energy Conservation in Buildings* (1974) and *The Systems Approach to School Construction* (1971).

RICHARD FRICKLAS is Technical Director of the Roofing Industry Educational Institute. He is a columnist for *Roofing, Siding and Insulation* magazine and recipient of the Voss Award from ASTM for "distinguished contributions to the advancement of knowledge in building technology...and pioneering, through education, the conversion of roofing technology from an art to disciplined science." Dick also received the James Q. McCawley Award from the MRCA "in recognition of and appreciation for outstanding service to the roofing industry," and the IRWC Award "for unceasing efforts on behalf of the roofing industry."